COLLEGE PHYSICS

HARCOURT BRACE JOVANOVICH COLLEGE OUTLINE SERIES

COLLEGE PHYSICS

Robert W. Stanley

Department of Physics
Purdue University
West Lafayette, Indiana

HBJ

Harcourt Brace Jovanovich College Publishers
Fort Worth Philadelphia San Diego New York Orlando Austin San Antonio
Toronto Montreal London Sydney Tokyo

Printed in the United States of America

Library of Congress Cataloging in Publication Data

Stanley, Robert W.
College Physics

(Harcourt Brace Jovanovich college outline series) (Books for professionals)
Includes index.
1. Physics. 2. Physics—Problems, exercises, etc. I. Title II. Series III. Series: Books for professionals.
OC21.2.S685 1987 530 86-32003

ISBN 0-15-601662-1

First edition

Harcourt Brace Jovanovich, Inc.
The Dryden Press
Saunders College Publishing

PREFACE

Do not *read* this Outline — *use* it. You can't learn physics simply by reading about it: You have to *do* it. Solving specific, practical problems is the best way to master — and to demonstrate your mastery of — the theories, laws, and definitions upon which the science of physics is based. Outside the laboratory, you need three tools to do physics: a pencil, paper, and a calculator. Add a fourth tool, this Outline, and you're all set.

This HBJ College Outline has been designed as a tool to help you sharpen your problem-solving skills in physics. Each chapter covers a single topic, whose fundamental principles are broken down in outline form for easy reference. The outline text is filled with worked-out examples, so you can see immediately how each new idea is applied in problem form. Each chapter also contains a Summary of all the principal equations presented and a Raise Your Grades section. Taken together, these two features give you an opportunity to review the primary principles of a topic and the problem-solving techniques implicit in those principles.

Most important, this Outline gives you plenty of problems to practice on. Work the Solved Problems, and check yourself against the step-by-step solutions provided. Test your mastery of the material in each chapter by doing the Supplementary Exercises. (In the Supplementary Exercises, you're given answers only — the details of the solutions are up to you.) Finally, you can review all the topics covered in the Outline by working the problems in the Midsemester and Semester Exams.

The level of difficulty is graded in the Examples, Solved Problems, and Exams. Most Examples involve the straightforward use of a single concept or equation. You should strive to understand the connection between the question asked and the solution presented. The Solved Problems are somewhat more complex, but again you should try to discover the reasoning behind the sequence of steps shown. (The reasoning is stated clearly as the parts of each problem are worked out: Try to anticipate where the solution should go next.) Everything is fair game in the Exams — the last problem in each Exam requires you to investigate a complex physical situation and determine the values of many physical quantities. Don't panic! Just apply the skills you learned in the Examples to the separate pieces of the problem.

Having the tools is one thing; knowing how to use them is another. The solution to any problem in physics requires six procedures: (1) UNDERSTANDING, (2) ANALYZING, (3) PLANNING, (4) EXECUTING, (5) CHECKING, (6) REPORTING. Let's look at each of these procedures in more detail.

1. **UNDERSTANDING:** Read over the problem carefully and be sure you understand every part of it. If you have difficulty with any of the terms or ideas in the problem, reread the text material on which the problem is based. (In this Outline, important ideas, principles, laws, and terms are printed in boldface type, so they will be easy to find.) Make certain that you understand what kind of answer will be required. If the problem is quantitative, make an estimate of the magnitude of the answer.

2. **ANALYZING:** Break the problem down into its components. Ask yourself

   - What are the data?
   - What is (are) the unknown(s)?
   - What equation, law, or definition connects the data to the unknowns?

3. **PLANNING:** Trace a connection between the data and the unknowns as a series of discrete operations (steps). This often involves manipulating one or more mathematical or physical expressions to isolate unknown quantities. Once you have a clear, stepwise path between data and solution, take note of any steps that require ancillary operations, such as substituting in equations or converting units. (Keep a sharp watch on units — they are often useful clues.)

4. **EXECUTION:** Follow your plan and execute any mathematical operations. It helps to work with symbols whenever possible: Substituting for variables should be the *last* thing you do. Make sure you've used the correct signs, exponents, and units.

5.- **CHECKING:** Never consider a problem solved until you have checked your work. Does your answer

- make sense?
- have the right units?
- answer the question?

Is your math right?

6. **REPORTING:** Make sure you have shown your reasoning and method clearly, and that your answer is readable. (It can't hurt to write the word "Answer" in front of your answer. That way, you—and your instructor—can find it at a glance, saving time and trouble all 'round.)

We have assumed that you own a calculator. Many hand-held calculators enable you to perform mathematical calculations that once required tables of logarithms and trigonometric functions. Calculators also permit you to eliminate the tedious methods of computation that arise from the properties of logarithms. Therefore, we provide neither trigonometric tables nor logarithm tables in this Outline. A hand-held calculator is essential for solving many of the problems. Even solutions that don't require the use of a calculator should be verified with a calculator.

The problem of precision and significant figures arises whenever something is measured. When you're applying your knowledge of physics to problems in the real world, you must be sure that your results do not imply more accuracy than your measurements warrant. In this Outline, which emphasizes principles and methods rather than measurements, you may assume that all numbers given in the Examples and Problems are precise—so you may express your answers in as many figures as your calculator allows. You may find it convenient, however, to give the final answers to the Examples and Problems to three significant figures, as we have done in the Outline. But avoid rounding errors! Many Problems contain several parts, and the answer to one part may be the starting point for calculations in the next part. Repeated approximations can introduce errors that accumulate over several steps. In this Outline, answers are given to more than three significant figures whenever necessary to prevent rounding errors in the next step. The problem of rounding errors arises often in physics applications, so in any problem based on measurements, round your answer to the proper number of significant figures only as the last step before reporting your result.

CONTENTS

1 VECTORS

THIS CHAPTER IS ABOUT

- ☑ Vectors and Scalars
- ☑ Components of a Vector
- ☑ Multiplication of a Vector by a Scalar
- ☑ Vector Addition
- ☑ Vector Multiplication

1-1. Vectors and Scalars

A. Vector quantities

Many of the quantities you work with in physics—e.g., velocity, displacement, acceleration, force—are vector quantities.

- A **vector quantity** has both *magnitude* and *direction*.

In printed material, **boldface** type represents a vector. Vector quantities may be represented graphically by arrows whose heads are pointed to give direction and whose lengths are drawn to scale to give magnitude.

note: The magnitude of a vector quantity is always a positive number.

A WORD ABOUT CALCULATIONS

Today most of us use pocket calculators, which commonly carry computations out to eight or ten figures. But in this book we will follow this procedure: All numbers given in an example or problem are assumed to be precise, and calculations will be carried out to three significant figures in the final answer.

EXAMPLE 1-1: When you're driving a car down a straight highway, you can measure your *velocity*: A speedometer registers your speed, which is the *magnitude* of your velocity; a compass registers the direction you're going, which is the *direction* of your velocity.

How would you represent graphically the velocity of a car moving (**a**) north at 50 miles per hour (mi/h)? (**b**) east at 25 mi/h? [Hint: Use some convenient unit to represent increments of 10 mi/h.]

Solution: Draw each velocity as a vector quantity represented by an arrow whose head points in the given direction and whose length represents the given speed.

(**a**) The top of a sheet of paper normally represents north, so your first arrow should point toward the top of the page as illustrated in Figure 1-1a. If you use 1 cm to represent 10 mi/h, the length of your arrow should be 5 cm.

(**b**) Your second arrow should point to the right and should be half as long as the first (Figure 1-1b).

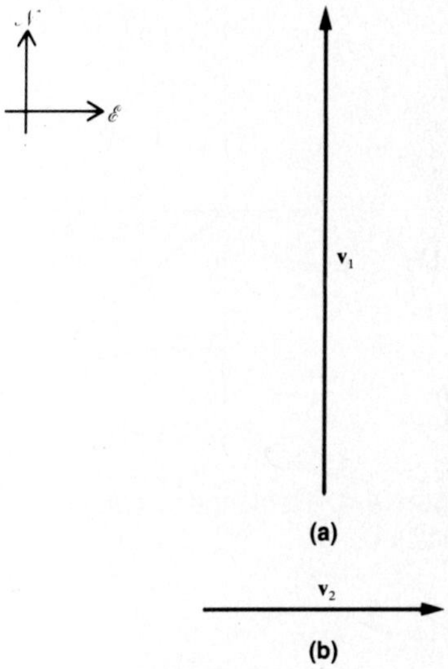

Figure 1-1

EXAMPLE 1-2: Two forces act on a small object. The first force $\mathbf{F}_1$ has a magnitude of 4 newtons (N) directed toward the right. The second force $\mathbf{F}_2$ has a magnitude of 3 N directed toward the left. (1 N = 1 kg • m/s^2, a measure of force we'll explore in detail in Chapter 3.)

Draw arrows to represent these two forces.

Solution: Use 1 cm to represent 1 N, and draw the two arrows to scale, as shown in Figure 1-2. The length of the arrow represents the magnitude of the vector, so the arrow directed toward the left should be three fourths as long as the arrow directed toward the right.

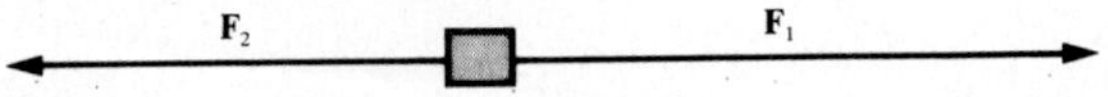

Figure 1-2

B. Scalars

You will also be working with many quantities in physics—e.g., time, temperature, volume—that are scalar quantities.

- A **scalar quantity** has *magnitude* only.

Scalar quantities are specified by a single number with units. In printed material, *italic* type represents a scalar so that the magnitude of a vector **F** may be written as

$$\text{mag of } \mathbf{F} \quad \text{or} \quad |\mathbf{F}| \quad \text{or} \quad F$$

note: Scalars may be positive or negative, and you add them using ordinary arithmetic.

C. Equality of vectors

- Two vectors are **equal** if they have the same magnitude and the same direction.

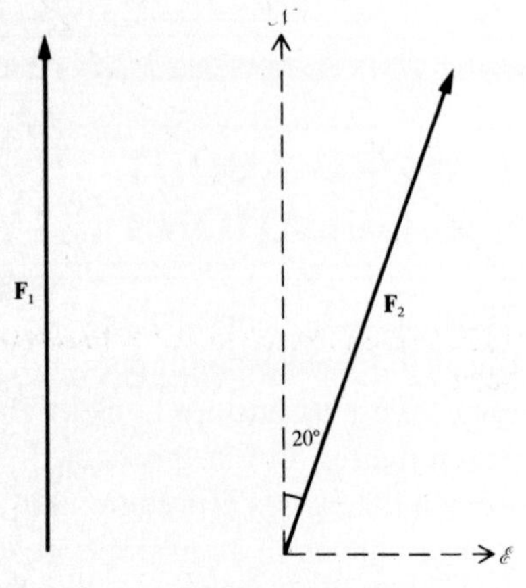

Figure 1-3

EXAMPLE 1-3: Two unequal forces act on an object. The first force $\mathbf{F}_1$ has a magnitude of 5 N directed toward the north. The second force $\mathbf{F}_2$ has a magnitude of 5 N and a direction 20° east of north.
Draw arrows to represent these two unequal forces.

Solution: The two arrows you draw to represent these forces should have the same length, but different directions (see Figure 1-3).

1-2. Components of a Vector

- The **component** of a vector is its projection along a given direction.

note: The components of a vector may be positive or negative, depending on the system of reference.

The process of representing a vector by two or more scalar components is called **resolving** the vector. The velocity vector **v** in Figure 1-4 has a magnitude v of 5 m/s. If we use the familiar Cartesian coordinate system, **v** has an x component v_x of 4 m/s and a y component v_y of -3 m/s. The component v_y is negative because it points downward in the direction of the negative y axis.

The vector **v** and its two rectangular components, v_x and v_y, form a right triangle, so you can use the Pythagorean theorem to determine the magnitude of the vector from its rectangular components:

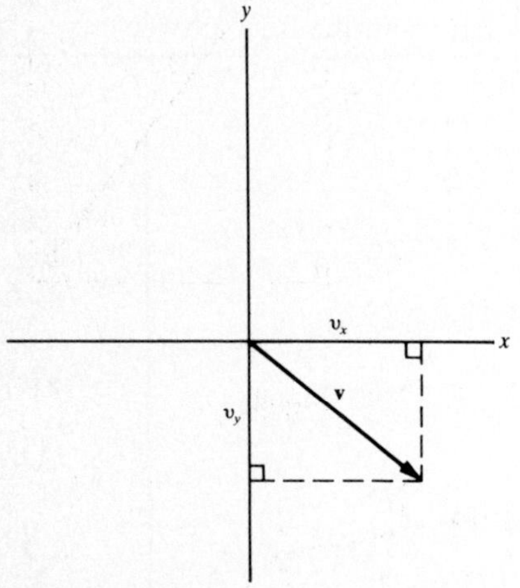

Figure 1-4. Rectangular components of a vector.

MAGNITUDE OF A VECTOR

$$|\mathbf{v}| = \sqrt{v_x^2 + v_y^2} \tag{1-1}$$

EXAMPLE 1-4: A force has a magnitude of 10 N directed 53° above the negative x axis (see Figure 1-5).
Find the x and y components of this force.

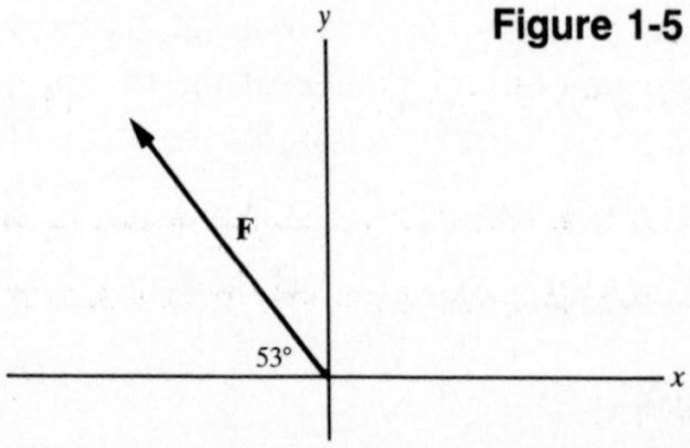

Figure 1-5

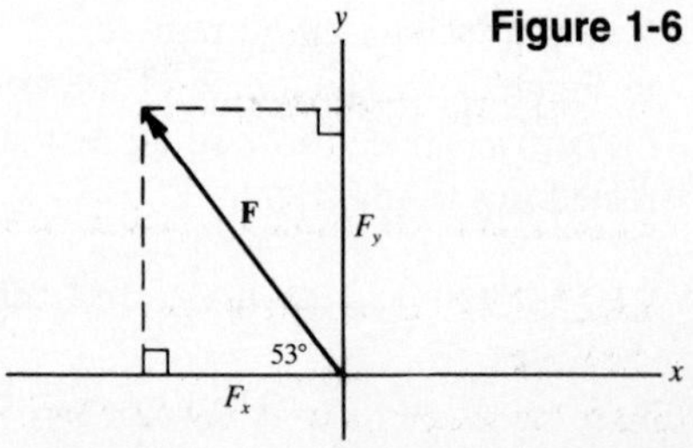

Figure 1-6

Solution: First, draw a straight line from the head of the arrow perpendicular to each axis, as shown in Figure 1-6. Now, use your knowledge of trigonometry to find the unknown side of each right triangle, adding negative signs as the quadrant requires:

$$\cos 53° = -\frac{F_x}{F} \qquad \sin 53° = \frac{F_y}{F}$$

$$F_x = -F \cos 53° \qquad F_y = F \sin 53°$$

$$= -(10\text{ N})(0.60) = -6.0\text{ N} \qquad = (10\text{ N})(0.80) = 8.0\text{ N}$$

note: F_x is negative since **F** is in the third quadrant.

1-3. Multiplication of a Vector by a Scalar

(1) Multiplying a vector by a *positive* scalar changes the magnitude of the vector, but not its direction. When you multiply the vector **F** in Figure 1-7 by the scalar 2, a new vector 2**F** is produced. 2**F** has twice the magnitude of **F**, but the same direction.
(2) Multiplying a vector by a *negative* scalar (other than −1) changes the magnitude of the vector and reverses its direction.
(3) Multiplying a vector by −1 reverses the direction of the vector, but its magnitude remains the same.

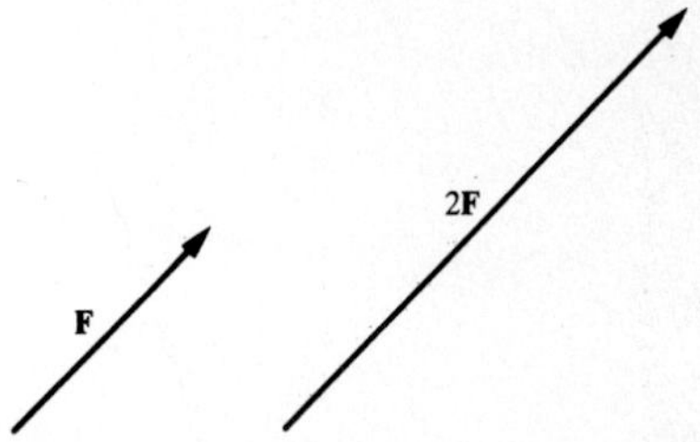

Figure 1-7. Multiplication of a vector by a scalar.

EXAMPLE 1-5: The force **F** from Example 1-4 (illustrated in Figure 1-5) has a magnitude of 10 N. Its x component F_x is −6.0 N and its y component F_y is 8.0 N. Find the x and y components of the force $\mathbf{F}_2$ such that $\mathbf{F}_2 = -2\mathbf{F}$.

Solution: Draw an arrow to represent the force $\mathbf{F}_2 = -2\mathbf{F}$. Since you are multiplying the vector **F** by a *negative* scalar −2, the force $\mathbf{F}_2$ will have a magnitude twice that of **F**, and a direction opposite that of **F** (see Figure 1-8).

$$F_{2_x} = |\mathbf{F}_2| \cos 53° = (20\text{ N})(0.60) = 12.0\text{ N}$$

$$F_{2_y} = -|\mathbf{F}_2| \sin 53° = -(20\text{ N})(0.80) = -16.0\text{ N}$$

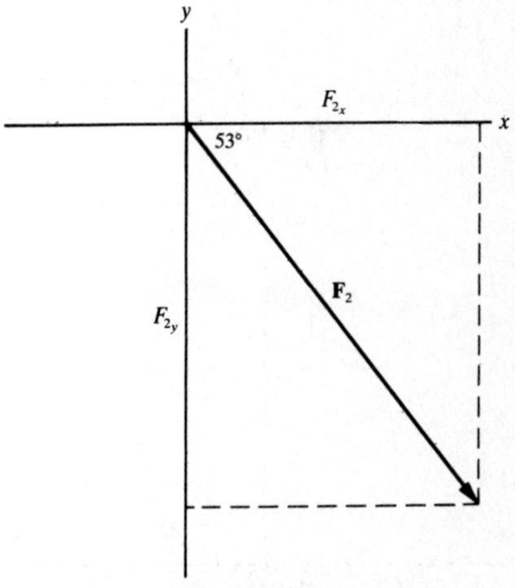

Figure 1-8

1-4. Vector Addition

A. Graphical method

Two vectors are added graphically by drawing the first vector and then placing the tail of the second vector at the head of the first, keeping the same magnitude and direction. The vector representing the sum of the two vectors begins at the tail of the first and extends to the head of the second. Figure 1-9 illustrates the graphical method of adding the vectors $\mathbf{F}_1$ and $\mathbf{F}_2$. First, place $\mathbf{F}_1$ and $\mathbf{F}_2$ head-to-tail. [Hint: The order of addition doesn't matter—$\mathbf{F}_1 + \mathbf{F}_2$ gives the same result as $\mathbf{F}_2 + \mathbf{F}_1$.] Now, draw a third vector from the tail of $\mathbf{F}_1$ to the head of $\mathbf{F}_2$. This vector is the vector sum or **resultant R** of $\mathbf{F}_1 + \mathbf{F}_2$.

If you know the angle θ opposite the unknown side of a triangle (see Figure 1-10), you can find the *magnitude* of the unknown side using the law of cosines:

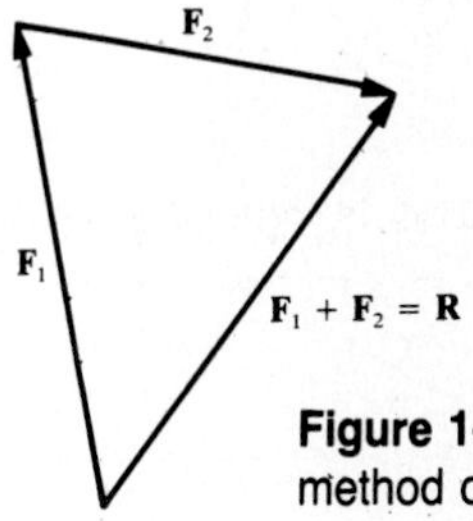

Figure 1-9. Graphical method of adding vectors.

LAW OF COSINES $$c^2 = a^2 + b^2 - 2ab \cos \theta \qquad \textbf{(1-2)}$$

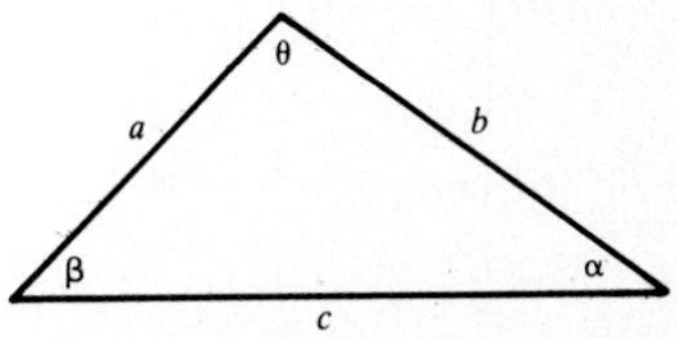

Figure 1-10

EXAMPLE 1-6: Jill walks 200 m toward the east and then 400 m toward the northeast.
Find the magnitude of her resultant displacement.

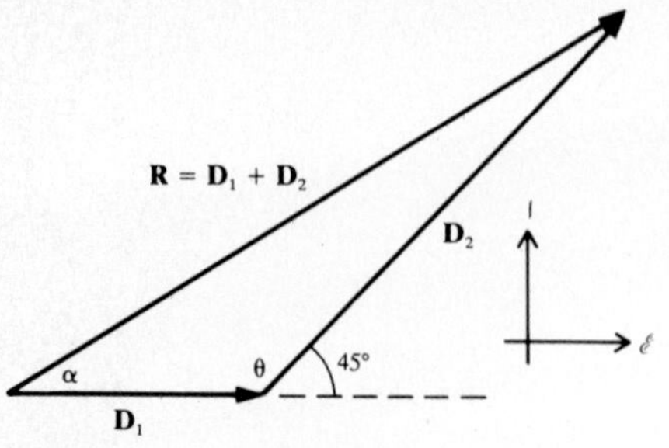

Figure 1-11

Solution: Draw arrows to represent the two displacement vectors, $\mathbf{D}_1$ and $\mathbf{D}_2$. Since you are going to add them, place the arrows head-to-tail as illustrated in Figure 1-11. The vector directed toward the northeast is 45° above the x axis, so you know that $\theta = 135°$. Now, use the law of cosines to calculate $|\mathbf{R}|$, the magnitude of the resultant displacement:

$$|\mathbf{R}| = |\mathbf{D}_1 + \mathbf{D}_2| = \sqrt{D_1^2 + D_2^2 - 2D_1D_2 \cos\theta}$$

$$= \sqrt{(200)^2 + (400)^2 - 2(200)(400)\cos 135°} = 560 \text{ m}$$

You can add several vectors together to find the resultant (or net) displacement of an object that has moved from point A to point B by some path other than a straight line. Just place the vectors head-to-tail and draw the net displacement vector from the tail of the first vector to the head of the last. Remember that the order of addition doesn't matter. The net displacement is a vector quantity whose magnitude is the linear distance from its starting point to its ending point.

You can find the *direction* of the net displacement vector—that is, the angle the vector makes with respect to a given system of reference—using the law of sines (see Figure 1-10):

LAW OF SINES

$$\frac{\sin\alpha}{a} = \frac{\sin\beta}{b} = \frac{\sin\theta}{c} \qquad \textbf{(1-3)}$$

EXAMPLE 1-7: Find the direction of Jill's resultant displacement in Example 1-6.

Solution: You want to find angle α, which will give you the direction of $\mathbf{R}$ with respect to the east. Use Eq. (1-3), the law of sines, to find the angle α:

$$\frac{\sin\alpha}{|\mathbf{D}_2|} = \frac{\sin\theta}{|\mathbf{D}_1 + \mathbf{D}_2|}$$

$$\sin\alpha = \frac{|\mathbf{D}_2|\sin\theta}{|\mathbf{D}_1 + \mathbf{D}_2|} = \frac{400 \sin 135°}{560} = 0.505$$

$$\alpha = \text{arc}\sin(0.505) = 30.3°$$

To complete the problem, note that compass headings are used as the system of reference. You should give the direction of Jill's resultant displacement as 59.7° east of north.

B. Component method

To add vectors using the component method, resolve each vector into its scalar components, add the x and y components individually, and then use the Pythagorean theorem to find the magnitude of the resultant. [Hint: The component method is useful for adding more than two vectors, particularly if you organize the work in tabular form, as shown in Example 1-8.]

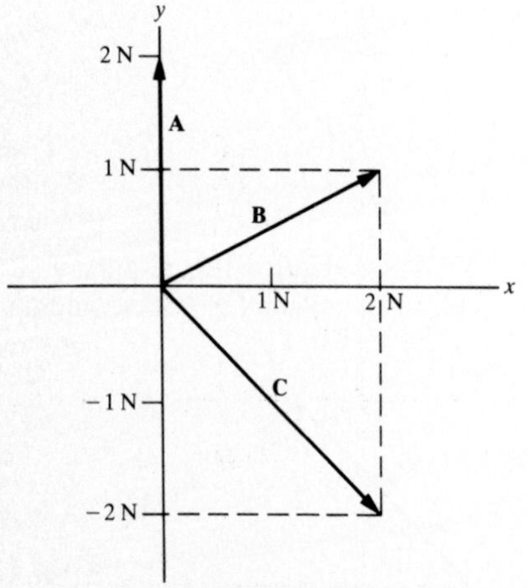

Figure 1-12

EXAMPLE 1-8: Find the magnitude and direction of the sum $\mathbf{R}$ of vectors $\mathbf{A}$, $\mathbf{B}$, and $\mathbf{C}$, shown in Figure 1-12.

Solution: You'll be less likely to make mistakes if you follow these steps:

(1) Draw each vector with its tail at the origin of the system of reference. (Assume the rectangular coordinate system as the system of reference.)

(2) Resolve the vector into components—that is, calculate the x and y components of each vector.
(3) Prepare a table containing all the vectors to be added, with their x and y components placed in separate columns. Remember that some components may be negative.
(4) Add all the x components to get the x component of **R**; then, add all the y components to get the y component of **R**.
(5) Use the Pythagorean theorem to get the magnitude of **R**.
(6) Use the tangent function to find the angle between **R** and the x axis.

For the three vectors shown in Figure 1-12, you can prepare the following table from steps 2-4:

Vector	x component	y component
A	0	2 N
B	2 N	1 N
C	2 N	−2 N
R	4 N	1 N

Because you are using the rectangular coordinate system, the x and y components of **R** are at right angles by definition, so you can use the Pythagorean theorem to find the magnitude of **R**:

$$|\mathbf{R}| = \sqrt{(4\text{ N})^2 + (1\text{ N})^2} = 4.12\text{ N}$$

Finally, use the tangent function to get the direction of **R**:

$$\tan\theta = \frac{y\text{ component}}{x\text{ component}} = \frac{1\text{ N}}{4\text{ N}} = 0.250$$

$$\theta = \arctan(0.250) = 14.0°$$

C. Vector subtraction

To solve a problem in which you must subtract vectors, first change it into an addition problem—then, you can use either the graphical or the component method.

EXAMPLE 1-9: You're driving an automobile due north at a speed of 22 m/s (about 50 mi/h). To take a curve in the highway, you reduce speed to 20 m/s and turn the car 30° east of north.
Calculate (**a**) the magnitude and (**b**) the direction of your change in velocity.

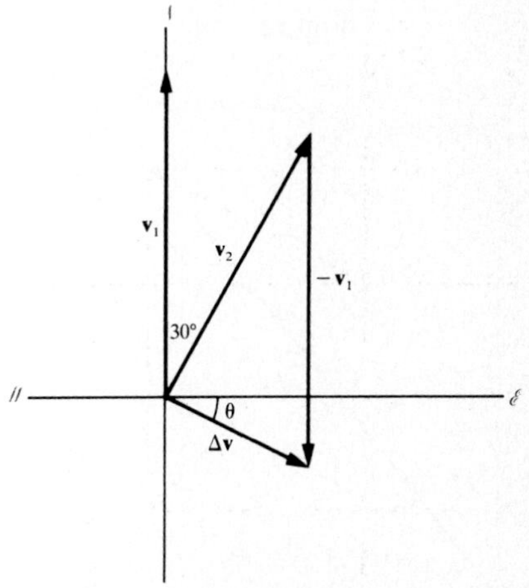

Figure 1-13

Solution:

(**a**) The change in any quantity is always the later value minus the earlier value. Use the symbol $\Delta\mathbf{v}$ to represent the change in $\mathbf{v}$:

$$\Delta\mathbf{v} = \mathbf{v}_2 - \mathbf{v}_1 = \mathbf{v}_2 + (-\mathbf{v}_1)$$

Draw arrows to represent the vectors $\mathbf{v}_1$ and $\mathbf{v}_2$, as illustrated in Figure 1-13. You'll also need to draw an arrow to represent the vector $-\mathbf{v}_1$. Place this arrow head-to-tail with $\mathbf{v}_2$, because this is the vector you add to $\mathbf{v}_2$ to find the resultant $\Delta\mathbf{v}$. Now, make a table of the vectors and their N and E components.

Vector	E component	N component
$\mathbf{v}_2$ $-\mathbf{v}_1$	$v_2 \sin 30° = 10$ m/s 0	$v_2 \cos 30° = 17.3$ m/s -22 m/s
$\Delta\mathbf{v}$	10 m/s	-4.7 m/s

You can use the Pythagorean theorem to find the magnitude of $\Delta\mathbf{v}$:

$$\Delta v = \sqrt{\Delta v_E^2 + \Delta v_N^2} = \sqrt{(10 \text{ m/s})^2 + (-4.7 \text{ m/s})^2} = 11.0 \text{ m/s}$$

(b) You can see from Figure 1-13 that the vector $\Delta\mathbf{v}$ lies below the E axis at an angle θ. The tangent of θ is

$$\tan\theta = \frac{\Delta v_N}{\Delta v_E} = \frac{-4.7 \text{ m/s}}{10 \text{ m/s}} = -0.470$$

$$\theta = \text{arc}\tan(-0.470) = -25.2°$$

The direction of your change in velocity is 25.2° S of E.

1-5. Vector Multiplication

A. Dot product

The **dot product**—or scalar product—of two vectors is a scalar equal to the product of the magnitudes of the two vectors and the cosine of the angle θ between them. If θ is less than 90°, the dot product will be positive; if θ is greater than 90°, the dot product will be negative. [Note: The dot product is written **A • B**.]

DOT PRODUCT

$$\mathbf{A} \cdot \mathbf{B} = |\mathbf{A}||\mathbf{B}| \cos\theta = AB \cos\theta \qquad \textbf{(1-4)}$$

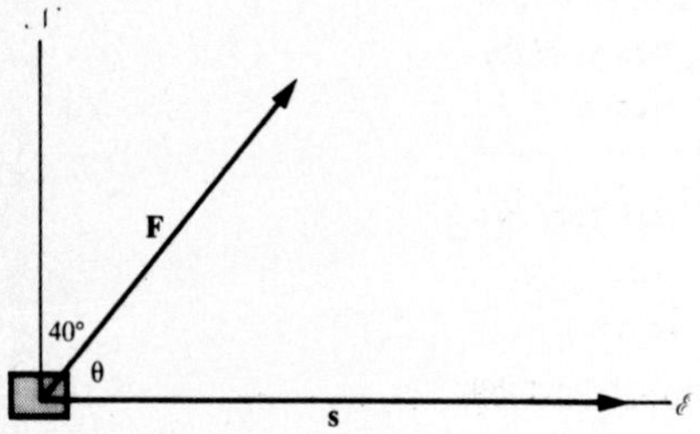

Figure 1-14

EXAMPLE 1-10: A force **F** of magnitude 3 N and direction 40° east of north acts on an object. The eastward displacement **s** of the object has a magnitude of 2 m. These two vectors are illustrated in Figure 1-14.
Calculate the scalar product of these two vectors.

Solution: From Figure 1-14 you can see that the angle θ between the two vectors is $(90 - 40)° = 50°$. Now, use Eq. (1-4) to calculate the dot product:

$$\mathbf{F} \cdot \mathbf{s} = |\mathbf{F}||\mathbf{s}| \cos\theta = (3 \text{ N})(2 \text{ m}) \cos 50° = 3.86 \text{ N m}$$

B. Cross product

The **cross product**—or vector product—of two vectors **A** and **B** is a new vector whose direction is perpendicular to both **A** and **B**. The magnitude of the cross product is equal to the product of the magnitudes of the two original vectors and the sine of the angle between them. [Note: The cross product is written **A × B**.]

CROSS PRODUCT MAGNITUDE

$$|\mathbf{A} \times \mathbf{B}| = |\mathbf{A}||\mathbf{B}| \sin\theta = AB \sin\theta \qquad \textbf{(1-5)}$$

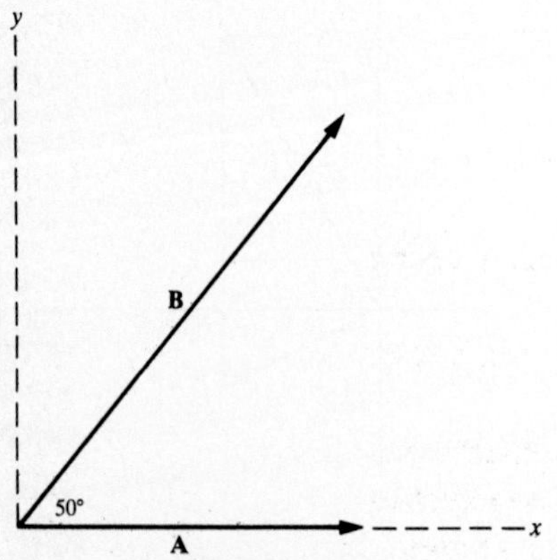

Figure 1-15

EXAMPLE 1-11: Vectors **A** and **B** have magnitudes of 2 cm and 3 cm, respectively, and the angle between them is 50° (see Figure 1-15).
Calculate the magnitude of the cross product **A × B**.

Solution: From Eq. (1-5):

$$|\mathbf{A} \times \mathbf{B}| = AB \sin \theta = (2 \text{ cm})(3 \text{ cm}) \sin 50° = 4.60 \text{ cm}^2$$

The cross product **A** × **B** is a vector quantity, so you must find the direction of **A** × **B**, as well as its magnitude. **A** and **B** lie in the *xy* plane and since their cross product is perpendicular to both of them, **A** × **B** must lie along the *z* axis. To determine the direction of **A** × **B**, imagine a right-hand screw placed along the *z* axis, as shown in Figure 1-16. Twist the screw in the direction that will rotate the first vector **A** toward the second vector **B**. If you view this rotation from above the *xy* plane, you can see that it is counterclockwise. A right-handed screw twisted in this way would move upward along the *z* axis, so you can conclude that **A** × **B** has the direction of the positive *z* axis.

note: Another method is to curl the fingers of your right hand so that your fingertips point from *A* to *B*. Your thumb indicates the direction of the cross product on the *z* axis.

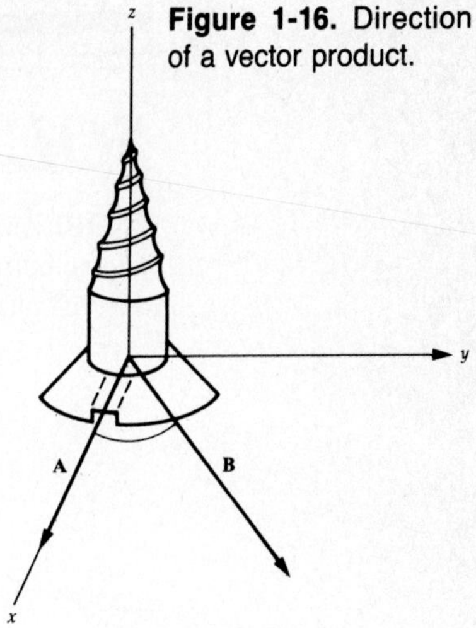

Figure 1-16. Direction of a vector product.

SUMMARY

Magnitude of a vector	$\lvert\mathbf{v}\rvert = \sqrt{v_x^2 + v_y^2}$	gives the magnitude of a vector from its rectangular components
Law of cosines	$c^2 = a^2 + b^2 - 2ab \cos \theta$	yields the magnitude of the unknown side *c* of a triangle, given two sides and the angle θ between them
Law of sines	$\frac{\sin \alpha}{a} = \frac{\sin \beta}{b} = \frac{\sin \theta}{c}$	the sides of a triangle are proportional to the sines of the opposite angles
Dot product	$\mathbf{A} \cdot \mathbf{B} = AB \cos \theta$	yields the scalar product of two vectors, given the cosine of the angle θ between them
Cross product magnitude	$\lvert\mathbf{A} \times \mathbf{B}\rvert = AB \sin \theta$	yields the vector product of two vectors, given the sine of the angle θ between them

RAISE YOUR GRADES

Can you define . . . ?

- [x] a vector
- [x] a scalar
- [x] a component
- [x] a resultant
- [x] a dot product
- [x] a cross product

Can you . . . ?

- [x] explain the difference between a vector and a scalar
- [x] tell if two vectors are equal
- [x] find a component of a vector
- [x] add two vectors graphically
- [x] add vectors using the method of components
- [x] explain the difference between vector −**A** and vector **A**
- [x] explain the difference between vector 2**B** and vector **B**
- [x] subtract one vector from another
- [x] calculate the dot product of two vectors
- [x] find the cross product of two vectors

SOLVED PROBLEMS

PROBLEM 1-1 An object moves at a speed of 30 m/s in the direction illustrated in Fig. 1-17. The velocity vector makes an angle of 25° with the positive y axis. Find the x and y components of this velocity.

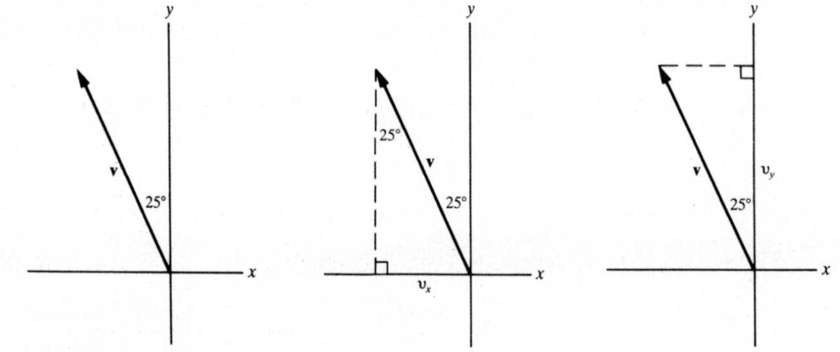

Figure 1-17 **Figure 1-18** **Figure 1-19**

Solution First, draw a perpendicular line from the head of the arrow to the x axis. When you do this, as shown in Fig. 1-18, the foot of the perpendicular lies to the left of the origin, so the x component is negative. Now, use your knowledge of trigonometry to find the unknown side v_x of the right triangle you have drawn:

$$\sin 25° = \frac{\text{side opposite}}{\text{hypotenuse}} = \frac{-v_x}{v}$$

$$v_x = -v \sin 25° = -(30\text{ m/s})(0.423) = -12.7\text{ m/s}$$

Find the y component of $\mathbf{v}$ in a similar way. When you draw the perpendicular from the head of the arrow to the y axis, as shown in Fig. 1-19, the foot of the perpendicular lies above the origin, so the y component of this vector is positive. Because the unknown side v_y is adjacent to the known angle, you'll need to use the cosine function:

$$\cos 25° = \frac{\text{side adjacent}}{\text{hypotenuse}} = \frac{v_y}{v}$$

$$v_y = v \cos 25° = (30\text{ m/s})(0.906) = 27.2\text{ m/s}$$

To check your answers, square the x and y components and then add them. The result should be equal to the square of the hypotenuse—900 m^2/s^2. Don't expect the result to be perfect, because you have kept only three significant figures:

$$(-12.7\text{ m/s})^2 + (27.2\text{ m/s})^2 = 901\text{ m}^2/\text{s}^2$$

PROBLEM 1-2 Two forces $\mathbf{F}_1$ and $\mathbf{F}_2$ with magnitudes of 3 N and 5 N, respectively, act on an object. The directions of the forces are shown in Fig. 1-20. Find **(a)** the magnitude and **(b)** the direction of the resultant $\mathbf{F}_{tot}$—that is, the sum $\mathbf{F}_1 + \mathbf{F}_2$.

Solution First place the two vectors head-to-tail and then draw the resultant vector from the tail of one vector to the head of the other. Make your drawing to scale, as shown in Fig. 1-21.
(a) You can calculate the magnitude of $\mathbf{F}_{tot}$ using the law of cosines:

$$|\mathbf{F}_{tot}| = \sqrt{F_1^2 + F_2^2 - 2F_1F_2 \cos 60°}$$

$$= \sqrt{(3\text{ N})^2 + (5\text{ N})^2 - 2(3\text{ N})(5\text{ N}) \cos 60°} = 4.36\text{ N}$$

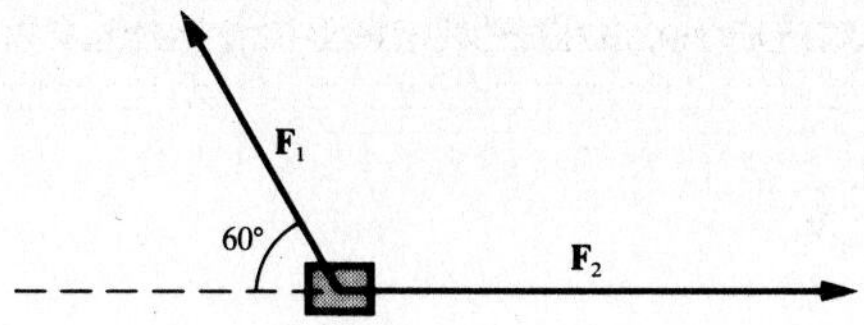

Figure 1-20

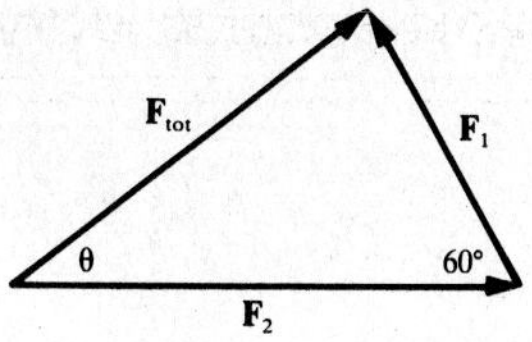

Figure 1-21

(**b**) Calculate the unknown angle θ by using the law of sines:

$$\frac{\sin\theta}{F_1} = \frac{\sin 60°}{F_{tot}}$$

$$\sin\theta = \frac{F_1 \sin 60°}{F_{tot}} = \frac{3\text{ N}}{4.36\text{ N}}(0.866) = 0.596$$

$$\theta = \text{arc}\sin(0.596) = 36.6°$$

PROBLEM 1-3 Find the resultant of the two forces in Problem 1-2 using the component method.

Solution Prepare a table listing the forces and their components, as illustrated in Example 1-9:

Vector	x component	y component
$\mathbf{F}_1$ $\mathbf{F}_2$	$-F_1 \cos 60° = -1.5$ N $+5$ N	$F_1 \sin 60° = +2.60$ N 0
$\mathbf{F}_{tot}$	3.50 N	2.60 N

[**recall**: Both components of $\mathbf{F}_{tot}$ are positive, which means that the resultant force lies in the first quadrant.]

Now, use the Pythagorean theorem to calculate the magnitude of the resultant:

$$F_{tot} = \sqrt{F_x^2 + F_y^2} = \sqrt{(3.50\text{ N})^2 + (2.60\text{ N})^2} = 4.36\text{ N}$$

You can find the angle that $\mathbf{F}_{tot}$ makes with the positive x axis from the tangent function:

$$\tan\theta = \frac{F_y}{F_x} = \frac{2.60\text{ N}}{3.50\text{ N}} = 0.743$$

$$\theta = \text{arc}\tan(0.743) = 36.6°$$

PROBLEM 1-4 The dashed line in Fig. 1-22 represents the path of an object that moves around a circle at a constant speed of 15 m/s. During the 3 seconds it takes for the object to move from point A to point B, the direction of the velocity changes by 30°. Calculate the magnitude of the change in velocity.

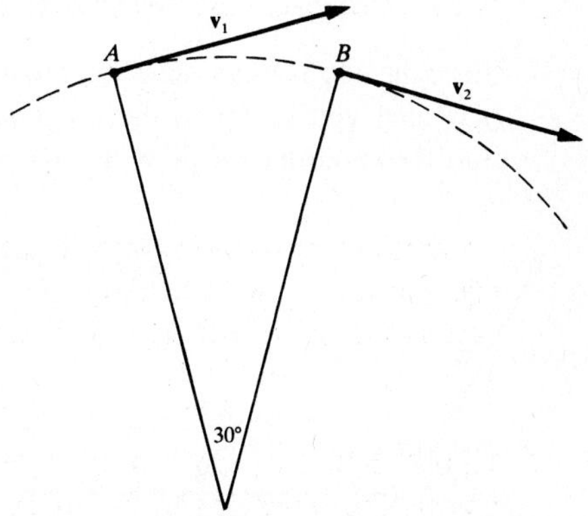

Figure 1-22. Velocities of an object moving along a circle.

Solution Remember that you always find the change in a quantity by subtracting the earlier value from the later value, so what you have here is a problem in vector subtraction. The symbol for the change in $\mathbf{v}$ is $\Delta\mathbf{v}$. Write the equation:

$$\Delta\mathbf{v} = \mathbf{v}_2 - \mathbf{v}_1 \quad \text{or} \quad \Delta\mathbf{v} = \mathbf{v}_2 + (-\mathbf{v}_1)$$

This problem is similar to the one you solved in Example 1-9. Next, draw the vector $-\mathbf{v}_1$ and add it to the vector $\mathbf{v}_2$ (see Fig. 1-23). You'll need to find the angle θ, which is opposite the unknown vector $\Delta\mathbf{v}$. Remember from geometry that since the two vectors $\mathbf{v}_1$ and $\mathbf{v}_2$ are mutually perpendicular to the radial lines, they necessarily intersect at the same angle as the radial lines. You can use the law of cosines to find the magnitude of $\Delta\mathbf{v}$:

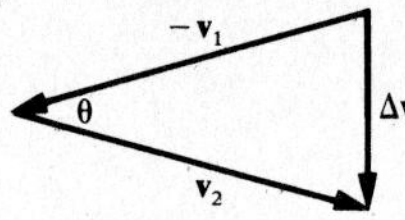

Figure 1-23

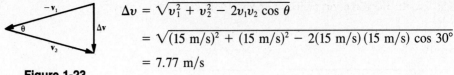

$$\Delta v = \sqrt{v_1^2 + v_2^2 - 2v_1v_2 \cos \theta}$$
$$= \sqrt{(15\ \text{m/s})^2 + (15\ \text{m/s})^2 - 2(15\ \text{m/s})(15\ \text{m/s}) \cos 30°}$$
$$= 7.77\ \text{m/s}$$

PROBLEM 1-5 Find the dot product of the dimensionless vectors **A** and **B**, illustrated in Fig. 1-24. Their magnitudes are 4 and 6, respectively.

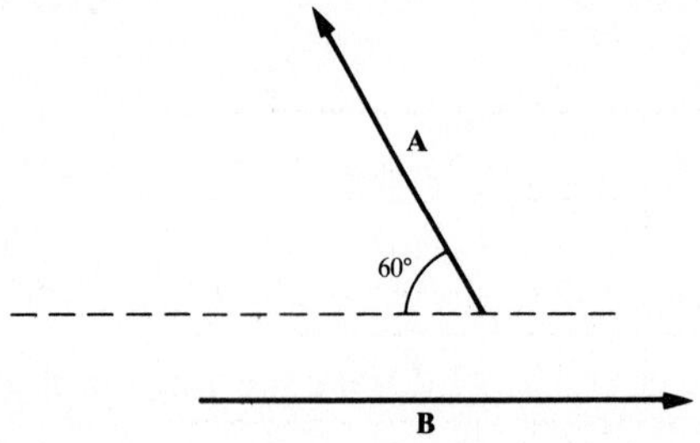

Figure 1-24

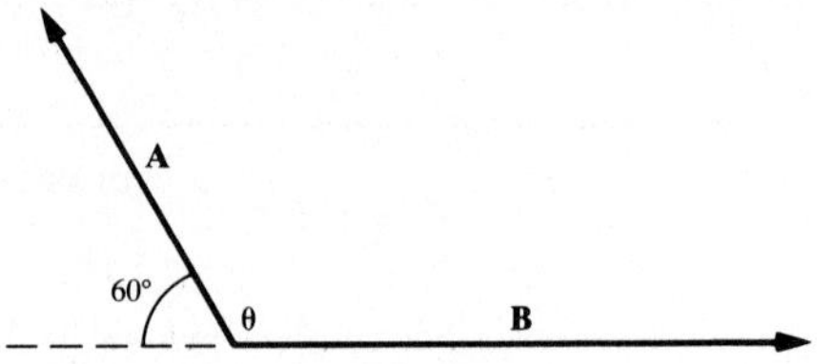

Figure 1-25

Solution Begin by writing down the equation that defines the dot product:

$$\mathbf{A} \cdot \mathbf{B} = AB \cos \theta$$

First, you have to find the angle θ, so place the two vectors with their tails together, as shown in Fig. 1-25. Now, you can see that

$$\theta = (180 - 60)° = 120°$$

and you can complete the calculation:

$$\mathbf{A} \cdot \mathbf{B} = (4)(6) \cos 120° = (24)(-0.5) = -12$$

The minus sign is important—don't omit it!

PROBLEM 1-6 Find the cross product of the vectors **A** and **B** in Problem 1-5.

Solution It's always a good idea to write down the equation you are going to use—in this case, the equation for the magnitude of the cross product:

$$|\mathbf{A} \times \mathbf{B}| = AB \sin \theta = (4)(6) \sin 120° = 20.8$$

Since the angle θ can never be greater than 180°, this calculation will *always* give a positive number. Now that you've found the magnitude of **A** × **B**, you're halfway through the problem. Now, just find the direction of **A** × **B**. [**recall:** **A** × **B** is perpendicular to both **A** and **B** (see Sec. 1-5).]

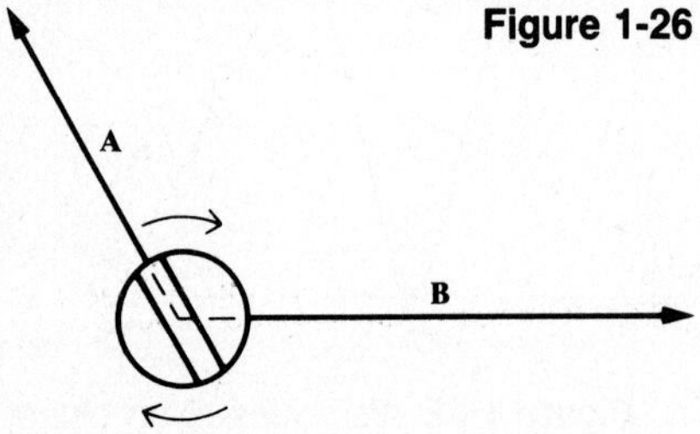

Figure 1-26

Since **A** and **B** lie in the horizontal plane of the drawing, their vector product must be perpendicular to this plane—that is, vertical. So, **A** × **B** points either upward or downward. Which is it?

Imagine a right-hand screw placed along this vertical axis. Turn the screw in the direction that will rotate the first vector **A** toward the second vector **B**. This is a clockwise rotation, as illustrated in Fig. 1-26. A screw twisted in this way will move downward into the plane of the drawing, so you can conclude that **A** × **B** points downward into the plane of the drawing.

PROBLEM 1-7 A small object is subject to three forces. The object is in equilibrium, which means that the sum of the three forces is zero. The two known forces $\mathbf{F}_1$ and $\mathbf{F}_2$, illustrated in Fig. 1-27, have magnitudes of 5 N and 7 N, respectively. The angle between them is 40°. Find (**a**) the magnitude and (**b**) the direction of the unknown force $\mathbf{F}_3$.

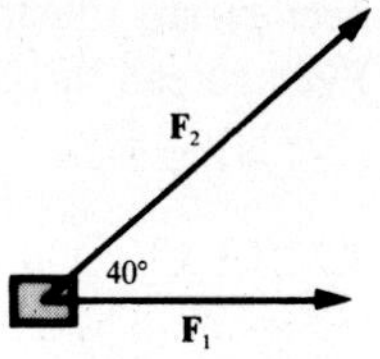

Figure 1-27

Solution

(**a**) Begin by writing the equation for an object in equilibrium:

$$\mathbf{F}_1 + \mathbf{F}_2 + \mathbf{F}_3 = 0$$

and solve for the unknown force $\mathbf{F}_3$:

$$\mathbf{F}_3 = -(\mathbf{F}_1 + \mathbf{F}_2)$$

Now, find the sum of $\mathbf{F}_1$ and $\mathbf{F}_2$ using the component method.

Forces	*x* component	*y* component
$\mathbf{F}_1$ $\mathbf{F}_2$	5 N (7 N) cos 40° = 5.36 N	0 N (7 N) sin 40° = 4.50 N
$\mathbf{F}_1 + \mathbf{F}_2$	10.36 N	4.50 N

Since the third force is equal to $-(\mathbf{F}_1 + \mathbf{F}_2)$, its x and y components are

$$F_{3_x} = -10.36\text{ N} \quad \text{and} \quad F_{3_y} = -4.50\text{ N}$$

and you can calculate the magnitude of $\mathbf{F}_3$:

$$F_3 = \sqrt{(-10.36\text{ N})^2 + (-4.50\text{ N})^2} = 11.3\text{ N}$$

(**b**) Both the x and y components of $\mathbf{F}_3$ are negative, so $\mathbf{F}_3$ lies in the third quadrant, as shown in Fig. 1-28. You can find the angle θ between $\mathbf{F}_3$ and the negative x axis by using the tangent function:

$$\tan\theta = \frac{F_{3_y}}{F_{3_x}} = \frac{4.50\text{ N}}{10.36\text{ N}} = 0.434$$

$$\theta = \arctan(0.434) = 23.5°$$

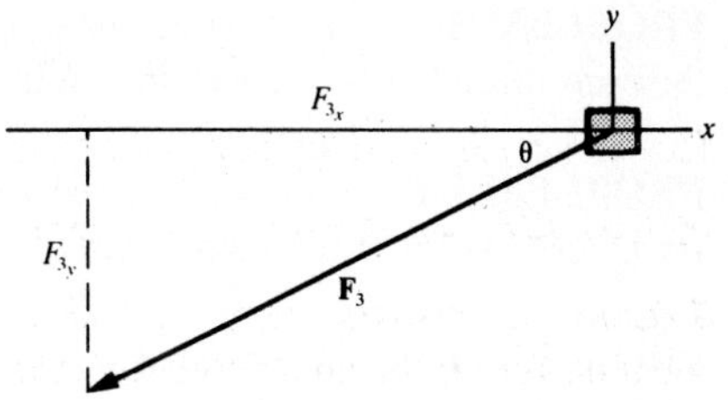

Figure 1-28

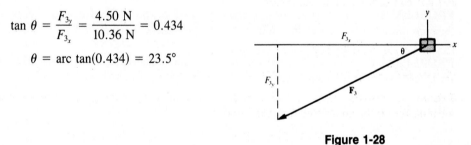

PROBLEM 1-8 The dimensionless vectors **A** and **B**, illustrated in Fig. 1-29, have magnitudes 4 and 6, respectively. Calculate the magnitude of the new vector **C**, which is equal to $\frac{3}{2}\mathbf{A} - \frac{2}{3}\mathbf{B}$.

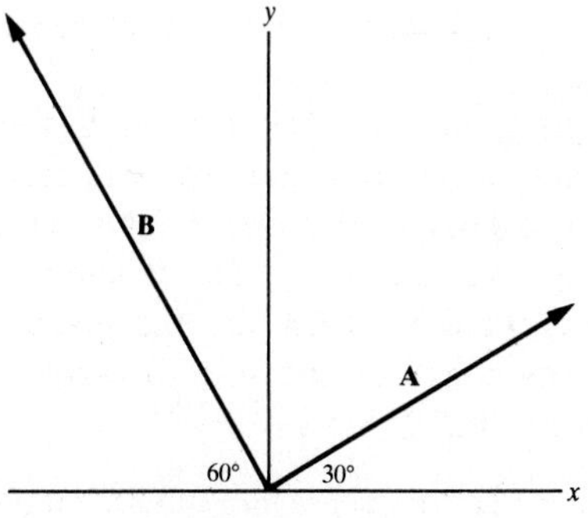

Figure 1-29

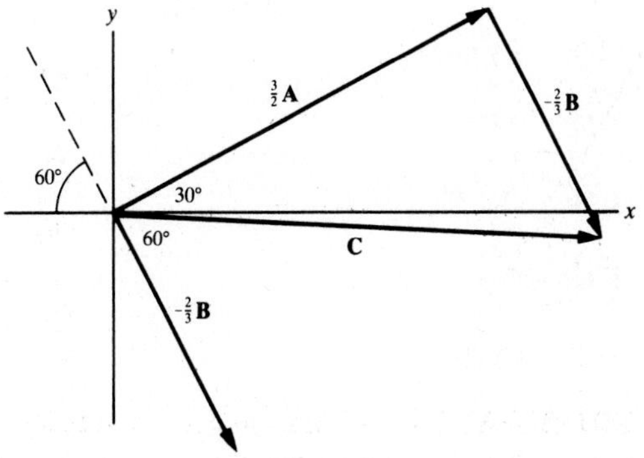

Figure 1-30

Solution Start by drawing the vectors $\frac{3}{2}\mathbf{A}$ and $-\frac{2}{3}\mathbf{B}$. Since **A** has a magnitude of 4, $\frac{3}{2}\mathbf{A}$ will have a magnitude of 6. The vector $-\frac{2}{3}\mathbf{B}$ will have a direction opposite to **B** and a magnitude of 4. These vectors are illustrated in Fig. 1-30. Since these two vectors are at right angles, you can use the Pythagorean theorem to find the magnitude of the sum:

$$|\mathbf{C}| = \sqrt{6^2 + 4^2} = 7.21$$

Supplementary Exercises

PROBLEM 1-9 A velocity **v** has a magnitude of 20 m/s. What is the magnitude of the vector 1.2**v**?

PROBLEM 1-10 A jet takes off at an angle of 20° with respect to the ground. The shadow of the jet, produced by the sun which is directly overhead, moves along the ground at a speed of 430 mi/h. Calculate the speed of the jet.

PROBLEM 1-11 A force has an x component of 3 N and a y component of −5 N. What is the magnitude of this force?

PROBLEM 1-12 A velocity **v** has an x component of 8 m/s and a y component of 5 m/s. What angle does this vector make with respect to the x axis?

PROBLEM 1-13 What property does a vector have that a scalar does not?

PROBLEM 1-14 Two forces have magnitudes of 8 N and 12 N. The angle between them when they are placed tail-to-tail is 40°. What is the magnitude of the sum of these two forces?

PROBLEM 1-15 Two velocity vectors have the same magnitude, 25 m/s, but their directions differ by 10°. What is the magnitude of the difference between these two velocities?

PROBLEM 1-16 A force has a magnitude of 12 N. It causes an object to be displaced a distance of 3 m. The angle between the vectors **F** and **s** is 20°. What is the dot product $\mathbf{F} \cdot \mathbf{s}$ of these two vectors?

PROBLEM 1-17 Vectors **A** and **B** have magnitudes of 3 m and 5 m. The angle between them when they are placed tail-to-tail is 25°. What is the magnitude of the cross product $\mathbf{A} \times \mathbf{B}$?

Answers to Supplementary Exercises

1-9: 24 m/s

1-10: 458 mi/h

1-11: 5.83 N

1-12: 32°

1-13: direction

1-14: 18.8 N

1-15: 4.36 m/s

1-16: 33.8 N m

1-17: 6.34 m^2

2 KINEMATICS: HOW THINGS MOVE

THIS CHAPTER IS ABOUT

☑ Displacement
☑ Velocity
☑ Acceleration
☑ Uniformly Accelerated Motion

Kinematics is the study of the motion of objects. It describes the path of an object using vector quantities—displacement, velocity, and acceleration.

2-1. Displacement

- The net **displacement** $\Delta\mathbf{r}$ of an object is the shortest straight line distance from the object's starting point to its end point—no matter what path the object follows.

Displacement is a vector quantity—it has both magnitude and direction. Consider an object that has moved along a curved path from point A to point B—the dashed line in Figure 2-1. The arrow drawn from A to B represents the net displacement $\Delta\mathbf{r}$ of the object.

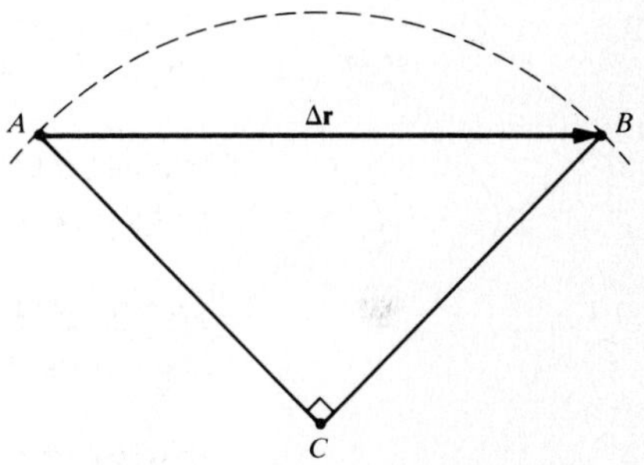

Figure 2-1. Net displacement.

note: The magnitude of the displacement is *not* the *total distance* the object has traveled. The net displacement of a round trip is zero—the object ends up right where it started.

2-2. Velocity

A. Average velocity

- **Average velocity** $\mathbf{v}_{ave}$ is the net displacement $\Delta\mathbf{r}$ divided by the time it takes for the displacement to occur, $\Delta t = t_2 - t_1$:

AVERAGE VELOCITY
$$\mathbf{v}_{ave} = \frac{\Delta\mathbf{r}}{\Delta t} \tag{2-1}$$

Average velocity is a vector quantity. Think of $\mathbf{v}_{ave}$ as the product of the vector $\Delta\mathbf{r}$ and the scalar $1/\Delta t$. Since multiplication by a positive number does not change the direction of a vector, you can see that the average velocity has the same direction as the displacement $\Delta\mathbf{r}$. [Note: The units of velocity are distance/time.]

EXAMPLE 2-1: The object illustrated in Figure 2-1 moves 90° along the arc of a circle whose radius is 20 cm. It moves from A to B in 2 s.
Calculate the magnitude of the average velocity $\mathbf{v}_{ave}$ over this 2 s interval.

Solution: First, calculate the magnitude of the net displacement $\Delta\mathbf{r}$. Since $\Delta\mathbf{r}$ is the hypotenuse of a right triangle, you can use the Pythagorean theorem:

$$|\Delta\mathbf{r}| = \sqrt{(20\text{ cm})^2 + (20\text{ cm})^2} = 28.28\text{ cm}$$

Now, divide by the time interval Δt to get the magnitude of $\mathbf{v}_{ave}$:

$$\mathbf{v}_{ave} = \frac{|\Delta\mathbf{r}|}{\Delta t} = \frac{28.28\text{ cm}}{2\text{ s}} = 14.1\text{ cm/s}$$

note: The magnitude of the average velocity is *not* always the *average speed,* which is defined as the total distance traveled divided by the total elapsed time. For example, the average speed of a boomerang might be 20 mi/h, but its average velocity, assuming it returns to its thrower, is always zero—there's no net displacement.

B. Instantaneous velocity

- **Instantaneous velocity v** or time rate of change of position, is essentially the average velocity over a very short time interval:

INSTANTANEOUS VELOCITY

$$\mathbf{v} = \lim_{\Delta t \to 0} \frac{\Delta \mathbf{r}}{\Delta t} \tag{2-2}$$

If you have studied calculus, you'll recognize this limit as the derivative of **r** with respect to t, which is $\frac{d\mathbf{r}}{dt}$.

note: The magnitude of the instantaneous velocity is called **speed.**

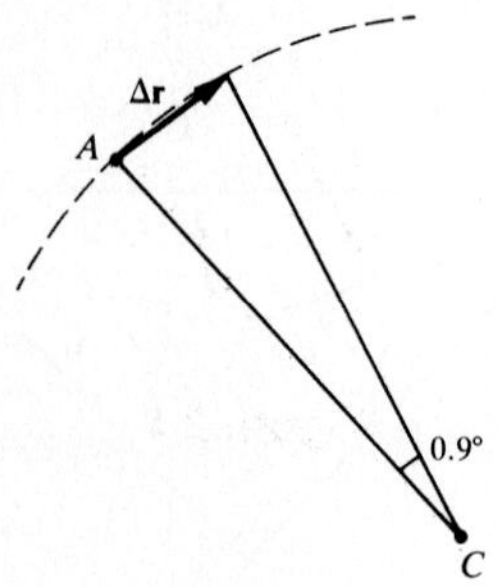

Figure 2-2

EXAMPLE 2-2: Using the data from Example 2-1, calculate the magnitude of the instantaneous velocity **v** of an object as it passes through point A in Figure 2-1.

Solution: If you assume that the object moves around the circle at a constant speed, you can estimate the instantaneous velocity **v** at point A by taking a very small time interval, say 0.02 s. In Figure 2-2, we have drawn this part of the motion on an enlarged scale and exaggerated the angle for clarity.

First, calculate the distance along the arc that the object travels in 2 s. Begin by finding the circumference C of the circle:

$$C = 2\pi r = 2\pi(20 \text{ cm}) = 126 \text{ cm}$$

Since the object moves through a 90° angle, the distance s from A to B that it travels in 2 s will be C divided by 4:

$$s = \frac{C}{4} = 31.4 \text{ cm}$$

You've assumed that the magnitude of the velocity (speed) is constant, so the distance traveled in 0.02 s is 0.02 s/2 s = $\frac{1}{100}$ of the distance from A to B, and

$$\Delta s = \frac{31.4 \text{ cm}}{100} = 0.314 \text{ cm}$$

Because the angle is very small ($\frac{90°}{100} = 0.9°$) the length of the chord (the net displacement $\Delta\mathbf{r}$) is nearly equal to the length of the arc Δs (the actual distance traveled). Assume $|\Delta\mathbf{r}| = \Delta s$, and you can calculate the magnitude of **v** at A. From Eq. (2-2),

$$|\mathbf{v}| = \frac{|\Delta\mathbf{r}|}{\Delta t} = \frac{0.314 \text{ cm}}{0.02 \text{ s}} = 15.7 \text{ cm/s}$$

2-3. Acceleration

A. Average acceleration

- **Average acceleration $\mathbf{a}_{\text{ave}}$** is the *change in velocity* $\Delta\mathbf{v}$ divided by the *time* Δt required for the change to occur:

AVERAGE ACCELERATION

$$\mathbf{a}_{\text{ave}} = \frac{\Delta \mathbf{v}}{\Delta t} \tag{2-3}$$

Since $\Delta\mathbf{v}$ represents a change in velocity and Δt a change in time, average acceleration may also be written as

$$\mathbf{a}_{\text{ave}} = \frac{\mathbf{v}_2 - \mathbf{v}_1}{t_2 - t_1} \tag{2-4}$$

note: The units of acceleration are (distance/time)/time or distance/(time)^2. Also keep in mind that the two vectors $\mathbf{a}_{ave}$ and $\Delta\mathbf{v}$ have the same direction.

EXAMPLE 2-3: In Figure 2-3, the arrow marked $\mathbf{v}_1$ represents the velocity of a moving object at time t_1. The arrow marked $\mathbf{v}_2$ represents the velocity of the same object 0.5 s later. The two velocities have the same magnitude, 50 m/s, but their directions differ by 20°.
Find the magnitude of the average acceleration $\mathbf{a}_{ave}$ of the object during this half-second interval.

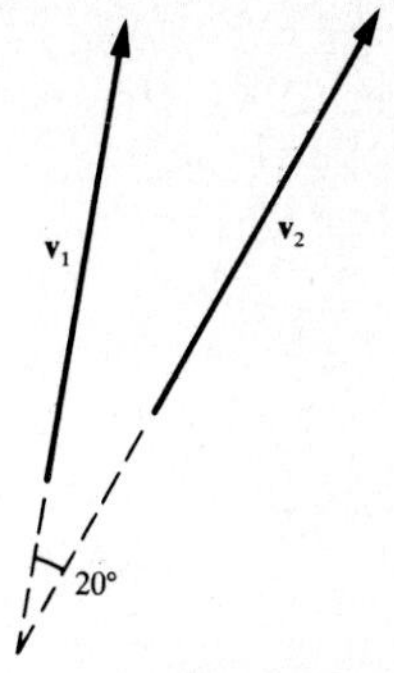

Figure 2-3

Solution: To find the average acceleration, you first have to calculate the change in velocity $\Delta\mathbf{v}$:

$$\Delta\mathbf{v} = \mathbf{v}_2 - \mathbf{v}_1 = \mathbf{v}_2 + (-\mathbf{v}_1)$$

As in Figure 2-4, place the vectors $\mathbf{v}_2$ and $-\mathbf{v}_1$ head-to-tail to obtain $\Delta\mathbf{v}$. Now, find the magnitude of $\Delta\mathbf{v}$ from Eq. (1-2), the law of cosines:

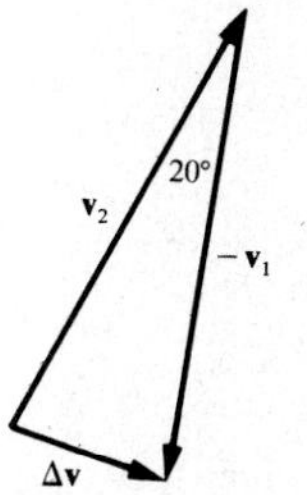

Figure 2-4

$$|\Delta\mathbf{v}| = \sqrt{(50)^2 + (50)^2 - 2(50)(50)\cos 20°} = 17.4 \text{ m/s}$$

From Eq. (2-3), the average acceleration during the half-second interval is

$$|\mathbf{a}_{ave}| = \frac{|\Delta\mathbf{v}|}{\Delta t} = \frac{17.4 \text{ m/s}}{0.5 \text{ s}} = 34.8 \text{ m/s}^2$$

B. Instantaneous acceleration

- **Instantaneous acceleration a,** or the time rate of change of velocity, is the average acceleration over a very short time interval:

INSTANTANEOUS ACCELERATION

$$\mathbf{a} = \lim_{\Delta t \to 0} \frac{\Delta\mathbf{v}}{\Delta t} \qquad \textbf{(2-5)}$$

You can approximate instantaneous acceleration by choosing very small time intervals—just as you can with instantaneous velocity.

EXAMPLE 2-4: An object moves in a circular path with a radius of 20 cm at a constant speed of 15.7 cm/s.
Calculate the instantaneous acceleration **a** for this object.

Solution: As in Example 2-2, a short time interval of 0.02 s will give you a good approximation to the instantaneous acceleration. During the 0.02 s, the object will move 0.9° around the circle. In Figure 2-5, we have exaggerated the angle for clarity. First, find the magnitude of $\Delta\mathbf{v}$ from the law of cosines:

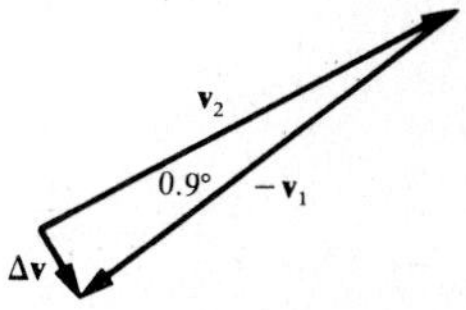

Figure 2-5

$$|\Delta\mathbf{v}| = \sqrt{(15.7)^2 + (15.7)^2 - 2(15.7)(15.7)\cos 0.9°} = 0.247 \text{ cm/s}$$

Now, divide $|\Delta\mathbf{v}|$ by the time interval Δt during which $\Delta\mathbf{v}$ took place:

$$|\mathbf{a}| = \frac{|\Delta\mathbf{v}|}{\Delta t} = \frac{0.247 \text{ cm/s}}{0.02 \text{ s}} = 12.3 \text{ cm/s}^2$$

The direction of **a** is the same as that of $\Delta\mathbf{v}$—toward the center of the circular path.

2-4. Uniformly Accelerated Motion

- **Uniformly accelerated motion** is the motion of an object whose acceleration is constant in magnitude and direction.

A freely falling body is a common example of uniformly accelerated motion. A stone dropped from the roof of a building picks up speed at a constant rate and falls downward vertically due to the earth's gravitational attraction. This acceleration—acceleration due to gravity—is represented by the vector **g**. Near the earth's surface, the magnitude of **g** is $|\mathbf{g}| = 9.8\ \text{m/s}^2 = 32\ \text{ft/s}^2$.

A. Motion in a straight line

- **Motion in a straight line** occurs when the path traveled by an object is linear.

For uniformly accelerated, straight-line motion, we have the following kinematic relationships between the scalar quantities speed v, distance s, and time t:

FOR CONSTANT ACCELERATION

$$v = v_0 + at \tag{2-6}$$

$$v^2 = v_0^2 + 2as \tag{2-7}$$

$$s = v_0 t + \frac{1}{2}at^2 \tag{2-8}$$

$$s = \frac{1}{2}(v_0 + v)t \tag{2-9}$$

$$v_{\text{ave}} = \frac{1}{2}(v_0 + v) \tag{2-10}$$

where v_0 represents the initial speed.

note: We insert negative signs only when we substitute numerically into these equations. The equations are valid in any reference system regardless of which direction we choose as + or −. The distance s is the distance from the starting point s_0 defined as the origin. If you are using a reference system in which the starting point is *not* the origin, that is, $s_0 \neq 0$, then replace s by $(s - s_0)$ in the three equations that contain s. In this chapter, $s_0 = 0$ in all examples and problems.

EXAMPLE 2-5: A small stone is dropped from a height of 4 m above the ground. (**a**) How long does it take the stone to fall to the ground? (**b**) How fast will the stone be moving when it strikes the ground? (**c**) What is the average speed of the stone over the distance of 4 m?

Solution: Let's choose the downward direction to be positive. [Hint: You may choose either direction to be positive—just be sure to stick with your choice throughout the problem.] You already know that $s = 4$ m, $a = 9.8\ \text{m/s}^2$, and $v_0 = 0$; something that is "dropped" or "released" has an initial speed of zero.
(**a**) From Eq. (2-8) with $v_0 = 0$,

$$s = \frac{1}{2}at^2$$

and you solve algebraically for the unknown t:

$$t = \sqrt{\frac{2s}{a}}$$

Putting in the known values, you have

$$t = \sqrt{\frac{2(4\ \text{m})}{9.8\ \text{m/s}^2}} = 0.904\ \text{s}$$

(**b**) You can calculate the final speed v from Eq. (2-7) with $v_0 = 0$:

$$v^2 = 2as = 2(9.8\ \text{m/s}^2)(4\ \text{m})$$

$$v = \sqrt{78.4\ \text{m}^2/\text{s}^2} = 8.85\ \text{m/s}$$

note: Since you have already found the time in part (**a**), you could also use Eq. (2-6) to find v:

$$v = v_0 + at = 0 + (9.8\ \text{m/s}^2)(0.903\ \text{s}) = 8.85\ \text{m/s}$$

(**c**) Now, use Eq. (2-10) to find v_{ave}:

$$v_{\text{ave}} = \frac{1}{2}(v_0 + v) = \frac{1}{2}(0 + 8.85\ \text{m/s}) = 4.43\ \text{m/s}$$

note: You'll get the same answer if you use the definition of average speed:

$$v_{\text{ave}} = \frac{\text{total distance traveled}}{\text{total time elapsed}} = \frac{4\ \text{m}}{0.903\ \text{s}} = 4.43\ \text{m/s}$$

Problems in uniformly accelerated, straight-line motion are more challenging if the initial speed is *not* zero, $v_0 \neq 0$.

EXAMPLE 2-6: A baseball is thrown straight upward with an initial speed of 20 m/s.
(**a**) How high will it go? (**b**) How long will it take to return?

Solution: Let's choose the upward direction as positive this time. You know that $a = -9.8\ \text{m/s}^2$, because the acceleration due to gravity is always downward, and that $v_0 = 20\ \text{m/s}$. You also know that at the top of the baseball's path, $v = 0\ \text{m/s}$, the ball is not moving up or down. Now, pick the equation that will most easily give the answer you're looking for.

(**a**) Since you want to find the height (distance s) the baseball will travel, use Eq. (2-7) because it contains only the one unknown quantity you're seeking. Solve algebraically for s, which in this case is the vertical height:

$$s = \frac{v^2 - v_0^2}{2a} = \frac{(0\ \text{m/s})^2 - (20\ \text{m/s})^2}{2(-9.8\ \text{m/s}^2)} = 20.4\ \text{m}$$

(**b**) Now, use Eq. (2-6) to find the time t_1 required to reach the highest point:

$$t_1 = \frac{v - v_0}{a} = \frac{0 - 20\ \text{m/s}}{-9.8\ \text{m/s}^2} = 2.04\ \text{s}$$

Since you found that $s = 20.4$ m in part (**a**), you could also use Eq. (2-9) to find t_1:

$$t_1 = \frac{2s}{v_0 + v} = \frac{2(20.4\ \text{m})}{20\ \text{m/s} + 0} = 2.04\ \text{s}$$

There's still more to do because you've found only the time required to reach the highest point. You still must calculate the time t_2 that the ball takes to fall back down to the starting point. Since the ball begins the return trip from rest, $v_0 = 0$, and you can write Eq. (2-8) as:

$$s = \frac{1}{2}at_2^2 \quad \text{or} \quad t_2 = \sqrt{\frac{2s}{a}}$$

Put in the known values:

$$t_2 = \sqrt{\frac{2(-20.4\ \text{m/s})}{-9.8\ \text{m/s}^2}} = 2.04\ \text{s}$$

The time for the return trip equals the time required to go up! This is always the case when the two distances are the same, so you can save some effort by making use of this fact:

$$t_{\text{total}} = 2t_1 = 4.08 \text{ s}$$

B. Graphical relationships between *a*, *v*, *s*, and *t*

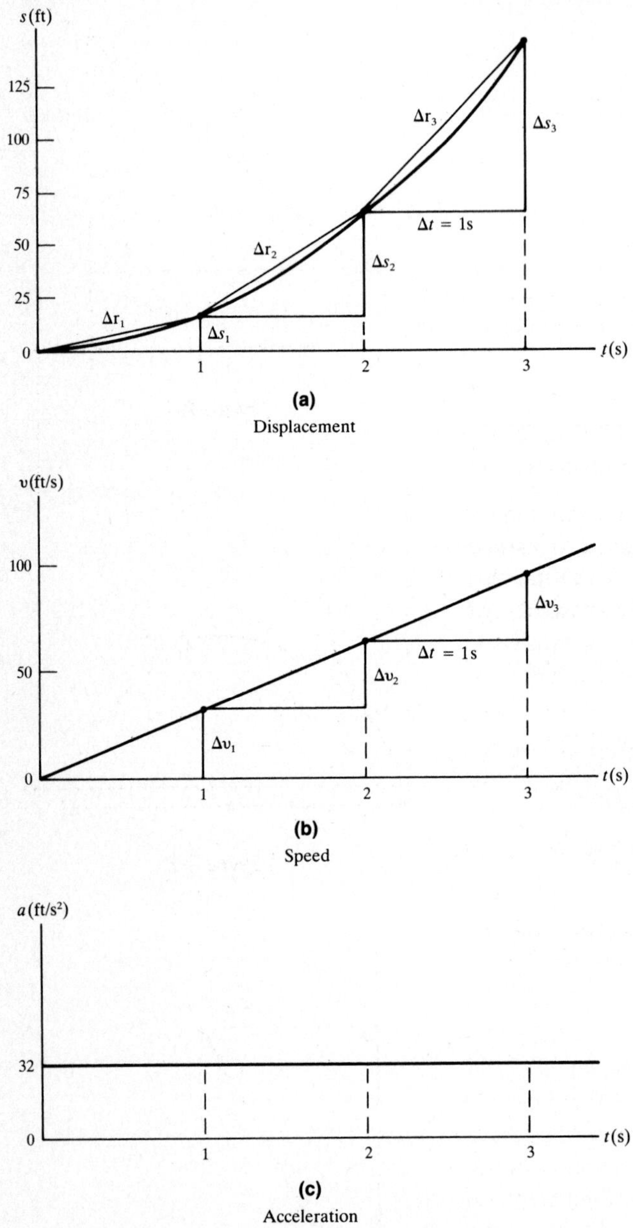

Figure 2-6. Acceleration, speed, and distance as functions of time.

Consider a small stone dropped from the top of a 144 ft cliff. You can use Eq. (2-8) to calculate the distance the stone has fallen at 1, 2, and 3 seconds, and then graph these points. Since $v_0 = 0$, Eq. (2-8) reduces to $s_t = \frac{1}{2}at^2$:

$$s_1 = \frac{1}{2}(32)(1)^2 = 16 \text{ ft}$$

$$s_2 = \frac{1}{2}(32)(2)^2 = 64 \text{ ft}$$

$$s_3 = \frac{1}{2}(32)(3)^2 = 144 \text{ ft}$$

These points are shown in Figure 2-6(a), connected by a smooth curve.

Note that the velocity (speed) of the stone for each second can be found from Eq. (2-6). Since $v_0 = 0$, the equation reduces to $v = at$:

$$v_1 = 32(1) = 32 \text{ m/s}$$

$$v_2 = 32(2) = 64 \text{ m/s}$$

$$v_3 = 32(3) = 96 \text{ m/s}$$

These points are shown in Figure 2-6(b), and they can be connected by a straight line.

Finally, the acceleration for each second is uniform at 32 ft/s²—the acceleration due to gravity—sketched in Figure 2-6(c).

There are *three* important relationships illustrated in Figure 2-6.

(1) The speed at any instant is the slope of the displacement/time curve at that instant: $\lim_{\Delta t \to 0} \Delta s/\Delta t = v$

(2) The area under the speed/time curve for any time interval gives the net displacement during that time. For example, the area under the curve between $t = 0$ and $t = 1$ is a right triangle:

$$A = \frac{1}{2}bh = \frac{1}{2}(1 \text{ s})(32 \text{ ft/s}) = 16 \text{ ft}$$

The area under the curve between t_1 and t_2 is a right triangle and a rectangle:

$$A = \frac{1}{2}bh + \ell w = \frac{1}{2}(1 \text{ s})(32 \text{ ft/s}) + (1 \text{ s})(32 \text{ ft/s}) = 48 \text{ ft}$$

(3) The acceleration at any instant is the slope of the speed/time curve. For uniformly accelerated linear motion, the acceleration is a constant.

note: If you're familiar with differentiation, you can see that these curves are related as follows (given $v_0 = 0$):

Displacement: $s(t) = \frac{1}{2}at^2$

Speed: $v = ds/dt = at$

Acceleration: $dv/dt = a$

EXAMPLE 2-7: The speed of a moving object as a function of time is illustrated in Figure 2-7.
How far did the object move during the first two seconds?

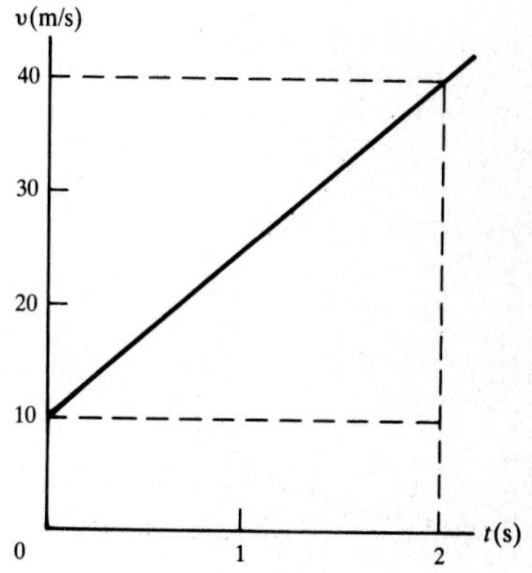

Figure 2-7

Solution: The area under the speed/time curve consists of a rectangle and a triangle, so the total area—the distance traveled s—is

$$s = \text{area of rectangle} + \text{area of triangle}$$

$$= (2\text{ s})(10\text{ m/s}) + \frac{1}{2}(2\text{ s})(30\text{ m/s}) = 50\text{ m}$$

C. Projectile motion

- **Projectile motion** is the path traced by an unpowered object launched at some angle θ above the horizontal and moving only under the influence of gravity.

In projectile motion, there are two motions occurring at the same time—*horizontal* motion, which is uniform motion in a straight line, and *vertical* motion, which is uniformly accelerated motion. Each motion is independent of the other, so we resolve the velocity into its horizontal and vertical components and treat each motion separately.

EXAMPLE 2-8: A stone is thrown at an angle of 60° above the horizontal with a speed of 20 m/s. Vector $\mathbf{v}_0$ in Figure 2-8 represents the initial velocity of the stone. Its subsequent motion will be confined to the xy plane.
How high will the stone go? [Hint: Use the vertical component to find distance in the air.]

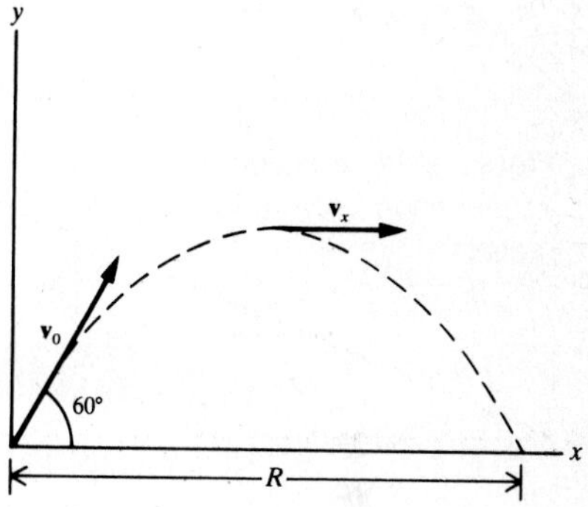

Figure 2-8

Solution: Your first step is to find the x and y components of v_0, which are drawn in Figure 2-9. The method for resolving a vector is described in Example 1-4.

$$v_{x_0} = v_0 \cos 60° = (20\text{ m/s})(0.5) = 10\text{ m/s}$$

$$v_{y_0} = v_0 \sin 60° = (20\text{ m/s})(0.866) = 17.3\text{ m/s}$$

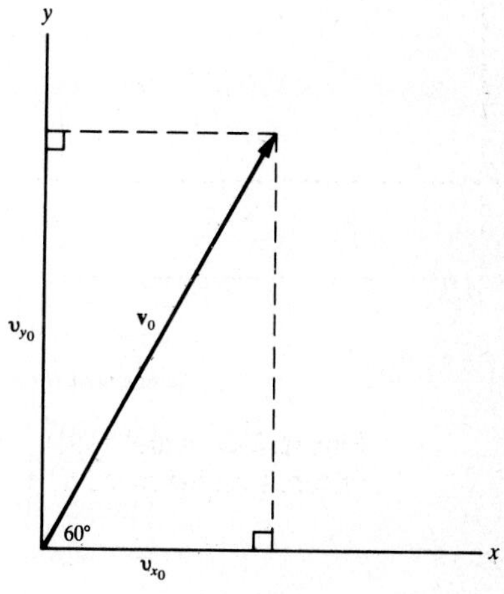

Figure 2-9

When the stone reaches the highest point of its path, it will be moving exactly horizontally and the vertical component of its velocity will be zero, $v_y = 0$. Since the vertical motion takes place at a constant acceleration of 9.8 m/s^2 downward, $a = -9.8\text{ m/s}^2$, you can use one of the Eqs. (2-6) through (2-10). Find the vertical height s (the distance the stone travels upwards) with Eq. (2-7):

$$v_y^2 = v_{y0}^2 + 2as$$

$$s = \frac{v_y^2 - v_{y0}^2}{2a} = \frac{0 - (17.3\text{ m/s})^2}{2(-9.8\text{ m/s}^2)} = 15.3\text{ m}$$

EXAMPLE 2-9: How far downrange will the stone in Example 2-8 (Figure 2-8) land on level ground? [Hint: You'll want to use the horizontal component to find the range of the projectile.]

Solution: The range R of the stone is the horizontal distance s it covers, so to find R you'll first have to find the total time of the stone's flight. You could double the time needed to reach the highest point, as you did in Example 2-6, or you could use the fact that the velocity of the stone just before it strikes the ground has the same magnitude as the stone's original velocity—except that the y components have reversed direction:

$$v_{y_0} = 17.3 \text{ m/s} \qquad v_y = -17.3 \text{ m/s}$$

Now, you can find t from Eq. (2-6):

$$v_y = v_{y_0} + at$$

$$t = \frac{v_y - v_{y_0}}{a} = \frac{-17.3 \text{ m/s} - 17.3 \text{ m/s}}{-9.8 \text{ m/s}^2} = 3.53 \text{ s}$$

You can assume that air resistance is negligible, so the horizontal motion takes place at a constant speed: $v_x = v_{x_0}$ = a constant. This means that the distance traveled R divided by the time t is equal to the horizontal speed v_x:

$$v_x = \frac{R}{t}$$

$$R = v_x t = (10 \text{ m/s})(3.53 \text{ s}) = 35.3 \text{ m}$$

warning: Don't assume that the direction of the position, velocity, and acceleration vectors are *necessarily* the same. For projectile motion, these three vectors generally have different directions:

- the position vector **r** varies with time;
- the velocity vector **v** is always tangent to the path traveled and points in the direction of travel;
- the acceleration vector **a** is vertical.

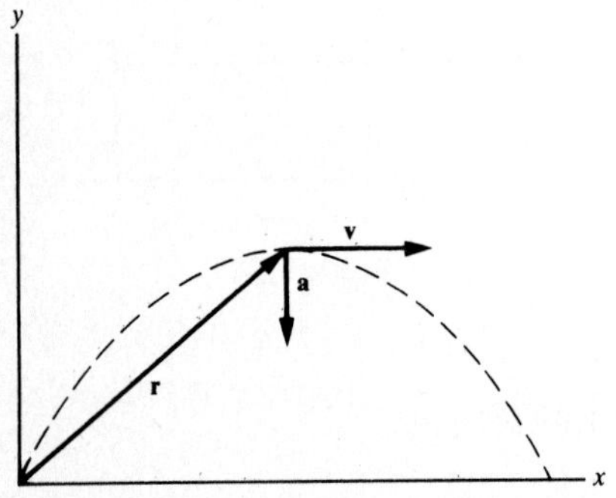

Figure 2-10. Position, velocity, and acceleration vectors for projectile motion.

Figure 2-10 shows these vectors for a projectile at the top of its trajectory. There is no point along this trajectory at which the three vectors have the same direction.

SUMMARY

Average velocity	$\mathbf{v}_{\text{ave}} = \dfrac{\Delta \mathbf{r}}{\Delta t}$	net displacement divided by the time involved
Instantaneous velocity	$\mathbf{v} = \lim\limits_{\Delta t \to 0} \dfrac{\Delta \mathbf{r}}{\Delta t}$	average velocity over a very short time interval
Average acceleration	$\mathbf{a}_{\text{ave}} = \dfrac{\Delta \mathbf{v}}{\Delta t}$ or $\dfrac{\mathbf{v}_2 - \mathbf{v}_1}{t_2 - t_1}$	change in velocity divided by the time required for the change
Instantaneous acceleration	$\mathbf{a} = \lim\limits_{\Delta t \to 0} \dfrac{\Delta \mathbf{v}}{\Delta t}$	average acceleration over a very short time interval
Straight-line motion with constant acceleration	$v = v_0 + at$ $v^2 = v_0^2 + 2as$ $s = v_0 t + \frac{1}{2}at^2$ $s = \frac{1}{2}(v_0 + v)t$ $v_{\text{ave}} = \frac{1}{2}(v_0 + v)$	kinematic relationships between acceleration a, speed v, distance s, and time t for uniformly accelerated, straight-line motion

RAISE YOUR GRADES

Can you define . . . ?

- ☑ displacement
- ☑ instantaneous velocity
- ☑ average acceleration
- ☑ speed
- ☑ uniformly accelerated motion
- ☑ projectile motion

Can you . . . ?

- ☑ find the average velocity of a moving object
- ☑ express the distinction between velocity and speed
- ☑ explain how the direction of **a** is related to the direction of $\Delta\mathbf{v}$
- ☑ define the conditions under which the equations

$$v = v_0 + at \qquad v^2 = v_0^2 + 2as \quad \text{and} \quad s = v_0t + \tfrac{1}{2}at^2$$

  are valid
- ☑ find the time for an object to fall a given distance
- ☑ find the initial speed required to throw a stone to a given height
- ☑ find the speed of an object that has fallen through a given distance
- ☑ find the acceleration of an object from a graph of v versus t
- ☑ find the distance traveled by an object from a graph of v versus t
- ☑ find the horizontal distance traveled by a projectile when the initial velocity is given

SOLVED PROBLEMS

Displacement

PROBLEM 2-1 During a time interval of 5 s, an object moves from the point (1 m, 3 m) in the xy plane to the point (4 m, −2 m). Find **(a)** the magnitude and **(b)** direction of the displacement.

Solution

(a) First, make a drawing that shows the initial position vector $\mathbf{r}_1$ and the later position vector $\mathbf{r}_2$, as we have done in Fig. 2-11. Next, write down the equation for the displacement $\Delta\mathbf{r}$, which is the change in position:

$$\Delta\mathbf{r} = \mathbf{r}_2 - \mathbf{r}_1 = \mathbf{r}_2 + (-\mathbf{r}_1)$$

Now, you'll need to make a table of the vectors to be added and their components. You worked a similar problem in Example 1-9. The x component of $\Delta\mathbf{r}$ is the sum of the x components of $\mathbf{r}_2$ and $-\mathbf{r}_1$, and the y component of $\Delta\mathbf{r}$ is the sum of the y components of $\mathbf{r}_2$ and $-\mathbf{r}_1$.

Vector	x component	y component
$\mathbf{r}_2$	4 m	−2 m
$-\mathbf{r}_1$	−1 m	−3 m
$\Delta\mathbf{r}$	3 m	−5 m

Now, you can find the magnitude of $\Delta\mathbf{r}$ from the Pythagorean theorem:

$$|\Delta\mathbf{r}| = \sqrt{(\Delta\mathbf{r}_x)^2 + (\Delta\mathbf{r}_y)^2} = \sqrt{(3\text{ m})^2 + (-5\text{ m})^2} = \sqrt{34}\text{ m} = 5.83\text{ m}$$

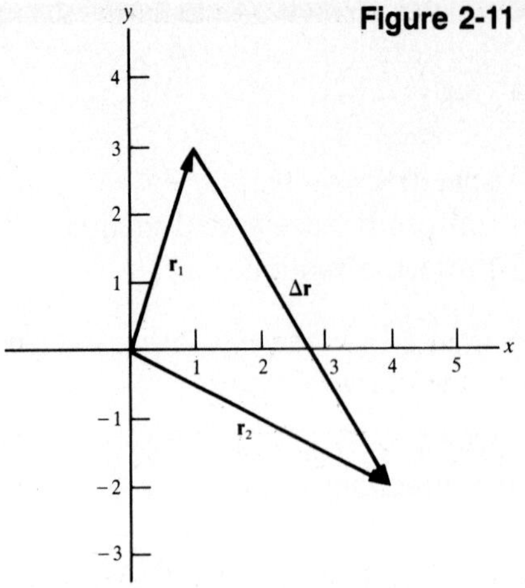

Figure 2-11

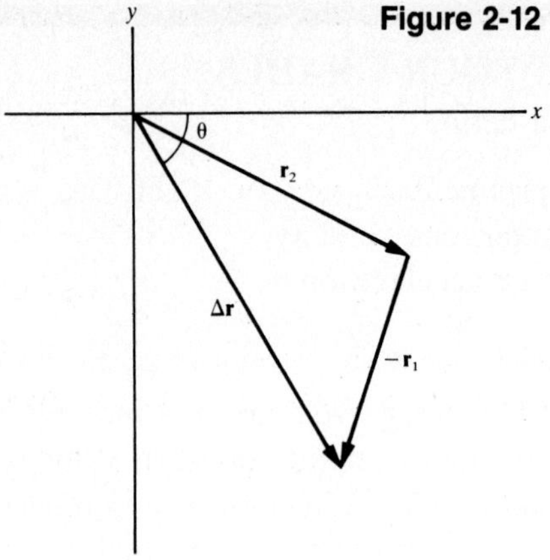

Figure 2-12

(**b**) In Fig. 2-12, we show how to obtain $\mathbf{\Delta r}$ by adding $\mathbf{r}_2$ and $-\mathbf{r}_1$. [**recall**: Find the angle θ that $\mathbf{\Delta r}$ makes with the x axis by using the tangent function (see Problem 1-3).]

$$\tan\theta = \frac{\Delta \mathbf{r}_y}{\Delta \mathbf{r}_x} = \frac{-5\text{ m}}{3\text{ m}} = -1.67$$

$$\theta = -59.0°$$

The negative sign indicates that θ is measured in the clockwise direction from the positive x axis.

Velocity

PROBLEM 2-2 Find the average velocity $\mathbf{v}_{\text{ave}}$ of the moving object described in Problem 2-1. [Hint: To find velocity, you must find direction as well as magnitude.]

Solution Start by writing down Eq. (2-1):

$$\mathbf{v}_{\text{ave}} = \frac{\Delta \mathbf{r}}{\Delta t} = \Delta \mathbf{r}(1/\Delta t)$$

Then you can easily find the magnitude of $\mathbf{v}_{\text{ave}}$ by substituting into the equation:

$$|\mathbf{v}_{\text{ave}}| = \frac{|\Delta \mathbf{r}|}{\Delta t} = \frac{5.83\text{ m}}{5\text{ s}}$$

$$|\mathbf{v}_{\text{ave}}| = 1.17\text{ m/s}$$

[**recall**: Multiplication of a vector by a scalar does not change the direction of the vector (see Sec. 2-2). The direction of $\mathbf{v}_{\text{ave}}$ is the same as that of $\mathbf{\Delta r}$: 59.0° below the positive x axis.]

Acceleration

PROBLEM 2-3 Find the average acceleration of an automobile that turns a sharp corner in 3 s, as illustrated in Fig. 2-13. The speed of the car is 11 m/s before and after the turn.

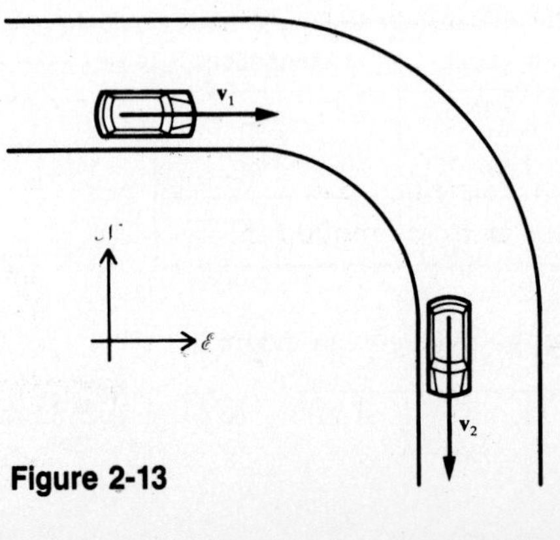

Figure 2-13

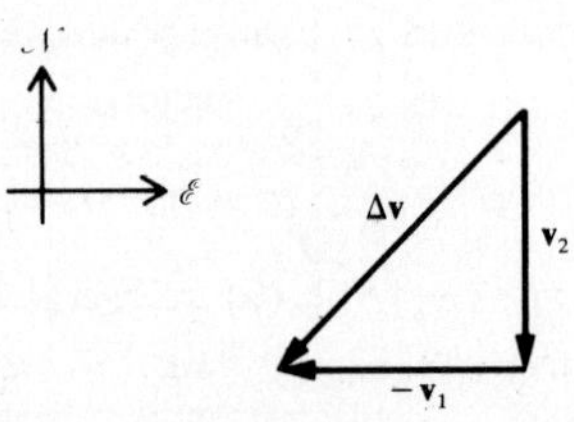

Figure 2-14

Solution Write down Eq. (2-3) or (2-4) defining average acceleration—the unknown:

$$\mathbf{a}_{\text{ave}} = \frac{\Delta \mathbf{v}}{\Delta t} \quad \text{or} \quad \mathbf{a}_{\text{ave}} = \frac{\mathbf{v}_2 - \mathbf{v}_1}{t_2 - t_1}$$

You are again faced with the subtraction of one vector from another, a problem you solved in Example 1-9.

$$\Delta \mathbf{v} = \mathbf{v}_2 + (-\mathbf{v}_1)$$

Make a drawing of the vectors to be added, which should look like Fig. 2-14. Since the vectors you're adding are at right angles, use the Pythagorean theorem to find $\Delta \mathbf{v}$:

$$|\Delta \mathbf{v}| = \sqrt{(\mathbf{v}_2)^2 + (-\mathbf{v}_1)^2} = \sqrt{(11\ \text{m/s})^2 + (11\ \text{m/s})^2}$$

$$|\Delta \mathbf{v}| = 15.56\ \text{m/s}$$

You can see from the figure that $\Delta \mathbf{v}$ makes an angle of 45° with the east-west axis, so the direction of $\Delta \mathbf{v}$ is southwest.

Next, calculate the magnitude of the average acceleration. Write down the equation,

$$|\mathbf{a}_{\text{ave}}| = \frac{|\Delta \mathbf{v}|}{\Delta t}$$

Then, substitute the known values, and solve:

$$|\mathbf{a}_{\text{ave}}| = \frac{15.56\ \text{m/s}}{3\ \text{s}} = 5.19\ \text{m/s}^2$$

The direction of $\mathbf{a}_{\text{ave}}$ is the same as that of $\Delta \mathbf{v}$—southwest.

Uniformly Accelerated Motion

PROBLEM 2-4 A rock is thrown vertically upward from the edge of a rooftop with an initial speed of 12 m/s. The rock misses the edge of the building and strikes the ground 2.994 s after being thrown. (**a**) Find the maximum height h_1, measured from the top of the building, and (**b**) the height h_2 of the roof above the ground.

Solution First, make a drawing of the situation, similar to Fig. 2-15. Since the acceleration has a constant magnitude and direction, you can use any of the equations for uniformly accelerated motion.

(**a**) Choose an origin and decide on a positive direction. Let's call the upward direction positive and measure distances from the initial position of the rock—the rooftop. Since you know that the speed v at the highest point is zero, you can use Eq. (2-7) to find the vertical distance h_1:

$$v^2 = v_0^2 + 2as$$

Now, solve algebraically for the unknown:

$$h_1 = s = \frac{v^2 - v_0^2}{2a}$$

Next, substitute all of the known quantities. Remember that the gravitational acceleration is downward and, therefore, negative.

$$h_1 = \frac{0 - (12\ \text{m/s})^2}{2(-9.8\ \text{m/s}^2)} = 7.35\ \text{m}$$

The rock rises to a height of about 7.35 m above the roof.

(**b**) You can use Eq. (2-8) to find the distance from the rooftop to the ground:

$$s = v_0 t + \frac{1}{2}at^2 = (12\ \text{m/s})(2.994\ \text{s})$$

$$+ \frac{1}{2}(-9.8\ \text{m/s}^2)(2.994\ \text{s})^2 = -8.00\ \text{m}$$

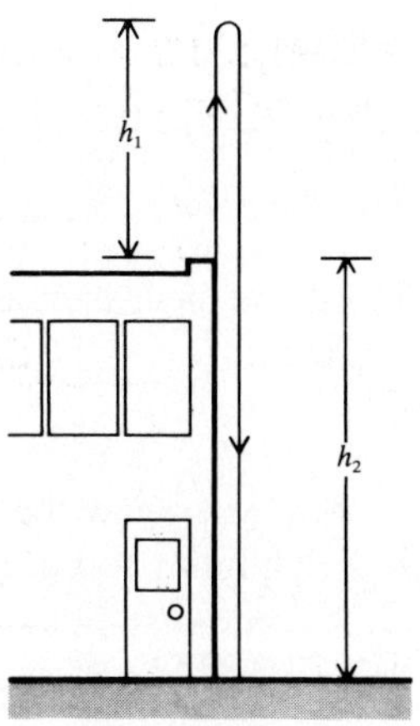

Figure 2-15

The negative sign is just what you should have expected. It means that the rock landed 8.00 m below the origin, which you placed at the top of the building, so the roof is about 8 m above the ground.

PROBLEM 2-5 Two marbles are thrown vertically upward from the same point. The second marble has a greater initial speed and reaches a maximum height 50% greater than the first. Find the ratio of their initial speeds v_2/v_1.

Solution Since you are considering two marbles, use subscript 1 for the first marble and subscript 2 for the second. The speed at the maximum height for both marbles is zero ($v = 0$), so use Eq. (2-7) for each marble to find the initial speed. Since the acceleration is downward, $a = -g$.

For the first marble:

$$v^2 = v_0^2 + 2as$$
$$0 = v_1^2 + 2as_1$$
$$v_1^2 = 2gs_1$$
$$v_1 = \sqrt{2gs_1}$$

For the second marble:

$$v^2 = v_0^2 + 2as$$
$$0 = v_2^2 + 2as_2$$
$$v_2^2 = 2gs_2$$
$$v_2 = \sqrt{2gs_2}$$

Then, divide the second equation by the first to get the ratio of their initial speeds:

$$\frac{v_2}{v_1} = \sqrt{\frac{s_2}{s_1}}$$

The maximum height of the second marble is 1.5 times that of the first, so

$$\frac{v_2}{v_1} = \sqrt{\frac{s_2}{s_1}} = \sqrt{\frac{1.5s_1}{s_1}} = 1.23$$

PROBLEM 2-6 A stone is thrown from the roof of a building 8 m in height at an angle of 40° above the horizontal. It strikes the ground 3.143 s later at a horizontal distance of 48.16 m from the base of the building. (**a**) Find the initial speed of the stone and (**b**) the maximum height that it reaches above the roof.

Solution Make a diagram showing the stone's path. Your drawing should look like Fig. 2-16.

(**a**) Since the horizontal motion v_x takes place at a constant speed, you can use the following relationship (see Example 2-9):

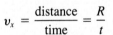

Figure 2-16

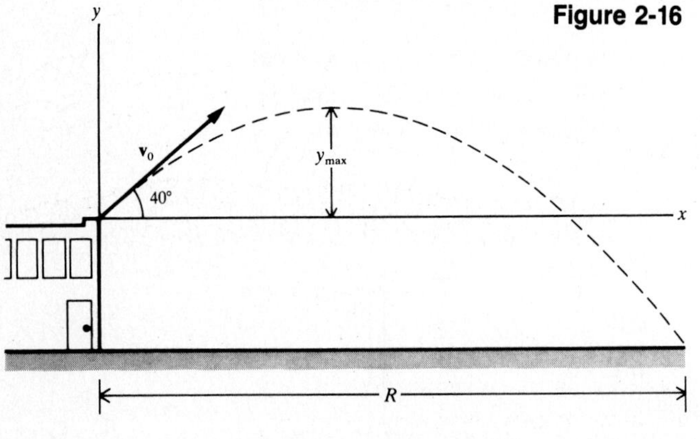

$$v_x = \frac{\text{distance}}{\text{time}} = \frac{R}{t}$$

Substituting the given values of distance and time:

$$v_x = \frac{48.16\ \text{m}}{3.143\ \text{s}} = 15.32\ \text{m/s}$$

This horizontal speed is the horizontal component of the initial velocity $\mathbf{v}_0$, so

$$v_x = v_0 \cos 40°$$

$$v_0 = \frac{v_x}{\cos 40°} = \frac{15.32}{0.7660} = 20.0\ \text{m/s}$$

(**b**) To find the maximum height the stone reaches, you first calculate the initial value of the vertical speed:

$$v_{y_0} = v_0 \sin 40° = (20\ \text{m/s})(0.6428) = 12.86\ \text{m/s}$$

Now you know the initial and final speeds, and the acceleration for the vertical motion, $g = -9.8\ \text{m/s}^2$, so Eq. (2-7) contains only one unknown—the vertical distance y:

$$v_y^2 = v_{y_0}^2 + 2ay \qquad y = \frac{v_y^2 - v_{y_0}^2}{2a} = \frac{0 - (12.86\ \text{m/s})^2}{2(-9.8\ \text{m/s}^2)} = 8.44\ \text{m}$$

The maximum height of the stone above the roof of the building is $y_{max} = 8.44$ m.

PROBLEM 2-7 An automobile travels due north for 20 min at a constant speed of 45 mi/h. The driver makes a right turn and drives due east for 30 min at a constant speed of 40 mi/h. Then the driver makes yet another right turn and drives due south for 6 min at a speed of 50 mi/h. (**a**) What is the total distance the automobile travels and (**b**) what is the magnitude of the net displacement?

Solution

(**a**) Since the speed is constant during each leg of the trip, you can use the equation $v = s/t$, which is valid only if v is constant. The magnitudes of the displacements are

$$s_1 = vt = (45\ \text{mi/h})(1/3\ \text{h}) = 15\ \text{mi}$$

$$s_2 = vt = (40\ \text{mi/h})(1/2\ \text{h}) = 20\ \text{mi}$$

$$s_3 = vt = (50\ \text{mi/h})(0.1\ \text{h}) = 5\ \text{mi}$$

The total distance traveled is $s_1 + s_2 + s_3 = 15 + 20 + 5 = 40$ mi.

(**b**) Now, make a drawing showing the three displacements and their resultant. Your drawing should look like Fig. 2-17. Use the method of components to find the net resultant:

Displacements	x components	y components
$\mathbf{s}_1$	0	15 mi
$\mathbf{s}_2$	20 mi	0
$\mathbf{s}_3$	0	−5 mi
$\mathbf{s}$	20 mi	10 mi

$$\mathbf{s} = \sqrt{(20\ \text{mi})^2 + (10\ \text{mi})^2} = 22.4\ \text{mi}$$

The magnitude of the net displacement is 22.4 mi.

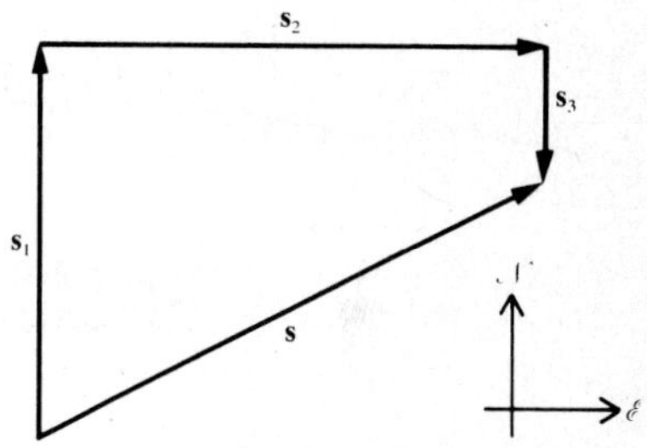

Figure 2-17

PROBLEM 2-8 Two Explorer Scouts paddle their canoe across a river at a steady speed of 3 mi/h. The river is flowing due west at a speed of 2 mi/h. The two scouts are heading their canoe due north, but much to their consternation, the canoe is moving roughly toward the northwest because of the flow of the river. What is the resultant velocity of the canoe with respect to the shore?

Solution Make a diagram, similar to Fig. 2-18, showing the three velocities involved. Since $\mathbf{v}_w$ (the velocity of the water) is perpendicular to $\mathbf{v}_{cw}$ (the velocity of the canoe with respect to the water), you can find the magnitude of the resultant velocity $\mathbf{v}$ from the Pythagorean theorem:

$$|\mathbf{v}| = \sqrt{v_w^2 + v_{cw}^2} = \sqrt{(2\ \text{mi/h})^2 + (3\ \text{mi/h})^2} = 3.61\ \text{mi/h}$$

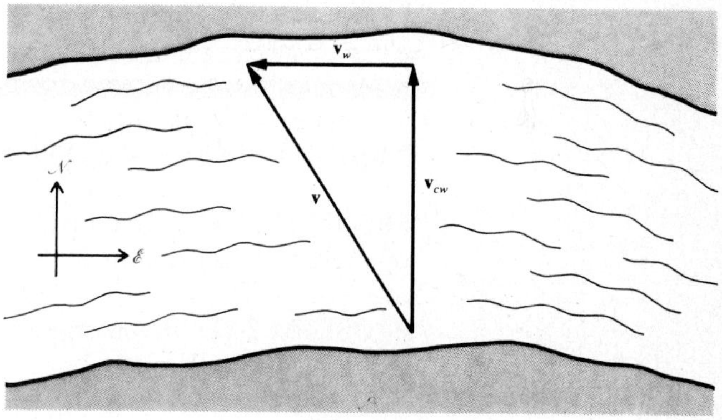

Figure 2-18

PROBLEM 2-9 In Problem 2-8, it is obvious that the scouts will land downstream from the point at which they launched their canoe. If the river is 0.1 mi wide, how far downstream will they be when they reach the opposite shore?

Solution You first have to know how long it takes to reach the opposite shore. [Hint: If the river banks are parallel, the time to cross the river is the same whether or not the river is flowing.]

$$v_{cw} = \frac{s}{t} \qquad t = \frac{s}{v_{cw}} = \frac{0.1\ \text{mi}}{3\ \text{mi/h}} = 0.033\ \text{h}$$

During this time the flowing water will carry the boat downstream at a speed of 2 mi/h, so that the distance traveled downstream will be

$$s = v_w t = (2\ \text{mi/h})(0.033\ \text{h}) = 0.066\ \text{mi}$$

PROBLEM 2-10 Suppose the scouts in Problem 2-8 want to cross the river without drifting downstream so they will land on the other side directly opposite their point of departure. (**a**) In what direction must they point their canoe and (**b**) how long will it take them to make the crossing?

Solution

(**a**) You'll need to make another vector diagram of this new situation. Note that the vector sum of $\mathbf{v}_w$ and $\mathbf{v}_{cw}$ must now be due north, as shown in Fig. 2-19. Since the vector diagram still forms a right triangle, you can use the Pythagorean theorem to find the resultant speed of the canoe with respect to the shore:

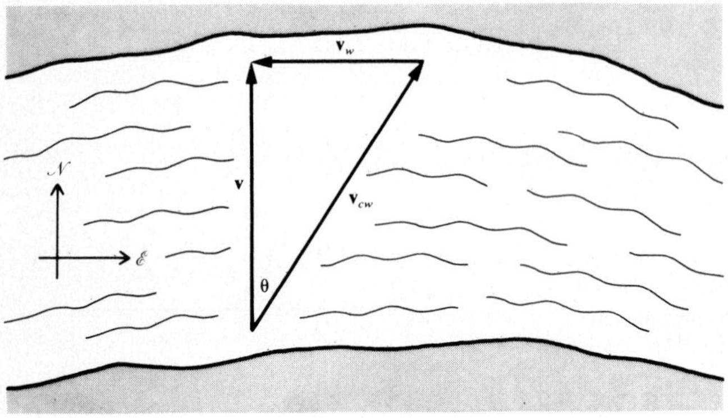

Figure 2-19

$$v = \sqrt{v_{cw}^2 - v_w^2}$$

$$= \sqrt{(3\ \text{mi/h})^2 - (2\ \text{mi/h})^2}$$

$$= 2.236\ \text{mi/h}$$

Now, you can use the sine function to determine the angle θ at which the canoe must be headed:

$$\sin\theta = \frac{v_w}{v_{cw}} = \frac{2\ \text{mi/h}}{3\ \text{mi/h}} = 0.667$$

$$\theta = \text{arc}\sin(0.667) = 41.8°$$

The scouts will have to point the canoe 41.8° east of north.

(**b**) Now, use the resultant speed v to calculate the time required for crossing the river:

$$t = \frac{s}{v} = \frac{0.1\ \text{mi}}{2.236\ \text{mi/h}} = (0.0447\ \text{h})(60\ \text{min/h})$$

$$= 2.68\ \text{min}$$

Supplementary Exercises

PROBLEM 2-11 If you drive 4 km due west and then 1 km due north, what is the magnitude of your resultant displacement?

PROBLEM 2-12 If you drive 5 km due west and then 1 km due north, what is the direction of your resultant displacement?

PROBLEM 2-13 If it takes a total of 7.5 min to drive 4 km due east and 6 km due south, what is the magnitude of the average velocity in km/h?

PROBLEM 2-14 An object moves along the arc of a circle whose radius is 12 cm. It travels half-way around the circle in 0.12 s. What is the average speed of this object?

PROBLEM 2-15 An automobile travels due north along a straight section of a highway. During a 6 s interval the driver increases her speed from 50 mi/h to 60 mi/h. What is the acceleration of the automobile in mi/h^2?

PROBLEM 2-16 An automobile travels due north at a speed of 80 km/h. Because of a smooth curve in the highway, its direction changes to 40° west of north during an 8 s time interval, but its speed remains at 80 km/h. Calculate the magnitude of the average acceleration in km/s^2 during this 8 s interval.

PROBLEM 2-17 A smooth steel ball is dropped from a height of 3.0 m. How long does it take to fall 3.0 m when released from rest?

PROBLEM 2-18 A small steel ball is dropped from a height of 4 m. Its initial speed is zero. What is the speed of the ball when it has fallen a distance of 4 m?

PROBLEM 2-19 An arrow is shot into the air at an angle of 30° with respect to the horizontal. What maximum height will it reach if its initial speed is 12 m/s?

PROBLEM 2-20 An arrow is shot into the air at an angle of 50° with respect to the horizontal. Its initial speed is 16 m/s and its initial height above the level ground is 2.0 m. How long does it take for the arrow to return to a height of 2.0 m above the ground?

PROBLEM 2-21 What is the horizontal distance traveled by the arrow in Problem 2-21 when it returns to a height of 2.0 m above the ground?

PROBLEM 2-22 The speed of a moving object as a function of time is illustrated in Fig. 2-20. How far did the object move during the first 7 s?

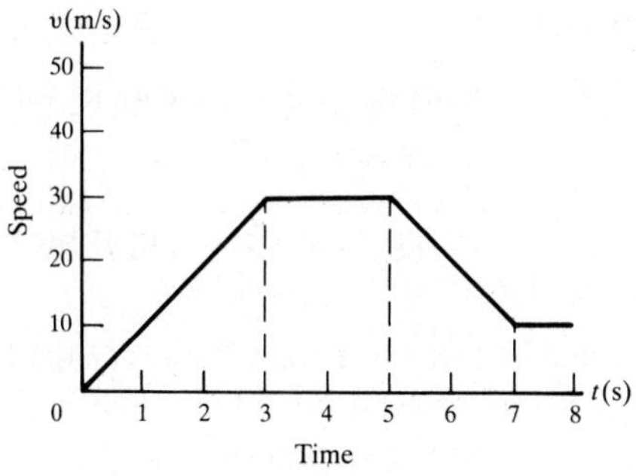

Figure 2-20

Answers to Supplementary Exercises

2-11: 4.12 km

2-12: 11.3° north of west

2-13: 57.7 km/h

2-14: 314 cm/s

2-15: 6000 mi/h^2

2-16: 1.90×10^{-3} km/s^2

2-17: 0.782 s

2-18: 8.85 m/s

2-19: 1.84 m

2-20: 2.50 s

2-21: 25.73 m

2-22: 145 m

3 DYNAMICS: WHY THINGS MOVE

THIS CHAPTER IS ABOUT

- ☑ Force
- ☑ Newton's Laws of Motion
- ☑ Weight, Mass, and Gravitation
- ☑ Friction
- ☑ Systems of Connected Bodies

3-1. Force

- **Force F** is what causes the *acceleration* of an object; that is, the technical term for what we commonly call a push, shove, or pull.

There must be a net force acting on an object in order to change its velocity in direction or magnitude.

- **Net force** $\Sigma \mathbf{F}$ is the unbalanced, or resultant, force that acts on an object.

note: The symbol Σ means to take the sum.

If there are *no forces* acting on an object, or if there are *equal and opposite forces* acting on it, the object will *not* accelerate.

note: The SI unit of force is a combination of the base units of *mass, length,* and *time* called the **newton** N, a force that will impart an acceleration of one meter per second per second to a one-kilogram mass:

$$1\text{ N} = \frac{1\text{ kg m}}{\text{s}^2}$$

Newton's laws of motion give the quantitative relationships between force, mass, and acceleration.

3-2. Newton's Laws of Motion

A. Newton's first law

Newton's first law describes the motion of an isolated object which, for our purposes, is an object on which no net force is acting—an object such as a small fragment of matter in interstellar space.

- If an isolated object is at rest, it will remain at rest; if it is in motion, it will continue moving along a straight line at a constant speed.

The first law applies to an object with several forces that balance one another to produce zero net force, as well as to an object that has no force acting on it.

note: Newton's first law is often called the law of inertia. **Inertia** is the property of an object that resists acceleration.

NEWTON'S FIRST LAW OF MOTION $$\text{If } \Sigma \mathbf{F} = 0, \text{ then } \mathbf{v} = \text{constant} \qquad \textbf{(3-1)}$$

An object with a constant velocity is said to be in a state of **equilibrium**; its direction remains unchanged (it moves along a straight line), and its speed (the magnitude of its velocity) is constant—because the sum of the forces acting on the object equals zero. Keep in mind that a speed of zero (object at rest) is a constant speed.

EXAMPLE 3-1: A small meteor fragment moves through interstellar space at a speed of 200 km/s directly toward the star Vega.
Predict the state of motion of this meteor fragment three days from now.

Solution: If you assume that there are no forces acting on the meteor—a reasonable assumption in interstellar space—then it will continue moving in a straight line at a constant speed. Three days from now it will still be moving toward the star Vega at a speed of 200 km/s.

B. Newton's second law

Newton's second law of motion states

- If the sum of all forces acting on an object is *not* zero, then the object will be accelerated.

The acceleration produced depends on the sum of the forces $\Sigma\mathbf{F}$ and on the mass m of the object:

NEWTON'S SECOND LAW OF MOTION

$$\text{net } \mathbf{F} = m\mathbf{a} \quad \text{or} \quad \Sigma\mathbf{F} = m\mathbf{a} \tag{3-2}$$

Both "net **F**" and $\Sigma\mathbf{F}$ stand for the vector sum of all forces acting on an object or system of objects whose mass is m. Mass also plays an important role in determining acceleration—you can see this more easily if you solve Eq. (3-2) for acceleration:

$$\mathbf{a} = \frac{\text{net } \mathbf{F}}{m}$$

If you think of inertia as the qualitative term for the property of a body that resists acceleration, then **mass** (a scalar quantity) is the quantitative measure of inertia. If the mass is large, the acceleration produced by a given force will be small.

note: Keep in mind that you must find *all* the forces acting on an object to determine its acceleration—not just the most obvious ones. Many times you'll discover that a force acting on an object is balanced by an equal and opposite force—such as weight balanced by the upward push of a tabletop, or of the floor—so that the forces cancel out. And if no forces act on an object, its velocity does not change. Just remember—the *net* force acting on an object is what causes it to accelerate.

EXAMPLE 3-2: A mass of 3 kg is subject to forces in the horizontal plane—a force of 2 N toward the north and a force of 3 N toward the east.
What is the magnitude of the acceleration produced by these forces?

Solution: First determine the resultant or net force $\Sigma\mathbf{F}$ acting on the object. You'll find it helps to make a drawing of the forces involved (see Figure 3-1). Since these forces are at right angles, you can easily find the magnitude of their resultant from the Pythagorean theorem:

$$|\Sigma\mathbf{F}| = \sqrt{(2\text{ N})^2 + (3\text{ N})^2} = 3.606\text{ N} = 3.606\text{ kg m/s}^2$$

Then you can use Newton's second law (Eq. 3-2) to find the magnitude of the acceleration produced by $\Sigma\mathbf{F}$:

$$|\Sigma\mathbf{F}| = \Sigma F = m|\mathbf{a}| = ma$$

$$a = \frac{\Sigma F}{m} = \frac{3.606\text{ kg m/s}^2}{3\text{ kg}} = 1.20\text{ m/s}^2$$

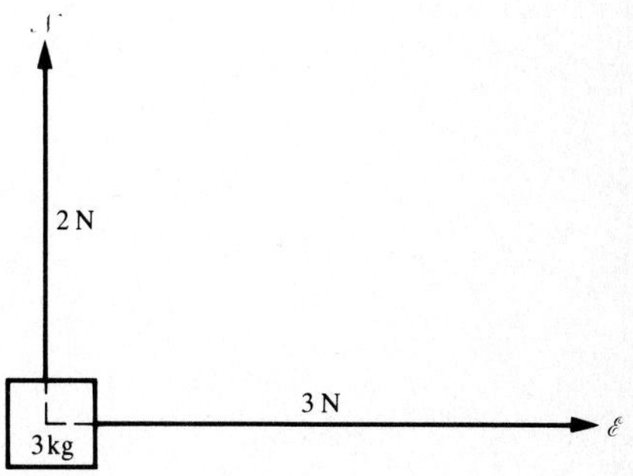

Figure 3-1

C. Newton's third law

Newton's third law states

- If object A exerts a force on object B, then object B exerts an equal and opposite force on object A.

Using mathematical symbolism, we write this statement

NEWTON'S THIRD LAW OF MOTION

$$\mathbf{F}_{A \text{ on } B} = -\mathbf{F}_{B \text{ on } A} \tag{3-3}$$

where the negative sign indicates the direction of the vector.

Forces always occur in pairs acting in opposite directions on two *different* objects. Newton called these forces *action* and *reaction*, which has led to this statement of the third law:

- For every action there is an equal and opposite reaction. This law is true for any type of force, including frictional, gravitational, electrical, and magnetic forces.

warning: Don't forget that you're investigating pairs of forces that always act on two different objects—not equal and opposite forces that act on one object—so the action/reaction forces never cancel out. Consider the case of a man pulling a train—difficult as that may seem! According to Newton's third law, the man exerts the same force on the train as the train exerts on the man. So, why does the train move—never mind the health of the poor guy pulling! Remember that you have to focus on the relevant object of the problem—the only forces that count in determining the motion of the train are those forces that act on the train. The fact that the train exerts a force on the man is irrelevant for solving a problem about the train's motion. Be sure when you draw your sketch that you can identify the action/reaction pair—and the objects they're acting on.

EXAMPLE 3-3: The earth, which has a mass of 5.98×10^{24} kg, exerts a force of 1.99×10^{22} N on the moon, which has a mass of 7.36×10^{22} kg. What is the magnitude of the force that the moon exerts on the earth?

Solution: From Newton's third law, you know that the force the moon exerts on the earth has exactly the same magnitude as the force the earth exerts on the moon. From Eq. (3-3),

$$\begin{aligned}\mathbf{F}_{\text{moon on earth}} &= -\mathbf{F}_{\text{earth on moon}}\\ &= -1.99 \times 10^{22}\ \text{N}\end{aligned}$$

Don't be misled by the fact that the moon is much smaller than the earth. Newton's third law is always true—no matter what the sizes or natures of the two objects may be.

3-3. Weight, Mass, and Gravitation

- The **weight w** of an object is the downward force the object experiences on or near the surface of the earth. Weight is a vector quantity—it has both magnitude and direction.

All freely falling objects, regardless of their mass, fall with the same acceleration due to gravity, written **g**. This means that at a given location, the weight of an object must be proportional to its mass. The relationship between weight, mass, and acceleration due to gravity is

WEIGHT AND MASS

$$\mathbf{w} = m\mathbf{g} \tag{3-4}$$

Gravitational acceleration **g** varies from one location to another, so the weight **w** of an object will vary according to its location. Mass m, which is the measure of an object's inertia, will remain constant no matter where the object is located. And, the relationship between weight, mass, and acceleration will always hold—no matter what the acceleration of the object is.

EXAMPLE 3-4: A nervous astronaut drops a wrench with a mass of 500 g from a height of 2 m above the surface of the moon. It takes the wrench 1.566 s to fall to the moon's surface.
How much does the wrench weigh (**a**) on the moon and (**b**) on the earth?

Solution:

(**a**) Before you can calculate the weight of the wrench on the moon $\mathbf{w}'$, you have to determine the gravitational acceleration on the surface of the moon. Since the acceleration is constant, you can use Eq. (2-8), $s = v_0t + \frac{1}{2}at^2$. The wrench was dropped, so its initial speed is $v_0 = 0$, and you have $s = \frac{1}{2}at^2$. The acceleration of the falling wrench is

$$a = \frac{2s}{t^2} = \frac{2(2\text{ m})}{(1.566\text{ s})^2} = 1.63\text{ m/s}^2$$

Let's call this acceleration of a freely falling body on the surface of the moon g'. Now you can calculate w':

$$w' = mg' = (0.50\text{ kg})(1.63\text{ m/s}^2) = 0.815\text{ N}$$

(**b**) Since you already know the acceleration due to gravity on the earth's surface, you can easily find the weight of the wrench w:

$$w = mg = (0.50\text{ kg})(9.8\text{ m/s}^2) = 4.90\text{ N}$$

You can see from this example that the weight of an object varies, but its mass does not.

EXAMPLE 3-5: If a smooth rock whose mass is 300 g is dropped from a second-story window, it has an acceleration of 9.8 m/s^2 toward the center of the earth. What is the acceleration of a rock with a mass of 600 g?

Solution: From Newton's second law (Eq. 3-2), you know that the acceleration of an object is the net force acting on it divided by its mass:

$$\mathbf{a} = \frac{\Sigma\mathbf{F}}{m}$$

Now the net force $\Sigma\mathbf{F}$ acting on the 600-g rock is its weight $m\mathbf{g}$, $\Sigma\mathbf{F} = m\mathbf{g}$, so you can solve for the acceleration of the falling rock:

$$\mathbf{a} = \frac{m\mathbf{g}}{m} = \mathbf{g}$$

Since the mass m cancels out, you can see that the acceleration of a freely falling object is independent of its mass—the rock whose mass is 600 g also has an acceleration of 9.8 m/s^2.

3-4. Friction

Two objects in contact exert force on each other. For example, a block resting on an inclined surface experiences an upward force as well as the downward force we call weight. The vector **F** in Figure 3-2 represents the upward force that the inclined surface exerts on the block. You may think of this upward force as the sum of two forces—one force *perpendicular* to the plane, which is called the **normal force N**; and a second force *parallel* to the plane, which is the **frictional force f**

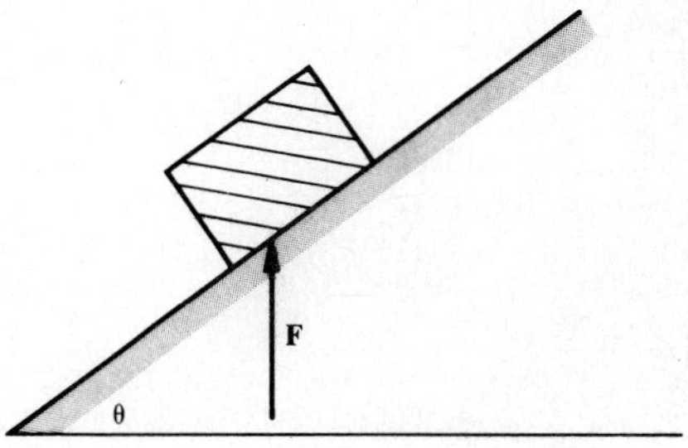

Figure 3-2. Total force **F** exerted on a block by an inclined plane.

between the two surfaces. These forces are shown in Figure 3-3.

- **Friction f** is the force that acts to oppose the tendency to move when one object slides over another. Frictional force is always parallel to the surfaces that are in contact.
- The force of **static friction $\mathbf{f}_s$** is the force parallel to the surface of contact when there is *no relative motion* between the two surfaces.

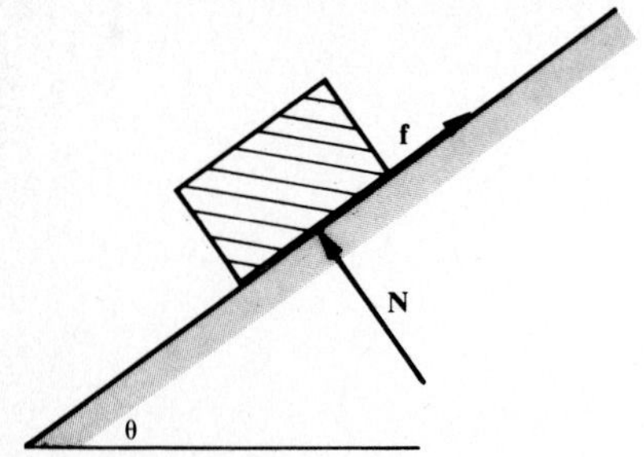

Figure 3-3. Parallel and perpendicular components of the force exerted by an inclined surface on a block.

EXAMPLE 3-6: A 5-kg block of aluminum is resting on a surface inclined at an angle of 10°.
What is the force of static friction $\mathbf{f}_s$ on the block?

Solution: First, draw a diagram (see Figure 3-4) and label the forces. You'll find the math is less cumbersome if you draw the x axis parallel to the inclined surface. Since the block is at rest, you know that the net force acting on it is zero. This means that the total upward force **F** acting on the block has the same magnitude as the block's weight—the downward force of gravity:

$$F = w = mg = (5\text{ kg})(9.8\text{ m/s}^2) = 49.0\text{ N}$$

Next, calculate the component of **F** parallel to the inclined surface—that is, the x component of **F**, which is the force of static friction:

$$f_s = F\sin 10° = (49.0\text{ N})(0.1736) = 8.51\text{ N}$$

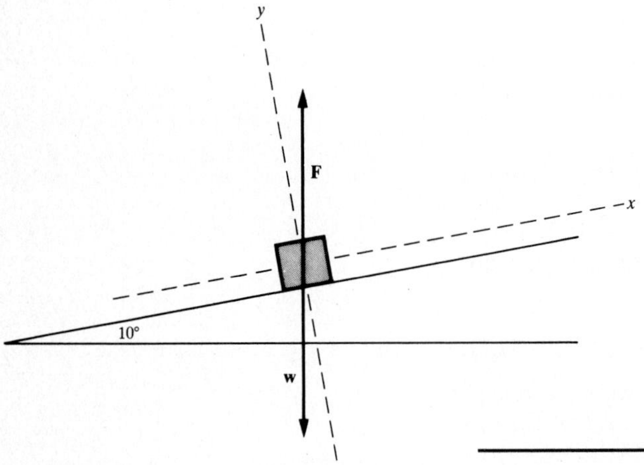

Figure 3-4

Increasing the angle of an inclined surface causes the force of static friction to increase. If the surface becomes too steep, the block starts to slide down the plane and the frictional force **f** acting on it becomes a force of kinetic friction $\mathbf{f}_k$.

- The force of **kinetic friction $\mathbf{f}_k$** is the force parallel to the surface of contact when there is *motion* between the two surfaces.

The force of kinetic friction is always opposite to the velocity of the object on which it acts.

EXAMPLE 3-7: The aluminum block of Example 3-6 remains at rest as long as the angle of inclination is less than 25°.
(a) Calculate the force of static friction $\mathbf{f}_s$ when the angle is 20°, and then
(b) Calculate the maximum force of static friction—the amount of force at the angle at which the block begins to slide.

Solution:
(a) At the angle of 20°, the force of static friction is

$$f_s = F\sin 20° = (49.0\text{ N})(0.342) = 16.8\text{ N}$$

(b) The block starts to slide when the angle of the plane is 25°, so the maximum force of static friction is

$$(f_s)_{\max} = F\sin 25° = (49.0\text{ N})(0.423) = 20.7\text{ N}$$

The maximum force of static friction $(\mathbf{f}_s)_{\max}$ between two surfaces is proportional to the normal force **N** that one object exerts against the other. This constant of proportionality is called the **coefficient of static friction** μ_s, so that $(\mathbf{f}_s)_{\max} = \mu_s\mathbf{N}$.

EXAMPLE 3-8: Calculate the coefficient of static friction μ_s between the aluminum block and the inclined surface described in Example 3-6.

Solution: First, find the magnitude of the normal force **N**—which is the component of **F** perpendicular to the surface of contact:

$$N = F \cos 10°$$

From Example 3-6, you know that the total upward force **F** = 49.0 N, so

$$N = (49.0 \text{ N})(0.985) = 48.3 \text{ N}$$

Now, you can calculate the coefficient of static friction μ_s:

$$(f_s)_{\text{max}} = \mu_s N \qquad \mu_s = \frac{(f_s)_{\text{max}}}{N}$$

and from Example 3-7, $(f_s)_{\text{max}} = 20.7$ N, so that

$$\mu_s = \frac{20.7 \text{ N}}{48.3 \text{ N}} = 0.429$$

The force of static friction can have any value up to the maximum determined by the normal force and the coefficient of static friction, so that the relationship between static friction and the normal force is best expressed by an inequality:

FORCE OF STATIC FRICTION

$$f_s \le \mu_s N \qquad \textbf{(3-5)}$$

Friction is also present when an object is sliding across a surface. If the surfaces are dry and the velocity is not too great, the magnitude of the force of kinetic friction $\mathbf{f}_k$ depends *only* on the nature of the two surfaces and the normal force **N**:

FORCE OF KINETIC FRICTION

$$f_k = \mu_k N \qquad \textbf{(3-6)}$$

The constant μ_k is the **coefficient of kinetic friction.**

- For a given pair of surfaces, the coefficient of kinetic friction is always less than the coefficient of static friction.

EXAMPLE 3-9: If you nudge the aluminum block of the last three examples, it will slide down the inclined surface at a constant speed when the angle of inclination is 22°.

Determine the coefficient of kinetic friction μ_k between the aluminum block and the inclined surface.

Solution: Draw a sketch of the situation and label the forces (see Figure 3-5). Since the speed is constant, you know that the acceleration is zero. This means that the x component of the downward force acting on the block must be equal to the force of kinetic friction $\mathbf{f}_k$ since $\mathbf{f}_k$ is always opposite to the velocity of the object:

$$f_k = w_x = w \sin 22°$$

You'll also need the normal force ($N = w \cos 22°$), then you can use Eq. (3-6) to find the coefficient of kinetic friction:

$$\mu_k = \frac{f_k}{N} = \frac{w \sin 22°}{w \cos 22°} = \tan 22° = 0.404$$

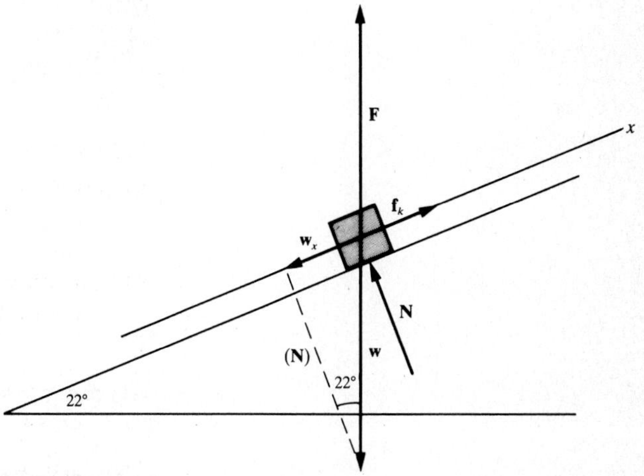

Figure 3-5

EXAMPLE 3-10: A granite block with a mass of 1.5 kg is placed on a smooth, horizontal surface. The coefficients of friction between these two surfaces are 0.35 and 0.28. A horizontal force of 3.0 N is applied to the granite block. Determine the magnitude of the frictional force(s) acting on the block.

Solution: First, you need to know whether the block is at rest or sliding along the surface. The surface is horizontal, so the normal force that the surface exerts against the block is equal in magnitude to the weight of the block:

$$N = w = mg = (1.5\ \text{kg})(9.8\ \text{m/s}^2) = 14.7\ \text{N}$$

Now you can calculate the force required to start the block moving. You know that $\mu_k < \mu_s$, so $\mu_k = 0.28$, and $\mu_s = 0.35$:

$$(f_s)_{\text{max}} = \mu_s N = (0.35)(14.7\ \text{N}) = 5.15\ \text{N}$$

Since the force applied to the block is only 3.0 N, the block remains at rest. Because an object at rest is in equilibrium—if $\Sigma\mathbf{F} = 0$, then $\mathbf{v}$ = a constant—you can write:

$$\Sigma\mathbf{F}_x = \mathbf{F} - \mathbf{f}_s = 0$$

and solve for the force of static friction:

$$f_s = F = 3.00\ \text{N}$$

3-5. Systems of Connected Bodies

Newton's laws also apply to a group of objects that are connected to one another. You find the acceleration of a system of connected objects by dividing the net force on the system by the total mass being accelerated:

ACCELERATION OF A SYSTEM

$$\mathbf{a} = \frac{\Sigma\mathbf{F}\ (\text{on system})}{m_{\text{tot}}\ (\text{of system})} \tag{3-7}$$

EXAMPLE 3-11: A single engine pulls a train of four identical freight cars, each with a mass of 1.5×10^3 kg. The engine exerts a constant force of 3×10^3 N on the car to which it is connected. Neglect frictional forces and determine the acceleration of the train.

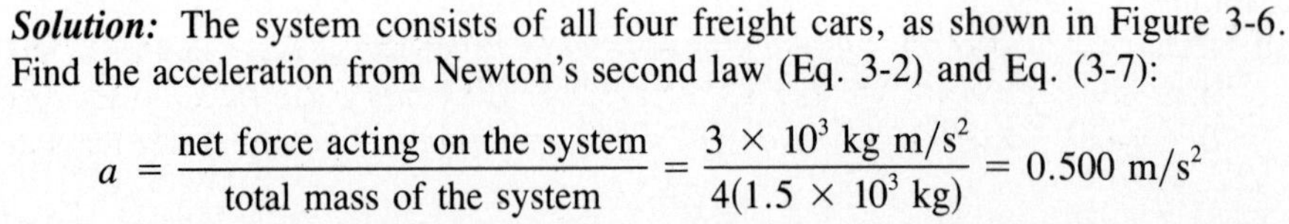

Figure 3-6

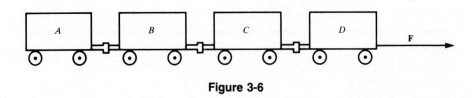

Solution: The system consists of all four freight cars, as shown in Figure 3-6. Find the acceleration from Newton's second law (Eq. 3-2) and Eq. (3-7):

$$a = \frac{\text{net force acting on the system}}{\text{total mass of the system}} = \frac{3 \times 10^3\ \text{kg m/s}^2}{4(1.5 \times 10^3\ \text{kg})} = 0.500\ \text{m/s}^2$$

You can also find the acceleration of any part of a system by isolating that particular part from the rest of the system.

- The **free-body method** is the technique of isolating a part of a system by considering only the mass of that part and the forces acting directly on it.

EXAMPLE 3-12: Determine the magnitude of the force that the second car C exerts on the third car B in the system of freight cars of Example 3-11.

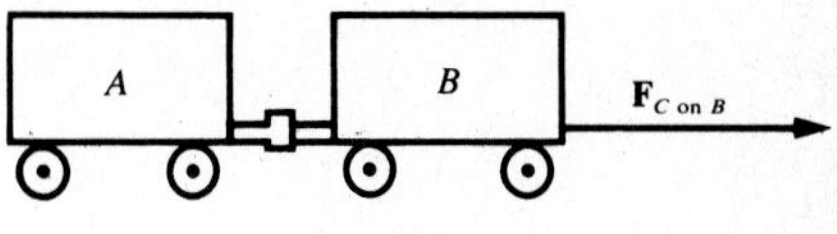

Figure 3-7

Solution: Isolate the part of the system you want to consider. Since the force acting on car B also accelerates car A, this part consists of the two cars, as shown in Figure 3-7. You can see that car B exerts a force on car A, and car A exerts a force on car B —an action/reaction pair. But, because both car A and car B are parts of the system you're considering, these are internal forces. According to Newton's third law, these internal forces are equal and opposite—their sum is zero.

$$\mathbf{F}_{A \text{ on } B} + \mathbf{F}_{B \text{ on } A} = 0$$

This is always the case with internal forces, which is why they cannot contribute to the acceleration of a system.

The force that does produce the acceleration of cars A and B is an external force, that is, a force caused by an agent outside the system under consideration—in this case, the force exerted by car C. Again, from Newton's second law:

$$\Sigma\mathbf{F} = m\mathbf{a}$$

$$\mathbf{F}_{(\text{on } B)} = (m_A + m_B)\mathbf{a}$$

$$= (3.0 \times 10^3 \text{ kg})(0.5 \text{ m/s}^2) = 1.50 \times 10^3 \text{ N}$$

Tension exists in a system when a cord or rod is used to connect the bodies. Whether this tension constitutes an internal or external force depends on whether you are considering all or only part of the system.

- **Tension** is a force that occurs when a flexible cord is used to transmit a *pull* from one point to another.

note: A rod can also transmit a tension or a **compression**—a *push* rather than a pull.

A cord that is pulled at opposite ends by forces of equal magnitude T is under a tension T. This tension is present uniformly at any point throughout the length of the cord—no matter what distance the point is from either end. If the cord passes around a frictionless, massless pulley, the direction of the tension changes, but the magnitude remains the same.

note: In order to simplify your calculations, you can assume that such cords have negligible mass, and ignore the force of gravity on them.

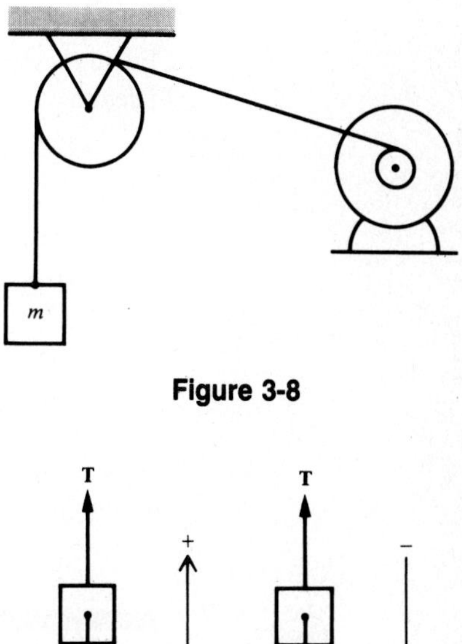

Figure 3-8

EXAMPLE 3-13: In Figure 3-8, a cord passes from a motor around a frictionless, massless pulley and is attached to a 4-kg mass.
Determine the tension in the cord if **(a)** the cord raises the mass with an acceleration of 5 m/s^2; **(b)** the cord lowers the mass with an acceleration of 5 m/s^2; and **(c)** the mass is raised at a constant velocity of 5 m/s.

Solution:

(a) Isolate the part of the system you want to consider and make a sketch of this part (see Figure 3–9a). The acceleration of the mass is upward, so let that direction be positive. Now, apply Newton's second law and solve for the tension T:

$$\Sigma\mathbf{F} = m\mathbf{a}$$

$$T - w = ma$$

$$T = w + ma = mg + ma = m(g + a)$$

$$= (4 \text{ kg})(9.8 \text{ m/s}^2 + 5 \text{ m/s}^2) = 59.2 \text{ N}$$

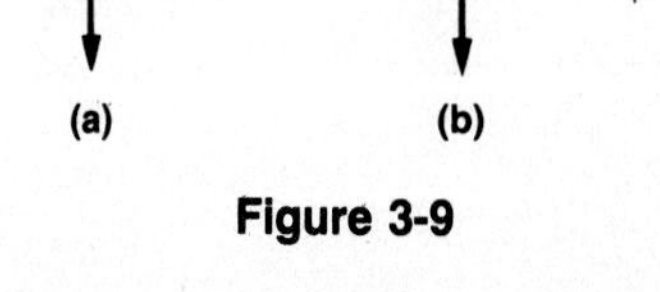

Figure 3-9

(b) Since the acceleration of the mass is downward, let that be the positive direction (see Figure 3-9b). Again, apply Newton's second law and solve for T:

$$\Sigma \mathbf{F} = m\mathbf{a}$$

$$w - T = ma$$

$$T = w - ma = mg - ma = m(g - a)$$

$$= (4 \text{ kg})(9.8 \text{ m/s}^2 - 5 \text{ m/s}^2) = 19.2 \text{ N}$$

(c) The mass is moving upward, so let that be the positive direction. Since the velocity is constant, the sum of the forces acting on the mass equals zero, $T - w = ma = 0$. So, when you apply Newton's second law,

$$T = w = mg = (4 \text{ kg})(9.8 \text{ m/s}^2) = 39.2 \text{ N}$$

note: When acceleration is upward, the tension is greater than the object's weight; when acceleration is downward, the tension is less than the weight.

SUMMARY

Newton's first law of motion	If $\Sigma \mathbf{F} = 0$, then $\mathbf{v}$ = a constant	describes the motion of an object on which no net force is acting
Newton's second law of motion	net $\mathbf{F} = m\mathbf{a}$ or $\Sigma \mathbf{F} = m\mathbf{a}$	states that an object is accelerated if the sum of all forces acting on it is *not* zero
Newton's third law of motion	$\mathbf{F}_{A \text{ on } B} = -\mathbf{F}_{B \text{ on } A}$	states that if object A exerts a force on object B, then object B exerts an equal and opposite force on object A
Weight and mass	$\mathbf{w} = m\mathbf{g}$	gives the relationship between weight, mass, and acceleration due to gravity
Force of static friction	$f_s \leq \mu_s N$ $f_{s\,max} = \mu_s N$	gives the relationship between the force of friction and the normal force when there is no relative motion between the two surfaces
Force of kinetic friction	$f_k = \mu_k N$	gives the relationship between the force of friction and the normal force when there is motion between the two surfaces
Acceleration of a system	$\mathbf{a} = \dfrac{\Sigma \mathbf{F} \text{ (on system)}}{m_{tot} \text{ (of system)}}$	yields the acceleration of a system by dividing the net force on the system by its total mass

RAISE YOUR GRADES

Can you define . . . ?

- [x] force
- [x] mass
- [x] inertia
- [x] a newton
- [x] weight
- [x] friction

Can you . . . ?

- ☑ predict what will happen to a moving object if the sum of all forces acting on it is zero
- ☑ express the relationship between the acceleration of an object and the net force acting on it
- ☑ predict how the acceleration of an object will change if the mass of the object is doubled and the force on it remains constant
- ☑ find the resultant of two forces acting on an object if the angle between the two forces is given
- ☑ describe the force that body *B* exerts on body *A* if you know the force that body *A* exerts on body *B*
- ☑ calculate the weight of an object whose mass is known and vice versa
- ☑ describe the direction of the gravitational force acting on an object near the earth
- ☑ explain why an object weighs much less on the surface of the moon than it weighs on the earth
- ☑ calculate the x and y components of a force if you know the angle between the force and the x axis
- ☑ write the relationship between the force of static friction and the normal force between the two surfaces
- ☑ calculate the force of kinetic friction between two surfaces, given the normal force and the coefficient of kinetic friction
- ☑ calculate the acceleration of a system of connected bodies, given the net force acting on the system

SOLVED PROBLEMS

Newton's laws of motion

PROBLEM 3-1 A loaded freight car weighing 12 tons (1 ton = 2000 lb) rests on a level track. A switch engine coupled to the freight car exerts a constant force of 1500 lb for 4 s. Determine (**a**) the displacement s of the freight car in 4 s, and (**b**) its speed v at the end of 4 s. (Neglect frictional forces.)

Solution

(**a**) You want to find the displacement s, and you already know the time, t = 4 s, and the initial velocity, $v_0 = 0$. The acceleration is constant in both magnitude and direction. So, if you calculate the magnitude of the acceleration, you can find s from Eq. (2-8):

$$s = v_0 t + \frac{1}{2}at^2$$

Because the net force in the horizontal direction is given, you can use Newton's second law to find the acceleration. (The mass of the car is equal to its weight **w** divided by its acceleration due to gravity **g**.)

$$a = \frac{\text{net } \mathbf{F}}{m} = \frac{\text{net } \mathbf{F}}{w/g} = \frac{(\text{net } \mathbf{F})(g)}{w} = \frac{1500 \text{ lb } (32 \text{ ft/s}^2)}{2400 \text{ lb}} = 2 \text{ ft/s}^2$$

Now, substitute the values into Eq. (2-8):

$$s = 0 + \frac{1}{2}(2 \text{ ft/s}^2)(4 \text{ s})^2 = 16.0 \text{ ft}$$

(**b**) Use Eq. (2-6) to determine the final speed:

$$v = v_0 + at = 0 + (2 \text{ ft/s}^2)(4 \text{ s}) = 8.00 \text{ ft/s}$$

PROBLEM 3-2 Two forces ($\mathbf{F}_1 = 10$ N; $\theta_1 = 37°$; $\mathbf{F}_2 = 8$ N, $\theta_2 = -30°$) act on a 5-kg object. Determine (**a**) the magnitude and (**b**) the direction of the acceleration produced by these forces.

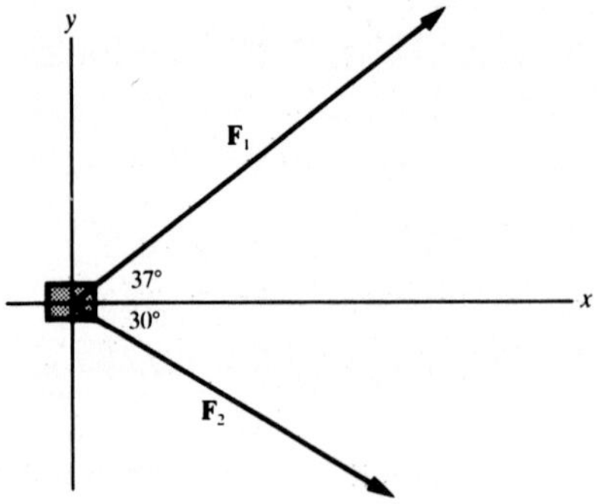

Figure 3-10

Solution

(**a**) Sketch the situation (see Fig. 3-10). You need to find the net force acting on the object, and since there are two forces you'll find it easier to use the method of components (see Sec. 1-4):

Force	*x* component	*y* component
$\mathbf{F}_1$	(10 N) cos 37°	(10 N) sin 37°
$\mathbf{F}_2$	(8 N) cos 30°	−(8 N) sin 30°

$$\Sigma F_x = 7.99 \text{ N} + 6.93 \text{ N} = 14.9 \text{ N}$$

$$\Sigma F_y = 6.02 \text{ N} - 4.00 \text{ N} = 2.02 \text{ N}$$

Let $\mathbf{R} = \mathbf{F}_1 + \mathbf{F}_2$ be the net force, so the magnitude of the net force is

$$R = \sqrt{R_x^2 + R_y^2} = \sqrt{(14.9 \text{ N})^2 + (2.02 \text{ N})^2} = 15.0 \text{ N}$$

Now use Newton's second law of motion (Eq. 3-2) to determine the magnitude of the acceleration:

$$\text{net } \mathbf{F} = m\mathbf{a}$$

$$\mathbf{a} = \frac{\text{net } \mathbf{F}}{m} = \frac{15 \text{ N}}{5 \text{ kg}} = 3.00 \text{ m/s}^2$$

(**b**) The acceleration is in the same direction as the net force producing it. You can find the direction of **R** with the tangent function. Let θ be the angle that **R** makes with the positive x axis—then,

$$\tan\theta = \frac{R_y}{R_x} = \frac{2.02 \text{ N}}{14.9 \text{ N}} = 0.136$$

$$\theta = \arctan(0.136) = 7.74°$$

PROBLEM 3-3 Block A with a mass of 200 g and block B with a mass of 500 g are connected by a spring of negligible mass, as shown in Fig. 3-11. The blocks move without friction over the horizontal surface of an air table. When the blocks are pulled apart and then released, they are accelerated toward each other by the spring. Calculate the ratio of their accelerations, a_A/a_B.

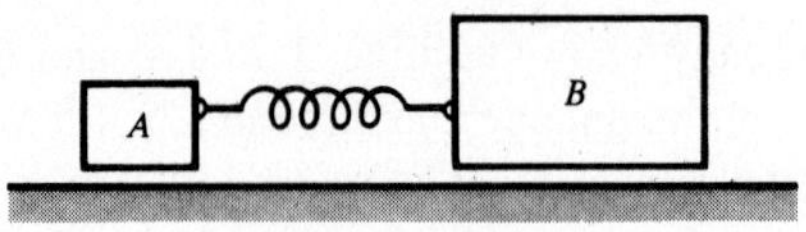

Figure 3-11

Figure 3-12

Solution You're considering a system that contains three parts: the two blocks and the spring connecting them. Since the spring is extended beyond its normal length, you know that block A is exerting a force on the spring toward the left and block B is exerting a force toward the right (see Fig. 3-12), so that the net force acting on the spring is $\mathbf{F}_{A \text{ on s}} + \mathbf{F}_{B \text{ on s}}$. From Newton's second law, you have

$$\mathbf{F}_{A \text{ on s}} + \mathbf{F}_{B \text{ on s}} = m\mathbf{a}$$

And, since the spring has a negligible mass, $m = 0$, so

$$\mathbf{F}_{A \text{ on s}} + \mathbf{F}_{B \text{ on s}} = 0 \quad \text{or} \quad \mathbf{F}_{A \text{ on s}} = -\mathbf{F}_{B \text{ on s}}$$

The two forces acting on the spring have the same magnitude.

Now consider block A — exerting the force $\mathbf{F}_{A \text{ on s}}$ on the left end of the spring. According to Newton's third law (Eq. 3-3), the spring exerts an equal and opposite force on the block (see Fig. 3-13), so that the magnitude of the force $\mathbf{F}_{\text{s on }A}$ has the same magnitude as $\mathbf{F}_{A \text{ on s}}$. Of course, Newton's third law also applies to the spring and block B, so you know that $F_{B \text{ on s}} = F_{\text{s on }B}$. Let's list the three relationships we have so far:

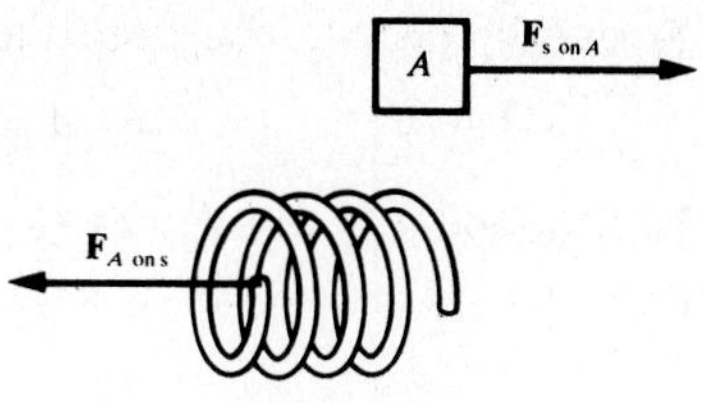

Figure 3-13

$$F_{A \text{ on s}} = F_{B \text{ on s}} \qquad F_{A \text{ on s}} = F_{\text{s on }A} \qquad F_{B \text{ on s}} = F_{\text{s on }B}$$

As you can see, the force acting on block A, $\mathbf{F}_{\text{s on }A}$, has the same magnitude as the force acting on block B, $\mathbf{F}_{\text{s on }B}$. Now from Newton's second law (Eq. 2-2) you can write expressions for the accelerations of the two blocks:

$$a_A = \frac{F_{\text{s on }A}}{m_A} \qquad a_B = \frac{F_{\text{s on }B}}{m_B}$$

When you divide a_A by a_B to get the ratio of their accelerations, the forces cancel out. This means that the accelerations of the two blocks are inversely proportional to their masses:

$$\frac{a_A}{a_B} = \frac{F_{\text{s on }A}/m_A}{F_{\text{s on }B}/m_B} = \frac{1/m_A}{1/m_B} = \frac{m_B}{m_A} = \frac{500 \text{ g}}{200 \text{ g}} = 2.50$$

Weight, mass, and gravitation

PROBLEM 3-4 A small bolt from the antenna of a planetary probe has a mass of 50 grams. On the surface of the planet Mercury this bolt weighs 0.186 N. When the probe lands on Mercury, the bolt comes loose and falls 1.5 m to the surface. How long does it take the bolt to fall?

Solution Since the time it takes to fall a known distance depends on the local acceleration of gravity, you'll have to calculate **g** on the planet Mercury. From Eq. (3-4),

$$g = \frac{w}{m} = \frac{0.186 \text{ N}}{0.050 \text{ kg}} = 3.72 \text{ m/s}^2$$

Since the initial velocity of the bolt is not given, assume that it is zero. Then the vertical distance s is given by Eq. (2-8):

$$s = \frac{1}{2}gt^2$$

Now solve algebraically for t, the unknown:

$$t = \sqrt{\frac{2\,s}{g}} = \sqrt{\frac{2(1.5 \text{ m})}{(3.72 \text{ m/s}^2)}} = 0.898 \text{ s}$$

Friction

PROBLEM 3-5 The coefficient of kinetic friction between smooth, dry surfaces of steel and wood is approximately 0.2. Calculate the force required to pull a 1.5-kg block of steel along a horizontal wooden surface at a steady speed of 0.2 m/s.

Solution Make a sketch of the forces acting on the block, (see Fig. 3-14). To find the force of kinetic friction, $f_k = \mu_k N$, you'll first have to find the normal force **N**. You've been told that the velocity is constant in both magnitude and direction (steady speed), so you can be sure that the net force acting on the block is zero. This means that the sum of the vertical forces **N** and **w** is zero, and the sum of the horizontal forces $\mathbf{f}_k$ and **F** is zero. Thus

$$|\mathbf{N}| = |\mathbf{w}| \quad \text{and} \quad |\mathbf{F}| = |\mathbf{f}_k|$$

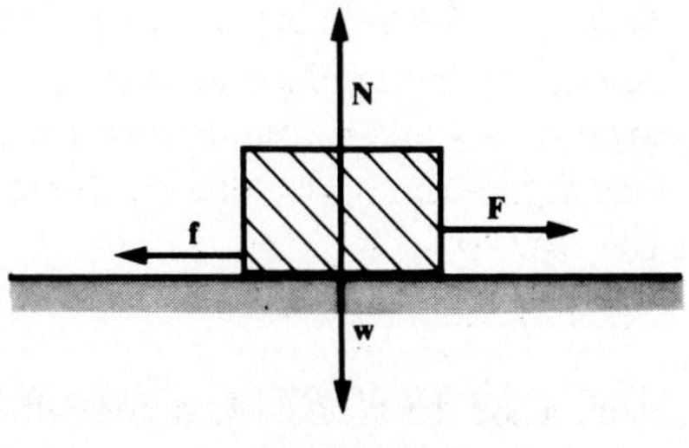

Figure 3-14

From Eq. (3-4),

$$N = w = mg = (1.5 \text{ kg})(9.8 \text{ m/s}^2) = 14.7 \text{ N}$$

Now substitute N into Eq. (3-6) and you can find the force of kinetic friction.

$$f_k = \mu_k N = (0.2)(14.7 \text{ N}) = 2.94 \text{ N}$$

Since $|\mathbf{F}| = |\mathbf{f}_k|$, $|\mathbf{F}| = 2.94$ N.

PROBLEM 3-6 Determine the force required to pull the steel block of Problem 3-5 up a 30° inclined wooden surface at a constant speed.

Solution First, make a diagram showing all the forces acting on the block. [Hint: Place the x axis parallel to the surface of the inclined plane.] Don't put the forces exerted by the block into your drawing—they'll only confuse the issue. Your drawing should resemble Fig. 3-15. You can see that three of the four forces acting on the block are parallel ($\mathbf{f}_k$ and $\mathbf{F}$) or perpendicular ($\mathbf{N}$) to the inclined surface. Only the force $\mathbf{w}$ has to be resolved into its x and y components.

Now, in order to find the force $\mathbf{F}$ required to pull the block up the inclined surface, you must calculate all the x components of the forces acting on the block, $\mathbf{f}_k$ and the x component of $\mathbf{w}$. You can find the force of kinetic friction $\mathbf{f}_k$ from Eq. (3-6), $f_k = \mu_k N$, but first you must find $\mathbf{N}$. Recall from Problem 3-5 that the sum of all y components is zero; that is, $\mathbf{N}$ plus the y component of $\mathbf{w}$ is zero, and

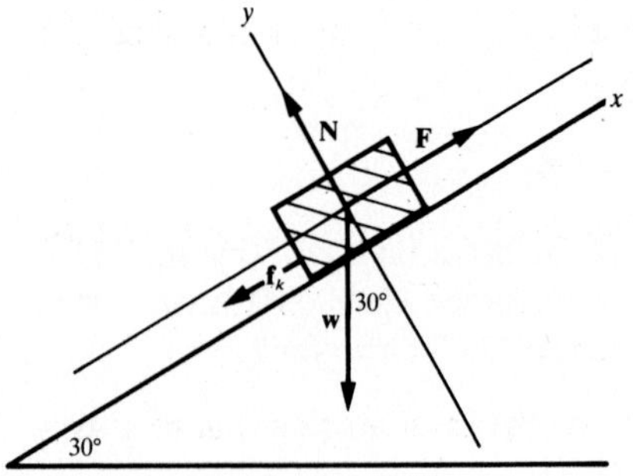

Figure 3-15

$$N = mg \cos 30° = (14.7 \text{ N})(0.866) = 12.73 \text{ N}$$

Then you can find the magnitude of $\mathbf{f}_k$ from Eq. (3-6):

$$f_k = \mu_k N = (0.2)(12.73 \text{ N}) = 2.55 \text{ N}$$

The block is moving at a constant velocity, so the forces acting along the x axis are in equilibrium, which means that $\Sigma \mathbf{F}_x = 0$. Now, you can set up the equality and solve for the unknown force $\mathbf{F}$:

$$F - f_k - mg \sin 30° = 0$$

$$F = f_k + mg \sin 30° = 2.55 \text{ N} + (14.7)(0.50)$$

$$= 9.90 \text{ N}$$

PROBLEM 3-7 For the steel block of Problems 3-5 and 3-6, determine the force that must be applied parallel to the inclined surface to prevent the block from accelerating as it slides down the surface.

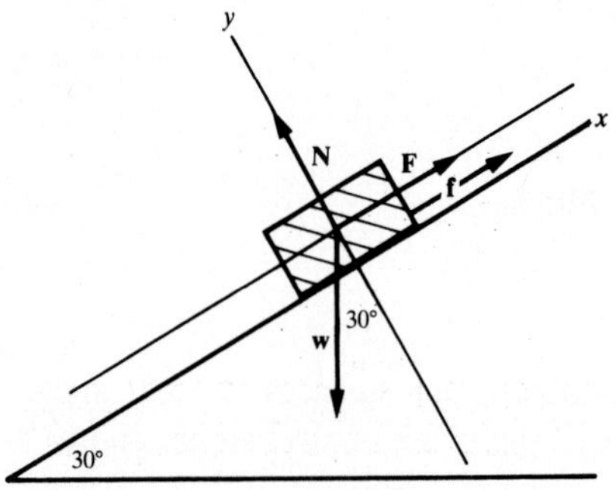

Figure 3-16

Solution This problem is similar to Problem 3-6 except that the frictional force $\mathbf{f}_k$ reverses when the direction of motion is changed, since $\mathbf{f}_k$ always opposes the direction of motion of the object upon which it acts. Your force diagram for this situation will look like Fig. 3-16. The normal force is the same as in Problem 3-6:

$$N = mg \cos 30° = 12.73 \text{ N}$$

The frictional force has the same magnitude:

$$f_k = \mu_k N = 2.55 \text{ N}$$

And the sum of the x components remains zero:

$$F + f_k - mg \sin 30° = 0$$

$$F = mg \sin 30° - f_k = (14.7 \text{ N})(0.50) - 2.55 \text{ N}$$

$$= 4.80 \text{ N}$$

PROBLEM 3-8 A wooden plank weighing 10 N is resting on the floor. A steel block weighing 5 N is placed on top of it, and a wire attached to the wall prevents the steel block from moving (see

Fig. 3-17). The coefficient of static friction between the steel block and the wood is 0.4; between the piece of wood and the floor, 0.3. Calculate the force required to start the wooden plank moving toward the right.

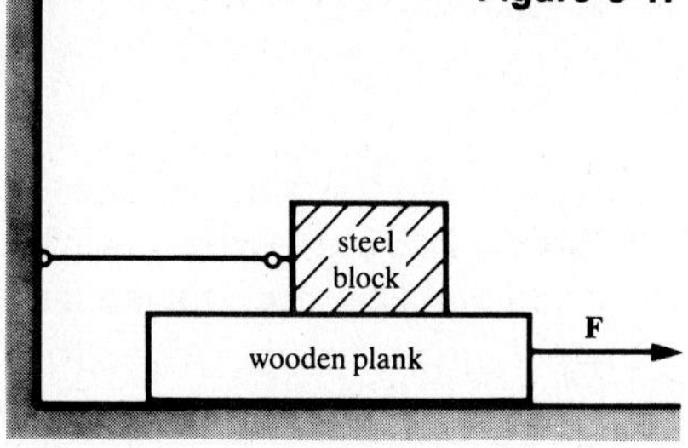

Figure 3-17

Solution Make a force diagram for the wooden plank. Note that there are forces acting at every point (or surface) where the plank makes contact with other objects. Your diagram should look like Fig. 3-18. To find the force **F** required to start the plank moving to the right, you'll have to find the maximum frictional forces acting on the plank— since **F** must be at least equal to them. You can see from Fig. 3-18 that

$$\mathbf{F} = (\mathbf{f}_1)_{\max} + (\mathbf{f}_2)_{\max}$$

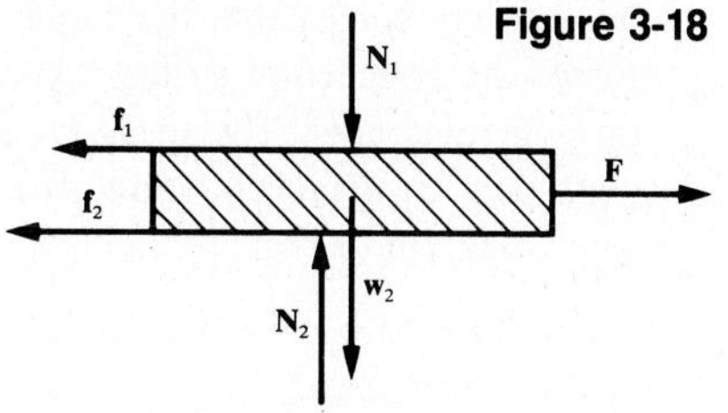

Figure 3-18

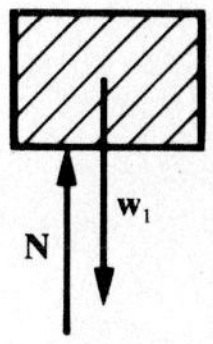

Figure 3-19

In order to determine $\mathbf{f}_1$, the frictional force acting on the upper surface, you must determine the normal force $\mathbf{N}_1$ that the steel block exerts on the upper surface of the plank (see Fig. 3-19). Because there is no acceleration in the vertical direction, $\Sigma\mathbf{F}_y = 0$. Thus,

$$\mathbf{N} + \mathbf{w}_1 = 0 \quad \text{and} \quad N = w_1 = 5 \text{ N}$$

You know from Newton's third law (Eq. 3-3) that the force exerted by the wooden plank on the block **N** is equal in magnitude to the force that the block exerts on the plank $\mathbf{N}_1$, so

$$N_1 = N = 5 \text{ N}$$

From Eq. (3-6), the maximum force of static friction that the steel block exerts on the upper surface of the plank is

$$(f_1)_{\max} = \mu_s N_1 = (0.4)(5 \text{ N}) = 2.0 \text{ N}$$

There is no acceleration of the plank in the vertical direction, so the vertical forces must have a sum of zero:

$$\mathbf{N}_1 + \mathbf{w}_2 + \mathbf{N}_2 = 0$$

Since $\mathbf{N}_1$ and $\mathbf{w}_2$ are downward forces, the magnitude of the upward force $\mathbf{N}_2$ is

$$N_2 = N_1 + w_2 = (5 + 10) \text{ N} = 15 \text{ N}$$

And, the maximum force of static friction against the lower surface of the plank is

$$(f_2)_{\max} = \mu_s N_2 = (0.3)(15 \text{ N}) = 4.5 \text{ N}$$

So, the force required to start the plank moving is

$$F = (f_1)_{\max} + (f_2)_{\max} = (2.0 + 4.5) \text{ N} = 6.50 \text{ N}$$

Systems of connected bodies

PROBLEM 3-9 The physical system shown in Fig. 3-20 consists of three masses connected by strings passing over two pulleys. The pulleys are frictionless, but there is a frictional force acting on mass m_2 (4 kg) as it slides along a horizontal tabletop. The suspended mass m_1 is 2 kg. When the suspended mass m_3 is 3 kg and the system is set in motion, it moves at a constant speed. If an additional 2-kg mass is added to the 3-kg mass, the system accelerates. Find the tension T_1 in the string that connects m_1 and m_2 when the system is accelerating.

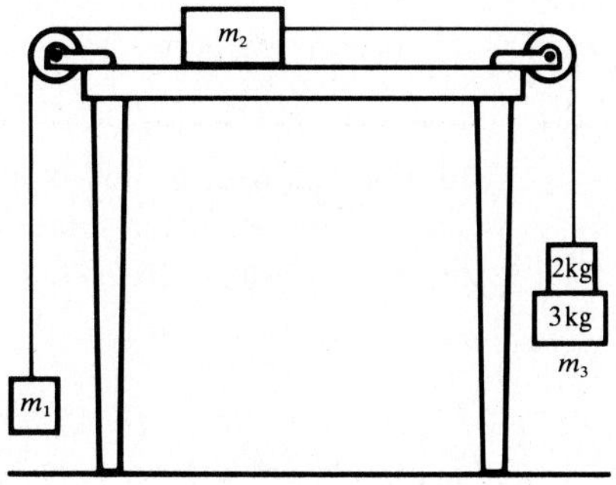

Figure 3-20

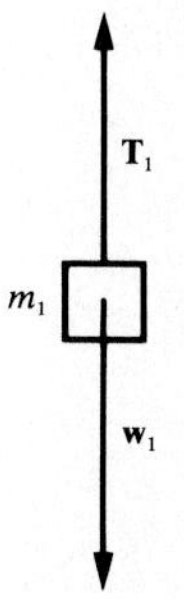

Figure 3-21

Solution You might begin this solution by drawing a force diagram for m_1 (see Fig. 3-21). Since the acceleration of

m_1 is clearly upward, choose that as the positive direction, so that the net force acting on m_1 is $T_1 - m_1g$. Applying Newton's second law, you get an expression for the unknown force of tension in the string:

$$T_1 - m_1g = m_1a$$

$$T_1 = m_1g + m_1a = m_1(g + a)$$

You can see that you can't get a numerical value for T_1 until you know the acceleration. So, first find the acceleration of m_1—which is also the acceleration of the entire system.

If you consider all three masses as a single system, the unknown forces of tension become *internal forces* that do not contribute to the acceleration of the system. Start with the system that moves at a constant speed—that is, the configuration that has an acceleration of zero. Now, make a diagram showing the *external forces* acting on this entire system. The table is not a part of the system, so the forces that the table exerts on the sliding block are external forces. [Hint: Since the forces the connecting strings exert on the masses are internal forces, you don't need to draw them on your diagram.] The mass m_2 is moving horizontally, so the vertical forces acting on it do not contribute to its acceleration. But, don't forget the frictional force **f** acting on m_2. Compare your force diagram to Fig. 3-22. Now, the net force acting on the entire system is

$$\text{net } \mathbf{F} = m_3\mathbf{g} - \mathbf{f} - m_1\mathbf{g}$$

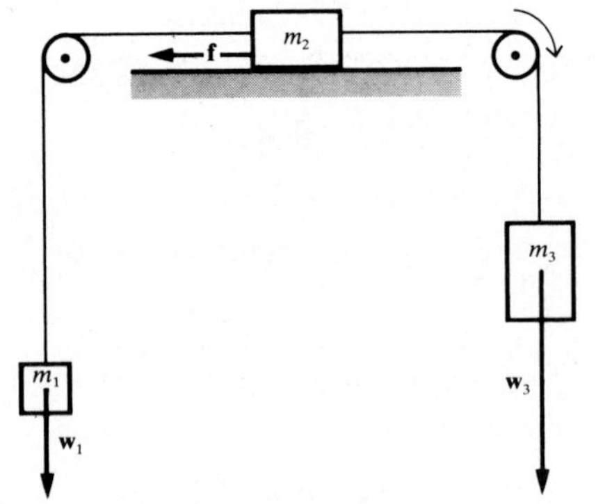

Figure 3-22

When m_3 is 3 kg, the system moves at a constant speed, so set the net force equal to zero and solve for **f**—the frictional force:

$$f = m_3g - m_1g = (m_3 - m_1)g$$

$$= (3 \text{ kg} - 2 \text{ kg})(9.8 \text{ m/s}^2) = 9.8 \text{ N}$$

Now that you know the value of **f**, you can find the acceleration of the system when m_3 is 5 kg. From Eq. (3-7):

$$\mathbf{a} = \frac{\text{net } \mathbf{F}}{m_{\text{total}}} = \frac{m_3g - m_1g - f}{m_1 + m_2 + m_3}$$

$$= \frac{(5 \text{ kg} - 2 \text{ kg})(9.8 \text{ m/s}^2) - 9.8 \text{ N}}{(2 + 4 + 5) \text{ kg}} = \frac{19.6 \text{ N}}{11 \text{ kg}}$$

$$= 1.782 \text{ m/s}^2$$

And just substitute the numerical values into the expression you began with—for the force of tension in the string:

$$T_1 = m_1(g + a) = (2 \text{ kg})(9.8 \text{ m/s}^2 + 1.782 \text{ m/s}^2) = 23.2 \text{ N}$$

PROBLEM 3-10 For the system described in Problem 3-9, find **(a)** the coefficient of kinetic friction μ_k between m_2 and the tabletop, and **(b)** the tension T_2 in the string connecting m_2 and m_3.

Solution

(a) You can find the coefficient of friction from Eq. (3-6):

$$\mu_k = \frac{f_k}{N}$$

But first you have to find **N**—the normal force. There are only two vertical forces acting on m_2 (see Fig. 3-23). Since there is no acceleration in the vertical direction, the sum of these two forces must equal zero—that is, their magnitudes must be equal:

$$N = w_2 = m_2g = (4 \text{ kg})(9.8 \text{ m/s}^2) = 39.2 \text{ N}$$

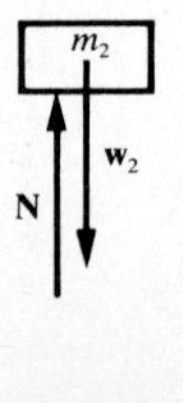

Figure 3-23

And from Problem 3-9, $\mathbf{f}_k = 9.8$ N. Now, plug the numerical values into Eq. (3-6):

$$\mu_k = \frac{f_k}{N} = \frac{9.8 \text{ N}}{39.2 \text{ N}} = 0.250$$

(b) To find T_2, choose a system acted on by this unknown force. You do have a choice! The string is connected to the sliding mass m_2 on the tabletop, and also to the mass m_3 hanging on the right. Mass m_3 is the better choice because only two forces are acting on it. Draw a force diagram for m_3 and then compare it to Fig. 3-24. Since the acceleration is downward, make that the positive direction so that the net force acting on m_3 is

$$\text{net } F = m_3g - T_2$$

From Newton's second law,

$$\text{net } F = m_3a$$

Now, substitute and solve for the unknown force T_2:

$$T_2 = m_3g - m_3a = m_3(g - a)$$

$$= (5 \text{ kg})(9.8 \text{ m/s}^2 - 1.78 \text{ m/s}^2) = 40.1 \text{ N}$$

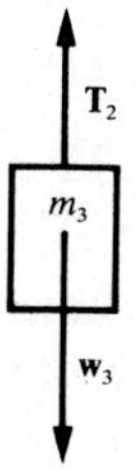

Figure 3-24

You can also find T_2 by isolating m_2 and applying Newton's second law to it. Use this to check your solution. Again, begin with a force diagram for m_2 (compare your diagram to Fig. 3-25). Since there are two forces of tension, you'll want to distinguish them by subscripts. The net force acting on this system is

$$\text{net } F = T_2 - f - T_1$$

Of course, the net force is equal to the product m_2a:

$$T_2 - f - T_1 = m_2a$$

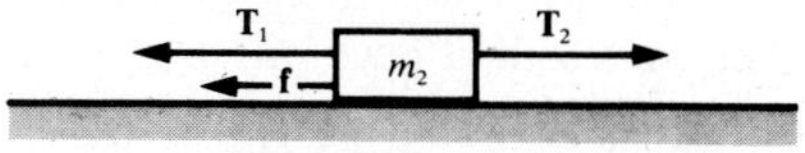

Figure 3-25

Solving for T_2 gives

$$T_2 = m_2a + T_1 + f = (4 \text{ kg})(1.782 \text{ m/s}^2) + 23.2 \text{ N} + 9.8 \text{ N} = 40.1 \text{ N}$$

PROBLEM 3-11 When a freight elevator that weighs 1600 lb moves, there is a frictional force of 100 lb acting on it. What force must be exerted on the elevator by its supporting cable if the maximum upward acceleration is to be 3 ft/s^2?

Solution Begin by listing all the forces acting on the elevator. There are two downward forces—the weight **w** and the friction **f**. The only upward force is that exerted by the cable, **T**. Thus the net force is **T** − **w** − **f**. Because $\Sigma\mathbf{F} = m\mathbf{a}$, you can write

$$T - w - f = ma$$

Now, solve for T—the unknown force:

$$T = ma + w + f = \left(\frac{w}{g}\right)a + w + f$$

$$= \left(\frac{1600 \text{ lb}}{32 \text{ ft/s}^2}\right)(3 \text{ ft/s}^2) + 1600 \text{ lb} + 100 \text{ lb} = 1850 \text{ lb}$$

PROBLEM 3-12 Two nearly equal masses are connected by a string passing over a massless, frictionless pulley. This physical system—called Atwood's machine—is shown in Fig. 3-26. For masses of $m_1 = 195$ g and $m_2 = 205$ g, determine the acceleration of the system in terms of g, the acceleration of gravity.

Solution The acceleration of a system of connected bodies is equal to the net force acting on the system divided by the total mass. Since the heavier mass will be moving downward, this is the natural choice for the positive direction. The weight of the smaller mass on the left acts in a direction opposite to the acceleration of the system, so you must regard it as a negative force. This means that the net external force on the system is

$$\text{net } F = m_2g - m_1g$$

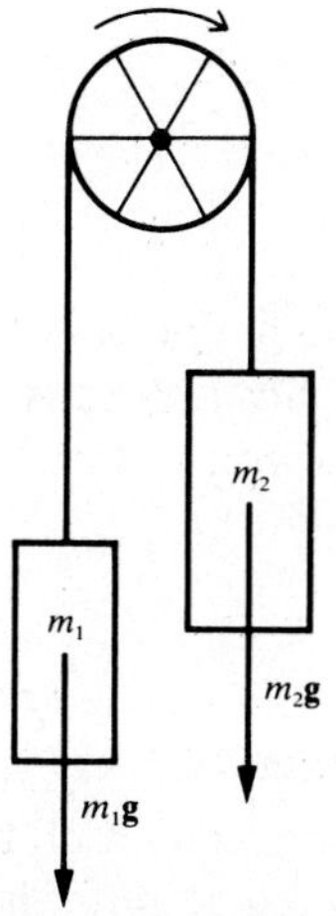

Figure 3-26. Atwood's machine.

The acceleration is

$$a = \frac{\text{net } F}{m_{\text{tot}}} = \frac{m_2 g - m_1 g}{m_1 + m_2} = \frac{(m_2 - m_1)}{m_1 + m_2} g$$

$$= \frac{0.01 \text{ kg}}{0.40 \text{ kg}} g = 0.025\ g$$

PROBLEM 3-13 A 2-kg mass is subject to two forces, only one of which is known. The acceleration produced by the two forces has a magnitude of 4.33 m/s² along the positive x axis. The known force $\mathbf{F}_1$ has a magnitude of 5 N and lies along the positive y axis. Find the unknown force $\mathbf{F}_2$.

Solution Since the unknown force $\mathbf{F}_2$ is a vector, you'll have to find both its *magnitude* and *direction*. The resultant force **R** (see Sec. 1-4) is the vector sum of the two forces acting on the mass—$\mathbf{F}_1$ and $\mathbf{F}_2$:

$$\mathbf{R} = \mathbf{F}_1 + \mathbf{F}_2$$

Solving for $\mathbf{F}_2$, you get

$$\mathbf{F}_2 = \mathbf{R} - \mathbf{F}_1 = \mathbf{R} + (-\mathbf{F}_1)$$

You already know the mass and acceleration of the object, so you can calculate **R** directly from Newton's second law:

$$\Sigma\mathbf{F} = m\mathbf{a}$$

$$|\mathbf{R}| = |\Sigma\mathbf{F}| = ma = (2 \text{ kg})(4.33 \text{ m/s}^2) = 8.66 \text{ N}$$

Now you can use the method of components (see Sec. 1-4) to find $\mathbf{F}_2$:

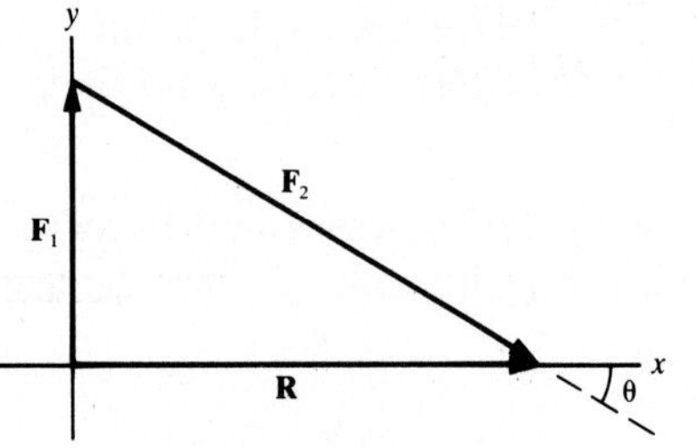

Figure 3-27

Forces	x components	y components
R	8.66 N	0
$-\mathbf{F}_1$	0	−5 N
$\mathbf{F}_2$	8.66 N	−5 N

And the magnitude of $\mathbf{F}_2$ is

$$F_2 = \sqrt{F_{2x}^2 + F_{2y}^2} = \sqrt{(8.66 \text{ N})^2 + (-5 \text{ N}^2)}$$

$$= 10.0 \text{ N}$$

The y component of $\mathbf{F}_2$ is negative, which means that $\mathbf{F}_2$ points downward, as shown in Fig. 3-27. You can find the angle between $\mathbf{F}_2$ and the positive x axis using the tan function:

$$\tan\theta = \frac{F_{2y}}{F_{2x}} = \frac{-5 \text{ N}}{8.66 \text{ N}} = -0.577$$

$$\theta = \text{arc}\tan(-0.577) = -30°$$

So, the direction of $\mathbf{F}_2$ is 30° south of east.

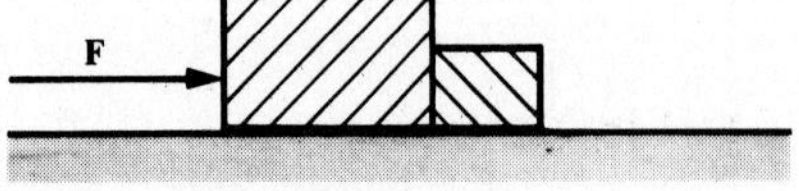

Figure 3-28

PROBLEM 3-14 Two rectangular blocks in contact with each other move without friction along a horizontal surface. The masses of the blocks are 1.8 kg and 0.2 kg. When a constant force is applied, as shown in Fig. 3-28, the blocks move 20 cm during a time interval of 0.283 s. **(a)** How fast is the smaller block moving at the end of the time interval and **(b)** what force does it exert against the larger block?

Solution

(a) First, find the acceleration of the system—the two blocks considered as a unit. From Eq. (2.8) and an initial speed of zero, the distance traveled is $s = \frac{1}{2}at^2$. And the acceleration is

$$a = \frac{2s}{t^2} = \frac{2(0.2 \text{ m})}{(0.283 \text{ s})^2} = 5.0 \text{ m/s}^2$$

The acceleration is constant, so the speed can be found from Eq. (2-6):

$$v = v_0 + at = 0 + (5.0 \text{ m/s}^2)(0.283 \text{ s}) = 1.42 \text{ m/s}$$

(**b**) According to Newton's third law, the force that the smaller block exerts on the larger one has the same magnitude as the force that the larger block exerts on the smaller one. Because there are two unknown forces acting on the larger block and only one unknown force on the smaller block, analyze the smaller block:

$$F = ma = (0.2 \text{ kg})(5 \text{ m/s}^2) = 1.00 \text{ N}$$

The smaller block exerts a force of 1.00 N against the larger one.

PROBLEM 3-15 An 80-lb crate is dragged across the floor of a warehouse by a rope that makes an angle of 30° with respect to the horizontal, as shown in Fig. 3-29. A force of 40 lb is required to pull the crate across the floor at a constant speed. What is the coefficient of kinetic friction between the crate and the floor?

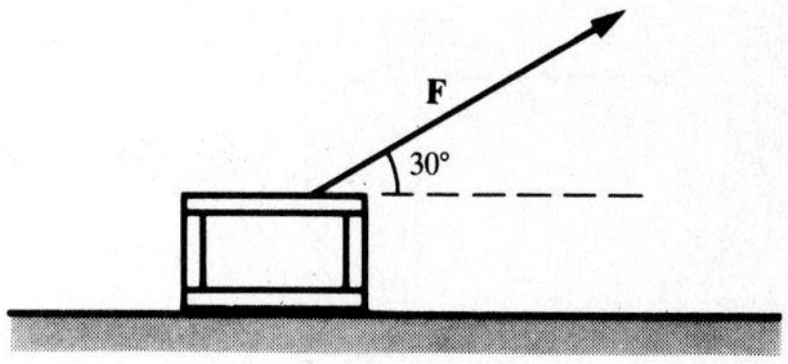

Figure 3-29

Solution Your first step, as always, is to make a sketch of the forces. Compare your diagram with Fig. 3-30 to be sure that you have correctly identified *all* of the forces that act *on* the crate. You'll use Eq. (3-6), $f_k = \mu_k N$, and solve for the coefficient of kinetic friction:

$$\mu_k = \frac{f_k}{N}$$

You'll have to find the normal force **N** from the y components of the forces acting on the crate, and the frictional force $\mathbf{f}_k$ from the x components. So, make a table of all the x and y components of the forces:

Forces	x components	y components
N	0	N
F	$F \cos 30°$	$F \sin 30°$
w	0	$-w$
$\mathbf{f}_k$	$-f_k$	0

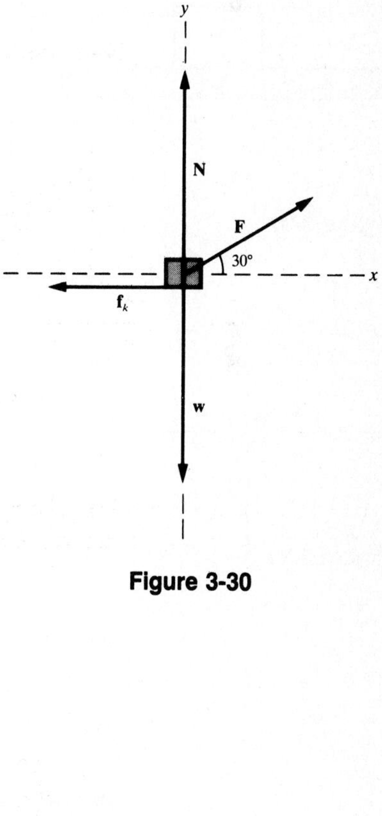

Figure 3-30

The crate is moving in a straight line at a constant speed, so

$$\Sigma \mathbf{F}_x = 0 \quad \text{and} \quad \Sigma \mathbf{F}_y = 0$$

Now, set the sum of the y components equal to zero and solve for N:

$$\Sigma F_y = N + F \sin 30° - w = 0$$

$$N = w - F \sin 30° = 80 \text{ lb} - (40 \text{ lb})(0.5) = 60 \text{ lb}$$

And set the sum of the x components equal to zero and solve for f_k:

$$\Sigma F_x = F \cos 30° - f_k = 0$$

$$f_k = F \cos 30° = (40 \text{ lb})(0.866) = 34.64 \text{ lb}$$

Now you can calculate the coefficient of kinetic friction:

$$\mu_k = \frac{f_k}{N} = \frac{34.64 \text{ lb}}{60 \text{ lb}} = 0.577$$

PROBLEM 3-16 Suppose that the crate of Problem 3-15 is *pushed* across the floor. If the force on the crate makes an angle of 30° downward with respect to the positive x axis, what force would be required to make the crate move at a constant speed?

Solution Because the force on the crate now has a downward component, **N** will be greater than before. You would be wise to make a force diagram for this case (see Fig. 3-31). The table of forces and their components will be almost the same as in Problem 3-15 except that the force **F** now has a negative y component. Start with the y components:

$$\Sigma F_y = N - F \sin 30° - w = 0$$

$$N = w + F \sin 30° = 80 \text{ lb} + F(0.5)$$

Then, the x components:

$$\Sigma F_x = F \cos 30° - f_k = 0$$

$$F \cos(30°) = f_k = \mu_k N = (0.577)[80 \text{ lb} + F(0.5)]$$

$$(0.866)F = 46.16 \text{ lb} + (0.2885)F$$

$$(0.866 - 0.2885)F = 46.16 \text{ lb}$$

$$F = 79.9 \text{ lb}$$

As you might have guessed, you'll need a considerably larger force to push the crate than to pull it.

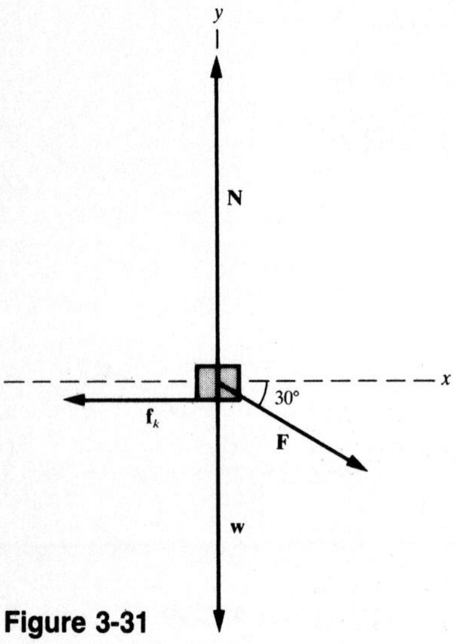

Figure 3-31

PROBLEM 3-17 A 2.0-kg block is attached to a string that passes over a pulley, as shown in Fig. 3-32. The other end of the string is fastened to a second block that hangs vertically. When the hanging block has a mass of 0.5 kg, the larger block moves across the horizontal surface at a constant speed. (**a**) What is the tension in the string and (**b**) what is the coefficient of friction between the block and the horizontal surface?

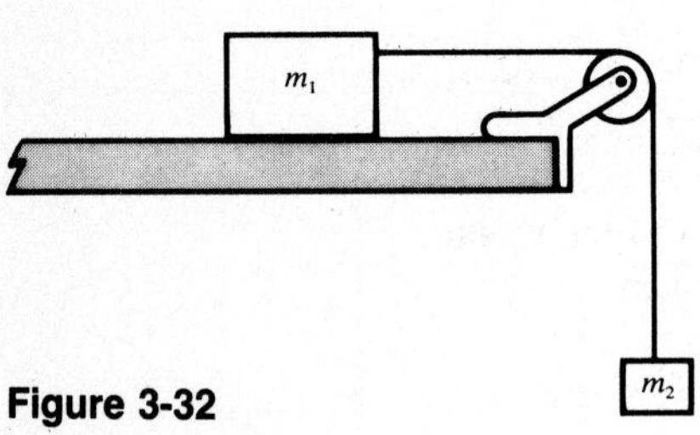

Figure 3-32

Solution

(**a**) There are only two forces acting on the hanging block m_2. Since it has no acceleration, the upward force T must have the same magnitude as the downward force mg. Thus, the tension is

$$T = mg = (0.5 \text{ kg})(9.8 \text{ m/s}^2) = 4.90 \text{ N}$$

(**b**) The sliding block is subject to four forces, as shown in Fig. 3-33. The two horizontal forces are the force of friction to the left and the pull of the string to the right. Since the sliding block is in equilibrium, the magnitude of the force of friction **f** must be equal to the magnitude of the tension in the string **T**:

$$f = T = 4.90 \text{ N}$$

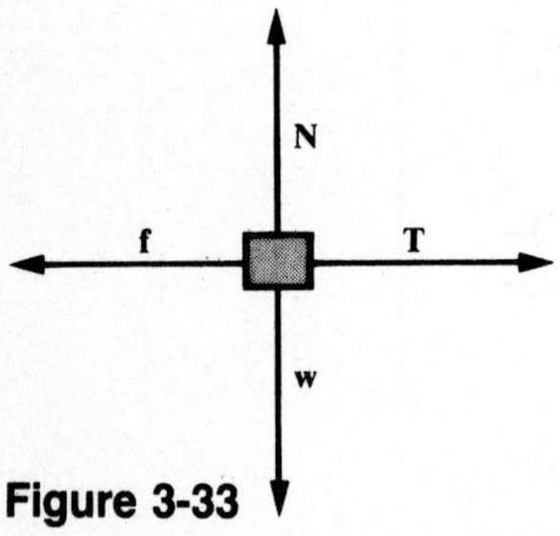

Figure 3-33

Since there is no acceleration in the vertical direction, you know that the sum of the y components is also zero, so the magnitude of the upward force on the block N is equal to the magnitude of the downward force mg.

$$N = mg = (2.0 \text{ kg})(9.8 \text{ m/s}^2) = 19.6 \text{ N}$$

Now, calculate the coefficient of kinetic friction:

$$\mu_k = \frac{f}{N} = \frac{4.90 \text{ N}}{19.6 \text{ N}} = 0.250$$

PROBLEM 3-18 If the mass of the hanging block in Problem 3-17 is doubled, calculate (**a**) the acceleration of the system and (**b**) the tension in the string.

Solution

(**a**) As you know, the acceleration is simply the net force on the system divided by the total mass being accelerated. There are two forces parallel to the motion of the system: the weight of the hanging block and the force of friction on the sliding block. Thus, the net force is

$$\Sigma F = m_2 g - f = (1 \text{ kg})(9.8 \text{ m/s}^2) - 4.90 \text{ N}$$

$$\text{net } F = 4.90 \text{ N}$$

Since the mass of the system is now 3 kg, the acceleration is

$$a = \frac{\Sigma F}{m} = \frac{4.90\ \text{N}}{3\ \text{kg}} = 1.63\ \text{m/s}^2$$

(b) Calculate the tension in the string by applying Newton's second law to the hanging block:

$$\Sigma F = m_2 a$$

Since this mass accelerates downward, choose this as the positive direction. Now, replace ΣF by the two forces acting on m_2:

$$m_2 g - T = m_2 a$$

And solve for the tension:

$$T = m_2(g - a) = (1.0\ \text{kg})(9.8\ \text{m/s}^2 - 1.63\ \text{m/s}^2) = 8.17\ \text{N}$$

Supplementary Exercises

PROBLEM 3-19 Two horizontal forces act on a 4-kg mass. One force has a magnitude of 8 N and is directed toward the north. The second force toward the east has a magnitude of 6 N. What is the magnitude of the acceleration produced by these two forces?

PROBLEM 3-20 A book resting on a table top exerts a downward force of 30 N on the surface of the table. What is the magnitude and direction of the force that the table exerts against the book?

PROBLEM 3-21 An object that weighs 24.5 N on the surface of the earth is taken to the moon where the acceleration of gravity is 1.63 m/s². What does this object weigh on the moon?

PROBLEM 3-22 When a steel block is released from rest on a steep inclined plane, it moves 50 cm in 0.404 s. What is the acceleration of the block?

PROBLEM 3-23 The steel block in Problem 3-22 has a weight of 8.00 N. What is the magnitude of the net force that produces the acceleration of the block?

PROBLEM 3-24 The inclined plane in Problem 3-22 makes an angle of 50° with respect to the horizontal. There is a constant force of kinetic friction between the steel block and the inclined surface. What is the magnitude of the frictional force?

PROBLEM 3-25 Calculate the normal force between the steel block of the preceding problems and the 50° inclined plane, then calculate the coefficient of kinetic friction between the two surfaces.

PROBLEM 3-26 A 4-kg steel block rests on a horizontal surface. The coefficients of friction between the steel block and the horizontal surface are 0.4 and 0.3. A horizontal force of 6.0 N is applied to the block. What is the magnitude of the force of friction acting on the block?

PROBLEM 3-27 Two cylinders having masses of 190 g and 210 g are tied to a piece of string that passes over a frictionless pulley. What is the acceleration of this system?

PROBLEM 3-28 What is the tension in the string that is fastened to the 190-g cylinder of the preceding problem?

Answers to Supplementary Exercises

3-19: 2.50 m/s²

3-20: 30 N, upward **3-21:** 4.08 N **3-22:** 6.13 m/s² **3-23:** 5.00 N **3-24:** 1.13 N

3-25: $N = 5.14$ N, $\mu_k = 0.220$ **3-26:** 6.00 N **3-27:** 0.490 m/s² **3-28:** 1.96 N

4 CIRCULAR MOTION AND ROTATION

THIS CHAPTER IS ABOUT

☑ Angular Quantities
☑ Linear and Angular Motion Compared
☑ Circular Motion
☑ Banked Curves
☑ Torque and Moment of Inertia
☑ Uniformly Accelerated Rotation

4-1. Angular Quantities

A. Angular displacement

- An object moving along a circular path passes through an angle θ measured at the center of the circle. The measurement of the angle θ through which the object passes is the **angular displacement**.

You can express angular displacement in degrees, but for equations in which both angular and linear quantities appear, you *must* use radians.

- A **radian** (abbreviated **rad**) is the angle whose arc equals the radius of the circle.

The radian measure of an angle is the length of the arc s measured on the circumference divided by the radius r (length/length), so it is a pure, dimensionless number.

ANGULAR DISPLACEMENT (θ IN RADIANS)

$$\theta = \frac{s}{r} \qquad s = r\theta \tag{4-1}$$

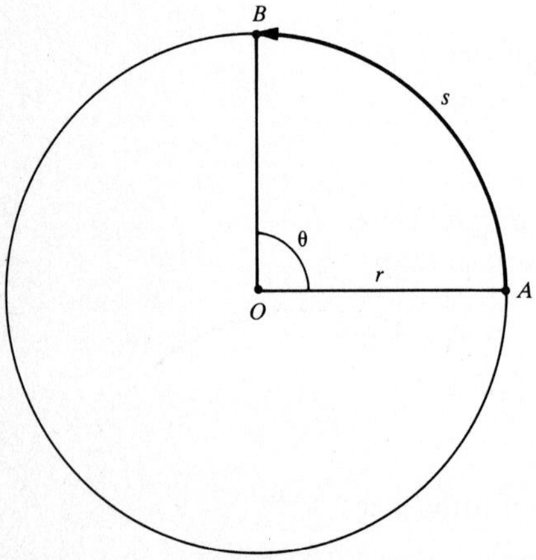

Figure 4-1

EXAMPLE 4-1: An object moves along a circle from point A to point B, as illustrated in Figure 4-1.
If the distance traveled is one quarter of the circumference, determine the angular displacement (**a**) in radians and (**b**) in degrees.

Solution:

(**a**) The circumference of a circle is $C = 2\pi r$, so the distance from A to B—the length of the arc—is $C/4 = 2\pi r/4 = \pi r/2$. From Eq. (4-1), the angular displacement is

$$\theta = \frac{s}{r} = \frac{\pi r/2}{r} = \frac{\pi}{2}\ \text{rad}$$

(**b**) The circumference of a circle is also $C = 360°$, so the angle corresponding to $C/4$ is 90°. The fact that $C = 2\pi r = 360°$ gives you a useful relationship between rads and degrees—the two dimensionless units we use to measure angles:

RADIAN–DEGREE CONVERSION

$$\frac{\pi}{2}\ \text{rad} = 90° \qquad 2\pi\ \text{rad} = 360°$$

$$\pi\ \text{rad} = 180° \qquad 1\ \text{rad} = \frac{360°}{2\pi} \cong 57.3°$$

B. Average angular speed

- The **average angular speed** ω_{ave} of a point that moves along a circle is the angular displacement divided by the time required for the displacement to occur:

AVERAGE ANGULAR SPEED

$$\omega_{ave} = \frac{\Delta\theta}{\Delta t} = \frac{\theta_2 - \theta_1}{t_2 - t_1} \quad \textbf{(4-2)}$$

EXAMPLE 4-2: An object moving along the circular path, shown in Figure 4-2, passes point C at $t_1 = 1.0$ s and point D at $t_2 = 1.5$ s. An arc of 30° separates point C and point D.
Calculate the average angular speed of this object in rad/s.

Solution: You're going to use an equation with both angular and linear quantities, so start by converting the angular displacement from degrees to radians:

$$\Delta\theta = 30°\frac{\pi \text{ rad}}{180°} = \frac{\pi}{6} \text{ rad}$$

Then, use Eq. (4-2) to calculate the average angular speed:

$$\omega_{ave} = \frac{\Delta\theta}{\Delta t} = \frac{\pi/6 \text{ rad}}{0.5 \text{ s}} = 1.05 \text{ rad/s}$$

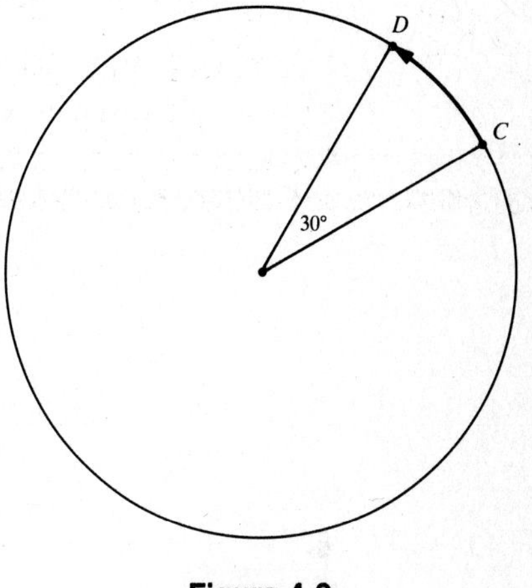

Figure 4-2

C. Instantaneous angular speed

- **Instantaneous angular speed** ω is the average angular displacement over a very short time interval:

INSTANTANEOUS ANGULAR SPEED

$$\omega = \lim_{\Delta t \to 0} \frac{\Delta\theta}{\Delta t} \quad \textbf{(4-3)}$$

D. Average angular acceleration

- **Average angular acceleration** α_{ave} is the change in angular speed divided by the time interval during which this change took place:

AVERAGE ANGULAR ACCELERATION

$$\alpha_{ave} = \frac{\Delta\omega}{\Delta t} = \frac{\omega_2 - \omega_1}{t_2 - t_1} \quad \textbf{(4-4)}$$

E. Instantaneous angular acceleration

- **Instantaneous angular acceleration** α is the average acceleration over a very short time interval:

INSTANTANEOUS ANGULAR ACCELERATION

$$\alpha = \lim_{\Delta t \to 0} \frac{\Delta\omega}{\Delta t} \quad \textbf{(4-5)}$$

EXAMPLE 4-3: Figure 4-3 shows the path of a moving object that passes point A at a speed of 2 rad/s. One second later it passes point B at a speed of 3 rad/s. Calculate the average angular acceleration that takes place.

Solution: From Eq. (4-4),

$$\alpha_{ave} = \frac{\Delta\omega}{\Delta t} = \frac{\omega_2 - \omega_1}{t_2 - t_1} = \frac{(3 - 2) \text{ rad/s}}{1 \text{ s}} = 1 \text{ rad/s}^2$$

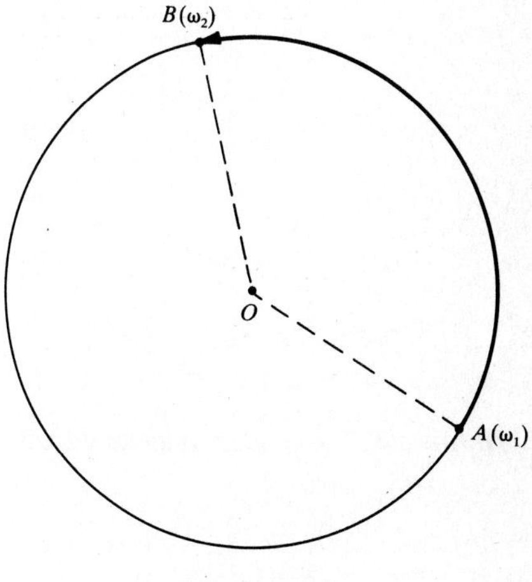

Figure 4-3

4-2. Linear and Angular Motion Compared

Take a moment to compare the definitions of angular speed and angular acceleration with the definitions of linear speed and linear acceleration from Chapter 2. (Also see Table 4-1.) There is a close analogy between linear and angular motion—to emphasize this, we replace the Roman letters of the linear equations by their Greek equivalents for the angular equations. You'll find this analogy extremely useful in solving problems.

TABLE 4-1: Linear and Angular Motion Compared

Symbols:	Linear	Angular	Definitions:	Linear	Angular
displacement	Δs or Δx	$\Delta\theta$	average speed	$v_{ave} = \frac{\Delta s}{\Delta t}$	$\omega_{ave} = \frac{\Delta\theta}{\Delta t}$
speed	v	ω			
acceleration	a	α	average acceleration	$a_{ave} = \frac{\Delta v}{\Delta t}$	$\alpha_{ave} = \frac{\Delta\omega}{\Delta t}$

Equations:	Linear	Angular	
	$v = v_0 + at$	$\omega = \omega_0 + \alpha t$	**(4-6)**
	$v_{ave} = \frac{1}{2}(v_1 + v_2)$	$\omega_{ave} = \frac{1}{2}(\omega_1 + \omega_2)$	**(4-7)**
	$v^2 = v_0^2 + 2as$	$\omega^2 = \omega_0^2 + 2\alpha\theta$	**(4-8)**
	$s = v_0t + \frac{1}{2}at^2$	$\theta = \omega_0t + \frac{1}{2}\alpha t^2$	**(4-9)**

note: In these examples and problems, θ represents the angular displacement from a starting point θ_0 defined as 0. We could define θ more broadly, as the angular displacement on the circle from any point designated 0 that is not the starting point, that is, where $\theta_0 \neq 0$. If you are using a reference system where θ_0 is *not* 0, then replace θ by $\theta - \theta_0$ in Eqs. (4-8) and (4-9). Notice the exact correspondence with the equations for linear motion, where a term s_0 must be added if linear position is being measured from an origin that is not the starting point.

4-3. Circular Motion

A. Tangential speed

The circular path of a moving mass is shown in Figure 4-4. Vector **v** represents the velocity of the mass as it passes through point *P*.

- **Tangential** (or **linear**) **speed** is the magnitude of the velocity vector of a mass moving through a particular point on a circular path.

Since the radius of the circle is constant, the velocity vector is necessarily *perpendicular* to the radius and *tangent* to the circular path.

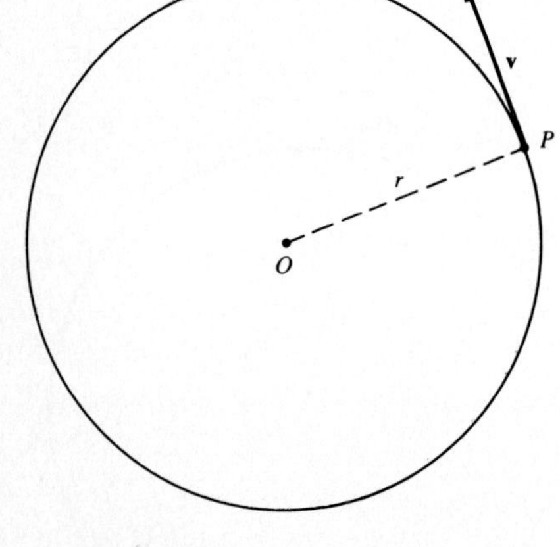

Figure 4-4. Tangential velocity vector.

EXAMPLE 4-4: A small body of mass *m* moves around a circle of radius r = 20 cm at a constant angular speed of 0.5 rad/s. Calculate the tangential speed of *m*.

Solution: From Eq. (4-1), you can write a relationship between angular and linear displacement—the change in linear displacement is equal to the radius times the change in angular displacement. [Note: Use delta Δ to represent change.]

$$\Delta s = r\Delta\theta$$

To find the speed of an object, you must divide its displacement by the time it takes for the object to travel that distance—so, divide both sides of the equation by Δt:

$$\frac{\Delta s}{\Delta t} = r\frac{\Delta\theta}{\Delta t}$$

The speed of m is constant, so the average speed is equal to the instantaneous speed: $v_{\text{ave}} = v$ and $\omega_{\text{ave}} = \omega$. And, from Table 4-1,

$$v_{\text{ave}} = \frac{\Delta s}{\Delta t} \quad \text{and} \quad \omega_{\text{ave}} = \frac{\Delta\theta}{\Delta t}$$

Now you have the relationship between tangential speed v_t and angular speed ω:

TANGENTIAL SPEED

$$v_t = r\omega \tag{4-10}$$

note: Although you have obtained Eq. (4-10) for the special case of uniform circular motion, it is valid even if the angular speed ω is not constant.

Simply substitute and solve. [Hint: Convert r from cm to m.]

$$v_t = r\omega = (0.2\text{ m})(0.5\text{ rad/s}) = 0.10\text{ m/s}$$

B. Tangential acceleration

Consider Eq. (4-10) again, and ask yourself what would happen if the angular speed ω were not constant. If ω is changing, then the tangential speed v_t must also be changing. Let Δv_t represent the change in the magnitude of the tangential speed during a short time interval, then $\Delta v_t = r\Delta\omega$. Now, divide both sides of this equation by the time interval Δt:

$$\frac{\Delta v_t}{\Delta t} = \frac{r\Delta\omega}{\Delta t}$$

You know that $\Delta\omega/\Delta t$ is the average angular acceleration, and that the change in the magnitude of v_t divided by the time interval is the average tangential acceleration (Eq. 4-4), so

$$a_{t(\text{ave})} = r\alpha_{\text{ave}}$$

Since the limiting value of $\Delta v/\Delta t$ as Δt approaches zero is the instantaneous tangential acceleration, you have the following relationship between tangential acceleration and angular acceleration:

TANGENTIAL AND ANGULAR ACCELERATION

$$a_t = r\alpha \tag{4-11}$$

note: The tangential acceleration a_t is present *only* if the angular speed ω is changing.

EXAMPLE 4-5: Calculate the tangential acceleration for the object of Example 4-3 if it follows a circular path of $r = 12$ cm.

Solution: You can calculate the tangential acceleration from Eq. (4-11) and the result of Example 4-3:

$$a_t = r\alpha = (12\text{ cm})(1.0\text{ rad/s}^2) = 12\text{ cm/s}^2$$

C. Centripetal acceleration

An object moving along a curved path has a changing velocity even if its speed is constant (see Ch. 2, Example 2-3). This is because the direction of the velocity vector changes even though its magnitude remains constant, so that an object moving in a circular path *always* has an acceleration.

- The linear acceleration of an object directed toward the center of its circular path is the **centripetal acceleration.** Centripetal means "seeking the center."

If the path is the arc of a circle of radius r, then the magnitude of the centripetal acceleration is given by

CENTRIPETAL ACCELERATION

$$a_c = \frac{v^2}{r} \quad \text{or} \quad a_c = r\omega^2 \tag{4-12}$$

note: In physics, you'll encounter two terms that refer to curved paths which are not perfectly circular:

- The **radius of curvature** is the radius of the circle that best approximates the curve of the path in which you're interested.
- The **center of curvature** is the center of the approximating circle.

The equations for the angular quantities in this chapter are exact for perfect circular paths, but less accurate for curved paths in general.

EXAMPLE 4-6: A car rounds a semicircular curve on a racetrack at a constant speed of 90 mi/h (132 ft/s), as shown in the aerial view of Figure 4-5. The turn's radius of curvature is 907.5 ft.
Determine (**a**) the magnitude and (**b**) the direction of the centripetal acceleration.

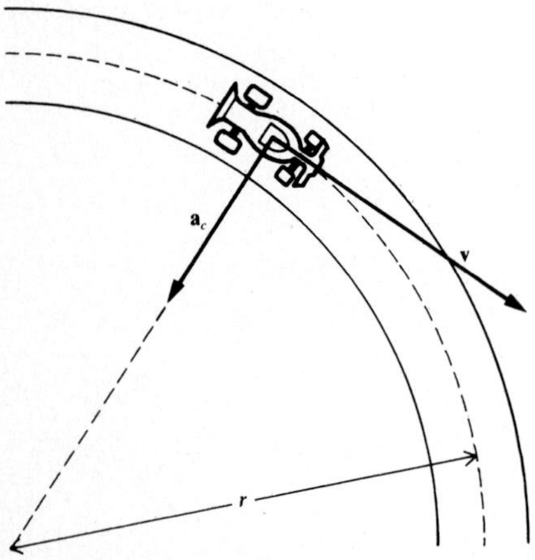

Figure 4-5

Solution:
(**a**) Use Eq. (4-12) to calculate the magnitude of the centripetal acceleration:

$$a_c = \frac{v^2}{r} = \frac{(132\ \text{ft/s})^2}{907.5\ \text{ft}} = 19.2\ \text{ft/s}^2$$

(**b**) The direction of the acceleration is toward the center of curvature.

D. Total acceleration

The tangential acceleration and the centripetal acceleration are at right angles to each other, so you can find the magnitude of the total acceleration from the Pythagorean theorem:

TOTAL ACCELERATION

$$(a_{tot})^2 = a_t^2 + a_c^2 \quad \text{or} \quad a_{tot} = \sqrt{a_t^2 + a_c^2} \tag{4-13}$$

E. Centripetal force

- The net force acting toward the center of a circle to keep an object moving in a circular path is the **centripetal force.**

EXAMPLE 4-7: The car in Example 4-6 has a weight of 2400 lb, and is not sliding in the curve.
Calculate (**a**) the magnitude of the centripetal force and (**b**) the minimum coefficient of static friction required to keep the car on the track.

Solution:
(**a**) Calculate the centripetal force by using Newton's second law of motion:

$$F_c = ma_c = \frac{w}{g}a_c = \frac{2400 \text{ lb}}{32 \text{ ft/s}^2}(19.2 \text{ ft/s}^2) = 1440 \text{ lb}$$

(**b**) In this situation, the force of static friction between the tire treads and the road surface provides the centripetal force which keeps the car from sliding off the track. If the car is to make the curve, the coefficient of static friction must be at least as great as the centripetal force divided by the normal force:

$$\mu_{s(\text{min})} = \frac{F_c}{N} = \frac{1440 \text{ lb}}{2400 \text{ lb}} = 0.6$$

F. Uniform circular motion

- An object moving along a circular path at a constant speed is in **uniform circular motion.**

In this special case, the angular acceleration α and the tangential acceleration a_t are both zero, so the total acceleration of the moving object is toward the center of the circular path—there is *no* tangential component. Eq. (4–6) through (4–9) are still valid even though the centripetal acceleration is the total acceleration, but you must set α equal to zero.

4-4. Banked Curves

A vehicle can make a sharp turn more safely if the roadway is banked, as illustrated in Figure 4-6. If the vehicle maintains the speed for which the curve is designed, no frictional force is needed to keep it on the road. The horizontal component of the normal force—the force the highway exerts against the vehicle—provides the necessary centripetal force.

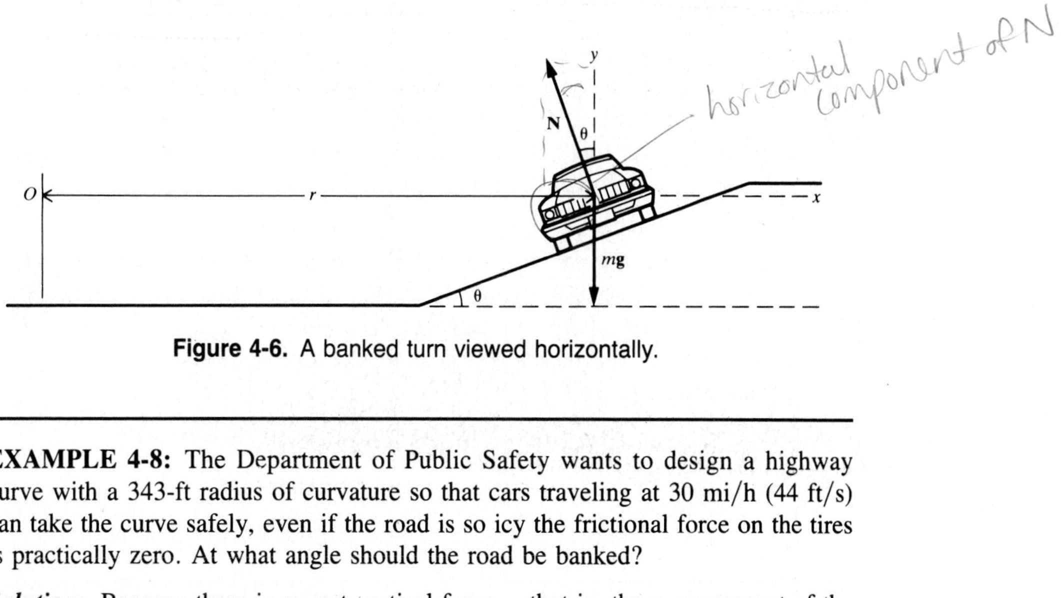

Figure 4-6. A banked turn viewed horizontally.

EXAMPLE 4-8: The Department of Public Safety wants to design a highway curve with a 343-ft radius of curvature so that cars traveling at 30 mi/h (44 ft/s) can take the curve safely, even if the road is so icy the frictional force on the tires is practically zero. At what angle should the road be banked?

Solution: Because there is no net vertical force—that is, the y component of the normal force **N** plus the weight mg of the car total zero—there is no acceleration in the vertical direction:

$$\Sigma F_y = 0 \qquad N \cos\theta = mg$$

The centripetal force needed to keep the cars on the road comes from the horizontal component of **N**:

$$N \sin\theta = \frac{mv^2}{r}$$

To get the angle at which the road should be banked, you'll need to use the tangent function, dividing ΣF_x by ΣF_y:

BANK ANGLE
$$\tan\theta = \frac{mv^2/r}{mg} = \frac{v^2}{rg} \qquad \textbf{(4-14)}$$

Now, substitute the known values:

$$\tan\theta = \frac{(44\ \text{ft/s})^2}{(343\ \text{ft})(32\ \text{ft/s}^2)} = 0.1764$$

The road must be banked at an angle of

$$\theta = \text{arc}\tan(0.1764) = 10.0°$$

4-5. Torque and Moment of Inertia

A. Torque

You know that a force is needed to give a mass linear acceleration. In the same way, a torque is needed to give a mass angular acceleration about a given axis—that is, torque takes the place of force in rotational motion.

- **Torque** τ is the product of a force and its moment arm.
- A **moment arm** (or lever arm) is the perpendicular distance between the direction line (or line of action) of the force and the axis of rotation.

TORQUE τ = moment arm × force **(4-15)**

Three factors determine how effective a torque is in altering the rotational state of an object—the magnitude of the force, its direction, and its point of application.

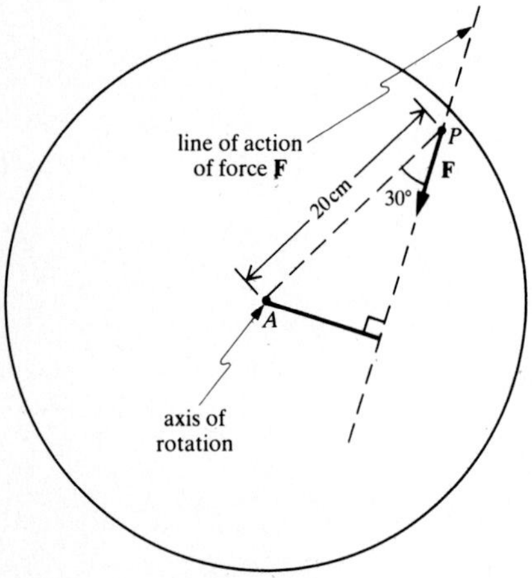

Figure 4-7

EXAMPLE 4-9: Figure 4-7 shows a force **F** of 8 N acting at point P on a disk free to rotate about an axis perpendicular to the plane of the drawing. The point of application of the force is 20 cm from the axis of rotation. The line of action of the force forms a 30° angle with a line from the axis of rotation A to the point P. Find (**a**) the moment arm of this force and (**b**) the torque about the axis of rotation.

Solution:

(**a**) First, draw the line of action of the force—the dotted line superimposed on vector **F** in Figure 4-7. Next, start at the axis of rotation—point A—and draw a line perpendicular to the line of action of **F**. The length of this perpendicular line is the moment arm of the force, which you can find using trigonometry:

$$\text{moment arm} = (20\ \text{cm})\sin 30° = 10\ \text{cm} = 0.10\ \text{m}$$

(**b**) The torque of **F** about the axis of rotation through A is

$$\tau = \text{moment arm} \times \text{force} = (0.10\ \text{m})(8\ \text{N}) = 0.80\ \text{N m}$$

note: You can consider torque as a vector quantity with the direction of the vector parallel to the axis of rotation. Use the right-hand screw analogy of Chapter 1 to define the direction. In this example, the direction is into the paper. In conventional notation, we usually designate **r** as the moment arm vector, so that $\tau = \mathbf{r} \times \mathbf{F}$, a cross product.

B. Moment of inertia

In linear motion, mass is the property of an object that resists acceleration. Newton's second law, $\Sigma\mathbf{F} = m\mathbf{a}$, expresses this relationship. Moment of inertia takes the place of mass in rotational motion.

- **Moment of inertia** I is the quantitative angular measure of the property of an object that resists acceleration.

The corresponding form of Newton's second law for angular motion is

NEWTON'S SECOND LAW (ANGULAR)

$$\Sigma\tau = I\alpha \tag{4-16}$$

Moment of inertia has dimensions of mass times length squared (kg m^2 in SI). When moment of inertia is in units of kg m^2 and angular acceleration has units of rad/s^2, then the torque is in units of N m.

The moment of inertia of a rigid body depends on the body's shape and the location of the axis of rotation. Table 4-2 gives the formulas for calculating the moment of inertia of several common objects.

TABLE 4-2: Moments of Inertia for Common Objects (r = radius, ℓ = length, m = mass)

Object	Moment of Inertia	
Thin ring, axis of rotation through the center or point mass, at distance r from axis of revolution	$I = mr^2$	**(4-17a)**
Solid uniform cylinder or disk, axis through center	$I = \frac{1}{2}mr^2$	**(4-17b)**
Uniform solid sphere, axis through center	$I = \frac{2}{5}mr^2$	**(4-17c)**
Thin uniform rod, axis through center	$I = \frac{1}{12}m\ell^2$	**(4-17d)**
Thin uniform rod, axis through one end	$I = \frac{1}{3}m\ell^2$	**(4-17e)**

EXAMPLE 4-10: The solid disk of Example 4-9 has a radius of 25 cm and a mass of 2.4 kg.
Calculate the moment of inertia of the disk about its axis.

Solution: Use Eq. (4-17b) from Table 4-2 to find the moment of inertia of a solid disk:

$$I = \frac{1}{2}mr^2 = \frac{1}{2}(2.4 \text{ kg})(0.25 \text{ m})^2 = 0.075 \text{ kg m}^2$$

EXAMPLE 4-11: Use the results of Examples 4-9 and 4-10 to determine the angular acceleration imparted to the disk shown in Figure 4-7.

Solution: You can find the disk's angular acceleration by solving the angular form of Newton's second law of motion (Eq. 4-16) for α:

$$\alpha = \frac{\Sigma\tau}{I} = \frac{0.8 \text{ N m}}{0.075 \text{ kg m}^2} = 10.7 \text{ rad/s}^2$$

4-6. Uniformly Accelerated Rotation

The solid disk illustrated in Figure 4-8 rotates freely about horizontal axis A through the center of the disk. A string wrapped around the rim of the disk attaches to mass m, which is hanging vertically as shown. The string exerts a constant tangential force on the disk—represented by the vector **F**—so the angular speed ω of the disk increases at a regular rate over time: a constant angular acceleration α. Because the disk is subject to the constant torque due to the constant tangential force, any point P on the disk undergoes **uniformly accelerated rotation**—that is, its acceleration increases linearly over time.

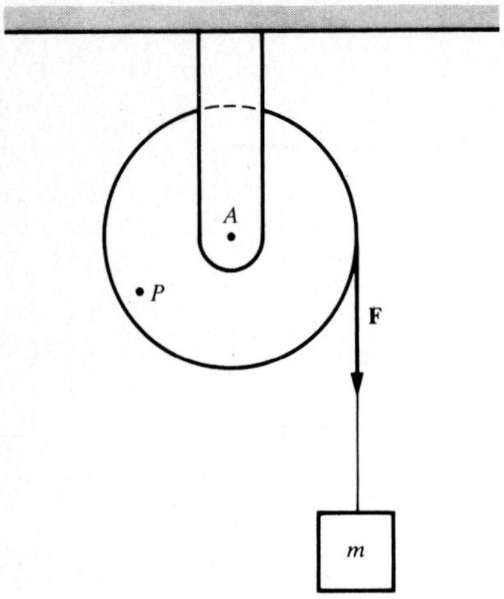

Figure 4-8

EXAMPLE 4-12: Starting from rest, the disk illustrated in Figure 4-8 reaches an angular speed of 2.5 revolutions (rev) per second in 2 s.
Calculate the angular acceleration of the disk.

Solution: You can calculate the angular acceleration α_{ave} from Eq. (4-4):

$$\alpha_{ave} = \frac{\Delta\omega}{\Delta t} = \frac{\omega_2 - \omega_1}{t_2 - t_1} = \frac{2.5 \text{ rev/s} - 0}{2 \text{ s} - 0 \text{ s}} = 1.25 \text{ rev/s}^2$$

In radians,

$$\alpha_{ave} = (1.25 \text{ rev/s}^2)(2\pi \text{ rad/rev}) = 2.5\pi \text{ rad/s}^2 = 7.85 \text{ rad/s}^2$$

Since α is constant, the average value you just calculated is also the instantaneous value: $\alpha_{ave} = \alpha = 7.85 \text{ rad/s}^2$.

EXAMPLE 4-13: What is the angular speed of the rotating disk of Example 4-12 at $t = 3$ s?

Solution: Let the initial angular speed of the disk be its speed at $t = 2$ s, $\omega_0 = 2.5$ rev/s, so you can find ω at $t = 3$ s from Eq. (4-6):

$$\omega = \omega_0 + \alpha t = (2.5 \text{ rev/s})(2\pi \text{ rad/rev}) + (7.85 \text{ rad/s}^2)(1 \text{ s}) = 23.6 \text{ rad/s}$$

EXAMPLE 4-14: Through how many revolutions does the disk of Examples 4-12 and 4-13 rotate during the first three seconds?

Solution: Calculate the angular displacement of the disk from Eq. (4-9). In this case, the initial speed ω_0 is zero. From Example 4-12, you know that the constant angular acceleration is 7.85 rad/s², so

$$\theta = \omega_0 t + \frac{1}{2}\alpha t^2 = 0 + \frac{1}{2}(7.85 \text{ rad/s}^2)(3 \text{ s})^2 = \frac{35.3 \text{ rad}}{2\pi \text{ rad/rev}} = 5.62 \text{ rev}$$

EXAMPLE 4-15: Calculate the constant torque that causes the disk of the preceding examples to have an angular acceleration of 7.85 rad/s², given that the mass of the solid disk is 8 kg and its radius is 0.25 m.

Solution: Since the torque is the product of the moment of inertia and the angular acceleration, you'll have to calculate the moment of inertia of the disk. Use Eq. (4-17b) from Table 4-2:

$$I = \frac{1}{2}mr^2 = \frac{1}{2}(8 \text{ kg})(0.25 \text{ m})^2 = 0.25 \text{ kg m}^2$$

Now, you can use Eq. (4-16) to calculate the torque:

$$\tau = I\alpha = (0.25 \text{ kg m}^2)(7.85 \text{ rad/s}^2) = 1.96 \text{ N m}$$

EXAMPLE 4-16: Calculate the tangential force F acting on the disk of the preceding examples.

Solution: You'll need to find the moment arm of the force acting on the disk. Since this force is perpendicular to the radius, the moment arm is equal to the radius of the disk. [**recall**: The moment arm is the perpendicular distance between the direction line of the force and the axis of rotation (see Sec. 4-5).] Use Eq. (4-15) to find the magnitude of F:

$$F = \frac{\tau}{\text{moment arm}} = \frac{1.96 \text{ N m}}{0.25 \text{ m}} = 7.84 \text{ N}$$

EXAMPLE 4-17: Find the mass m that is fastened to the string wrapped around the edge of the disk of the preceding examples.

Solution: Since *both* the disk and the hanging mass are being accelerated, the tension in the string is *not* equal to the weight of the mass. Figure 4-9 shows the two forces acting on the hanging mass. The mass is being accelerated downward, so choose this as the positive direction. You can calculate the mass from Newton's second law:

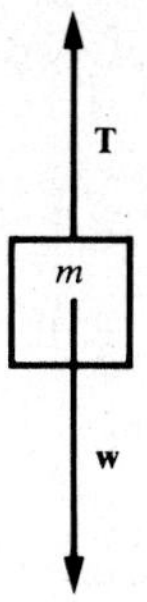

Figure 4-9

$$\Sigma F = ma \qquad mg - T = ma \qquad m = \frac{T}{g - a}$$

But, first you have to find the acceleration. Use Eq. (4-11):

$$a = r\alpha = (0.25\ \text{m})(7.84\ \text{rad/s}^2) = 1.96\ \text{m/s}^2$$

Because the string has a negligible mass, the tension T is equal to the force F acting on the disk. Substitute the values:

$$m = \frac{F}{g - a} = \frac{7.84\ \text{N}}{(9.8 - 1.96)\ \text{m/s}^2} = 1.00\ \text{kg}$$

SUMMARY

Angular displacement	$\theta = \frac{s}{r}$ or $s = r\theta$	the length of the arc divided by the radius
Average angular speed	$\omega_{\text{ave}} = \frac{\Delta\theta}{\Delta t} = \frac{\theta_2 - \theta_1}{t_2 - t_1}$	angular displacement divided by the time needed for the displacement to occur
Instantaneous angular speed	$\omega = \lim_{\Delta t \to 0} \frac{\Delta\theta}{\Delta t}$	average angular displacement over a very short time interval
Average angular acceleration	$\alpha_{\text{ave}} = \frac{\Delta\omega}{\Delta t} = \frac{\omega_2 - \omega_1}{t_2 - t_1}$	change in angular speed divided by the time needed for the change to occur
Instantaneous angular acceleration	$\alpha = \lim_{\Delta t \to 0} \frac{\Delta\omega}{\Delta t}$	average angular acceleration over a very short time interval
Angular relationships	$\omega = \omega_0 + \alpha t$ $\omega_{\text{ave}} = \frac{1}{2}(\omega_1 + \omega_2)$ $\omega^2 = \omega_0^2 + 2\alpha\theta$ $\theta = \omega_0 t + \frac{1}{2}\alpha t^2$	give angular relationships between speed, displacement, acceleration, and time
Tangential speed	$v_t = r\omega$	the radius times angular speed
Tangential and angular acceleration	$a_t = r\alpha$	the radius times angular acceleration
Centripetal acceleration	$a_c = \frac{v^2}{r}$ or $a_c = r\omega^2$	the linear acceleration of an object directed toward the center of curvature

Total acceleration	$(a_{tot})^2 = a_t^2 + a_c^2$ or $a_{tot} = \sqrt{a_t^2 + a_c^2}$	magnitude of the sum of tangential acceleration and centripetal acceleration
Bank angle	$\tan\theta = \dfrac{v^2}{rg}$	angle at which object remains on path without frictional force
Torque	τ = moment arm × force	the force causing angular acceleration about a given axis
Newton's second law (angular)	$\Sigma\tau = I\alpha$	the relationship between torque, moment of inertia, and angular acceleration for rotational motion
Moments of inertia for common objects	$I = mr^2$	for a thin ring, axis through center
	$I = \frac{1}{2}mr^2$	for a solid uniform cylinder or disk, axis through center
	$I = \frac{2}{5}mr^2$	for a uniform solid sphere, axis through center
	$I = \frac{1}{12}m\ell^2$	for a thin uniform rod, axis through center
	$I = \frac{1}{3}m\ell^2$	for a thin uniform rod, axis through one end

RAISE YOUR GRADES

Can you define . . . ?

☑ angular displacement
☑ tangential speed
☑ centripetal acceleration
☑ radius of curvature
☑ torque
☑ moment arm

Can you . . . ?

☑ state the relationship between an angle θ expressed in radians and the length of the subtended arc
☑ state the definition of instantaneous angular speed
☑ write the equation that defines average angular acceleration
☑ name the quantity in angular motion analogous to mass in linear motion
☑ name the angular quantity analogous to linear speed
☑ write the relationship between tangential speed and angular speed
☑ write the relationship between tangential acceleration and angular acceleration
☑ calculate the centripetal acceleration of a point moving in a circular path at an angular speed ω
☑ state the angle between tangential acceleration and centripetal acceleration
☑ calculate the moment arm of a force if you know where its line of action and the axis of rotation are
☑ calculate the torque produced by a force
☑ write the angular form of Newton's second law of motion
☑ calculate the moment of inertia of a long thin rod about an axis through one end
☑ write the equation for the moment of inertia of a solid cylinder with an axis of rotation through its center

SOLVED PROBLEMS

Angular quantities

PROBLEM 4-1 The turntable of a record player rotates at a constant speed of $33\frac{1}{3}$ rev/min. How long (in seconds) does it take the turntable to rotate through an angle of 60°?

Solution You have the angular displacement and angular speed, so begin by writing down Eq. (4-2)—the relationship between angular speed, angular displacement, and time:

$$\omega = \frac{\Delta\theta}{\Delta t}$$

Now, solve for Δt, the time interval, in seconds:

$$\Delta t = \frac{\Delta\theta}{\omega} = \left(\frac{60°}{33.3\ \text{rev/min}}\right)\left(\frac{1\ \text{rev}}{360°}\right)\left(\frac{60\ \text{s}}{1\ \text{min}}\right) = 0.300\ \text{s}$$

PROBLEM 4-2 After a record player is turned on, the turntable reaches its correct operating speed of $33\frac{1}{3}$ rev/min (or 3.49 rad/s) in 1.5 s. (**a**) What is the average acceleration of the turntable and (**b**) through what angle does it rotate during the first 1.5 s?

Solution

(**a**) You can calculate the average angular acceleration from Eq. (4-4):

$$\alpha_{\text{ave}} = \frac{\omega_2 - \omega_1}{t_2 - t_1} = \frac{3.49\ \text{rad/s} - 0}{1.5\ \text{s}} = 2.33\ \text{rad/s}^2$$

(**b**) Assume that the angular acceleration is constant during the 1.5 s interval, and use Eq. (4-9) to calculate the angular displacement:

$$\theta = \omega_0 t + \frac{1}{2}\alpha t^2 = 0 + \frac{1}{2}(2.327\ \text{rad/s}^2)(1.5\ \text{s})^2$$

$$= 2.618\ \text{rad} = (2.618\ \text{rad})(180°/\pi\ \text{rad}) = 150°$$

Circular motion

PROBLEM 4-3 For the record player of Problem 4-2, calculate the tangential acceleration of a point on the rim of the platter, given a radius of 15 cm.

Solution Since the point experiences an angular displacement, all you need is the relationship between tangential acceleration and angular acceleration, Eq. (4-11):

$$a_t = r\alpha = (15\ \text{cm})(2.327\ \text{rad/s}^2) = 34.9\ \text{cm/s}^2$$

PROBLEM 4-4 What is the magnitude of the total acceleration of a point on the rim of the turntable of Problem 4-2 1.2 s after the turntable has been turned on?

Solution The total acceleration is a vector having two components—the tangential acceleration, which you found in Problem 4-3, and the centripetal acceleration, $a_c = r\omega^2$. To calculate a_c, you must first determine the angular speed, Eq. (4-6):

$$\omega = \omega_0 + \alpha t = 0 + (2.327\ \text{rad/s}^2)(1.2\ \text{s}) = 2.792\ \text{rad/s}$$

Then, you can use Eq. (4-12) to calculate the centripetal acceleration:

$$a_c = r\omega^2 = (15\ \text{cm})(2.792\ \text{rad/s})^2 = 116.9\ \text{cm/s}^2$$

The total acceleration is the vector sum of the tangential and centripetal accelerations:

$$\mathbf{a}_{\text{tot}} = \mathbf{a}_t + \mathbf{a}_c$$

You can use the Pythagorean theorem to find the magnitude of $\mathbf{a}_{tot}$ because the two components of **a** are at right angles to each other:

$$a = \sqrt{a_t^2 + a_c^2} = \sqrt{(34.9\text{ cm/s}^2)^2 + (116.9\text{ cm/s}^2)^2} = 122\text{ cm/s}^2$$

PROBLEM 4-5 The needle of a record player is placed 12 cm from the center of a record, which is rotating at an angular speed of $33\frac{1}{3}$ rev/min. Determine the tangential speed of the groove in which the needle is placed.

Solution Use Eq. (4-10) to determine the tangential speed:

$$v_t = r\omega = (12\text{ cm})\left(\frac{33.33\text{ rev}}{1\text{ min}}\right)\left(\frac{2\pi\text{ rad}}{1\text{ rev}}\right)\left(\frac{1\text{ min}}{60\text{ s}}\right) = 41.9\text{ cm/s}$$

PROBLEM 4-6 A small object moves around a circle in a vertical plane at a constant speed of 5 m/s. The radius of the circular path is 8 cm. (**a**) Calculate the angular speed of this object and (**b**) the magnitude of its centripetal acceleration.

Solution

(**a**) You can find the angular speed from Eq. (4-10):

$$\omega = \frac{v_t}{r} = \frac{5\text{ m/s}}{0.08\text{ m}} = 62.5\text{ rad/s}$$

(**b**) Now, calculate the centripetal acceleration from Eq. (4-12):

$$a_c = r\omega^2 = (0.08\text{ m})(62.5\text{ rad/s})^2 = 313\text{ m/s}^2$$

or

$$a_c = \frac{v^2}{r} = \frac{(5\text{ m/s})^2}{0.08\text{ m}} = 313\text{ m/s}^2$$

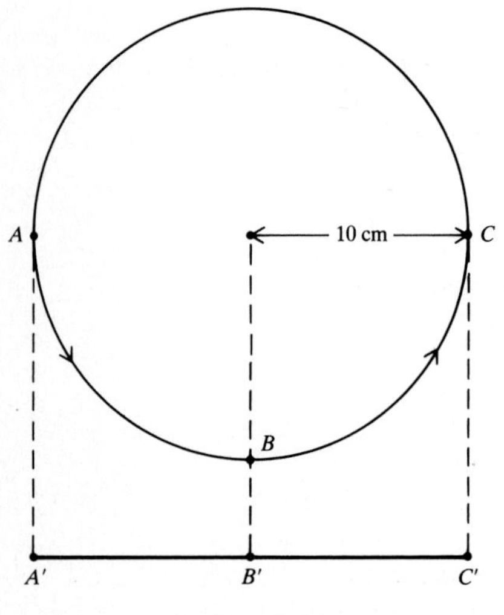

Figure 4-10

PROBLEM 4-7 A small object moves around a circle in a vertical plane at a constant angular speed of 2 rad/s. The radius of the circle is 10 cm. The sun, which is directly overhead, casts a shadow of the moving object on the ground. The shadow moves back and forth between points A' and C', as illustrated in Fig. 4-10. How long does it take the shadow to move from A' to C'?

Solution While the shadow moves from A' to C', the object itself moves from A to C in the circle—an angular displacement of 180° or π radians. Since the angular speed is constant, you can use Eq. (4-2) and solve for the time interval:

$$\Delta t = \frac{\Delta\theta}{\omega} = \frac{\pi\text{ rad}}{2\text{ rad/s}} = 1.57\text{ s}$$

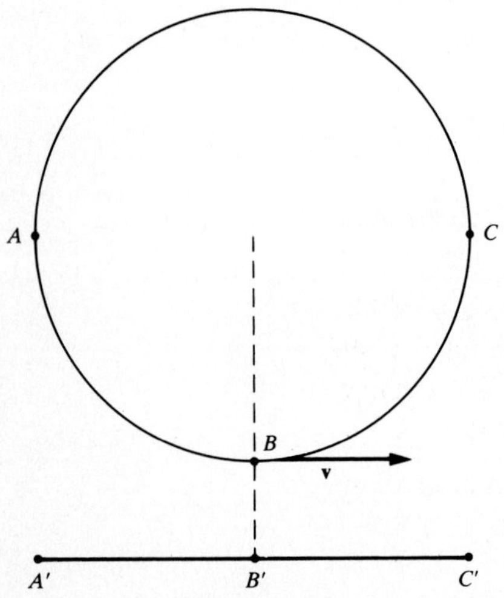

Figure 4-11

PROBLEM 4-8 The shadow of the object described in Problem 4-7 moves from A' to C', a distance of 20 cm, in 1.57 s. What is the speed of the shadow as it passes point B', the midpoint of line $A'C'$?

Solution As you can see in Fig. 4-11, the direction of the velocity vector of the moving object is horizontal as it passes point B in its circular path. In this case, the speed of the shadow—which also moves horizontally—is equal to the speed of the object itself, so you can find the tangential speed from Eq. (4-10):

$$v_t = r\omega = (10\text{ cm})(2\text{ rad/s}) = 20\text{ cm/s}$$

PROBLEM 4-9 The shadow of the object described in Problems 4-7 and 4-8 has its maximum speed at point B', the midpoint of its path. The speed of the shadow necessarily becomes zero at the two end points, A' and C'. Find the point P' at which the shadow will be moving at half of its maximum speed.

Solution: Draw a diagram of the situation, showing the relationship between the velocity of the object in the circle and its shadow (or projection) on the ground. It should be similar to Fig. 4-12. Can you see that the speed of the shadow as it moves along the horizontal line $B'C'$ is the horizontal component of the velocity **v** of the moving object? This is the key to the problem. Since this horizontal component is half of **v**, you can find the angle θ from its cosine:

$$\cos\theta = \frac{v_x}{v} = \frac{v/2}{v} = \frac{1}{2}$$

$$\theta = \arccos(0.5) = 60°$$

Since you know from trigonometry that this same angle appears in the larger triangle whose hypotenuse is r, you can now find the distance from B' to P':

$$x = r\sin\theta$$

$$= (10\text{ cm})\sin 60° = 8.66\text{ cm}$$

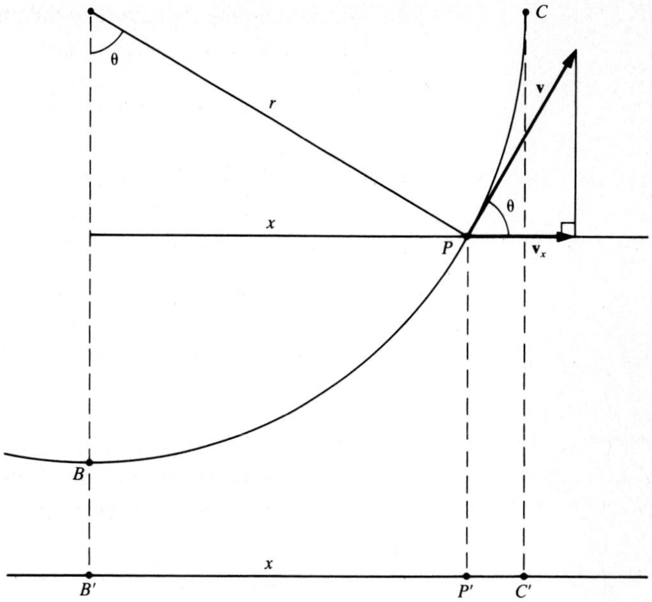

Figure 4-12

PROBLEM 4-10 A cyclist rides her racing bicycle at a speed of 16 mi/h. The bicycle wheels have a diameter of 27 in. What is the angular speed of rotation of the wheels in rad/s?

Solution You know the linear speed of the wheels and you can calculate their radius in SI units:

$$r = \frac{1}{2}(27\text{ in})(2.54\text{ cm/in})(1\text{ m}/100\text{ cm}) = 0.343\text{ m}$$

So, you can solve this problem with Eq. (4-10)—the relationship between linear speed and angular speed:

$$\omega = \frac{v}{r} = \left(\frac{16\text{ mi/h}}{0.343\text{ m}}\right)\left(\frac{1.609\times 10^3\text{ m}}{1\text{ mi}}\right)\left(\frac{1\text{ h}}{3600\text{ s}}\right) = 20.9\text{ rad/s}$$

Torque and moment of inertia

PROBLEM 4-11 A one-cylinder diesel engine has a flywheel that weighs 445 N. The moment of inertia of the flywheel is 1.50 kg m^2. What torque is required to bring this flywheel from rest to a speed of 50 rev/s in 5 minutes?

Solution From Eq. (4-16), you know that $\tau = I\alpha$, so first calculate the angular acceleration of the flywheel from Eq. (4-4):

$$\alpha = \frac{\Delta\omega}{\Delta t} = \left(\frac{50\text{ rev/s} - 0}{(5\text{ min})(60\text{ s/min})}\right)\left(\frac{2\pi\text{ rad}}{1\text{ rev}}\right) = 1.047\text{ rad/s}^2$$

Then, you can use Eq. (4-16) to calculate the torque:

$$\tau = I\alpha = (1.50\text{ kg m}^2)(1.047\text{ rad/s}^2) = 1.57\text{ N m}$$

PROBLEM 4-12 If you place a small coin close to the center of a record rotating on a turntable, the coin remains where you put it. However, if you place the coin close to the outer edge of the record, it slides off. If the coefficient of static friction between the coin and the record is 0.4, calculate the maximum radius at which the coin will stay on a record rotating at 78 rev/min.

Solution Since it is the force of static friction that produces the centripetal acceleration of the penny, you can set the frictional force equal to the centripetal force:

$$f = \mu N = \mu mg$$

$$F_c = mr\omega^2$$

So that:

$$\mu mg = mr\omega^2$$

And, solve for the radius:

$$r = \frac{\mu g}{\omega^2}$$

Find the angular speed in rad/s:

$$\omega = (78 \text{ rev/min})(2\pi \text{ rad/rev})(1 \text{ min}/60 \text{ s}) = 8.17 \text{ rad/s}$$

Now, solve for the maximum radius:

$$r = \frac{\mu g}{\omega^2} = \frac{(0.4)(9.8 \text{ m/s}^2)}{(8.17 \text{ rad/s})^2} = 0.0587 \text{ m} = 5.87 \text{ cm}$$

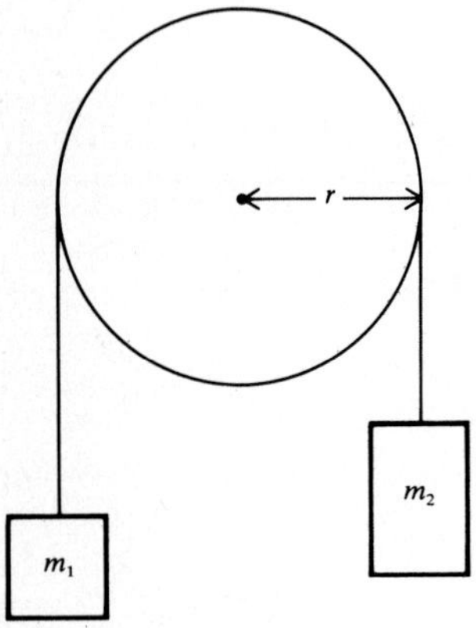

Figure 4-13

PROBLEM 4-13 A thin steel rod, 60 cm in length has a mass of 500 g. The rod rotates about an axis that is perpendicular to the rod and that passes through its center. What torque is required to bring the stationary rod to an angular speed of 4 rev/s in 20 s?

Solution Torque is equal to the product of moment of inertia and angular acceleration. You can calculate the moment of inertia of the rod from Eq. (4-17d) in Table 4-2:

$$I = \frac{1}{12}m\ell^2 = \frac{1}{12}(0.50 \text{ kg})(0.60 \text{ m})^2 = 0.015 \text{ kg m}^2$$

Use Eq. (4-4) to calculate the constant angular acceleration:

$$\alpha = \frac{\omega_2 - \omega_1}{\Delta t} = \frac{(4 \text{ rev/s} - 0)}{20 \text{ s}}\left(\frac{2\pi \text{ rad}}{1 \text{ rev}}\right) = 1.26 \text{ rad/s}^2$$

And then use Eq. (4-16) to calculate the torque:

$$\tau = I\alpha = (0.015 \text{ kg m}^2)(1.26 \text{ rad/s}^2) = 0.0189 \text{ N m}$$

Uniformly accelerated rotation

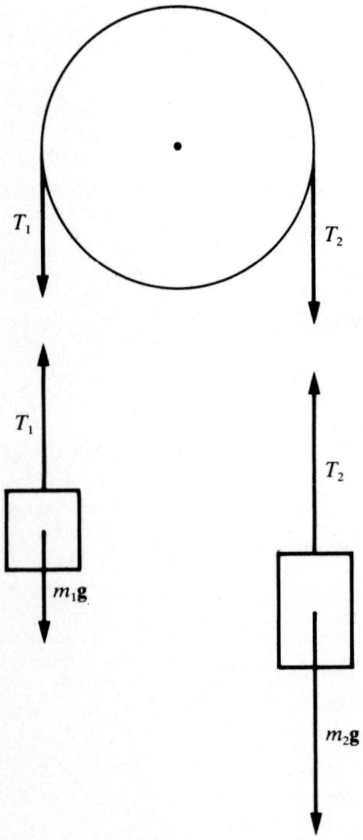

Figure 4-14

PROBLEM 4-14 A solid disk with a radius of 6 cm rotates without friction about an axis through its center. Two masses are fastened to a piece of string that passes over the rim of the disk (see Fig. 4-13). The string does not slip on the disk. The disk has a mass of 400 g and the two hanging blocks have masses of 200 g and 300 g. When released from rest, how long will it take the 300-g block to move 40 cm?

Solution Eq. (4-11) gives the relation of the tangential acceleration of the hanging blocks to the angular acceleration of the disk: $a = r\alpha$. As you can see from Fig. 4-14, there are two tangential forces acting on the disk—one caused by the string on the right side, and the other caused by the string on the left side. You want to find the net torque acting on the disk.

$$\text{net } \tau = \Sigma(\text{moment arm} \times \text{force})$$

The forces are the tensions in the strings, and the moment arms are both equal to the radius of the disk. You can calculate the tensions from the weights of the blocks. Notice that the blocks are accelerating at the same rate since they are connected by a common string. Use Newton's second law of motion, $\Sigma F = ma$, and choose the direction in which the blocks are being accelerated as the positive direction of rotation.

$$T_1 - m_1g = m_1a \qquad T_1 = m_1g + m_1a$$

$$m_2g - T_2 = m_2a \qquad T_2 = m_2g - m_2a$$

Now find the net torque from these equations:

$$\tau = T_2r - T_1r = m_2(g - a)r - m_1(g + a)r$$

The tangential acceleration of the disk is also a, since the string doesn't slip. It is also equal to αr, where $r = 6$ cm is the same as the moment arm of the torque.

$$\begin{aligned}\tau &= m_2(g - \alpha r)r - m_1(g + \alpha r)r \\ &= (m_2 - m_1)gr - (m_2 + m_1)\alpha r^2 \\ &= (0.30\text{ kg} - 0.20\text{ kg})(9.8\text{ m/s}^2)(0.06\text{ m}) - (0.30\text{ kg} + 0.20\text{ kg})(0.06\text{ m})^2\alpha \\ &= 5.88 \times 10^{-2}\text{ kg m}^2/\text{s}^2 - (1.8 \times 10^{-3}\text{ kg m}^2)\alpha\end{aligned}$$

You can use Eq. (4-16)—the angular form of Newton's second law—to find α:

$$\Sigma\tau = I\alpha$$

but first you'll have to calculate the moment of inertia of the disk from Eq. (4-17b), Table 4-2:

$$I = \frac{1}{2}mr^2 = \frac{1}{2}(0.40\text{ kg})(0.06\text{ m})^2 = 7.2 \times 10^{-4}\text{ kg m}^2$$

Then, from Eq. (4-16):

$$\begin{aligned}\tau &= I\alpha \\ I\alpha &= (5.88 \times 10^{-2}\text{ kg m}^2/\text{s}^2) - (1.8 \times 10^{-3}\text{ kg m}^2)\alpha \\ \alpha[7.2 \times 10^{-4}\text{ kg m}^2 + 1.8 \times 10^{-3}\text{ kg m}^2] &= 5.88 \times 10^{-2}\text{ kg m}^2/\text{s}^2 \\ \alpha &= \frac{5.88 \times 10^{-2}\text{ kg m}^2/\text{s}^2}{2.52 \times 10^{-3}\text{ kg m}^2} = 23.3\text{ rad/s}^2\end{aligned}$$

Now, you can calculate the linear acceleration of the 300 g block:

$$a = r\alpha = (0.06\text{ m})(23.3\text{ rad/s}^2) = 1.40\text{ m/s}^2$$

Since the block is initially at rest, $v_0 = 0$, and from Eq. (2-8), the distance that it moves is

$$s = \frac{1}{2}at^2$$

Solve for the unknown time:

$$t = \sqrt{\frac{2s}{a}} = \sqrt{\frac{2(0.40\text{ m})}{1.40\text{ m/s}^2}} = 0.756\text{ s}$$

Supplementary Exercises

PROBLEM 4-15 The turntable of a record player rotates at a speed of 45 rev/min. How long does it take for this turntable to rotate 90°?

PROBLEM 4-16 The turntable of another record player rotates at a speed of 33.3 rev/min. The phonograph needle is placed in a groove that is 14 cm from the axis of rotation. At what speed does the groove move under the needle?

PROBLEM 4-17 An airplane, flying at a speed of 250 km/h, moves along a horizontal circular path. The radius of the circle is 800 m. What is the magnitude of the centripetal acceleration of the plane?

PROBLEM 4-18 The curve of a race track is banked at an angle of 25°. The radius of curvature is 675 m. What is the safe speed in km/h for a race car to round this curve if the track is very slippery?

PROBLEM 4-19 A solid sphere is mounted on an axis that passes through the center of the sphere. The mass of the sphere is 4.0 kg and its radius is 5 cm. A motor attached to the axis produces a constant torque of 0.012 N m. What is the angular acceleration of the sphere?

PROBLEM 4-20 A disk whose radius is 5 cm has an angular acceleration of 40 rad/s^2. A point on the circumference of the disk has both centripetal acceleration and tangential acceleration. What is the magnitude of the centripetal acceleration and the tangential acceleration when the angular speed is 7.746 rad/s?

PROBLEM 4-21 What is the total acceleration of the point on the circumference of the disk described in Problem 4-20?

PROBLEM 4-22 A disk has an angular acceleration of 40 rad/s^2. If the initial angular speed of the disk is 5 rad/s, how long does it take to reach an angular speed of 20 rad/s?

Answers to Supplementary Exercises

4-15: 0.333 s

4-16: 48.8 cm/s

4-17: 6.03 m/s^2

4-18: 200 km/h

4-19: 3.00 rad/s^2

4-20: a_c = 3.00 m/s^2, a_t = 2.00 m/s^2

4-21: 3.61 m/s^2

4-22: 0.375 s

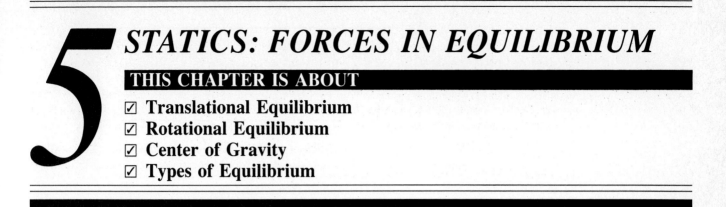

5 STATICS: FORCES IN EQUILIBRIUM

THIS CHAPTER IS ABOUT

☑ Translational Equilibrium
☑ Rotational Equilibrium
☑ Center of Gravity
☑ Types of Equilibrium

A particle or point object is in **equilibrium** if the vector sum of all forces acting on it is zero.

5-1. Translational Equilibrium

- A particle is said to be in **translational equilibrium** if its linear acceleration is zero; that is, $\mathbf{a} = 0$.

There are only two situations in which this condition is satisfied:

- A particle is said to be in **dynamic equilibrium** if it moves along a straight line at a constant speed; that is, its velocity $\mathbf{v}$ is constant so that $\mathbf{a} = 0$.
- A particle is said to be in **static equilibrium** if it is at rest; that is, $\mathbf{v} = 0$ and $\mathbf{a} = 0$.

An object does not accelerate unless there is a net force acting on it. [**recall:** Newton's second law of motion (Eq. 3-2) gives the relationship between forces and acceleration: $\Sigma\mathbf{F} = m\mathbf{a}$ or $\mathbf{a} = \Sigma\mathbf{F}/m$.] This means that the acceleration of an object is zero if the sum of all forces acting on it is zero. Since an object in equilibrium has zero acceleration, the condition for translational equilibrium—either static or dynamic—is

TRANSLATIONAL EQUILIBRIUM (VECTOR)

$$\Sigma\mathbf{F} = 0 \qquad \textbf{(5-1)}$$

A net force can be zero only if each of its rectangular components is zero, so the vector equation for translational equilibrium (Eq. (5-1)) is equivalent to three scalar equations:

TRANSLATIONAL EQUILIBRIUM (SCALAR)

$$\Sigma F_x = 0 \qquad \Sigma F_y = 0 \qquad \Sigma F_z = 0 \qquad \textbf{(5-2)}$$

You'll be using these three equations to solve problems in translational equilibrium.

EXAMPLE 5-1: The two forces $\mathbf{F}_1 = 4$ N and $\mathbf{F}_2 = 2$ N act on a particle as shown in Figure 5-1.
Is this object in translational equilibrium?

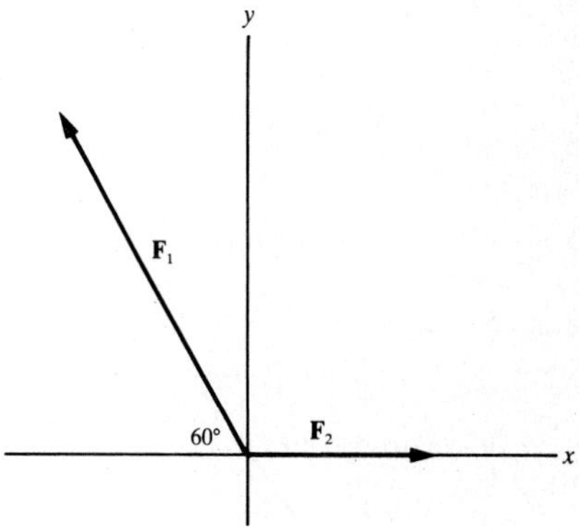

Figure 5-1

Solution: The object is in translational equilibrium only if Eqs. (5-2) are satisfied, so that's what you'll have to determine. $\mathbf{F}_1$ and $\mathbf{F}_2$ lie in the xy plane, which means that their z components are zero, so the equation $\Sigma F_z = 0$ is clearly satisfied. To check the other two equations, you'll have to find the x and y components of the resultant $\mathbf{R}$ of $\mathbf{F}_1$ and $\mathbf{F}_2$. The x component of $\mathbf{R}$ (R_x) is the sum of the x components of $\mathbf{F}_1$ and $\mathbf{F}_2$, and $R_y = \Sigma F_y$:

Forces	x components	y components
$\mathbf{F}_1$ $\mathbf{F}_2$	$-(4\text{ N})\cos 60°$ 2 N	$(4\text{ N})\sin 60°$ 0
$\mathbf{R} = \mathbf{F}_1 + \mathbf{F}_2$	0	3.46 N

You can see that the condition of equilibrium for the x components ($\Sigma \mathbf{F}_x = 0$) is satisfied, but $\Sigma \mathbf{F}_y = 0$ is not—so the object in question is not in equilibrium.

EXAMPLE 5-2: Find a third force **F** that can be added to the forces of Example 5-1 to satisfy the condition of equilibrium.

Solution: Let $\mathbf{F}_x$ and $\mathbf{F}_y$ represent the components of the third force, and revise your table to show all three forces and their components:

Forces	x components	y components
$\mathbf{F}_1$	$-(4\text{ N})\cos 60°$	$(4\text{ N})\sin 60°$
$\mathbf{F}_2$	2 N	0
$\mathbf{F}$	F_x	F_y

Now, set the sum of the x components equal to zero, and the sum of the y components equal to zero to satisfy the conditions of equilibrium:

$$-(2\text{ N}) + 2\text{ N} + F_x = 0 \qquad 3.46\text{ N} + 0 + F_y = 0$$

$$F_x = 0 \qquad F_y = -(3.46\text{ N})$$

The direction of **F** is along the negative y axis, as shown in Figure 5-2. With the addition of **F** to $\mathbf{F}_1$ and $\mathbf{F}_2$, the object is in equilibrium.

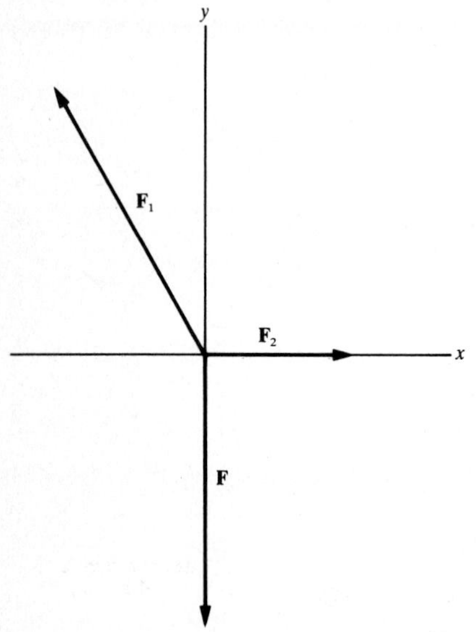

Figure 5-2

EXAMPLE 5-3: Three forces act on a point P in Figure 5-3. Downward force **w** has a magnitude of 80 lb, and the rope pulling upward and toward the right exerts a force **F** of 100 lb.
Find the magnitude of the force of tension **T** in the horizontal rope.

Solution: Since the point P where the three ropes are joined is in equilibrium, the sum of **w**, **F**, and **T** must be zero. This means that if you place the arrows representing the three forces head-to-tail, they will form a closed triangle: $\Sigma\mathbf{F} = 0$, so there is no resultant force. (See Figure 5-4.) Since this is a right triangle, you can use the Pythagorean theorem to find the magnitude of **T**:

$$T = \sqrt{(100\text{ lb})^2 - (80\text{ lb})^2} = 60\text{ lb}$$

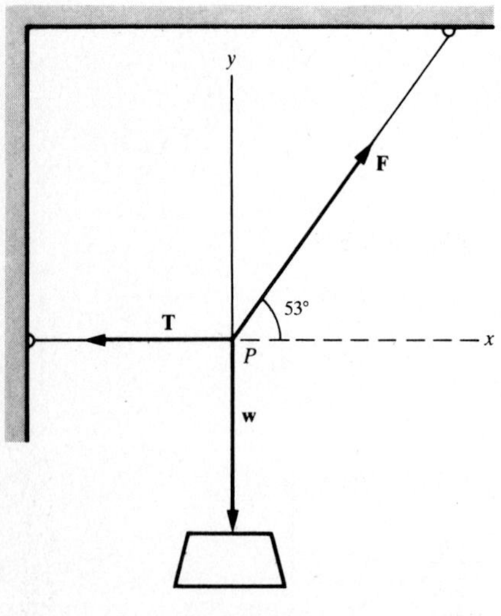

Figure 5-3

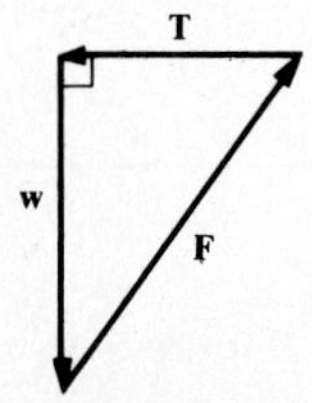

Figure 5-4

5-2. Rotational Equilibrium

- An object is said to be in **rotational equilibrium** if its angular acceleration is zero; that is, $\alpha = 0$.

There are also two situations in which this condition is satisfied:

- An object is said to be in **dynamic rotational equilibrium** if it is rotating about a fixed axis at a constant speed; that is, its angular speed ω is constant so that $\alpha = 0$.
- An object is said to be in **static rotational equilibrium** if it is not rotating at all; that is, $\omega = 0$ and $\alpha = 0$.

A net torque acting on an object causes its angular acceleration. [**recall:** the angular form of Newton's second law (Eq. 4-16) gives the relationship between torque and angular acceleration: $\Sigma\tau = I\alpha$ or $\alpha = \Sigma\tau/I$. (Sec. 4-5)] This means that the angular acceleration of an object is zero if the sum of the torques acting about its axis of rotation is zero. An object in equilibrium has zero acceleration, so the condition for rotational equilibrium—either static or dynamic—is

ROTATIONAL EQUILIBRIUM

$$\Sigma\tau = 0 \tag{5-3}$$

EXAMPLE 5-4: The horizontal rod shown in Figure 5-5 can rotate about a horizontal axis perpendicular to the drawing. A load of 8 N is attached to the left end of the rod, 0.5 m from the axis. An unknown force **T** is acting on the other end of the rod at an angle of 60°, as shown.
Find the value of **T** that will keep the rod in rotational equilibrium.

Solution: You have to find the sum of the torques acting about the axis of rotation. And, since τ = moment arm × force (Eq. 4-15), you need the moment arm of each force. [**recall:** A moment arm is the perpendicular distance between the line of action of the force and the axis of rotation. (Sec. 4-5)] The torque associated with **w** is positive because it acts in a counterclockwise direction. Since **T** would cause the rod to rotate in a clockwise direction if it were the only force present, the torque associated with **T** is negative. It will help if you tabulate the results:

Force	Magnitude	Moment arm	Torque
w	8 N	0.5 m	4.0 N m
T	T	(1.5 m) sin 60°	$-T$(1.30 m)

Now, set the sum of the torques equal to zero:

$$\Sigma\tau = 4.0 \text{ N m} - T(1.30 \text{ m}) = 0$$

$$T(1.30 \text{ m}) = 4.0 \text{ N m}$$

$$T = 3.08 \text{ N}$$

The rod will remain in rotational equilibrium if **T** has a value of 3.08 N exerted in the direction shown in Figure 5-5.

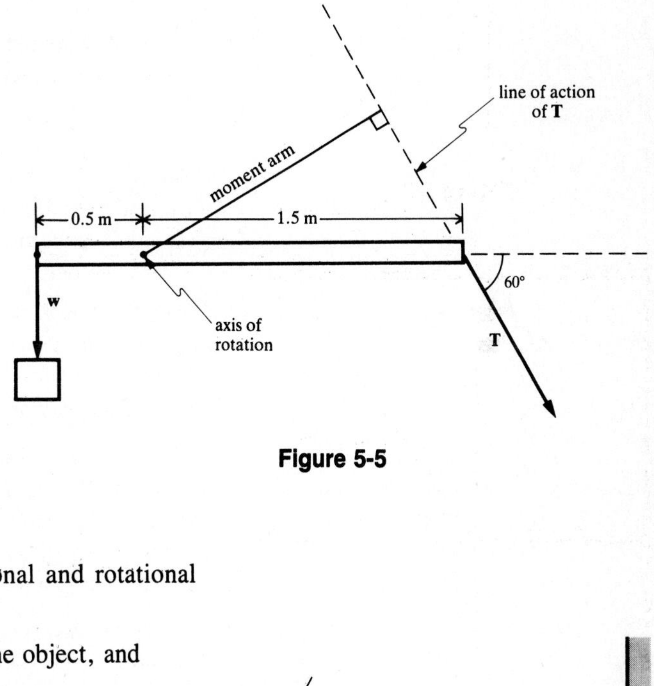

Figure 5-5

To solve problems in statics that involve both translational and rotational equilibrium:

- sketch a diagram showing all the external forces acting on the object, and
- apply the conditions of equilibrium: Eqs. (5-2) and Eq. (5-3).

note: You can simplify the mathematics in many problems by a careful choice of axes. Try to find a point through which more than one force acts. The moment arms of these forces equal zero, so the forces will not appear in the equation for torque.

EXAMPLE 5-5: The rod of negligible mass illustrated in Figure 5-6 is supported by a hinge at the wall and by a horizontal wire fastened at a 30° angle one third of the way along the rod. A weight of 12 N hangs from the left end of the rod.
Find (**a**) the tension T in the horizontal wire and (**b**) the magnitude and direction of the force that the hinge exerts on the end of the rod.

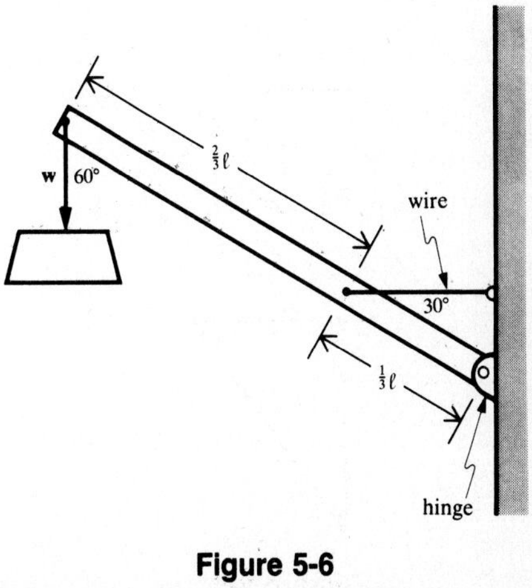

Figure 5-6

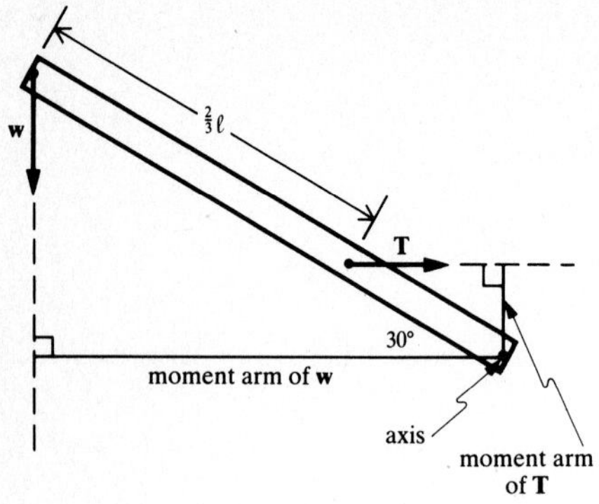

Figure 5-7

Solution:

(a) Place the axis of rotation at the hinge and then draw the moment arms of the forces acting on the rod (see Figure 5-7). [Hint: The unknown force **F** that the hinge exerts on the rod has a moment arm of zero, so it will not appear in your equation for torque.] Use ℓ to represent the length of the rod since it is not given, and prepare a table of the forces acting on the rod, their magnitudes, moment arms, and torques:

Force	Magnitude	Moment arm	Torque
w	12 N	$\ell \cos 30°$	$(12\text{ N})\ell \cos 30°$
T	T	$\frac{1}{3}\ell \sin 30°$	$-\frac{1}{3}T\ell \sin 30°$

Now, apply Eq. (5-3) for rotational equilibrium, $\Sigma\tau = 0$, and set the sum of the torques equal to zero:

$$(12\text{ N})\ell \cos 30° - \frac{1}{3}T\ell \sin 30° = 0$$

You can see that ℓ cancels out, so that

$$\frac{1}{3}T \sin 30° = (12\text{ N}) \cos 30°$$

And, you can solve for the magnitude of **T**, the unknown force of tension:

$$T = \frac{(12\text{ N}) \cos 30°}{\frac{1}{3} \sin 30°} = 62.4\text{ N}$$

(b) Now you know two of the three forces acting on the rod, and you can find the force the hinge exerts on the rod by applying Eqs. (5-2) for translational equilibrium in the x and y directions. Again, it will help to prepare a table of the forces and their components:

Forces	x components	y components
w	0	−12 N
T	62.4 N	0
F	F_x	F_y

Since the rod is in equilibrium, the x components must add up to zero, so $F_x = -62.4$ N. The y components must also add up to zero, so $F_y = 12$ N. Now, find the magnitude of **F** from the Pythagorean theorem:

$$F = \sqrt{F_x^2 + F_y^2} = \sqrt{(-62.4\text{ N})^2 + (12\text{ N})^2} = 63.5\text{ N}$$

Use the tangent function to find the angle between **F** and the negative x axis:

$$\tan\theta = \frac{12\text{ N}}{62.4\text{ N}} = 0.1923$$

$$\theta = \arctan(0.1923) = 10.9°$$

5-3. Center of Gravity

The weight of an object is the gravitational force the earth exerts on the object. Although every atom or molecule of an object is subject to a gravitational attraction toward the center of the earth, we may consider the total gravitational force on the object to act at a single point for purposes of calculating torque.

- The **center of gravity** of an object is that point at which we may consider the total force of gravity to act.

The center of gravity is located at the *geometrical center* of a symmetrical object, such as a solid uniform sphere or a right circular cylinder.

- The **center of mass** of an object is that point at which we may consider the total mass to be concentrated.

In general, the center of gravity and the center of mass are located at the same point—but, if an object is so large that the acceleration of gravity varies from point to point, the center of mass and the center of gravity will have slightly different locations. The coordinates of the center of mass for a two-dimensional structure are given by

CENTER OF MASS

$$x_{cm} = \frac{x_1 m_1 + x_2 m_2 + \cdots + x_n m_n}{m_1 + m_2 + \cdots + m_n}$$
$$y_{cm} = \frac{y_1 m_1 + y_2 m_2 + \cdots + y_n m_n}{m_1 + m_2 + \cdots + m_n} \qquad \textbf{(5-4)}$$

EXAMPLE 5-6: A two-dimensional rigid body consists of three small masses ($m_1 = 2$ g, $m_2 = 3$ g, $m_3 = 1$ g) connected by rods of negligible mass (see Figure 5-8).
Find the x and y components of the center of gravity of this body.

Solution: Let M represent the total mass of the body, so that $M = m_1 + m_2 + m_3$. Since m_1, m_2, and m_3 lie in the xy plane, the center of gravity will also lie in the xy plane. The locations of the point masses are $m_1 = (0, 0)$, $m_2 = (1, 3)$, and $m_3 = (3, 0)$. Use Eq. (5-4) to find the x and y coordinates of the center of mass, which in this case coincides with the center of gravity:

$$x_{cm} = \frac{(0\text{ cm})(2\text{ g}) + (1\text{ cm})(3\text{ g}) + (3\text{ cm})(1\text{ g})}{6\text{ g}} = 1\text{ cm}$$

$$y_{cm} = \frac{(0\text{ cm})(2\text{ g}) + (3\text{ cm})(3\text{ g}) + (0\text{ cm})(1\text{ g})}{6\text{ g}} = 1.5\text{ cm}$$

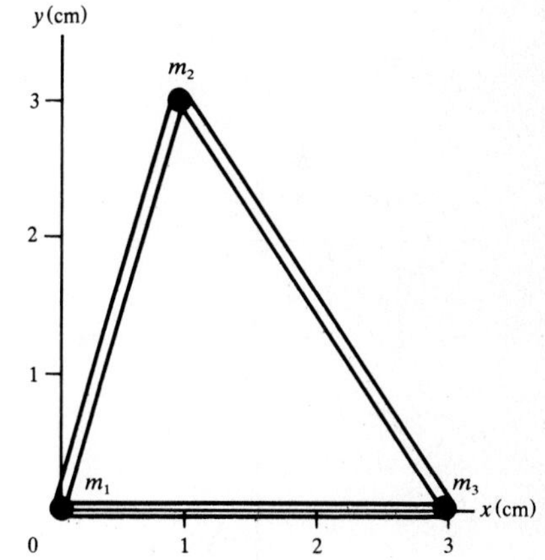

Figure 5-8

You might also think of an object's center of gravity as the point at which the object balances. Consider, for example, a beam of uniform cross section. The beam's center of gravity is located at its geometrical center, as shown in Figure 5-9. If a pivot is placed halfway between the two ends of the beam and a little above the center of gravity, there will be no tendency for the beam to rotate. The downward force **w** acting on the beam has a line of action passing through the pivot point. This means that the moment arm of **w** is zero for the torque about the pivot point, so there is no torque present to cause rotation—the beam remains in rotational equilibrium.

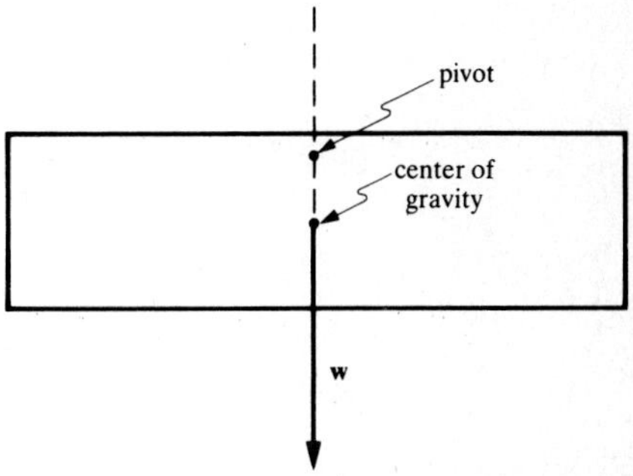

Figure 5-9. The center of gravity of a uniform beam.

EXAMPLE 5-7: A meter stick with a mass of 0.15 kg has two masses ($m_1 = 0.2$ kg; $m_2 = 0.1$ kg) fastened to it, as shown in Figure 5-10.
Find the balance point of the system.

Solution: The meter stick will balance if the total torque about the pivot point is zero. Let a represent the distance from the left end of the meter stick to the pivot, and let the direction of the torque associated with m_1 (counterclockwise) be positive. The magnitude of this torque is the weight of m_1 multiplied by its moment arm:

$$\tau_1 = (0.2\text{ kg})ga$$

The torque associated with m_2 is negative (clockwise):

$$\tau_2 = -(0.1 \text{ kg})g(100 \text{ cm} - a)$$

The weight of the meter stick causes a clockwise torque about the pivot point. Its line of action passes through the center of gravity. The moment arm of **w** is 50 cm − a, and the torque is negative:

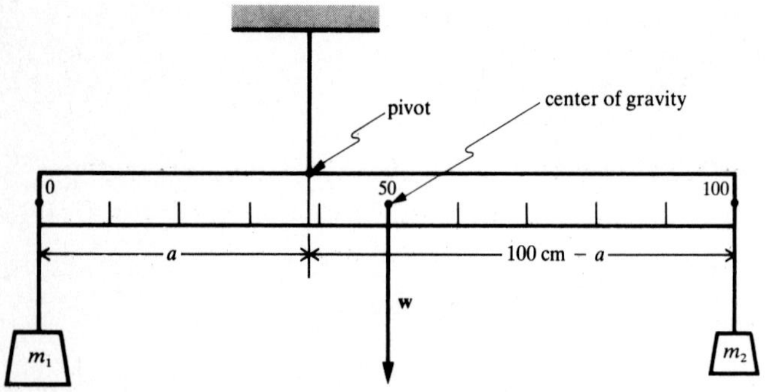

Figure 5-10

$$\tau_3 = -(0.15 \text{ kg})g(50 \text{ cm} - a)$$

Now, apply Eq. (5-4):

$$\tau_1 + \tau_2 + \tau_3 = 0$$

$$(0.2 \text{ kg})ga - (0.1 \text{ kg})g(100 \text{ cm} - a) - (0.15 \text{ kg})g(50 \text{ cm} - a) = 0$$

Cancel the common factor g and the common unit kg, and you get

$$0.2a + 0.1a + 0.15a = (10 + 7.5) \text{ cm}$$

$$0.45a = 17.5 \text{ cm}$$

$$a = 38.9 \text{ cm}$$

The point of support must be placed 38.9 cm from the left end to balance the meter stick and its two masses.

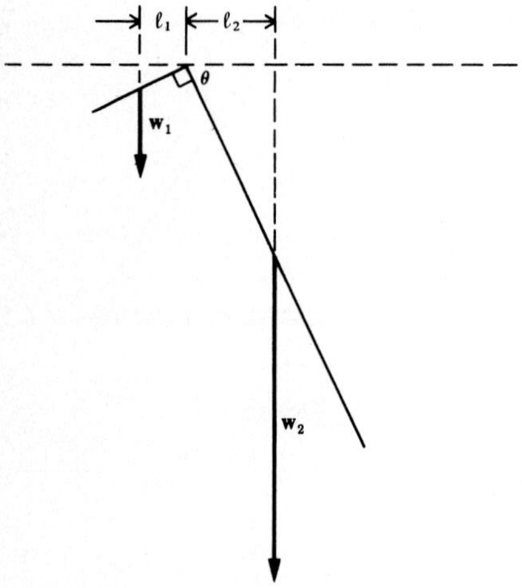

Figure 5-11

EXAMPLE 5-8: A thin uniform piece of wire 1.00 m long and mass 100 g is bent at a right angle 20.0 cm from one end and hung from the bend as shown in Figure 5-11. What angle does the long side of the wire make with the horizontal?

Solution: Think of the wire as two uniform pieces. The short wire has a mass of 20 g and the long wire has a mass of 80 g. Since the wire hangs in equilibrium, the sum of the torques is zero.

$$\Sigma\tau = 0 = \ell_1 w_1 - \ell_2 w_2$$

The center of gravity of each wire is at its geometric center, which is where the weight acts. With this information, calculate the moment arms:

$$\ell_1 = (0.10 \text{ m}) \cos(90° - \theta) = (0.10 \text{ m}) \sin\theta \qquad \ell_2 = (0.40 \text{ m}) \cos\theta$$

Then

$$\Sigma\tau = (0.10 \text{ m})(0.020 \text{ kg}) \sin\theta - (0.40 \text{ m})(0.080 \text{ kg}) \cos\theta = 0$$

$$\frac{\sin\theta}{\cos\theta} = \tan\theta = \frac{0.032 \text{ kg m}}{0.002 \text{ kg m}} = 16.0$$

$$\theta = \text{arc} \tan(16.0) = 86.4°$$

EXAMPLE 5-9: A construction crane lifts a cement bucket weighing 3500 N (see Figure 5-12). The supporting boom weighs 2500 N and makes a 25° angle with the support cable.
Calculate the tension in the horizontal cable that holds the boom in position.

Solution: First, draw the forces acting on the boom (see Figure 5-13). Since the force acting at the pivot has a moment arm of zero, it does not contribute to the torques acting on the boom. Choose the counterclockwise direction as positive. Now, make a table listing all the forces, their moment arms, and their torques:

Forces	Magnitude	Moment arm	Torque
$\mathbf{w}_1$	2500 N	$-(\ell/2) \cos 25°$	$-(1133 \text{ N})\ell$
$\mathbf{w}_2$	3500 N	$-\ell \cos 25°$	$-(3172 \text{ N})\ell$
T	T	$(\ell/2) \sin 25°$	$(0.2113)T\ell$

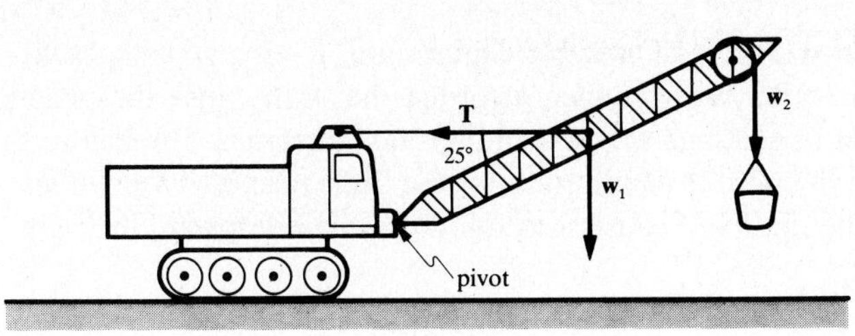

Figure 5-12

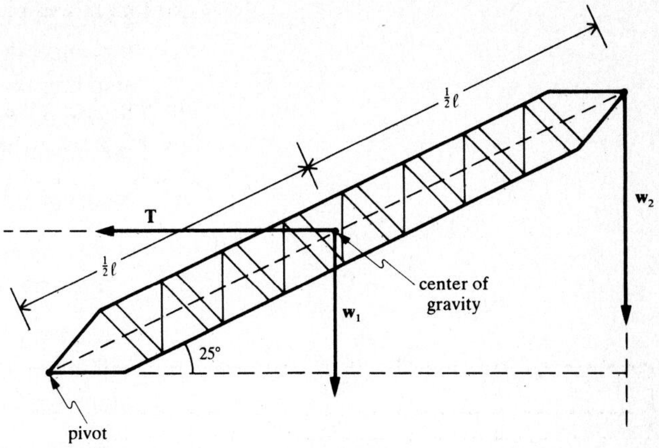

Figure 5-13

And, apply Eq. (5-3) and set the sum of the torques equal to zero:

$$0.2113T\ell - (1133\text{ N})\ell - (3172\text{ N})\ell = 0$$

$$T = \frac{4305\text{ N}}{0.2113} = 2.04 \times 10^4\text{ N}$$

5-4. Types of Equilibrium

A. Stable equilibrium

- An object in static equilibrium that is displaced slightly and then returns to its original position is in a state of **stable equilibrium.**

Figure 5-14 shows two objects in stable equilibrium. When the simple pendulum is displaced from its original position, it is subject to a force, $mg \sin \theta$, that pulls it back to its original position. When the solid, rectangular block is tilted slightly, its weight **w** produces a torque that will pull it back to its original position.

B. Unstable equilibrium

- An object that is displaced slightly from its original position and continues to move away because of its displacement is in a state of **unstable equilibrium.**

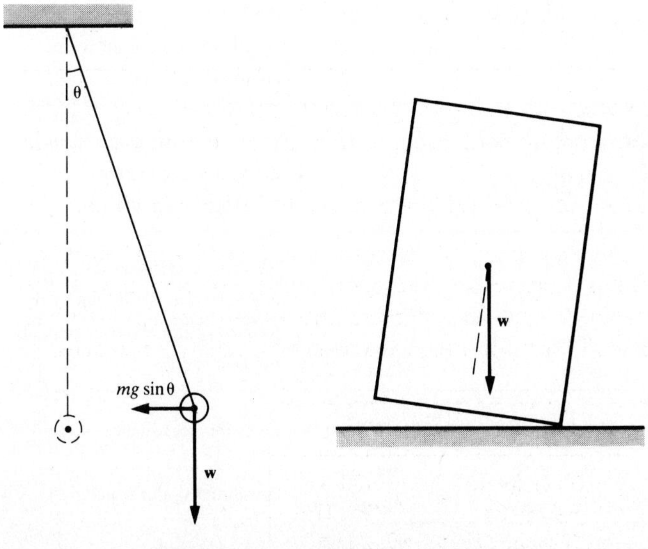

Figure 5-14. Stable equilibrium.

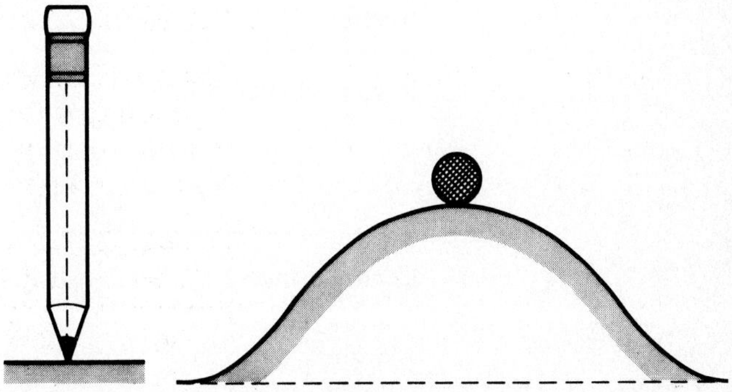

Figure 5-15. Unstable equilibrium.

Figure 5-15 shows two objects in unstable equilibrium. Even a very slight disturbance in any direction will produce a torque that will cause the pencil standing on its point to continue to move in the same direction. The ball resting on the hilltop is in unstable equilibrium for the same reason—a small displacement toward the right will cause it to roll down the hill toward the right.

C. Neutral equilibrium

- An object that is displaced slightly from its original position and neither returns nor moves farther away is in a state of **neutral equilibrium.**

Consider a marble on a horizontal surface. If the marble is nudged slightly, it will move a slight distance and then stop—neither returning to its original position nor moving farther away.

note: You can also determine the type of equilibrium from the direction in which a slight displacement moves an object's center of gravity:

(1) If a slight displacement *raises* the center of gravity, the equilibrium is *stable*.
(2) If a slight displacement *lowers* the center of gravity, the equilibrium is *unstable*.
(3) If a slight displacement has *no effect* on the height of the center of gravity, the equilibrium is *neutral*.

EXAMPLE 5-10: A rectangular safe for small valuables is 20 cm tall. Its base is 10 cm × 10 cm.
How far can you tilt the safe before it falls over?

Solution: You can tilt the safe through an angle θ so that the line of action of its weight **w** passes through the pivot, as shown in Figure 5-16—but no farther before it will fall over. Use your knowledge of trigonometry to find this angle:

$$\tan \theta = \frac{5 \text{ cm}}{10 \text{ cm}} = 0.5$$

$$\theta = \arctan(0.5) = 26.6°$$

When you tilt the safe at a 26.6° angle, the moment arm of **w** is zero, so there is no torque acting on the safe. Tilt the safe any farther and it will make enough noise to wake up the owners!

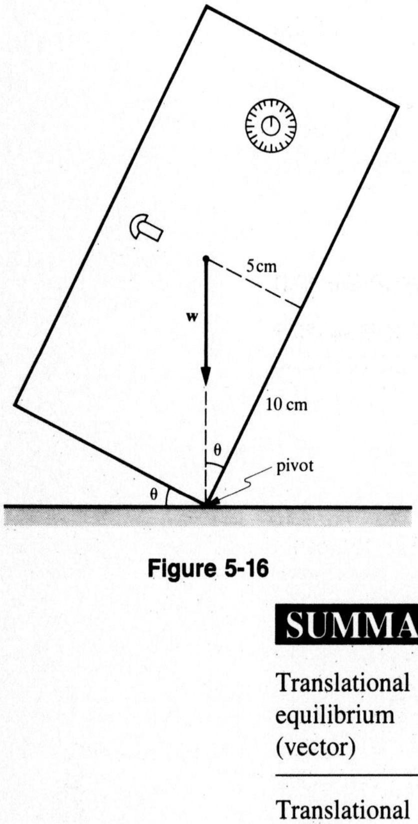

Figure 5-16

SUMMARY

Translational equilibrium (vector)	$\Sigma \mathbf{F} = 0$	the acceleration of an object is zero if the sum of all forces acting on it is zero
Translational equilibrium (scalar)	$\Sigma F_x = 0 \quad \Sigma F_y = 0 \quad \Sigma F_z = 0$	gives the scalar components of the vector equation for translational equilibrium
Rotational equilibrium	$\Sigma \tau = 0$	the angular acceleration of an object is zero if the sum of the torques acting about its axis of rotation is zero
Center of mass	$x_{cm} = \dfrac{x_1 m_1 + x_2 m_2 + \cdots + x_n m_n}{m_1 + m_2 + \cdots + m_n}$ $y_{cm} = \dfrac{y_1 m_1 + y_2 m_2 + \cdots + y_n m_n}{m_1 + m_2 + \cdots + m_n}$	gives the coordinates of the center of mass for a two-dimensional structure

RAISE YOUR GRADES

Can you define . . . ?

☑ static translational equilibrium
☑ dynamic rotational equilibrium
☑ center of gravity
☑ center of mass
☑ unstable equilibrium
☑ neutral equilibrium

Can you . . . ?

☑ state the technical term that describes a point particle moving along a straight line at a constant speed
☑ state the speed of an object that is in static equilibrium
☑ write the three scalar equations for translational equilibrium
☑ state the vector sum of the three forces acting on a point that is in translational equilibrium
☑ state the technical term that describes an object rotating about a fixed axis at a constant angular speed
☑ state the value of the angular acceleration of an object in rotational equilibrium
☑ write the equation that describes the condition of rotational equilibrium in terms of the torques acting on the object
☑ express the relationship between torque and moment arm
☑ state the technical term for the point at which gravitational force acts on a rigid body
☑ state the location of the center of gravity of a steel beam having a uniform cross section
☑ describe an object that is in stable equilibrium
☑ describe the type of equilibrium of a basketball resting on the floor of a gym

SOLVED PROBLEMS

Translational equilibrium

PROBLEM 5-1 A tightrope walker balances one third of the way across a steel cable. The cable supports are 30 m apart and the weight of the tightrope walker causes the cable to sag 1 m. Calculate the ratio of the tensions in the cable on either side of the tightrope walker.

Solution First, draw the situation (Fig. 5-17). Now, find the angles from the tangent function:

$$\tan \theta_1 = \frac{1 \text{ m}}{20 \text{ m}} = 0.05$$

$$\theta_1 = \arctan(0.05) = 2.86°$$

$$\tan \theta_2 = \frac{1 \text{ m}}{10 \text{ m}} = 0.10$$

$$\theta_2 = \arctan(0.10) = 5.71°$$

You can find the ratio of the tensions T_1/T_2 by using Eqs. (5-2) for the x components:

$$\Sigma F_x = 0$$

$$T_1 \cos \theta_1 - T_2 \cos \theta_2 = 0$$

$$\frac{T_1}{T_2} = \frac{\cos \theta_2}{\cos \theta_1} = \frac{\cos 5.71°}{\cos 2.86°} = \frac{0.9950}{0.9989} = 0.996$$

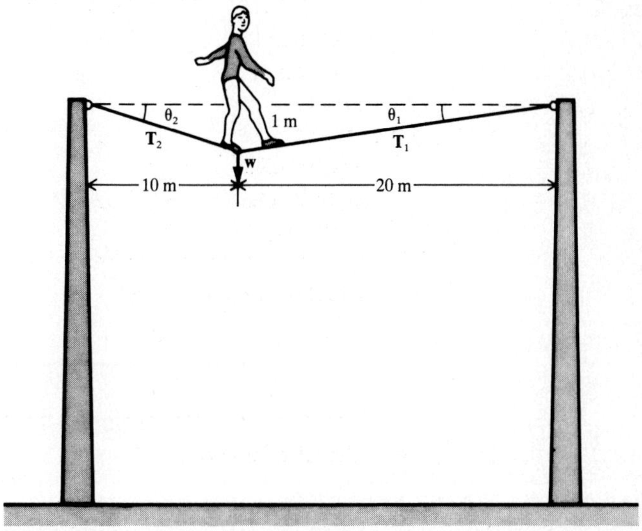

Figure 5-17

PROBLEM 5-2 If the tightrope walker in Problem 5-1 weighs 800 N, calculate the tension in the longer section of cable.

Solution This time, use Eqs. (5-2) for the vertical direction:

$$\Sigma F_y = 0$$

$$T_1 \sin \theta_1 + T_2 \sin \theta_2 - w = 0$$

To eliminate T_2, express it in terms of T_1, and substitute: $T_2 = T_1/0.996$. So,

$$0.0499T_1 + T_1\left(\frac{0.0995}{0.996}\right) = 800 \text{ N}$$

$$T_1 = \frac{800 \text{ N}}{0.1498} = 5340 \text{ N}$$

PROBLEM 5-3 A frictionless cart is being pushed up a 20° incline at a constant speed. The cart is in dynamic equilibrium and weighs 50 N. Calculate the magnitude of the horizontal force.

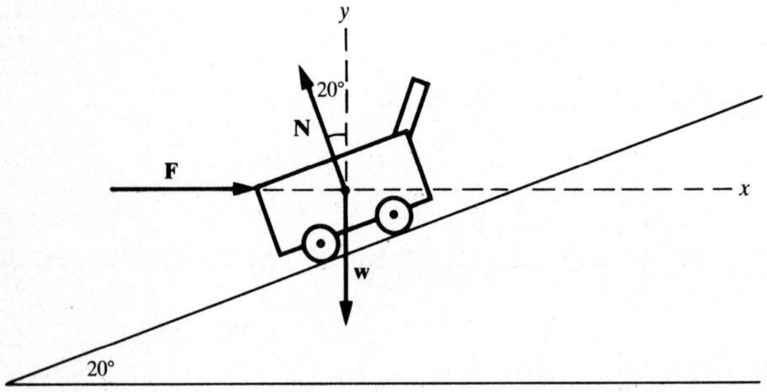

Figure 5-18

Solution Start with a diagram (Fig. 5-18). Use Eqs. (5-2) in the x and y directions:

$$\Sigma F_y = 0$$

$$N \cos 20° - w = 0$$

$$N = \frac{50 \text{ N}}{\cos 20°} = 53.2 \text{ N}$$

$$\Sigma F_x = 0$$

$$F - N \sin 20° = 0$$

$$F = N \sin 20° = (53.2 \text{ N})(0.342) = 18.2 \text{ N}$$

Rotational equilibrium

PROBLEM 5-4 A 100-lb ladder of length ℓ leans against a vertical wall, as shown in Fig. 5-19. The upper end of the ladder has two rollers so that the wall exerts no friction in the vertical direction, but the force of friction that the ground exerts on the ladder holds the ladder in place. If the coefficient of friction between the ladder and the ground is 0.4, what is the minimum angle θ at which the ladder will remain in equilibrium?

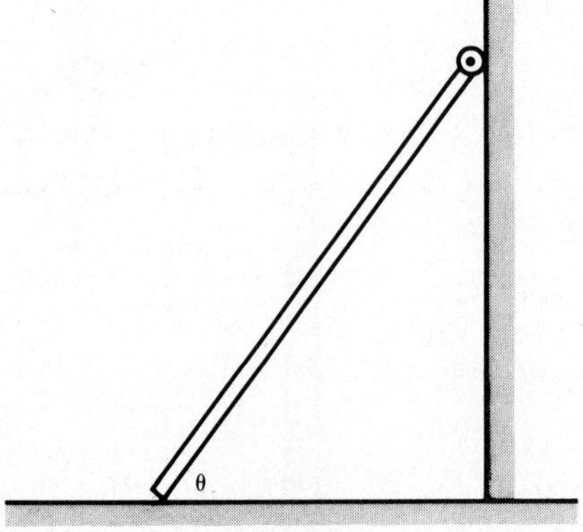

Figure 5-19

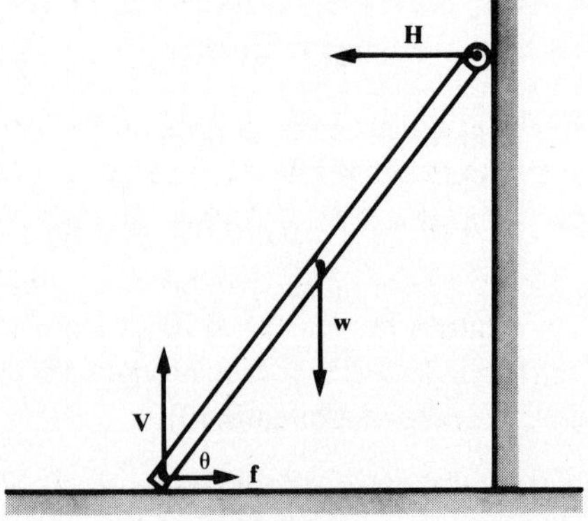

Figure 5-20

Solution First, make a drawing showing all the forces acting on the ladder. Placing the axis of rotation at the upper end of the ladder will simplify the final calculations (see Fig. 5-20). Make counterclockwise the positive direction, and list all the torques acting on the ladder:

Forces	Magnitude	Moment arm	Torque
w	100 lb	$(\ell/2)\cos\theta$	$(50\text{ lb})\ell\cos\theta$
V	V	$\ell\cos\theta$	$-V\ell\cos\theta$
f	f	$\ell\sin\theta$	$f\ell\sin\theta$

Apply the equations for translational equilibrium. Since there are only two vertical forces, **V** and **w**, acting in opposite directions, they must have the same magnitude, so $V = w$. Now, use Eq. (5-3) for rotational equilibrium, and Eq. (3-6), $\mathbf{f} = \mu V$:

$$\Sigma\tau = 0 = (50\text{ lb})\ell\cos\theta - V\ell\cos\theta + \mu V\ell\sin\theta$$

$$(100\text{ lb})\ell\cos\theta = (50\text{ lb})\ell\cos\theta + \mu(100\text{ lb})\ell\sin\theta$$

Cancel the common factor, $\ell(50\text{ lb})$, and solve for the angle θ:

$$(2-1)\cos\theta = (0.4)(2\sin\theta)$$

$$\tan\theta = \frac{1}{0.8} = 1.25$$

$$\theta = \text{arc}\tan(1.25) = 51.3°$$

PROBLEM 5-5 A heavy steel beam 4 m long is supported by a column at each end. The beam weighs 200 N and supports a load of 700 N that is 1 m from the left end, as shown in Fig. 5-21. What force does the beam exert on each supporting column?

Solution For simplicity, choose the left end of the beam as the axis of rotation and calculate the torque produced by each force acting on the beam:

$$\tau_A = F_2(4\text{ m})$$

$$\tau_B = -w(2\text{ m}) = -400\text{ N m}$$

$$\tau_C = -L(1\text{ m}) = -700\text{ N m}$$

Now, set the sum of the torques equal to zero:

$$\Sigma\tau = 0 = \tau_A + \tau_B + \tau_C$$

$$F_2(4\text{ m}) = 400\text{ N m} + 700\text{ N m}$$

$$F_2 = \frac{1100\text{ N m}}{4\text{ m}} = 275\text{ N}$$

Since the beam is in equilibrium, you know that $\Sigma F_y = 0$, so

$$F_1 + 275\text{ N} = 900\text{ N}$$

$$F_1 = 625\text{ N}$$

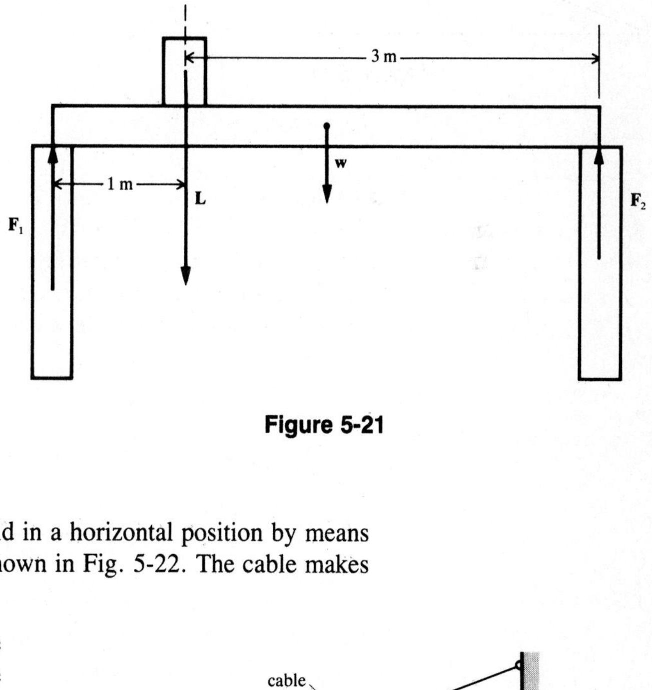

Figure 5-21

PROBLEM 5-6 A uniform steel beam weighing 2000 N is held in a horizontal position by means of a hinge at one end and a cable fastened to the other end, as shown in Fig. 5-22. The cable makes a 20° angle with the beam. Calculate the tension in the cable.

Solution Use the condition of rotational equilibrium and place the axis of rotation at the end of the beam supported by the hinge. Then, calculate the torques produced by the weight of the beam and the tension in the cable:

$$\tau_1 = w(\ell/2) \qquad \tau_2 = -T\ell\sin 20°$$

Now, set the sum of the torques equal to zero:

$$\tau_1 + \tau_2 = 0$$

$$T\ell\sin 20° = w(\ell/2)$$

$$T = \frac{w/2}{\sin 20°} = \frac{1000\text{ N}}{0.342} = 2924\text{ N}$$

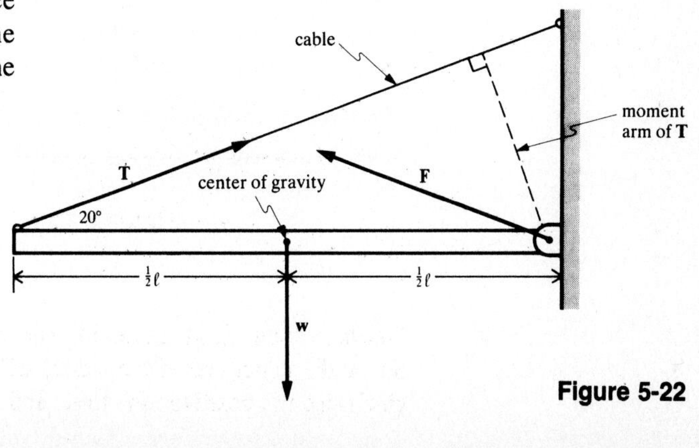

Figure 5-22

PROBLEM 5-7 Calculate the magnitude and the direction of the force that the hinge exerts on the horizontal beam described in Problem 5-6.

Solution Find the x and y components of force **F** by using the equations of translational equilibrium (Eqs. 5-2):

Forces	x components	y components
w **T** **F**	0 (2924 N) cos 20° F_x	−2000 N (2924 N) sin 20° F_y

$$\Sigma F_x = 0 \qquad \Sigma F_y = 0$$

$$F_x = -(2924\text{ N})\cos 20° \qquad F_y = 2000\text{ N} - (2924\text{ N})\sin 20°$$

$$= -2748\text{ N} \qquad = 2000\text{ N} - 1000\text{ N} = 1000\text{ N}$$

You can see from the x and y components that the force on the hinge is directed upward and to the left. Find the angle that **F** makes with respect to the x axis by using the tangent function:

$$\tan\theta = \frac{1000\text{ N}}{2748\text{ N}} = 0.3639$$

$$\theta = \text{arc}\tan(0.3639) = 20°$$

Finally, use the Pythagorean theorem to find the magnitude of **F**:

$$F = \sqrt{F_x^2 + F_y^2}$$

$$= \sqrt{(-2748\text{ N})^2 + (1000\text{ N})^2} = 2924\text{ N}$$

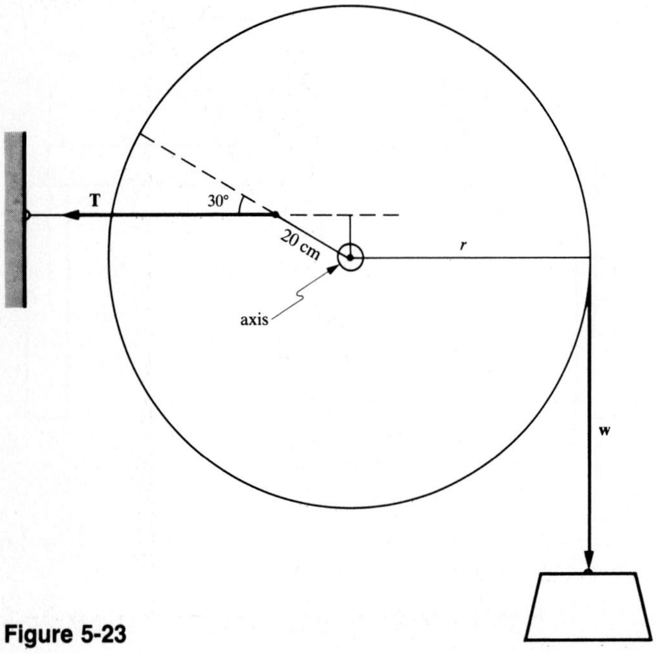

Figure 5-23

PROBLEM 5-8 Consider the two forces acting on the solid disk illustrated in Fig. 5-23. The 50-N weight is tied to the rim of the disk, which has a radius of 50 cm. The unknown force **T** acts at a point that is 20 cm from the axis rotation of the disk at an angle of 30° and keeps the disk from rotating. Calculate the magnitude of **T**.

Solution Calculate the two torques acting on the disk:

$$\tau_1 = T(20\text{ cm})\sin 30°$$

$$\tau_2 = -(50\text{ N})(50\text{ cm})$$

Now, set the sum of the torques equal to zero:

$$T(0.2\text{ m})\sin 30° - (50\text{ N})(0.5\text{ m}) = 0$$

$$T = \frac{25\text{ N m}}{0.1\text{ m}} = 250\text{ N}$$

PROBLEM 5-9 A door 1 m wide and weighing 200 N is supported by two hinges. The hinges are 1.8 m apart and are designed so that the vertical force provided by each hinge is equal to half the weight of the door. Find the horizontal force exerted on the door by each hinge.

Solution Your sketch of the forces should look like Fig. 5-24. You can find the horizontal force exerted by the upper hinge by placing the axis of rotation at the lower hinge, so the moment arms for $\mathbf{V}_1$, $\mathbf{V}_2$, and H_2 are then each zero. First, calculate the torques produced by the forces $\mathbf{H}_1$ and **w**:

$$\tau_1 = H_1(1.8\text{ m})$$

$$\tau_2 = -w(0.50\text{ m})$$

Next, set the sum of the torques equal to zero:

$$\tau_1 + \tau_2 = 0$$

$$H_1(1.8\text{ m}) = w(0.50\text{ m})$$

$$H_1 = \frac{(200\text{ N})(0.50\text{ m})}{1.8\text{ m}} = 55.6\text{ N}$$

Now, use Eq. (5-2) in the horizontal direction:

$$\Sigma F_x = 0$$

$$H_2 - H_1 = 0$$

$$H_2 = H_1 = 55.6\text{ N}$$

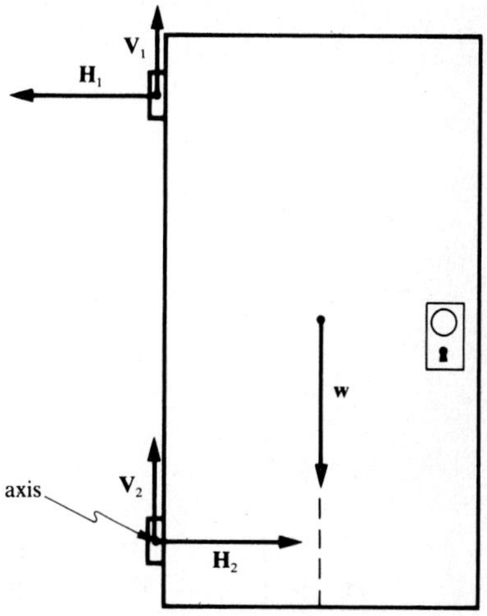

Figure 5-24

PROBLEM 5-10 A 100-lb ladder, 20 ft long, leans against the side of a building. Because of rollers at the top of the ladder, there is no frictional force on its upper end. The coefficient of friction between the bottom of the ladder and the horizontal surface is 0.4. The angle between the ladder and the horizontal surface is 60°. Calculate the maximum distance that a 200-lb house painter can climb before the ladder falls down.

Solution Your sketch should resemble Fig. 5-25. Let x be the distance from the top of the ladder to the maximum height the hefty painter can climb. Place the axis of rotation at the upper end of the ladder and then calculate the four torques:

$$\tau_1 = w_1(\ell/2)\cos 60^\circ \qquad \tau_3 = f\ell \sin 60^\circ$$

$$\tau_2 = w_2 x \cos 60^\circ \qquad \tau_4 = -V\ell \cos 60^\circ$$

Set the sum of the torques equal to zero:

$$V\ell \cos 60^\circ = w_1(\ell/2)\cos 60^\circ + w_2 x \cos 60^\circ + f\ell \sin 60^\circ$$

Divide each term by $\ell \cos 60^\circ$ and substitute μV for f:

$$V = (w_1/2) + w_2(x/\ell) + \mu V \tan 60^\circ$$

Since $\Sigma F_y = 0$, the magnitude of $\mathbf{V}$ is 300 lb. Thus,

$$300\text{ lb} = 50\text{ lb} + (200\text{ lb})(x/20\text{ ft}) + (120\text{ lb})(1.732)$$

Now you can solve for x:

$$(10\text{ lb/ft})x = 250 - 207.8\text{ lb}$$

$$x = \frac{42.2\text{ lb}}{10\text{ lb/ft}} = 4.22\text{ ft}$$

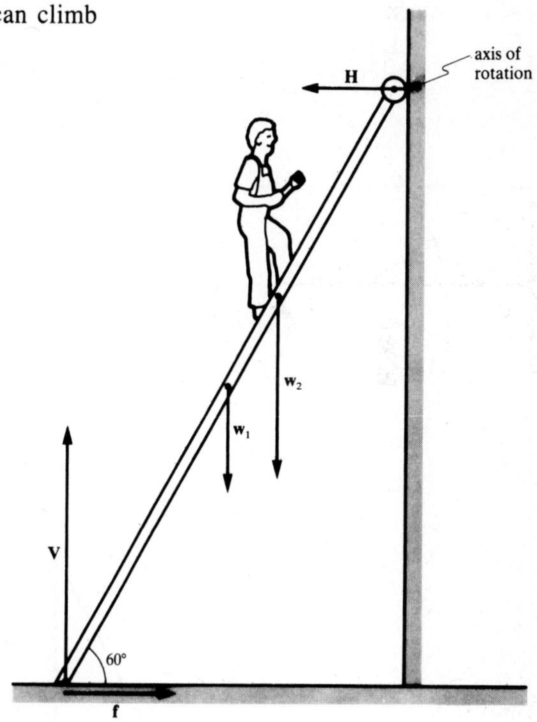

Figure 5-25

So, the maximum distance the painter can climb the ladder is (20 − 4.22) ft = 15.8 ft.

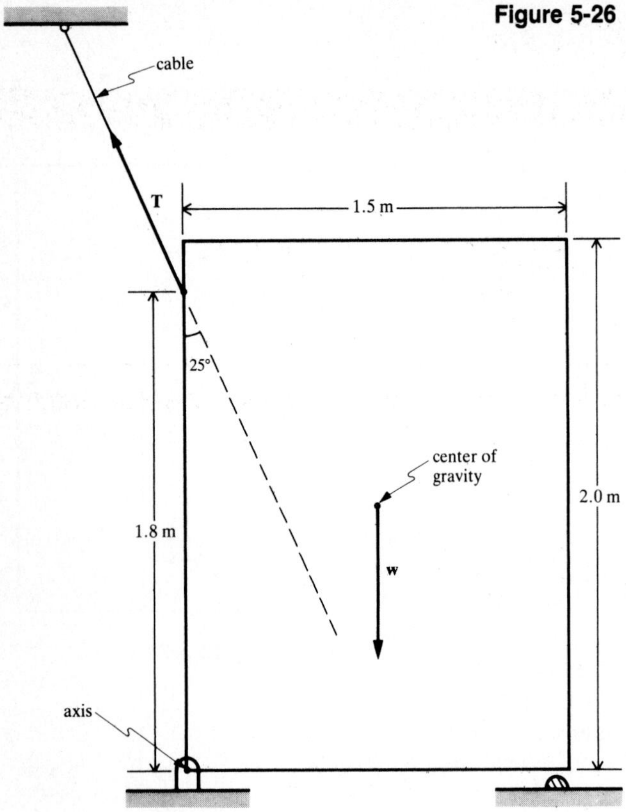

Figure 5-26

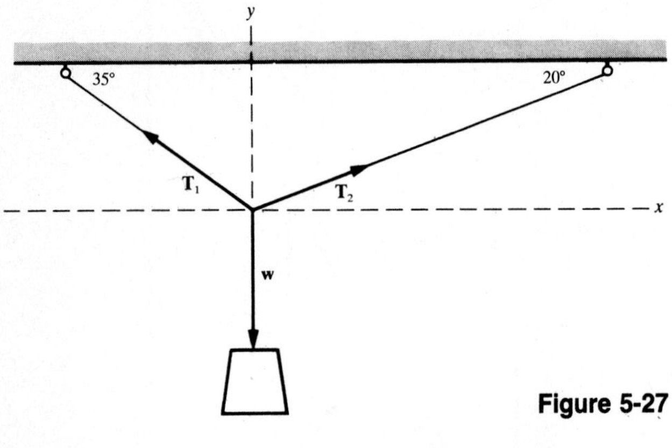

Figure 5-27

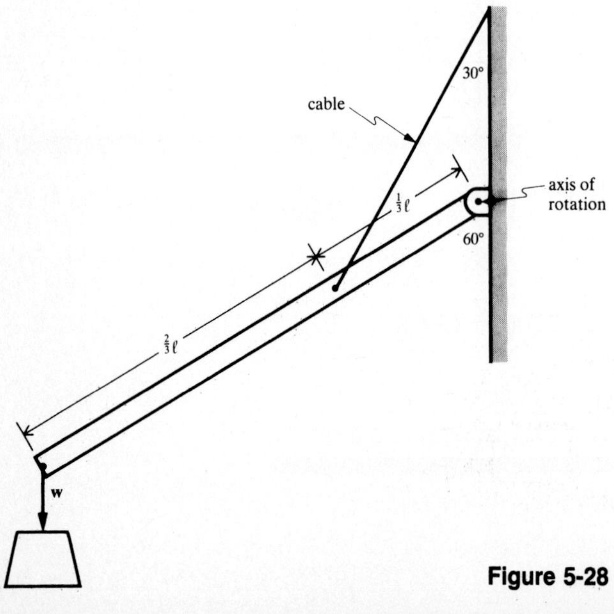

Figure 5-28

Supplementary Exercises

PROBLEM 5-11 A sheet of steel weighing 800 N is supported by a bolt at the lower left-hand corner and a cable tied to the left edge, as illustrated in Fig. 5-26. What is the moment arm of the force exerted by the cable?

PROBLEM 5-12 What is the torque produced by the weight of the sheet of steel illustrated in Fig. 5-26?

PROBLEM 5-13 What is the magnitude of the force exerted by the cable in Fig. 5-26?

PROBLEM 5-14 If the tension in the cable is reduced to 500 N, an additional vertical force will be supplied by the block at the lower right-hand corner of the sheet of steel. Calculate the magnitude of this third force.

PROBLEM 5-15 A 200-N weight is supported by two ropes that are tied to the ceiling as shown in Fig. 5-27. Calculate the ratio of the tensions in the two ropes, T_2/T_1.

PROBLEM 5-16 Calculate the force of tension in the rope on the left, illustrated in Fig. 5-27.

PROBLEM 5-17 What is the tension in the rope on the right-hand side of Fig. 5-27?

PROBLEM 5-18 A light rod is supported by a hinge and a steel cable, as illustrated in Fig. 5-28. A load of 100 N is fastened to the left end of the rod. What is the moment arm of the force exerted by the cable?

PROBLEM 5-19 What is the torque produced by the 100-N weight hanging from the end of the rod shown in Fig. 5-28?

PROBLEM 5-20 Calculate the force of tension in the cable shown in Fig. 5-28.

Answers to Supplementary Exercises

5-11: 0.761 m

5-12: −600 N m

5-13: $T = 788$ N

5-14: 146 N

5-15: $T_2/T_1 = 0.872$

5-16: $T_1 = 229$ N

5-17: $T_2 = 200$ N

5-18: $\ell/6$

5-19: $0.866\ \ell$

5-20: $T = 520$ N

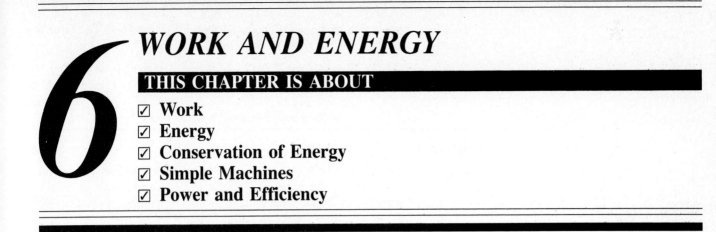

6 WORK AND ENERGY

THIS CHAPTER IS ABOUT

☑ Work
☑ Energy
☑ Conservation of Energy
☑ Simple Machines
☑ Power and Efficiency

6-1. Work

A. Work done by a constant force

- **Work** is the product of the magnitude of an applied force and the distance through which the force acts, i.e., the dot product of the force **F** acting on an object and the object's displacement **s**.

[**recall:** The dot product—or scalar product—of two vectors is a scalar equal to the product of the two vectors and the cosine of the angle θ between them. (Sec. 1-5)] To calculate work, you multiply the component of **F** parallel to the direction of **s** times the magnitude of **s**. This means that if **F** is at an angle θ with respect to **s**, the component of **F** parallel to **s** will be $F \cos \theta$, so that we can define work as

WORK $$W = \mathbf{F} \cdot \mathbf{s} \quad \text{or} \quad W = Fs \cos \theta \tag{6-1}$$

note: In physics, no work is done if there is not a net displacement (if $s = 0$, $Fs \cos \theta = 0$), or if the line of action of a force is at a right angle to the displacement (if $\theta = 90°$, $Fs \cos \theta = 0$ since $\cos 90° = 0$).

In SI, the newton-meter, called the **joule** and abbreviated J, is the unit associated with work. You can think of one joule as the work done when an object is moved one meter against an opposing force of one newton.

EXAMPLE 6-1: A barge is towed through a canal by an electric engine on a track parallel to the canal. The tow rope exerts a force of 6500 N, and the angle between this force and the direction the barge is moving is 20°.
How much work is done on the barge for each kilometer it is towed?

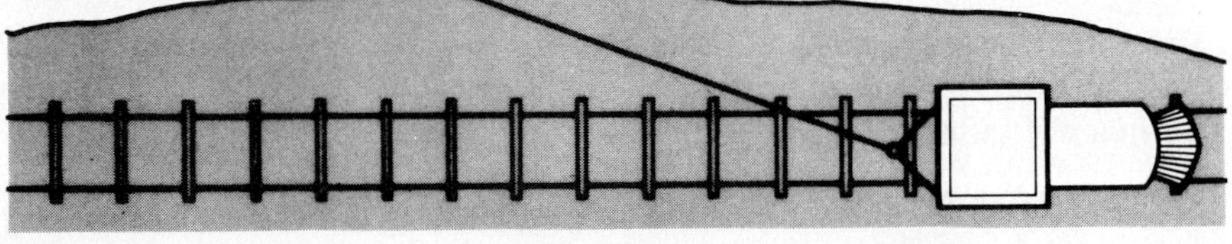

Figure 6-1

Solution: Your sketch of the situation should look like Figure 6-1. The force is constant, so use Eq. (6-1) to calculate the work:

$$W = Fs \cos \theta$$
$$= (6500 \text{ N})(1000 \text{ m}) \cos 20°$$
$$= 6.11 \times 10^6 \text{ N m} = 6.11 \times 10^6 \text{ J}$$

B. Work done by a variable force

If the value of the force **F** changes as the displacement **s** increases, the work done is the area under the graph of the force plotted as a function of the distance. Figure 6-2 shows the force required to stretch a spring. As you can see in the diagram, the force required increases linearly as the spring gets longer.

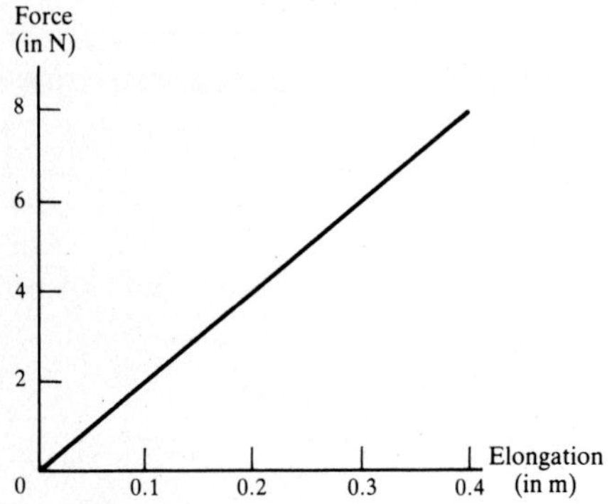

Figure 6-2. The force required to stretch a spring.

EXAMPLE 6-2: If a spring exhibits the properties shown in Figure 6-2, how much work is required to stretch it a distance of 0.3 m?

Solution: When the force is not constant, you can find the work from the graph of the force as a function of distance. The work required to stretch this spring is the triangular area under the graph from 0.0 m to 0.3 m, so

$$\text{area} = \frac{1}{2}(\text{base})(\text{altitude}) = \frac{1}{2}(0.3 \text{ m})(6 \text{ N}) = 0.9 \text{ N m} = 0.900 \text{ J}$$

6-2. Energy

A. Potential energy

- The **potential energy** PE of an object is the amount of work "stored" by the object due to its position, shape, or configuration relative to other objects.

For example, a book has more gravitational potential energy when it is on top of a table than when it is lying on the floor.

- **Gravitational potential energy** is the product of the weight of an object and its height above a reference level that can be chosen arbitrarily:

GRAVITATIONAL POTENTIAL ENERGY $$\text{PE} = wh \quad \text{or} \quad \text{PE} = mgh \qquad (6\text{-}2)$$

Energy, like work, is expressed in joules.

EXAMPLE 6-3: Suppose you lift a book weighing 20 N from the floor and place it on a table that is 0.8 m high.
What is the increase in PE of the book?

Solution: From Eq. (6-2),

$$\text{PE} = wh = (20 \text{ N})(0.8 \text{ m}) = 16 \text{ J}$$

In addition to gravitational potential energy, there are other common forms of PE, such as chemical and electrical PE. Mechanical PE, such as the energy associated with the elongation of a spring, always refers to work done against gravitational or elastic forces. When you stretch the spring described in Example 6-2 from 0.0 m to 0.3 m, the increase in its potential energy is equal to the work required to stretch it. You can calculate this potential energy from the spring constant k and the elongation x:

POTENTIAL ENERGY OF A SPRING $$\text{PE} = \frac{1}{2}kx^2 \qquad (6\text{-}3)$$

In Figure 6-3, the elongation of the spring at x is equal to the length of the stretched spring (c) minus the length of the relaxed spring (b). The compression of the spring at $-x'$ is equal to the length of the compressed spring (a) minus the length of the relaxed spring (b).

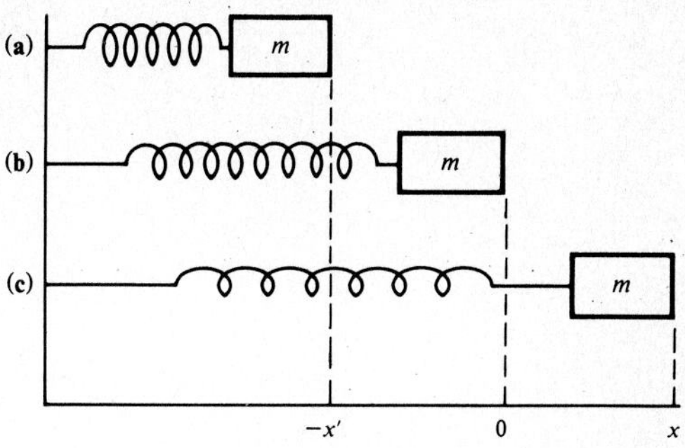

Figure 6-3

note: Since $x = -x'$, whether the spring is at position x or $-x'$ the spring has the same potential energy: $\text{PE} = 1/2kx^2 = 1/2k(-x')^2$.

The spring constant k is the ratio of the change in force divided by the elongation produced. Geometrically, k is the slope of the graph of F versus x, or

SPRING CONSTANT

$$k = \frac{\Delta F}{\Delta x} \qquad \textbf{(6-4)}$$

note: The force in Eq. (6-4) is the *external* force that produces the elongation of the spring. The spring itself produces a **restoring force** that has the same magnitude but opposite direction.

EXAMPLE 6-4: Calculate the PE of the spring of Example 6-2 when the spring is stretched a distance of 0.3 m.

Solution: First, you'll have to calculate the spring constant. Use Eq. (6-4) and the graph of Figure 6-2 from 0.2 m to 0.3 m to determine the slope of the line:

$$k = \frac{\Delta F}{\Delta x} = \frac{6\text{ N} - 4\text{ N}}{0.3\text{ m} - 0.2\text{ m}} = \frac{2\text{ N}}{0.1\text{ m}} = 20\text{ N/m}$$

Now, you can use Eq. (6-3) to calculate the increase in PE of the spring:

$$\text{PE} = \frac{1}{2}kx^2 = \frac{1}{2}(20\text{ N/m})(0.3\text{ m})^2 = 0.9\text{ J}$$

B. Kinetic energy

- The **kinetic energy** KE of an object is the amount of work "stored" by the object due to its motion.

A moving object has the capacity to do work on another object. The kinetic energy of an object is the object's capacity to perform work due to its mass and speed:

KINETIC ENERGY

$$\text{KE} = \frac{1}{2}mv^2 \qquad \textbf{(6-5a)}$$

A rotating object is also in motion and has kinetic energy. For rotation, the moment of inertia I plays the role of the mass and ω is the velocity, so that

KINETIC ENERGY OF ROTATION

$$\text{KE} = \frac{1}{2}I\omega^2 \qquad \textbf{(6-5b)}$$

EXAMPLE 6-5: A constant force of 6 N is applied parallel to a 3-kg mass initially at rest. The mass moves through a horizontal distance of 0.5 m. Determine (**a**) the work done by the force and (**b**) the KE acquired by the mass.

Solution:

(**a**) Since the force is constant, you can calculate the work from Eq. (6-1). The force is parallel to the mass' motion, so $\theta = 0$ and $\cos 0° = 1$, which means that $W = Fs$:

$$W = Fs = (6\text{ N})(0.5\text{ m}) = 3\text{ J}$$

(b) To find the KE of the mass, you'll first have to calculate its acceleration—from Newton's second law of motion (Eq. 3-2):

$$a = \frac{F}{m} = \frac{6\ \text{N}}{3\ \text{kg}} = 2.0\ \text{m/s}^2$$

A constant force necessarily produces a constant acceleration, so you can calculate the final speed from Eq. (2-7):

$$v^2 = v_0^2 + 2as$$
$$= 0 + 2(2.0\ \text{m/s}^2)(0.5\ \text{m}) = 2\ \text{m}^2/\text{s}^2$$
$$v = \sqrt{2}\ \text{m/s}$$

Now, you can calculate the KE from Eq. (6-5):

$$\text{KE} = \frac{1}{2}mv^2 = \frac{1}{2}(3\ \text{kg})(\sqrt{2}\ \text{m/s})^2 = 3\ \text{J}$$

EXAMPLE 6-6: A sphere with a radius of 5 cm and a mass of 1.4 kg has a rotational kinetic energy of 7 J. A cylinder of the same radius and mass is rotating at the same angular speed.
What is the cylinder's kinetic energy?

Solution: You can find the moment of inertia for the sphere from Eq. (4-17c):

$$I = \frac{2}{5}mr^2 = \frac{2}{5}(1.4\ \text{kg})(0.05\ \text{m})^2 = 1.4 \times 10^{-3}\ \text{kg m}^2$$

You can figure its angular speed from Eq. (6-5b):

$$\text{KE} = \frac{1}{2}I\omega^2$$

$$\omega = \sqrt{\frac{2\ \text{KE}}{I}} = \sqrt{\frac{2(7\ \text{J})}{1.4 \times 10^{-3}\ \text{kg m}^2}} = 100\ \text{rad/s}$$

The moment of inertia for the cylinder (Eq. 4-17b) is

$$I = \frac{1}{2}mr^2$$

So you can substitute for I in Eq. (6-5b) to find the cylinder's kinetic energy:

$$\text{KE} = \frac{1}{4}mr^2\omega^2 = \frac{1}{4}(1.4\ \text{kg})(0.05\ \text{m})^2(100\ \text{rad/s})^2 = 8.75\ \text{J}$$

6-3. Conservation of Energy

Work and energy are closely related concepts. Both must be accounted for—they can't simply disappear. When work is done on a system, the system acquires potential energy or kinetic energy or both. In fact, when there is neither friction nor an external force acting on a system, the total mechanical energy of the system is conserved. In the absence of friction, the sum of the PE and the KE acquired by a system is exactly equal to work done on the system by an external force. If there is friction, some of the mechanical energy of the system is converted into heat energy. For example, if you compare the results of Examples 6-2 and 6-4, you'll see that the work done by the external force that stretched the spring is equal to the potential energy acquired by the spring. Note also that the work done on the mass in Example 6-5 is equal to the KE acquired by the mass. These are examples of an extremely important principle in physics—the conservation of energy.

- According to the law of the **conservation of energy,** the energy of a system cannot be created or destroyed, but it can be converted from one kind to another.

EXAMPLE 6-7: A constant force of 32 N pushes a block weighing 16 N up an inclined plane (see Figure 6-4). The force is parallel to the plane, which is inclined 30°. The block starts from rest and moves a distance of $\frac{2}{3}$ m along the plane. Friction between the block and the inclined plane is negligible.
Calculate the total energy (PE + KE) acquired by the block.

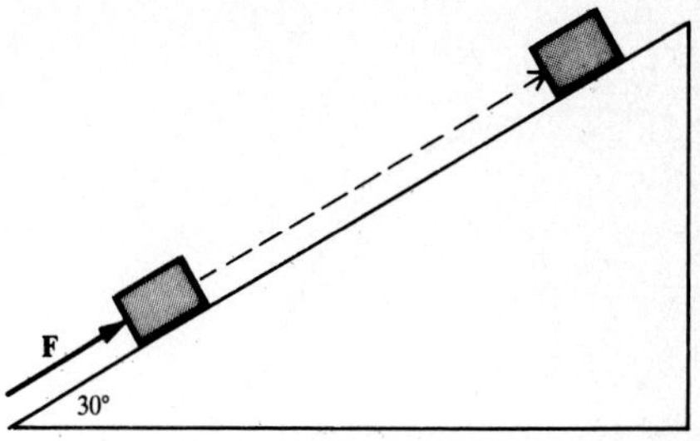

Figure 6-4

Solution: Start with the KE. To find the speed of the block, you'll have to calculate its acceleration from Newton's second law. Since you need the net force acting in the direction of motion, make a force diagram and compare it to Figure 6-5. The component of w along the x axis is

$$w_x = -w \sin 30° = -(16 \text{ N})(0.5) = -8 \text{ N}$$

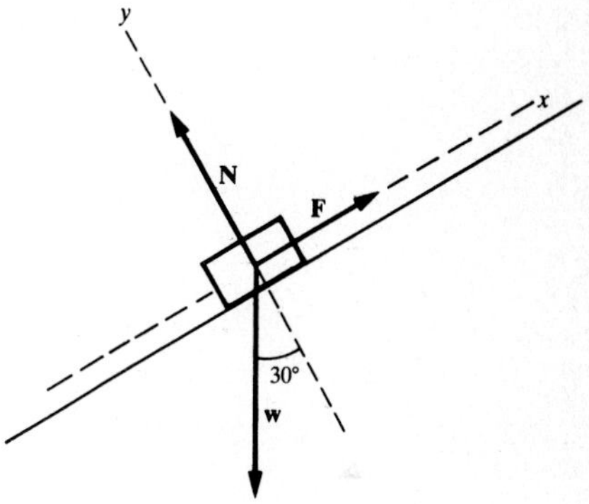

Figure 6-5

So, the net force acting in the direction of motion is

$$F_x = F + w_x = 32 \text{ N} - 8 \text{ N} = 24 \text{ N}$$

And, you can calculate the constant acceleration of the block:

$$a = \frac{\Sigma F_x}{m} = \frac{24 \text{ N}}{(16 \text{ N})/g} = 1.5g = (1.5)(9.8 \text{ m/s}^2) = 14.7 \text{ m/s}^2$$

Now, calculate the square of the final speed of the block from Eq. (2-7). [Hint: Since the speed is squared in Eq. (6-5) for KE, you don't need to figure the square root of v^2.]

$$v^2 = v_0^2 + 2as = 0 + 2(14.7 \text{ m/s}^2)\left(\frac{2}{3} \text{ m}\right) = 19.6 \text{ m}^2/\text{s}^2$$

From Eq. (6-5), the final KE of the block is

$$\text{KE} = \frac{1}{2}mv^2 = \frac{1}{2}\left(\frac{16 \text{ N}}{9.8 \text{ m/s}^2}\right)(19.6 \text{ m}^2/\text{s}^2) = 16 \text{ J}$$

Now for the PE. You'll need to find the vertical distance h through which the block has moved, so use your knowledge of trigonometry:

$$h = (0.667 \text{ m}) \sin 30° = 0.333 \text{ m}$$

Now, calculate the increase in gravitational PE from Eq. (6-2):

$$\text{PE} = wh = (16 \text{ N})(0.333 \text{ m}) = 5.33 \text{ J}$$

The total energy acquired by the block is

$$\text{E} = \text{KE} + \text{PE} = 16 \text{ J} + 5.33 \text{ J} = 21.3 \text{ J}$$

To check your math, calculate the work done by the external force and verify that energy has been conserved:

$$W = \mathbf{F} \cdot \mathbf{s} = Fs \cos 0° = (32 \text{ N})(0.667 \text{ m}) = 21.3 \text{ J}$$

You can see from these calculations that the work done by the external force is equal to the kinetic energy plus the potential energy acquired by the block. This is just what you'd expect since no friction is present.

EXAMPLE 6-8: A roller coaster car starts at the top of a 12-m crest (point A in Fig. 6-6) with a speed $v_0 = 0$ and climbs to the next crest (9 m).
How fast will it be moving when it passes point B if the track is frictionless?

Solution: In the absence of friction, the total mechanical energy of the system is constant, so the total energy of the cart (PE + KE) has the same value at A as it has at B:

CONSERVATION OF ENERGY

$$\text{PE}_A + \text{KE}_A = \text{PE}_B + \text{KE}_B \qquad \textbf{(6-6)}$$

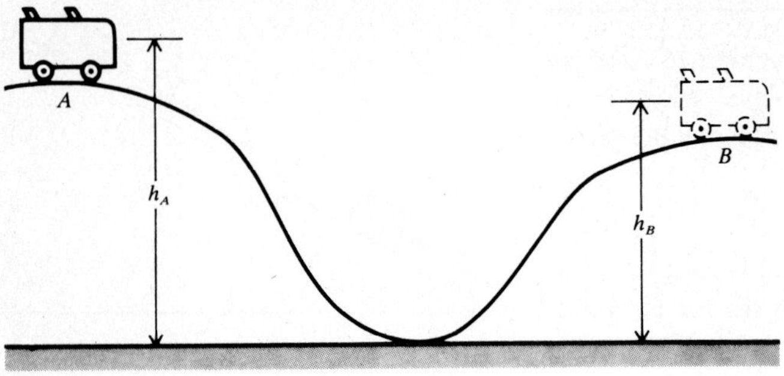

Figure 6-6

Since the cart starts from rest, its initial KE is zero. Its potential energy is a result of its position and gravitational acceleration:

$$mgh_A + 0 = mgh_B + \frac{1}{2}mv^2$$

Solve for v:

$$v^2 = 2g(h_A - h_B)$$
$$= 2(9.8\ \text{m/s}^2)(12\ \text{m} - 9\ \text{m})$$
$$= 58.8\ \text{m}^2/\text{s}^2$$
$$v = \sqrt{58.8\ \text{m}^2/\text{s}^2} = 7.67\ \text{m/s}$$

note: Equation (6-6) for conservation of energy is the most important formula in this chapter.

EXAMPLE 6-9: A metal hoop with a radius of 40 cm and a mass of 1.6 kg rolls along a horizontal surface at an initial speed of 3 m/s before it rolls up a 20° incline. The hoop slows down, stops, and then rolls back down the incline. Because the hoop rolls without slipping, there is no energy loss due to friction. Calculate the maximum height that the hoop's center of mass reaches above its initial height.

Solution: Your sketch of the situation should look like Fig. 6-7. Conservation of energy plays a key role in this problem because the total energy of the hoop remains constant. Before the hoop starts up the inclined surface, its energy is kinetic energy due to the linear motion of its center of mass plus the energy associated with its rotation. When the hoop comes to rest on the incline, its energy is entirely potential energy. So, you can express the principle of conservation of energy as follows:

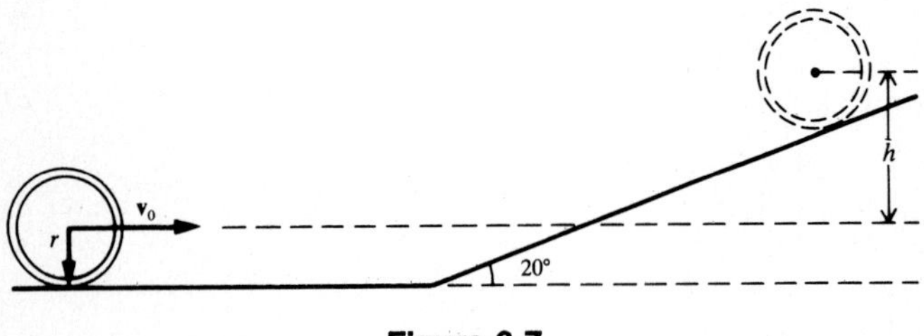

Figure 6-7

$$\frac{1}{2}mv^2 + \frac{1}{2}I\omega^2 = mgh \qquad \textbf{(6-7)}$$

Use the original position of the hoop's center of mass as the reference level from which to measure h. The relationship between linear speed v and angular speed ω, Eq. (4-10), is

$$v = r\omega \quad \text{or} \quad \omega = \frac{v}{r}$$

The entire mass of the hoop is at a distance r from the axis of rotation, so the moment of inertia of the hoop is

$$I = mr^2$$

Substitute the values for ω and I into Eq. (6-7) and solve for h:

$$mgh = \frac{1}{2}mv^2 + \frac{1}{2}\left(mr^2\left(\frac{v}{r}\right)^2\right) = \frac{1}{2}mv^2 + \frac{1}{2}mv^2 = mv^2$$

$$h = \frac{v^2}{g} = \frac{(3\ \text{m/s})^2}{9.8\ \text{m/s}^2} = 0.918\ \text{m}$$

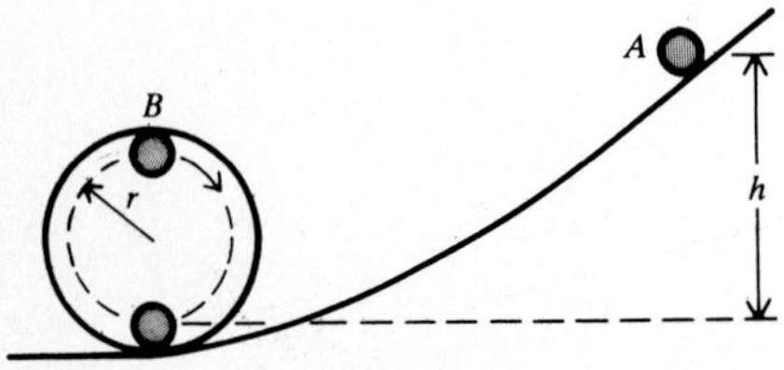

Figure 6-8

EXAMPLE 6-10: A child's toy called "loop-the-loop" consists of a grooved track containing a loop with a radius of 12 cm as shown in Fig. 6-8. When you place a marble high enough on the curved track, it will travel around the loop without falling off the track. The marble rolls along the track without slipping. Find the minimum height from which the marble can be released without falling off the track.

Solution: Because the marble rolls along the track, it has both rotational kinetic energy and translational kinetic energy—its total kinetic energy is

$$\text{KE (total)} = \frac{1}{2}mv^2 + \frac{1}{2}I\omega^2$$

Linear speed is related to angular speed as follows:

$$v = r\omega \quad \text{or} \quad \omega = \frac{v}{r}$$

A sphere's moment of inertia about an axis through its center (Eq. 4-17c) is

$$I = \frac{2}{5}mr^2$$

Now, you can write the kinetic energy of rotation in terms of v and m:

$$\frac{1}{2}I\omega^2 = \frac{1}{2}\left(\frac{2}{5}mr^2\right)\left(\frac{v}{r}\right)^2 = \frac{1}{5}mv^2$$

And the total kinetic energy of the rolling sphere is

$$\text{KE (total)} = \frac{1}{2}mv^2 + \frac{1}{5}mv^2 = 0.7mv^2$$

You know that the total energy of the marble remains constant, therefore,

$$\text{E}_A = \text{E}_B \qquad \text{PE}_A = \text{PE}_B + \text{KE}_B \qquad mgh = mg(2r) + (0.7)mv^2$$

Find the minimum speed of the marble at the top of the loop by setting the centripetal force (Eq. 4-12) equal to the gravitational force:

$$\frac{mv^2}{r} = mg \qquad v^2 = rg$$

Now, use this result to eliminate v in the energy equation:

$$gh = 2rg + (0.7)rg = (2.7)rg$$

Then, solve for the minimum height:

$$h = (2.7)r = (2.7)(12 \text{ cm}) = 32.4 \text{ cm}$$

EXAMPLE 6-11: A solid sphere and a solid cylinder with radii of 3 cm are released simultaneously from the same height on an inclined surface (see Fig. 6-9). The sphere has a mass of 1.2 kg and the cylinder has a mass of 1.0 kg. The distance to the bottom of the inclined plane is 50 cm. (**a**) Which object arrives at the bottom first and (**b**) what is the difference in their arrival times?

Solution:

(**a**) You can solve this problem most easily by using the principle of energy conservation. If you place the reference level at the center of the cylinder when it reaches the bottom of the inclined plane, the total energy of the cylinder will be the sum of kinetic energy of translation and kinetic energy of rotation. At the height from which the cylinder is released, the energy is entirely potential:

$$\text{E (at bottom)} = \text{E (at top)}$$

$$\frac{1}{2}m_cv^2 + \frac{1}{2}I\omega^2 = m_cgh$$

For a cylinder, $I = \frac{1}{2}mr^2$ (Eq. 4-17b), so

$$\frac{1}{2}m_cv^2 + \frac{1}{2}\left(\frac{1}{2}m_cr^2\right)\left(\frac{v}{r}\right)^2 = m_cgh$$

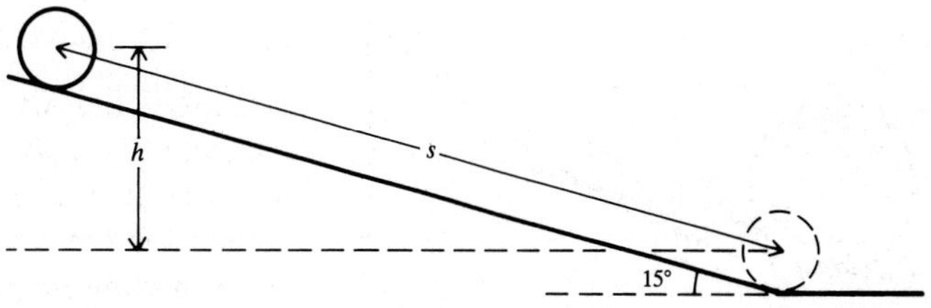

Figure 6-9

Cancel the mass and combine the two kinetic energy terms:

$$\left(\frac{1}{2} + \frac{1}{4}\right)v^2 = gh \qquad v^2 = \frac{4}{3}gh$$

Now you need the vertical distance:

$$h = s \sin 15° = (0.5 \text{ m})(0.259) = 0.1294 \text{ m}$$

So you can find the linear speed of the cylinder at the bottom of the incline:

$$v = \sqrt{\frac{4}{3}gh} = \sqrt{\frac{4}{3}(9.8 \text{ m/s}^2)(0.1294 \text{ m})} = 1.30 \text{ m/s}$$

Repeat this calculation for the solid sphere. The moment of inertia of a solid sphere is $I = \frac{2}{5}m_s r^2$:

$$\frac{1}{2}m_s v^2 + \frac{1}{2}\left(\frac{2}{5}m_s r^2\right)\left(\frac{v}{r}\right)^2 = m_s gh \qquad \left(\frac{1}{2} + \frac{1}{5}\right)v^2 = gh \qquad v^2 = \frac{gh}{0.7}$$

$$v = \sqrt{\frac{gh}{0.7}} = \sqrt{\frac{(9.8 \text{ m/s}^2)(0.1294 \text{ m})}{0.7}} = 1.35 \text{ m/s}$$

The sphere is going faster at the bottom of the inclined plane, so you conclude that it will win the race.

(b) To find the time required to move 50 cm along the plane, you'll have to calculate the acceleration. Since the acceleration is constant, you can use Eq. (2-7):

$$v^2 = v_0^2 + 2as \qquad a = \frac{v^2 - v_0^2}{2s}$$

For the cylinder:

$$a = \frac{(4/3)gh - 0}{2s} = \frac{4(9.8 \text{ m/s}^2)(0.1294 \text{ m})}{(3)(2)(0.50 \text{ m})} = 1.691 \text{ m/s}^2$$

And, for the sphere:

$$a = \frac{(gh/0.7) - 0}{2s} = \frac{(9.8 \text{ m/s}^2)(0.1294 \text{ m})}{(0.7)(2)(0.50 \text{ m})} = 1.812 \text{ m/s}^2$$

Now that you have the acceleration of the two rolling objects, you can calculate the time from Eq. (2-8):

$$s = v_0 t + \frac{1}{2}at^2$$

Since $v_0 = 0$, the time for the cylinder to reach the bottom is

$$t = \sqrt{\frac{2s}{a}} = \sqrt{\frac{2(0.50 \text{ m})}{1.691 \text{ m/s}^2}} = 0.769 \text{ s}$$

And the time required for the sphere is

$$t = \sqrt{\frac{2(0.50 \text{ m})}{1.812 \text{ m/s}^2}} = 0.743 \text{ s}$$

The sphere arrives at the bottom of the inclined plane 0.026 s before the cylinder.

6-4. Simple Machines

- A **machine** is a mechanical device that transmits or changes the magnitude or direction of an applied force.
  The most common simple machines are levers, pulleys, inclined planes, and gears.

A. The lever

Figure 6-10 illustrates the use of a **lever** to lift a heavy rock. The **fulcrum** is the point about which the lever pivots. Since the system is in rotational equilibrium, the sum of the torques about the fulcrum is zero.

EXAMPLE 6-12: Suppose you want to use the lever shown in Figure 6-10 to lift a 600-lb stone. The lever is 10 ft long and the maximum force you can exert is 50 lb.
Where should you place the fulcrum?

Solution: Begin by writing Eq. (5-3) for rotational equilibrium about an axis of rotation through the fulcrum:

$$\Sigma\tau = 0$$

$$F_1 d_1 - F_2 d_2 = 0$$

$$\frac{d_2}{d_1} = \frac{F_1}{F_2} = \frac{600 \text{ lb}}{50 \text{ lb}} = 12$$

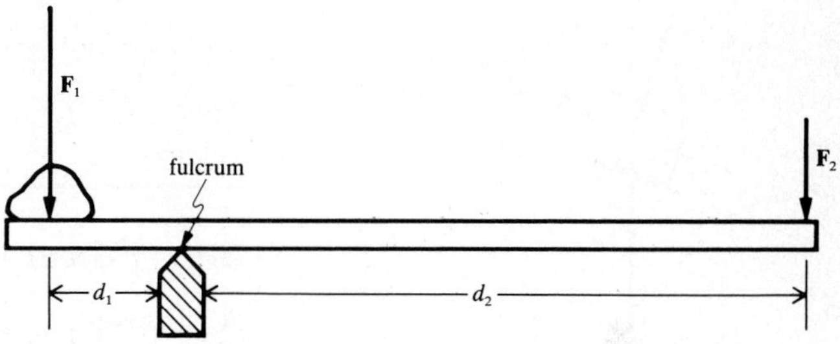

Figure 6-10

Since the length of the lever ($d_1 + d_2$) is 10 ft, you can replace d_2 by 10 ft $- d_1$, so that the equation becomes

$$10 \text{ ft} - d_1 = 12d_1$$

$$13d_1 = 10 \text{ ft}$$

$$d_1 = 0.769 \text{ ft}$$

You should place the fulcrum 0.769 ft from the left end of the lever.

Although the input force in Example 6-12 is multiplied by a factor of 12, you can't expect to get more work out of this simple machine than you put into it. If you assume that the lever rotates about the fulcrum without friction, then the work input should be exactly equal to the work output.

EXAMPLE 6-13: Calculate the work input and the work output for the simple lever in Example 6-12 if the stone is lifted 2 in.

Solution: You'll need to make a drawing similar to Figure 6-11 showing the rotation of the lever about the fulcrum. The two positions of the lever form a pair of similar triangles, and the distance the rock is lifted is s_1. From these similar triangles, you have

$$\frac{s_1}{d_1} = \frac{s_2}{d_2}$$

$$\frac{s_2}{s_1} = \frac{d_2}{d_1} = \frac{(10 - 0.769) \text{ ft}}{0.769 \text{ ft}} = 12$$

$$s_2 = 12s_1 = 12(2 \text{ in}) = 24 \text{ in}$$

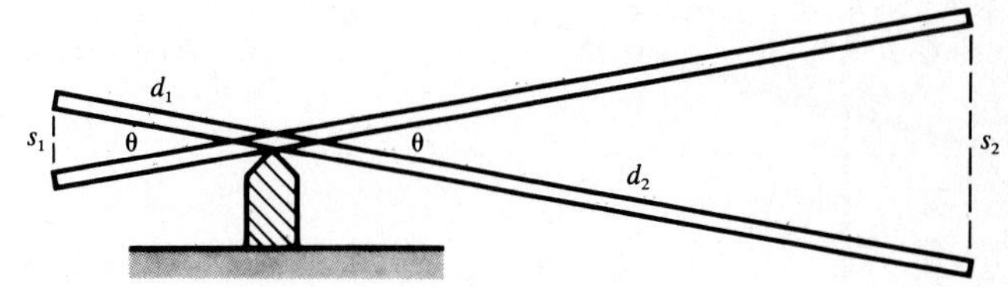

Figure 6-11

Now, you can calculate the work input and output:

$$W_{\text{in}} = \mathbf{F}_2 \cdot \mathbf{s}_2 = (50 \text{ lb})(2 \text{ ft}) \cos 0° = 100 \text{ ft lb}$$

$$W_{\text{out}} = \mathbf{F}_1 \cdot \mathbf{s}_1 = (600 \text{ lb})(0.167 \text{ ft}) \cos 0° = 100 \text{ ft lb}$$

Since there are no frictional losses, the work output is equal to the work input. (The **foot-pound** is the unit of work in the British system of units.)

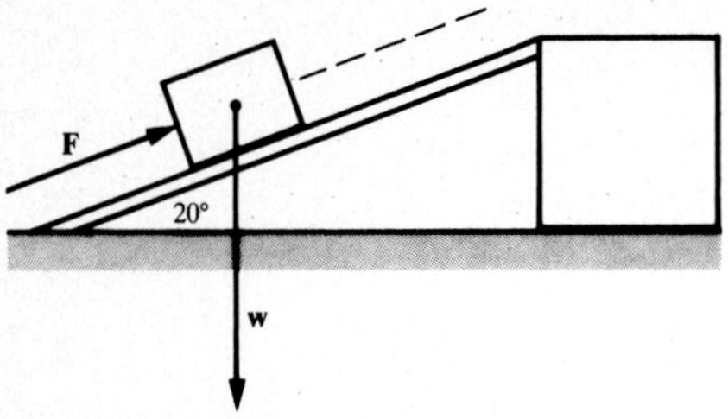

Figure 6-12

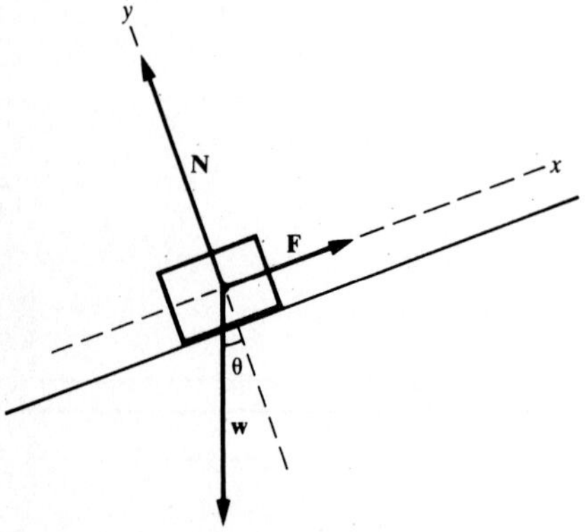

Figure 6-13

B. The inclined plane

- The **inclined plane** is another device that allows you to lift a very heavy load by means of a smaller force.

EXAMPLE 6-14: Calculate the force required to push a 400-lb crate up the ramp shown in Figure 6-12. You may assume that friction is negligible.

Solution: First, draw a force diagram, placing the x axis parallel to the inclined plane. Your diagram should look like Figure 6-13. From Eq. (5-2) for translational equilibrium in the x direction,

$$\Sigma F_x = 0$$

$$F - w \sin \theta = 0$$

$$F = w \sin \theta = (400 \text{ lb}) \sin 20° = 137 \text{ lb}$$

This is the force required to move the crate up the inclined plane at constant velocity.

6-5. Power and Efficiency

A. Power

- **Power** is the rate at which work is done or work per unit time.

In SI, the unit of power is the **watt** W, which is equal to one joule per second.

Since power is the rate of doing work, you'll have to take the limit of the ratio $\Delta W/\Delta t$ if the applied force isn't constant:

DEFINITION OF POWER

$$P = \lim_{\Delta t \to 0} \frac{\Delta W}{\Delta t} \tag{6-8}$$

When the applied force is constant, you can calculate power as follows:

$$P = \frac{W}{t} \qquad [\text{If } \mathbf{F} \text{ is constant}] \tag{6-9}$$

EXAMPLE 6-15: A winch pulls a hopper car filled with coal out of a mine at a speed of 7 m/s (17.7 mi/h). The mass of the loaded car is 4 metric tons, and the track is inclined at an angle of 15°. Friction can be neglected. Determine the minimum power requirement of the winch.

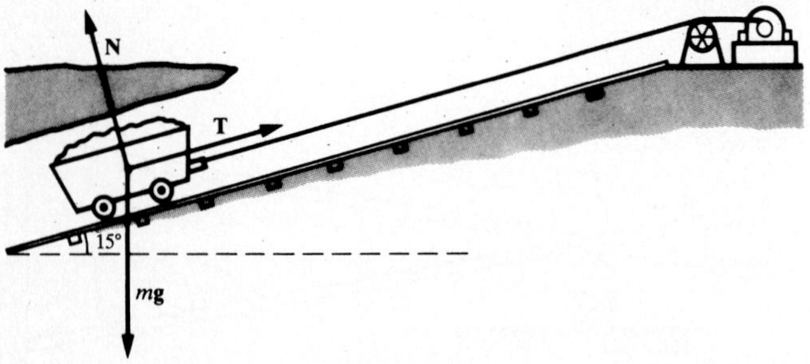

Figure 6-14

Solution: Draw the forces acting on the coal car (see Figure 6-14). First, you have to find the amount of work done. Since the force is constant and has the same direction as the displacement,

$$W = \mathbf{F} \cdot \mathbf{s} = Fs$$

Then, find how much power you need to do that amount of work:

$$P = \frac{W}{t} = F\left(\frac{s}{t}\right)$$

Since the constant speed of the car is the distance traveled divided by the time, $v = s/t$, you can express the power associated with a constant force as

POWER OF A CONSTANT FORCE

$$P = Fv \tag{6-10}$$

The coal car is in dynamic equilibrium, so from Eq. (5-2):

$$\Sigma F_x = 0$$

$$T - mg \sin \theta = 0$$

$$T = mg \sin \theta$$

$$= (4 \times 10^3 \text{ kg})(9.8 \text{ m/s}^2) \sin 15° = 1.015 \times 10^4 \text{ N}$$

The force is supplied by the tension T. Because the force is constant, you can use Eq. (6-10) to calculate the power needed:

$$P = Fv = (1.015 \times 10^4 \text{ N})(7 \text{ m/s})$$

$$= 7.10 \times 10^4 \text{ J/s} = 7.10 \times 10^4 \text{ W}$$

You can express the required power of the engine in either kilowatts or horsepower (1 hp = 0.746 kW):

$$P = 71.0 \text{ kW} = 71.0 \text{ kW}\left(\frac{1 \text{ hp}}{0.746 \text{ kW}}\right) = 95.2 \text{ hp}$$

B. Efficiency

- The **efficiency** e of a machine is the work output divided by the work input, usually expressed as a percent:

EFFICIENCY OF A MACHINE

$$e = \frac{\text{Work output}}{\text{Work input}} \times 100\% \qquad \textbf{(6-11)}$$

EXAMPLE 6-16: A crate of mass 250 kg is pushed 2 m up a 20° ramp at a constant speed. The coefficient of kinetic friction between the crate and the inclined surface is 0.1.

What is the efficiency of this inclined plane?

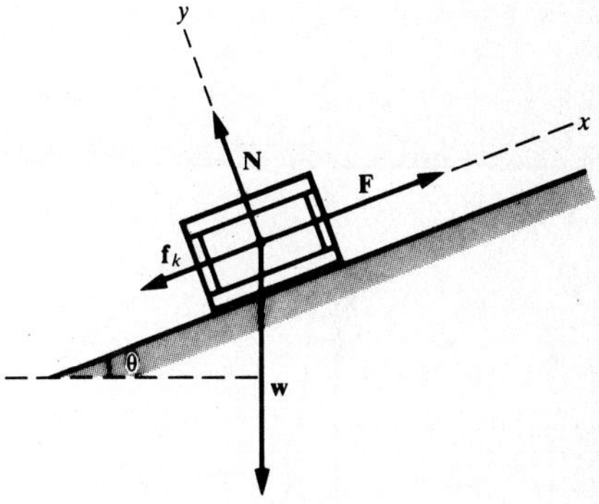

Figure 6-15

Solution: As always, make a force diagram (see Figure 6-15). Since the crate is moving at a constant speed, you can use Eq. (5-2) for the x and y components of the four forces acting on the crate:

Forces	x components	y components
w	$-w \sin \theta$	$-w \cos \theta$
N	0	N
F	F	0
$\mathbf{f}_k$	$-f_k$	0

To find the work input, you have to know the total force acting parallel to the direction of motion. Since there is friction present, you'll have to calculate the frictional force $f_k = \mu_k N$. First, find N from the y components:

$$\Sigma F_y = 0$$

$$N - w \cos \theta = 0$$

$$N = w \cos \theta = mg \cos \theta = (250 \text{ kg})(9.8 \text{ m/s}^2) \cos 20° = 2.30 \times 10^3 \text{ N}$$

Then, you can calculate the frictional force from Eq. (3-6):

$$f_k = \mu_k N$$

$$= (0.1)(2.30 \times 10^3 \text{ N})$$

$$= 2.30 \times 10^2 \text{ N}$$

Now, from Eq. (5-2) for equilibrium along the x axis:

$$\Sigma F_x = 0 \qquad F - f_k - w \sin \theta = 0$$

$$F = f_k + w \sin \theta = 2.30 \times 10^2 \text{ N} + (250 \text{ kg})(9.8 \text{ m/s}^2) \sin 20°$$

$$= 1.068 \times 10^3 \text{ N}$$

And, you can calculate the work input:

$$W_{\text{in}} = \mathbf{F} \cdot \mathbf{s} = Fs \cos 0°$$

$$= (1.068 \times 10^3 \text{ N})(2 \text{ m}) = 2.136 \times 10^3 \text{ J}$$

To figure the work output, you need to know the vertical height through which the crate is lifted:

$$h = s \sin 20° = (2 \text{ m})(0.342) = 0.684 \text{ m}$$

[**recall:** The potential energy PE of an object is the amount of work "stored" by the object, and PE $= wh$. (Sec. 6-2)] The work output is equal to the potential energy stored in the crate, so

$$W_{\text{out}} = wh = mgh = (250 \text{ kg})(9.8 \text{ m/s}^2)(0.684 \text{ m}) = 1.676 \times 10^3 \text{ J}$$

And, the efficiency of the inclined plane is

$$e = \frac{W_{\text{out}}}{W_{\text{in}}} \times 100\% = \frac{1.676 \times 10^3 \text{ J}}{2.136 \times 10^3 \text{ J}} \times 100\% = 78.5\%$$

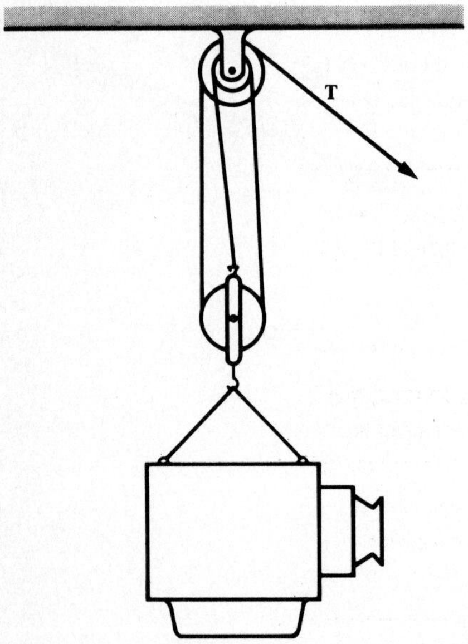

Figure 6-16

EXAMPLE 6-17: A 600-lb motor block is lifted by the system of pulleys illustrated in Figure 6-16. There is some friction in the pulleys. The input force required is 220 lb. When the cable is pulled 6 ft, the load is lifted 2 ft. What is the efficiency of this machine?

Solution: Calculate the work input and the work output:

$$W_{\text{in}} = Ts = (220 \text{ lb})(6 \text{ ft}) = 1320 \text{ ft lb}$$

$$W_{\text{out}} = wh = (600 \text{ lb})(2 \text{ ft}) = 1200 \text{ ft lb}$$

Then, calculate the efficiency of the system of pulleys:

$$e = \frac{W_{\text{out}}}{W_{\text{in}}} \times 100\% = \frac{1200 \text{ ft lb}}{1320 \text{ ft lb}} \times 100\% = 90.9\%$$

C. Mechanical advantage

In general, simple machines provide a **mechanical advantage** for moving objects from one place to another. Because of friction, the work done BY the machine is generally less than the work done ON the machine. We use the terms **actual mechanical advantage** (AMA) and **ideal mechanical advantage** (IMA) to express the ratio of the output force to the input force and the ratio of input distance to output distance, respectively. The ratio of AMA to IMA is the efficiency of the machine.

ACTUAL MECHANICAL ADVANTAGE

$$\text{AMA} = \frac{\text{output force}}{\text{input force}} \qquad \textbf{(6-12)}$$

IDEAL MECHANICAL ADVANTAGE

$$\text{IMA} = \frac{\text{input distance}}{\text{output distance}} \qquad \textbf{(6-13)}$$

EFFICIENCY

$$e = \frac{\text{AMA}}{\text{IMA}} \times 100\% \qquad \textbf{(6-14)}$$

EXAMPLE 6-18: What is the efficiency of the machine described in Example 6-17?

Solution:

$$\text{AMA} = \frac{600 \text{ lb}}{220 \text{ lb}} = 2.727$$

$$\text{IMA} = \frac{6 \text{ ft}}{2 \text{ ft}} = 3.00$$

$$e = \frac{\text{AMA}}{\text{IMA}} \times 100\% = \frac{2.727}{3.00} \times 100\% = 90.9\%$$

You can see this answer agrees with the answer you calculated from the work in Example 6-17.

SUMMARY

Work	$W = \mathbf{F} \cdot \mathbf{s}$ or $W = Fs \cos \theta$	the dot product of a constant force acting on an object, and the displacement of the object
Gravitational potential energy	$\text{PE} = wh$ or $\text{PE} = mgh$	product of the weight of an object and its height above a reference level
Potential energy of a spring	$\text{PE} = \frac{1}{2}kx^2$	yields the PE from the spring constant and its elongation
Spring constant	$k = \frac{\Delta F}{\Delta x}$	ratio of the change in force divided by the elongation produced
Kinetic energy	$\text{KE} = \frac{1}{2}mv^2$	capacity of an object to perform work due to its mass and speed
	$\text{KE} = \frac{1}{2}I\omega^2$	capacity of a rotating object to do work due to its moment of inertia and angular speed

Conservation of energy	$PE_A + KE_A = PE_B + KE_B$	energy is conserved if there is no friction
Definition of power	$P = \lim_{\Delta t \to 0} \frac{\Delta W}{\Delta t}$	rate at which work is done for a variable force
	$P = \frac{W}{t}$	rate at which work is done for a constant force
Power of a constant force	$P = Fv$	power associated with a constant force
Efficiency of a machine	$e = \frac{\text{Work output}}{\text{Work input}} \times 100\%$	work output divided by work input expressed as a percent
Actual mechanical advantage	$\text{AMA} = \frac{\text{output force}}{\text{input force}}$	ratio of the output force to the input force
Ideal mechanical advantage	$\text{IMA} = \frac{\text{input distance}}{\text{output distance}}$	ratio of input distance to output distance
Efficiency	$e = \frac{\text{AMA}}{\text{IMA}} \times 100\%$	ratio of AMA to IMA expressed as a percent

RAISE YOUR GRADES

Can you define . . . ?

☑ work
☑ potential energy
☑ kinetic energy
☑ a machine
☑ power
☑ efficiency of a machine

Can you . . . ?

☑ calculate the work done by a constant force **F** whose direction differs from that of the displacement **s**
☑ calculate the work done by a variable force when a graph of F versus s is available
☑ calculate the potential energy of an object that has been lifted above the reference level
☑ calculate the potential energy of a spring
☑ calculate the kinetic energy of a moving object
☑ calculate the energy acquired by an object as a result of work done by an external agent
☑ calculate the power expended by a force that is doing work on an object
☑ calculate the efficiency of a machine

SOLVED PROBLEMS

PROBLEM 6-1 A steel block weighing 12 N is pulled up an inclined surface by a constant force of 7.35 N, which makes an angle of 10° with respect to the surface. The block starts from rest and is pulled 2.0 m along the inclined plane whose angle is 20° with respect to the horizontal (see Fig. 6-17). The coefficient of friction between the block and the inclined plane is 0.2. Determine (**a**) the work done on the block by the force; (**b**) the increase in potential energy of the block; (**c**) the increase in kinetic energy of the block; and (**d**) the amount of work required to overcome the frictional force.

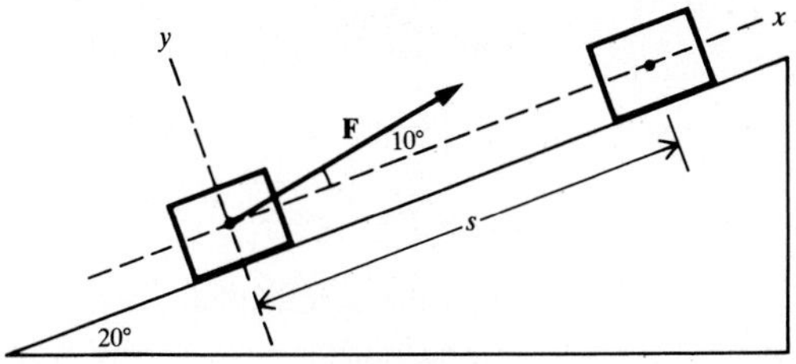

Figure 6-17

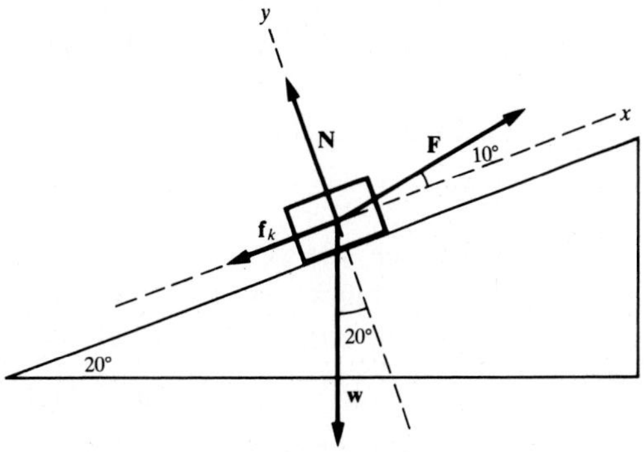

Figure 6-18

Solution

(**a**) Because the force is constant, you can calculate the work from Eq. (6-1). The displacement s is along the x axis:

$$W = \mathbf{F} \cdot \mathbf{s} = Fs \cos \theta = (7.35 \text{ N})(2.0) \cos 10° = 14.48 \text{ J}$$

(**b**) You can calculate the potential energy from Eq. (6-2), PE = wh, but first you must find the vertical distance h:

$$\sin 20° = \frac{h}{s} \qquad h = s \sin 20° = (2.0 \text{ m}) \sin 20° = 0.684 \text{ m}$$

Then,

$$\text{PE} = wh = (12 \text{ N})(0.684 \text{ m}) = 8.21 \text{ J}$$

(**c**) The kinetic energy acquired by the block depends on its final speed, which you can find by calculating the net force acting on the block in the direction of its movement and the acceleration produced by this net force. Make a drawing showing all forces acting on the block (see Fig. 6-18), and a table of all forces and their components:

Forces	x components	y components
F	$F \cos 10°$	$F \sin 10°$
w	$-w \sin 20°$	$-w \cos 20°$
N	0	N
$\mathbf{f}_k$	$-\mu_k N$	0

You know that there is no acceleration in the y direction, so the sum of the y components is zero:

$$\Sigma F_y = F \sin 10° - w \cos 20° + N = 0$$

$$N = w \cos 20° - F \sin 10° = (12 \text{ N})(0.9396) - (7.35 \text{ N})(0.1736) = 10.0 \text{ N}$$

Then, find the net force in the x direction:

$$\Sigma F_x = F \cos 10° - w \sin 20° - \mu_k N$$

$$= (7.35 \text{ N})(0.985) - (12 \text{ N})(0.342) - (0.2)(10.0 \text{ N}) = 1.134 \text{ N}$$

You can get the acceleration from Newton's second law:

$$a = \frac{\Sigma F_x}{m} = \frac{\Sigma F_x}{w/g} = \frac{1.134 \text{ N}}{12 \text{ N}}(9.8 \text{ m/s}^2) = 0.926 \text{ m/s}^2$$

and the square of the final speed of the block from Eq. (2-7):

$$v^2 = v_0^2 + 2as = 0 + 2(0.926 \text{ m/s}^2)(2.0 \text{ m}) = 3.705 \text{ m}^2/\text{s}^2$$

Now, calculate the kinetic energy:

$$\text{KE} = \frac{1}{2}mv^2 = \frac{1}{2}\left(\frac{w}{g}\right)v^2 = \frac{1}{2}\left(\frac{12 \text{ N}}{9.8 \text{ m/s}^2}\right)(3.705 \text{ m}^2/\text{s}^2) = 2.27 \text{ J}$$

(**d**) The total mechanical energy acquired by the block is

$$E = \text{PE} + \text{KE} = 8.21 \text{ J} + 2.27 \text{ J} = 10.48 \text{ J}$$

As you can see, the work done on the block, 14.48 J, is greater than the mechanical energy acquired by the block, 10.48 J. Because of friction between the two surfaces, some of the work done on the moving block is transformed into heat energy. The amount of heat energy is

$$q = 14.48 \text{ J} - 10.48 \text{ J} = 4.00 \text{ J}$$

Since the force of friction acting on the block is 2.0 N in the negative x direction, a force of 2.0 N in the positive x direction is needed to overcome it. So, the work done to overcome friction is

$$W = f_k s \cos 0° = (2.0 \text{ N})(2.0 \text{ m}) = 4.00 \text{ J}$$

This portion of the work done by the external force is converted into heat energy. We can write the equation for conservation of energy in this situation as:

$$\text{Work done} = \text{PE} + \text{KE} + q$$

where q stands for the heat energy produced.

PROBLEM 6-2 A variable force, illustrated in Fig. 6-19, acts on an object which has a displacement of 60 cm in the direction of the force. How much work is done by this force?

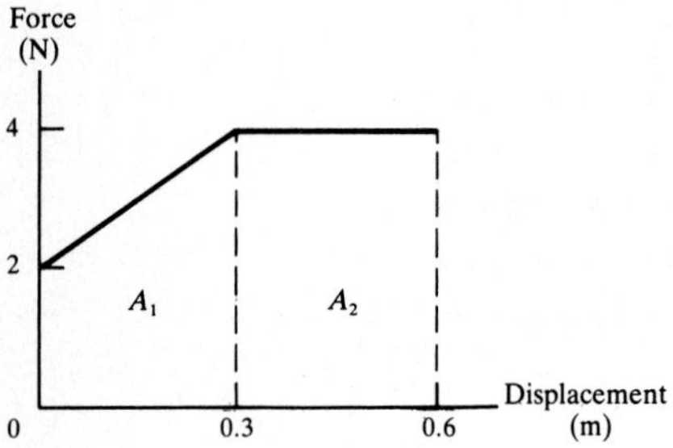

Figure 6-19

Solution You can calculate the work done by this variable force by determining the area under the graph. The total area consists of a trapezoid and a rectangle. The area of the trapezoid is

$$A_1 = \frac{1}{2}(2 \text{ N} + 4 \text{ N})(0.3 \text{ m}) = 0.9 \text{ J}$$

The area of the rectangle is

$$A_2 = (4 \text{ N})(0.6 \text{ m} - 0.3 \text{ m}) = 1.2 \text{ J}$$

And, the total work done is

$$W = A_1 + A_2 = 0.9 \text{ J} + 1.2 \text{ J} = 2.1 \text{ J}$$

PROBLEM 6-3 Fig. 6-20 shows an 80-cm pendulum that swings through an arc of 120°. The pendulum bob is pulled 60° from its lowest position C and released from rest. It reaches its maximum speed v_{max} as it passes through point C. Friction may be neglected. Calculate (**a**) the maximum speed of the pendulum bob and (**b**) the height at which the pendulum bob will be moving at one half its maximum speed.

Solution

(**a**) You'll find it convenient to choose a horizontal line passing through point C as the reference level. The height at A is $h_A = (0.8 \text{ m}) - (0.8 \text{ m})(\sin 30°) = 0.4$ m. Without friction, the total mechanical energy is conserved, so

$$\text{E}_A = \text{E}_C \qquad \text{PE}_A + \text{KE}_A = \text{PE}_C + \text{KE}_C$$

$$mgh_A + 0 = 0 + \frac{1}{2}mv_C^2$$

$$v_C^2 = 2gh_A = 2(9.8 \text{ m/s}^2)(0.4 \text{ m}) = 7.84 \text{ m}^2/\text{s}^2$$

$$v_C = \sqrt{7.84 \text{ m}^2/\text{s}^2} = 2.80 \text{ m/s}$$

(**b**) Since the total mechanical energy of the pendulum remains constant,

$$\mathrm{PE}_B + \mathrm{KE}_B = \mathrm{PE}_C + \mathrm{KE}_C$$

$$mgh_B + \frac{1}{2}mv_B^2 = 0 + \frac{1}{2}mv_C^2$$

Now, set v_B equal to $\frac{1}{2}v_C$:

$$mgh_B + \frac{1}{2}m\left(\frac{v_C}{2}\right)^2 = \frac{1}{2}mv_C^2$$

$$mgh_B = \frac{1}{2}mv_C^2 - \frac{1}{4}\left(\frac{1}{2}mv_C^2\right)$$

$$= \frac{3}{4}\left(\frac{1}{2}mv_C^2\right)$$

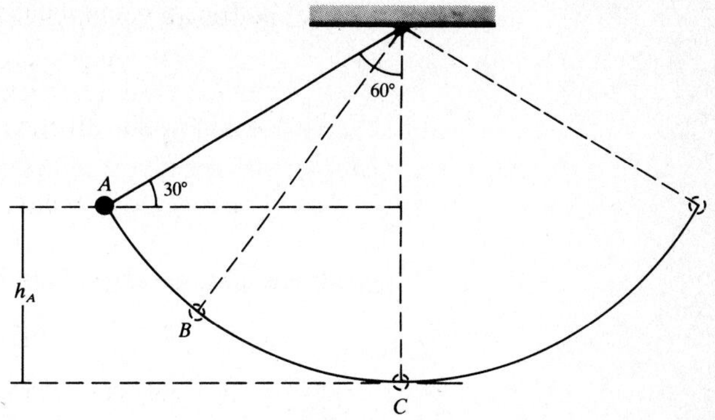

Figure 6-20

In part (**a**) you found that $\frac{1}{2}mv_C^2$ is equal to mgh_A, so you have

$$mgh_B = \frac{3}{4}(mgh_A)$$

$$h_B = \frac{3}{4}h_A = \frac{3}{4}(0.4 \text{ m}) = 0.3 \text{ m}$$

The pendulum bob reaches one half its maximum speed when its height is three fourths of the original height.

PROBLEM 6-4 A child and sled, weighing 290 N (approximately 65 lb), slide down an ice-covered (frictionless) hill that is 10 m high. At the bottom of the hill there is an ice-free, horizontal surface that brings the sled to a stop within a distance of 25 m. What is the coefficient of friction between the sled runners and the horizontal surface?

Solution Your sketch of the situation should look like Fig. 6-21. You need to calculate the speed of the sled when it starts across the horizontal surface. Since the icy slope is frictionless, the total mechanical energy remains constant while the sled is on the slope. Choose the horizontal surface as the reference level.

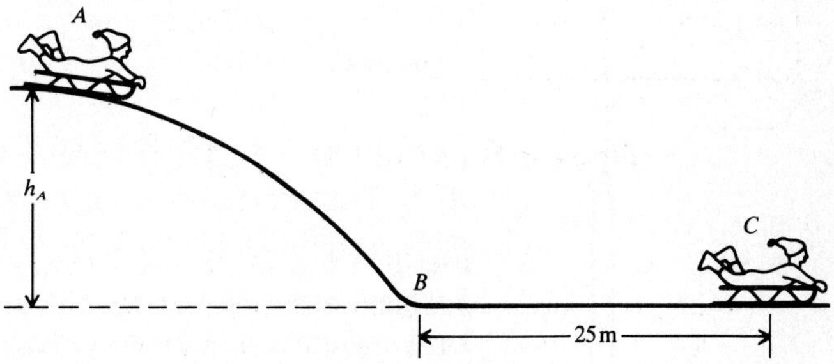

Figure 6-21

$$\mathrm{E}_B = \mathrm{E}_A$$

$$\frac{1}{2}mv_B^2 + 0 = 0 + mgh_A$$

$$v_B^2 = 2gh_A$$

The deceleration of the sled along the horizontal surface is produced by a constant force of friction, $f_k = -\mu_k N$. The minus sign indicates that the frictional force is toward the left, the negative x direction. You can find the constant deceleration of the sled from Newton's second law:

$$a = \frac{\Sigma F}{m} = \frac{f_k}{m} = \frac{-\mu_k N}{m}$$

For an object on a horizontal surface, the magnitude of the normal force is equal to the magnitude of the weight, $N = w = mg$, so the deceleration is

$$a = \frac{-\mu_k mg}{m} = -\mu_k g$$

Because the acceleration is constant, you can use Eq. (2-7):

$$v^2 - v_0^2 = 2as = -2\mu_k gs$$

The final speed of the sled v is zero and the initial speed is v_B. The frictional force acts toward the left—in the negative x direction—so the acceleration is negative:

$$0 - v_B^2 = -2\mu_k gs$$

Now, solve for the coefficient of kinetic friction:

$$\mu_k = \frac{v_B^2}{2gs} = \frac{2gh_A}{2gs} = \frac{h_A}{s} = \frac{10\text{ m}}{25\text{ m}} = 0.4$$

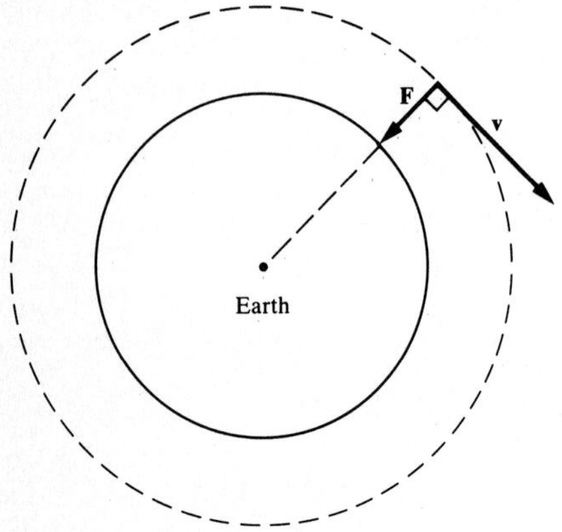

Figure 6-22

PROBLEM 6-5 A weather satellite maintains a circular orbit with a radius of 6500 km, measured from the center of the earth. The centripetal force which keeps the satellite in its circular path has a magnitude of 3.75×10^3 N. The direction of this force, produced by the gravitational attraction of the earth, is toward the center of the earth. The speed of the satellite is 7.83 km/s. How much work is done on the satellite by the gravitational force during a 2 s interval?

Solution Your first step is to make a drawing that shows the path of the satellite and the direction of the force acting on it (see Fig. 6-22). The satellite is moving along the circular path, so its displacement $\Delta\mathbf{s}$ is exactly perpendicular to the centripetal force **F**. Since cos 90° is equal to zero, the work done by the centripetal force is exactly zero:

$$W = \mathbf{F} \cdot \Delta\mathbf{s} = F\,\Delta s \cos 90^\circ = 0$$

PROBLEM 6-6 A steel girder to be used in the construction of a skyscraper weighs 1.80×10^4 N and is 18 m long. The girder is initially lying on the ground in a horizontal position, as shown in Fig. 6-23a. The cable of a construction crane is fastened to one end of the girder and lifts the girder to a vertical position with its lower end 6 m above the ground, as illustrated in Fig. 6-23b. How much work is required to lift the girder to this position?

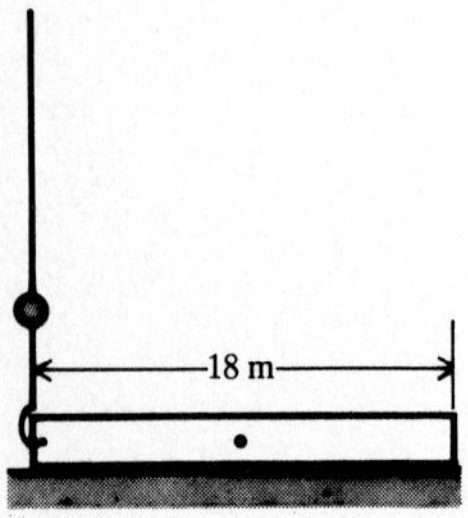

(a)

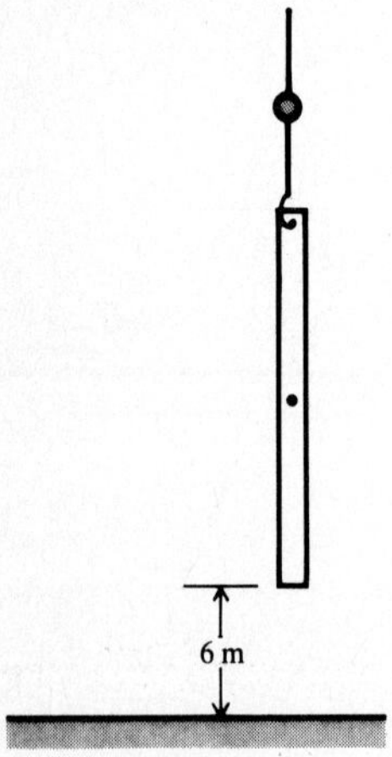

(b)

Figure 6-23

Solution Because the girder has a uniform cross section, its center of gravity is located at the center. You can see from the figure that the center of gravity of the girder is lifted a vertical distance of 15 m. You can use Eq. (6-2) to calculate the increase in PE of the girder:

$$\text{PE} = wh = (1.80 \times 10^4\text{ N})(15\text{ m}) = 2.70 \times 10^5\text{ J}$$

Since there is no friction, the work done by the crane is equal to the PE of the girder: $W = 2.70 \times 10^5$ J.

PROBLEM 6-7 A diesel engine weighing 2.4×10^5 N pulls 6 freight cars, each weighing 1.8×10^5 N. The train moves along a straight, horizontal section of track at 70 km/h. The total drag force acting to slow the train is 2.64×10^4 N. What is the power output of the engine?

Solution Since the velocity of the train is constant, the sum of the horizontal forces must be zero. Let **F** represent the forward force exerted by the engine and let **f** represent the total drag force, then $F - f = 0$, or $F = f$. Because the speed is given, you can use Eq. (6-10) to calculate the output power of the engine:

$$P = Fv = (2.64 \times 10^4\text{ N})\left(70\,\frac{\text{km}}{\text{h}}\right)\left(\frac{1\text{ h}}{3.6 \times 10^3\text{ s}}\right)\left(\frac{10^3\text{ m}}{1\text{ km}}\right) = 5.13 \times 10^5\text{ W}$$

PROBLEM 6-8 Suppose that the train described in Problem 6-7 starts from rest and reaches a speed of 70 km/h in 40 s. What is the total energy the engine expends during this 40 s interval?

Solution Because the train is accelerating, you'll have to use Newton's second law of motion to calculate the force exerted by the engine:

$$\Sigma F = ma$$

$$F - f = ma = \left(\frac{w}{g}\right)a$$

Find the constant acceleration from Eq. (2-3):

$$a = \frac{\Delta v}{\Delta t} = \frac{(70\text{ km/h})\left(\dfrac{1\text{ h}}{3600\text{ s}}\right)\left(\dfrac{10^3\text{ m}}{1\text{ km}}\right)}{40\text{ s}} = 0.486\text{ m/s}^2$$

Then, calculate the total force exerted by the engine:

$$F = \left(\frac{w}{g}\right)a + f$$

$$= \left(\frac{1.32 \times 10^6\text{ N}}{9.8\text{ m/s}^2}\right)(0.486\text{ m/s}^2) + 2.64 \times 10^4\text{ N}$$

$$= 9.19 \times 10^4\text{ N}$$

Next, calculate the distance the train travels during the 40 s interval (use Eq. (2-8), which is applicable in this situation because the acceleration is constant):

$$s = v_0 t + \frac{1}{2}at^2$$

$$= 0 + \frac{1}{2}(0.486\text{ m/s}^2)(40\text{ s})^2 = 389\text{ m}$$

Now, you can calculate the work from Eq. (6-1):

$$W = Fs\cos\theta$$

$$= (9.19 \times 10^4\text{ N})(389\text{ m})(1)$$

$$= 3.57 \times 10^7\text{ J}$$

PROBLEM 6-9 An arrow is shot from the top of an 8-m building with an initial speed of 6 m/s at an angle of 40° above the horizontal. Determine the velocity of the arrow (both magnitude and direction) when it reaches the ground.

Solution Your drawing should look like Fig. 6-24. Although you could solve this problem by using the equations of Chapter 2, you can get the solution more easily and quickly by using the principle of energy conservation. The total energy of the arrow at point B, just before it strikes the ground, is equal to its total energy at point A when it is shot from the bow. Thus,

$$E_B = E_A$$

$$\frac{1}{2}mv_B^2 = mgh + \frac{1}{2}mv_0^2$$

$$v_B^2 = 2gh + v_0^2$$

$$= 2(9.8\text{ m/s}^2)(8\text{ m}) + (6\text{ m/s})^2 = 192.8\text{ m}^2/\text{s}^2$$

$$v_B = \sqrt{192.8\text{ m}^2/\text{s}^2} = 13.9\text{ m/s}$$

Since the horizontal component of the velocity does not change, you can use it to obtain the direction of v_B:

$$\cos\theta = \frac{v_x}{v_B} = \frac{v_0\cos 40°}{v_B}$$

$$= \frac{(6\text{ m/s})\cos 40°}{13.89\text{ m/s}} = 0.331$$

$$\theta = \arccos(0.331) = 70.7°$$

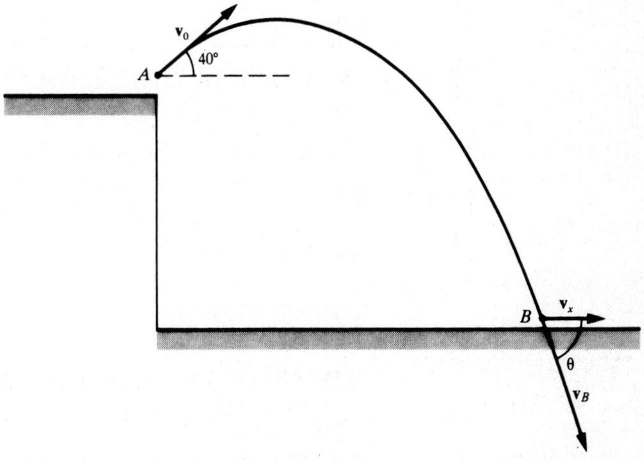

Figure 6-24

Supplementary Exercises

PROBLEM 6-10 A constant force of 6 N is applied to an object. The object moves along a straight line a distance of 4 m. The work done by the force is 22.4 J. What is the angle between the force and the displacement?

PROBLEM 6-11 A small rock is thrown straight upward. It rises to a maximum height of 15 m. What was the initial speed of the rock?

PROBLEM 6-12 A sandbag weighing 150 N (33.7 lb) is dropped from a height of 4 m. What is the KE of the sandbag when it is 1 m above the ground?

PROBLEM 6-13 The original length of a certain spring is 16 cm. The amount of work required to stretch it to a length of 20 cm is 0.05 J. What is the spring constant?

PROBLEM 6-14 When the spring described in Problem 6-13 is stretched to a length of 22 cm, what force is exerted by the spring?

PROBLEM 6-15 A frictionless cart is released on an inclined plane whose angle with respect to the horizontal is 10°. What is the speed of the cart when it has moved 2 m along the inclined surface?

PROBLEM 6-16 A trunk weighing 20 N is pushed up an inclined plane whose angle is 30°. The amount of work required to move the trunk a distance of 3 m along the inclined plane is 40 J. What is the efficiency of this inclined plane?

PROBLEM 6-17 A heavily loaded truck is moving along a level highway at a constant speed of 59.7 mi/h (96 km/h). The power output of the engine is 120 hp. What is the magnitude of the force required to keep the truck moving at this speed?

PROBLEM 6-18 A flywheel, spinning at 500 rev/s, has a moment of inertia of 4 kg m^2. What torque is required to bring this flywheel to rest in 5 s?

PROBLEM 6-19 A solid disk, rotating about an axis through its center, has a mass of 1.5 kg and a radius of 20 cm. How much energy is required to bring this disk from rest to an angular speed of 1000 rev/s?

PROBLEM 6-20 A solid disk is released from rest on a 15° incline. How long will it take for the disk to roll 50 cm along the inclined surface?

Answers to Supplementary Exercises

6-10: 21°

6-11: 17.15 m/s

6-12: 450 J

6-13: 62.5 N/m

6-14: 3.75 N

6-15: 2.61 m/s

6-16: 75%

6-17: 3.36×10^3 N

6-18: 2.51×10^3 Nm

6-19: 5.92×10^5 J

6-20: 0.769 s

7 MOMENTUM

THIS CHAPTER IS ABOUT

☑ Linear and Angular Momentum
☑ Newton's Second Law Revisited
☑ Conservation of Momentum

7-1. Linear and Angular Momentum

A. Linear momentum

- **Linear momentum p** is the product of an object's mass and velocity.

The concept of momentum is one of the most singularly useful concepts in mechanics. With it, you can analyze processes in terms of position and velocity, rather than in terms of the second-law concepts of force and acceleration. In vector notation,

LINEAR MOMENTUM $$\mathbf{p} = m\mathbf{v} \tag{7-1}$$

Since mass is a positive scalar, momentum is a vector that has the same direction as the velocity. [**recall:** Multiplying a vector by a positive scalar changes the magnitude of the vector, but not its direction. (Sec. 1-3)]

B. Angular momentum

- **Angular momentum L** is the cross product of position vector **r** and linear momentum **p**. [**recall:** A cross product of two vectors is a new vector whose direction is perpendicular to both of the original vectors. (Sec. 1-5)]

ANGULAR MOMENTUM $$\mathbf{L} = \mathbf{r} \times \mathbf{p} \quad \text{or} \quad L = rp \sin\theta \tag{7-2}$$

There are no special names for the units of momentum—you may use any mass unit multiplied by a velocity unit.

EXAMPLE 7-1: A 3-kg point mass moves parallel to the x axis along the line $y = 2$ m. The speed of the object is 4 m/s.
Find (**a**) the linear momentum and (**b**) the angular momentum of this object.

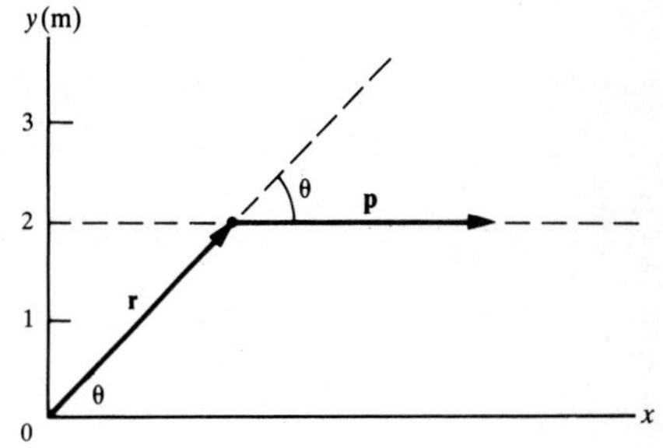

Figure 7-1

Solution:

(**a**) Begin by drawing the x and y axes and the path of the moving object (see Figure 7-1). From Eq. (7-1), the magnitude of the linear momentum is

$$p = mv = (3\ \text{kg})(4\ \text{m/s}) = 12\ \text{kg m/s}$$

As you can see from the diagram, **p** has a y component of zero, so the direction of **p** is the same as that of the positive x axis.

(**b**) You'll need to draw the position vector **r**, as shown in the figure. Since $r \sin\theta = 2$ m, the magnitude of **L** is

$$L = rp \sin\theta$$
$$= (2\ \text{m})(12\ \text{kg m/s}) = 24\ \text{kg m}^2/\text{s}$$

Since the cross product of two vectors is perpendicular to each of them, you know that the vector **L** is perpendicular to the drawing. Mentally place the vectors **r** and **p** with their tails together, and then imagine a right-hand screw twisted in the direction that will rotate **r** toward **p**. (If you can't visualize this process, make a quick sketch.) A clockwise rotation will produce the desired result. As you know, a right-hand screw twisted clockwise will move down into the drawing, so the angular momentum vector **L** is directed downward along the direction of the negative *z* axis. You might find it easier to determine the direction of **L** if you first place the vectors **r** and **p** with their tails together, and then place the fingers of your right hand so they point in the same direction as the vector **r**. Now, curl your fingers in the direction of the vector **p**. The thumb of your right hand now points in the direction of $\mathbf{r} \times \mathbf{p} = \mathbf{L}$.

- The **angular momentum of a rigid body** rotating about a fixed axis is the product of the moment of inertia of the body and its angular speed:

ANGULAR MOMENTUM OF A RIGID BODY

$$\mathbf{L} = I\omega \tag{7-3}$$

If a point mass m moves in a circular path of radius r, its moment of inertia is $I = mr^2$ (Eq. 4-17). Its velocity vector **v** is exactly perpendicular to the position vector **r**. Therefore, the magnitude of the angular momentum, from Eq. (7-2), is

$$\begin{aligned} L &= rp \sin 90° \\ &= r(mv)(1) = rm(r\omega) \\ &= mr^2\omega \\ &= I\omega \end{aligned}$$

You can see that Eqs. (7-2) and (7-3) lead to the same result.

7-2. Newton's Second Law Revisited

Newton's second law states that the sum of forces on an object equals the object's mass multiplied by its acceleration (see Sec. 3-1). If the mass is constant, we can express the second law in terms of the velocity's rate of change:

$$\Sigma F = \lim_{\Delta t \to 0} m\frac{\Delta \mathbf{v}}{\Delta t}$$

Since $\Delta \mathbf{v}$ represents the change in velocity,

$$m\,\Delta\mathbf{v} = m(\mathbf{v}_2 - \mathbf{v}_1) = m\mathbf{v}_2 - m\mathbf{v}_1 = \mathbf{p}_2 - \mathbf{p}_1$$

Of course, $\mathbf{p}_2 - \mathbf{p}_1$ is the change in momentum ($\Delta\mathbf{p}$). Now, we'll write Newton's second law as

NEWTON'S SECOND LAW

$$\Sigma \mathbf{F} = \lim_{\Delta t \to 0} \frac{\Delta \mathbf{p}}{\Delta t} \tag{7-4}$$

If you've studied calculus, you'll recognize the limit of $\Delta\mathbf{p}/\Delta t$ as the derivative of **p** with respect to t. So, you can also express Newton's second law as

$$\Sigma \mathbf{F} = \frac{d\mathbf{p}}{dt} \tag{7-5}$$

An important advantage in expressing Newton's second law in terms of momentum rather than velocity is that Eqs. (7-4) and (7-5) are valid for situations in which the mass is changing, such as a rocket using fuel (mass) to achieve thrust. Newton's second law expressed in terms of the velocity's rate of change is valid *only* if the mass is constant.

EXAMPLE 7-2: A horizontal conveyor belt moves coal from a storage facility to a dump truck, as shown in Figure 7-2. The belt moves at a constant speed of 0.5 m/s. Because of friction in the drive mechanism and the rollers that support

the belt, a force of 20 N is required to keep the belt moving even when no coal is falling onto it.
What additional force is needed to keep the belt moving when coal is falling onto it at the rate of 80 kg/s?

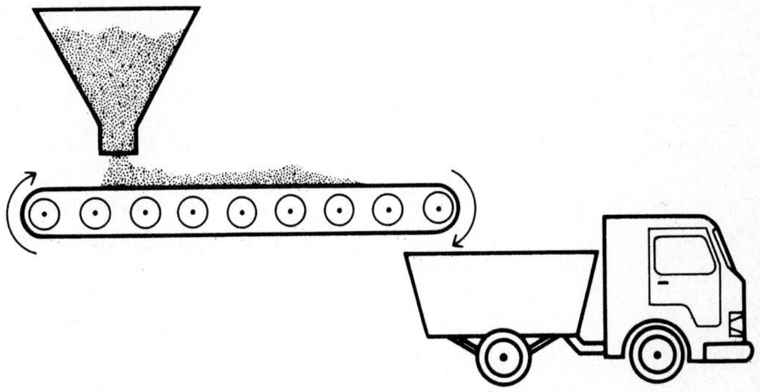

Figure 7-2

Solution: The speed is constant, so you can write the change in momentum as

$$\Delta p = \Delta(mv) = m_2 v - m_1 v$$
$$= v(m_2 - m_1) = v\,\Delta m$$

Then, you can write Newton's second law as

$$\Sigma F = \lim_{\Delta t \to 0} v \frac{\Delta m}{\Delta t}$$

The mass on the belt increases at the rate of $\Delta m/\Delta t =$ 80 kg/s, so you can calculate the additional force required:

$$F = v\frac{\Delta m}{\Delta t} = (0.5 \text{ m/s})(80 \text{ kg/s}) = 40 \text{ N}$$

7-3. Conservation of Momentum

A. Conservation of linear momentum

You know that the rate of change of an object's (or system's) momentum is equal to the net force acting on the object (or system). If the net force is zero, then the rate of change of the momentum is also zero; that is, if $\Sigma\mathbf{F} = 0$, then $\Delta\mathbf{p} = 0$. This result leads directly to the

- **Law of conservation of linear momentum:** If the net force acting on a system is zero, the total linear momentum of the system will remain constant.

The law of conservation of momentum is true if (a) the mechanical energy of an isolated system is conserved:

- In an **elastic collision,** no kinetic energy is lost. The total kinetic energy of the particles before collision is equal to the total kinetic energy after collision.

or if (b) the mechanical energy of the system has been transformed into some other form of energy:

- In an **inelastic collision,** some kinetic energy is "lost" by transformation into other forms of energy such as heat or sound. A collision is **completely inelastic** if the particles stick together after the collision.

The law of conservation of momentum is also true for an explosion, in which any number of bodies make up a system. For example, when a grenade explodes in the air, the vector sum of the momentum of all the fragments—regardless of their various directions and various speeds—is equal to the momentum of the grenade before it explodes.

note: In analyzing the effects of collisions, remember that you can *always* apply the principle of conservation of momentum. If the collision is elastic, you can also apply the principle of conservation of energy.

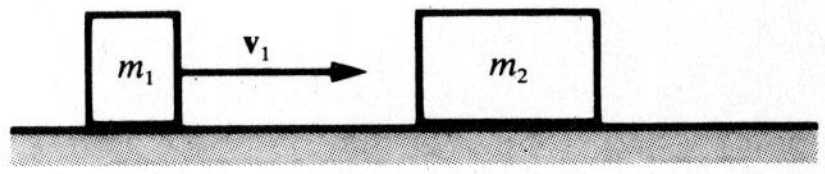

Figure 7-3

EXAMPLE 7-3: A system consists of two blocks that move without friction along a horizontal line, as shown in Figure 7-3. The smaller block m_1 has a mass of 0.5 kg and moves toward the right at a speed of 2 m/s. The larger block m_2

has a mass of 1.0 kg and is at rest. (The blocks are made of an elastic material so there is no loss of energy when they collide. This is called a **perfectly elastic collision**.)
Determine the velocities of the two blocks after m_1 collides with m_2.

Solution: There are external forces acting on this system—for example, m_1 is subject to a downward force of m_1g (0.5 kg $\times$ 9.8 m/s^2 = 4.9 N), and an upward force that the horizontal surface exerts. But, there is no acceleration in the vertical direction, so the net force in that direction is necessarily equal to zero.

During the collision, m_1 exerts a force on m_2, and m_2 exerts a force on m_1. You know from Newton's third law of motion (Eq. 3-3) that these forces are equal and opposite, so that

$$\mathbf{F}_{1\text{ on }2} = -\mathbf{F}_{2\text{ on }1}$$

The internal forces have a sum of zero and, therefore, cannot change the momentum of the system. The total momentum of this system comes from the mechanical energy of m_1 and is constant, so you can write

CONSERVATION OF LINEAR MOMENTUM

$$\begin{array}{ccc} \text{Before} & & \text{After} \\ p_1 + p_2 & = & p_1' + p_2' \\ m_1v_1 + 0 & = & m_1v_1' + m_2v_2' \end{array} \quad \textbf{(7-6)}$$

Since $m_2 = 2m_1$, you can write

$$m_1v_1 = m_1v_1' + 2m_1v_2' \quad \textbf{(7-6a)}$$

As you can see, there are two unknowns v_1' and v_2', but only one equation. You can obtain a second equation from the equation for kinetic energy (Eq. 6-5) and the principle of conservation of energy, which states that energy can be neither created nor destroyed, only transformed (see Eq. 6-6), so that the total energy before the collision must be equal to the total energy after the collision, $KE_1 + PE_1 = KE_2 + PE_2$:

$$\begin{array}{ccc} \text{Before} & & \text{After} \\ \frac{1}{2}m_1v_1^2 & = & \frac{1}{2}m_1(v_1')^2 + \frac{1}{2}(2m_1)(v_2')^2 \end{array} \quad \textbf{(7-7)}$$

Cancel out the mass by dividing Eqs. (7-6a) and (7-7) by m_1:

$$v_1 = v_1' + 2v_2' \quad \textbf{(7-8)}$$

$$v_1^2 = (v_1')^2 + 2(v_2')^2 \quad \textbf{(7-9)}$$

Now, use Eq. (7-8) to eliminate v_1' from Eq. (7-9):

$$v_1^2 = (v_1 - 2v_2')^2 + 2(v_2')^2$$

$$= v_1^2 - 4v_1v_2' + 4(v_2')^2 + 2(v_2')^2$$

$$4v_1v_2' = 6(v_2')^2$$

And, divide by $2v_2'$:

$$2v_1 = 3v_2'$$

When you substitute the numerical values, you'll find that the speed of m_2 after the collision is

$$v_2' = \frac{2}{3}v_1 = \frac{2}{3}(2\text{ m/s}) = 1.33\text{ m/s}$$

You can find the speed of m_1 from Eq. (7-8):

$$v_1 = v_1' + 2v_2'$$

$$v_1' = v_1 - 2v_2' = 2\text{ m/s} - 2(1.33\text{ m/s}) = -0.667\text{ m/s}$$

The negative sign is important—it tells you that m_1 is moving toward the left. Now, verify your results by calculating the total energy of the system before and after the collision. The total energy before the collision is

$$\text{KE} = \frac{1}{2}m_1v_1^2 = \frac{1}{2}(0.5 \text{ kg})(2 \text{ m/s})^2 = 1.0 \text{ J}$$

After the collision, the total energy is

$$\text{KE}' = \frac{1}{2}m_1(v_1')^2 + \frac{1}{2}m_2(v_2')^2$$

$$= \frac{1}{2}(0.5 \text{ kg})\left(-\frac{2}{3}\text{ m/s}\right)^2 + \frac{1}{2}(1.0 \text{ kg})\left(\frac{4}{3}\text{ m/s}\right)^2 = \frac{1}{9}\text{ J} + \frac{8}{9}\text{ J} = 1.0 \text{ J}$$

EXAMPLE 7-4: Reconsider the system of blocks of Example 7-3, given the same initial conditions, but for a **completely inelastic collision**—that is, the two blocks stick together and move as a single unit after the collision. (Mechanical energy is not conserved in an inelastic collision.)
Find (**a**) the common speed of the two blocks after the collision and (**b**) the fraction of kinetic energy "lost" during the collision.

Solution:

(**a**) Because the net external force including friction is zero, the total momentum of the system is conserved, so that

Before After

$$p_1 + p_2 = p_1' + p_2'$$

$$m_1v_1 = (m_1 + m_2)v'$$

$$v' = \frac{m_1v_1}{m_1 + m_2}$$

$$= \frac{(0.5 \text{ kg})(2 \text{ m/s})}{(0.5 + 1.0) \text{ kg}} = 0.667 \text{ m/s}$$

(**b**) The total energy before the collision is

$$\text{KE} = \frac{1}{2}m_1v_1^2 = \frac{1}{2}(0.5 \text{ kg})(2 \text{ m/s})^2 = 1.0 \text{ J}$$

The total energy after the collision is

$$\text{KE}' = \frac{1}{2}(m_1 + m_2)(v')^2 = \frac{1}{2}(0.5 \text{ kg} + 1.0 \text{ kg})(0.667 \text{ m/s})^2 = 0.333 \text{ J}$$

Two-thirds of the original kinetic energy was "lost" in the collision. (In actuality, the missing kinetic energy reappears as frictional heating, sound energy, etc.)

EXAMPLE 7-5: A 3-kg puck slides on a frictionless, straight, horizontal rail at 40 cm/s and strikes a 4-kg puck at rest. The collision is elastic and restricted to one-dimensional motion.
Find the speeds of the two pucks after the collision.

Solution: Let the direction in which the 3-kg puck moves before the collision be the positive direction. Apply the principle of the conservation of momentum:

$$m_1v_1 = m_1v_1' + m_2v_2'$$

And, solve for v_1':

$$v_1' = v_1 - \frac{m_2}{m_1}v_2'$$

The collision is elastic, so you can also apply the principle of conservation of energy:

$$\frac{1}{2}m_1v_1^2 = \frac{1}{2}m_1(v_1')^2 + \frac{1}{2}m_2(v_2')^2$$

Divide by $\frac{1}{2}m_1$:

$$v_1^2 = (v_1')^2 + \frac{m_2}{m_1}(v_2')^2$$

Substitute for v_1':

$$v_1^2 = v_1^2 - 2\frac{m_2}{m_1}v_1v_2' + \frac{m_2^2}{m_1^2}(v_2')^2 + \frac{m_2}{m_1}(v_2')^2$$

And solve for v_2':

$$v_2' = \frac{2v_1}{1 + \dfrac{m_2}{m_1}} = \frac{2(0.4\ \text{m/s})}{1 + \dfrac{4\ \text{kg}}{3\ \text{kg}}} = 0.343\ \text{m/s}$$

The 4-kg puck moves in the positive x direction at 34.3 cm/s. Now, you can solve for the speed of the 3-kg puck:

$$v_1' = v_1 - \frac{m_2}{m_1}v_2' = 0.4\ \text{m/s} - \frac{4\ \text{kg}}{3\ \text{kg}}(0.343\ \text{m/s}) = -5.73 \times 10^{-2}\ \text{m/s}$$

The 3-kg puck moves in the negative x direction at 5.73 cm/s.

EXAMPLE 7-6: A 3-kg puck slides on a frictionless surface at 40 cm/s and strikes a 4-kg puck at rest. The first puck moves off at 30 cm/s at an angle of 35° from the incident direction. Find (**a**) the angle at which the 4-kg puck moves off and (**b**) the speed of the 4-kg puck after the impact.

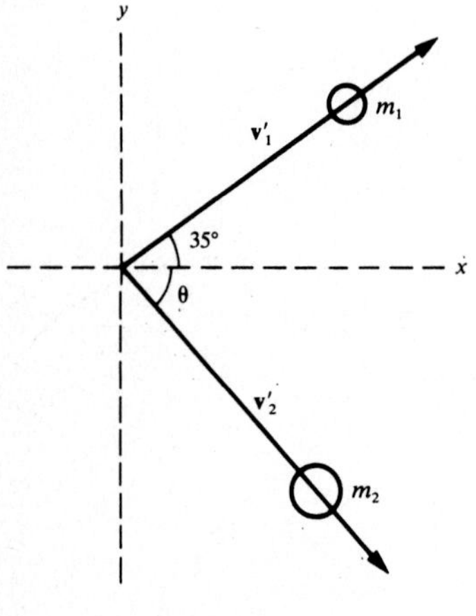

Figure 7-4

Solution:

(**a**) Draw a diagram of the collision (see Fig. 7-4), and apply the principle of the conservation of momentum to the x and y components of $\mathbf{v}_2$:

x component

$$m_1v_1 + 0 = m_1v_1' \cos 35° + m_2v_2' \cos\theta$$

$$v_2' \cos\theta = \frac{m_1v_1 - m_1v_1' \cos 35°}{m_2}$$

$$= \frac{(3\ \text{kg})(0.4\ \text{m/s}) - (3\ \text{kg})(0.3\ \text{m/s})\cos 35°}{4\ \text{kg}} = 0.116\ \text{m/s}$$

y component

$$0 = m_1v_1' \sin 35° - m_2v_2' \sin\theta$$

$$v_2' \sin\theta = \frac{m_1v_1' \sin 35°}{m_2} = \frac{(3\ \text{kg})(0.3\ \text{m/s})\sin 35°}{4\ \text{kg}} = 0.129\ \text{m/s}$$

Now, obtain the value of θ from the tangent function:

$$\tan\theta = \frac{v_2' \sin\theta}{v_2' \cos\theta} = \frac{0.129\ \text{m/s}}{0.116\ \text{m/s}} = 1.12$$

$$\theta = \text{arc}\tan(1.12) = 48.2°$$

(**b**) Then you can find v_2—the speed after impact:

$$v_2' = \frac{0.129\ \text{m/s}}{\sin 48.2°} = 0.173\ \text{m/s}$$

B. Conservation of angular momentum

The angular momentum **L** of an object rotating about a fixed axis is given by $\mathbf{L} = I\omega$, where I is the body's moment of inertia about the axis of rotation. [**recall:** The angular form of Newton's second law is $\Sigma\tau = I\alpha$ (Eq. 4-16).] Since $\Delta\omega$ represents the change in the magnitude of the angular velocity, you can write

$$I\,\Delta\omega = I(\omega_2 - \omega_1) = I\omega_2 - I\omega_1$$
$$= L_2 - L_1 = \Delta L$$

You can also express Newton's second law in terms of the angular momentum's rate of change:

NEWTON'S SECOND LAW (ANGULAR FORM)

$$\Sigma\tau = \lim_{\Delta t \to 0} \frac{\Delta \mathbf{L}}{\Delta t} \qquad \textbf{(7-10)}$$

Since the limit of $\Delta\mathbf{L}/\Delta t$ as t approaches zero is the derivative of **L** with respect to t, you can express the second law in calculus notation:

$$\Sigma\tau = \frac{d\mathbf{L}}{dt} \qquad \textbf{(7-11)}$$

Only an external torque can cause the angular momentum to change, so the angular momentum will remain constant if the net torque acting on the system is zero. This result leads directly to the

- **Law of conservation of angular momentum:** If the net torque acting on a system is zero, the total angular momentum of the system will remain constant.

EXAMPLE 7-7: A record turntable with a mass of 800 g and a diameter of 35 cm accelerates from rest to 33.3 rpm in 3 seconds.
(**a**) What is the change in angular momentum of the turntable, and (**b**) what torque is necessary to accelerate the disk?

Solution:
(**a**) The turntable is a disk so its angular momentum is (see Sec. 6-2B and Eq. 4-17b)

$$\mathbf{L} = I\omega = \frac{1}{2}mr^2$$

$$\Delta\mathbf{L} = \mathbf{L}_f - \mathbf{L}_i = \frac{1}{2}mr^2\omega - 0$$

$$= \frac{1}{2}(0.800 \text{ kg})(0.125 \text{ m})^2\left(\frac{33.3 \text{ rev}}{1 \text{ min}}\right)\left(\frac{1 \text{ min}}{60 \text{ sec}}\right)\left(\frac{2\pi \text{ rad}}{1 \text{ rev}}\right)$$

$$= 2.18 \times 10^{-2} \text{ kg m}^2/\text{s}$$

(**b**) The torque is constant to give a smooth acceleration, so from Eq. (7-11),

$$\tau = \frac{\Delta\mathbf{L}}{\Delta t} = \frac{2.18 \times 10^{-2} \text{ kg m}^2/\text{s}}{3 \text{ s}} = 7.27 \times 10^{-3} \text{ kg m}^2/\text{s}^2$$

EXAMPLE 7-8: A record, originally at rest, drops onto a turntable rotating at a speed of 33.3 rev/min. The turntable, disconnected from the driving motor, rotates about the center spindle without friction, as illustrated in Figure 7-5. The turntable has a mass of 0.8 kg and a radius of 12 cm. The record has a mass of 0.2 kg and a radius of 15 cm. After the record drops onto the rotating turntable, the two disks rotate at a common speed.

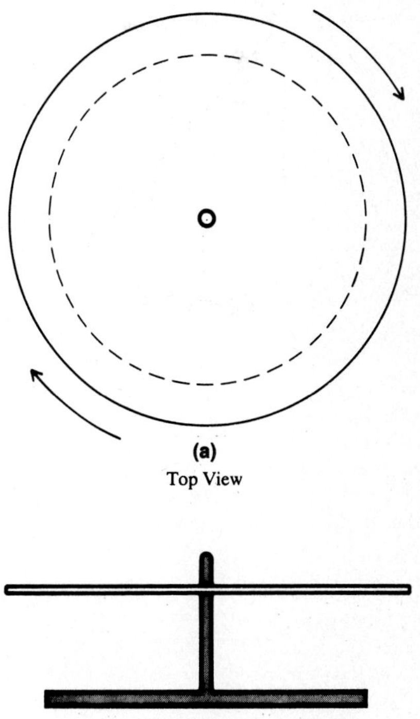

(a) Top View

(b) Side View

Figure 7-5

Calculate
(**a**) the final angular speed of the two disks and
(**b**) the energy loss of this "collision."

Solution:

(**a**) Since there is no external torque acting on the system, you know that its total angular momentum does not change because of the collision—the angular momentum before the collision is equal to the angular momentum after the collision. [Hint: Let ω represent the angular velocity before the collision and ω' represent angular velocity after the collision.]

$$I_1\omega = I_1\omega' + I_2\omega'$$
$$= \omega'(I_1 + I_2)$$

The moment of inertia of a solid disk (Eq. 4-17b) is

$$I = \frac{1}{2}mr^2$$

so that the moments of inertia of the two disks are

$$I_1 = \frac{1}{2}(0.8\text{ kg})(0.12\text{ m})^2 = 5.76 \times 10^{-3}\text{ kg m}^2$$

$$I_2 = \frac{1}{2}(0.2\text{ kg})(0.15\text{ m})^2 = 2.25 \times 10^{-3}\text{ kg m}^2$$

Now, you can calculate the final angular speed:

$$\omega' = \omega\frac{I_1}{I_1 + I_2}$$
$$= (33.3\text{ rev/min})\frac{5.76 \times 10^{-3}\text{ kg m}^2}{(5.76 + 2.25) \times 10^{-3}\text{ kg m}^2}$$
$$= 23.9\text{ rev/min}$$

(**b**) The initial kinetic energy due to the rotation of the turntable is

$$\text{KE} = \frac{1}{2}I_1\omega^2$$
$$= \frac{1}{2}(5.76 \times 10^{-3}\text{ kg m}^2)[(33.3\text{ rev/min})(2\pi\text{ rad/rev})/(1\text{ min}/60\text{ s})]^2$$
$$= 3.50 \times 10^{-2}\text{ J}$$

After the record has fallen onto the turntable, the kinetic energy is

$$\text{KE}' = \frac{1}{2}(I_1 + I_2)(\omega')^2$$
$$= \frac{1}{2}(8.01 \times 10^{-3}\text{ kg m}^2)[(23.9\text{ rev/min})(2\pi\text{ rad/rev})/(1\text{ min}/60\text{ s})]^2$$
$$= 2.51 \times 10^{-2}\text{ J}$$

The collision's energy loss is

$$\text{KE} - \text{KE}' = 0.99 \times 10^{-2}\text{ J}$$

SUMMARY

Linear momentum	$\mathbf{p} = m\mathbf{v}$	gives linear momentum as the product of an object's mass and velocity
Angular momentum	$\mathbf{L} = \mathbf{r} \times \mathbf{p}$ or $L = rp \sin\theta$	gives angular momentum as the cross product of an object's position vector and linear momentum
Angular momentum of a rigid body	$\mathbf{L} = I\omega$	gives angular momentum as the product of the body's moment of inertia and its angular speed
Newton's second law	$\Sigma\mathbf{F} = \lim_{\Delta t \to 0} \frac{\Delta\mathbf{p}}{\Delta t}$	expressed as the rate of change of linear momentum
Conservation of linear momentum	$m_1v_1 + m_2v_2 = m_1v_1' + m_2v_2'$	expressed in terms of the constant linear momentum of two colliding objects
Newton's second law (angular form)	$\Sigma\tau = \lim_{\Delta t \to 0} \frac{\Delta\mathbf{L}}{\Delta t}$	expressed as the rate of change of angular momentum

RAISE YOUR GRADES

Can you define . . . ?

- ☑ linear momentum
- ☑ angular momentum of a moving point mass
- ☑ angular momentum of a rotating rigid body

Do you know . . . ?

- ☑ the relationship between net force and rate of change of linear momentum
- ☑ the conditions under which the total linear momentum of a system will be conserved
- ☑ the conditions under which the total energy of a system will be conserved
- ☑ the relationship between net torque and the rate of change of the angular momentum of a system
- ☑ the difference between an elastic collision and an inelastic collision

SOLVED PROBLEMS

PROBLEM 7-1 A 400-g racquetball is dropped onto an inclined concrete ramp from a height of 1.837 m (see Fig. 7-6). The ball bounces off the ramp horizontally at a speed of 5 m/s. During the brief period of impact with the ramp, a force acts on the ball and causes its momentum to change. Calculate this change in the ball's momentum.

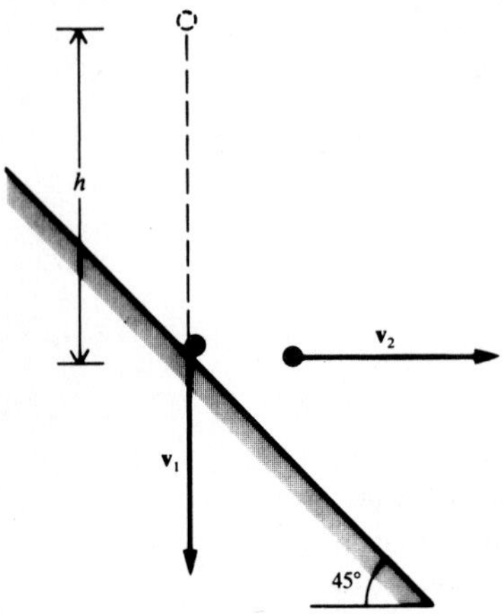

Figure 7-6

Solution Before you can figure the ball's momentum, you need to know its speed before it hits the ramp. Make the impact point the reference level for the potential energy, and use the principle of energy conservation to find the speed of the ball before impact. As the ball falls vertically, its total energy remains constant:

$$E \text{ (at release)} = E \text{ (just before impact)}$$

$$mgh = \frac{1}{2}mv^2$$

$$v^2 = 2gh$$

$$v = \sqrt{2gh} = \sqrt{2(9.8 \text{ m/s}^2)(1.837 \text{ m})}$$

$$= 6.00 \text{ m/s}$$

Now, from Eq. (7-1), calculate the magnitude of the momentum before impact:

$$p_1 = mv_1 = (0.40 \text{ kg})(6.0 \text{ m/s}) = 2.4 \text{ kg m/s}$$

The magnitude of the momentum after impact is

$$p_2 = mv_2 = (0.40 \text{ kg})(5.0 \text{ m/s}) = 2.0 \text{ kg m/s}$$

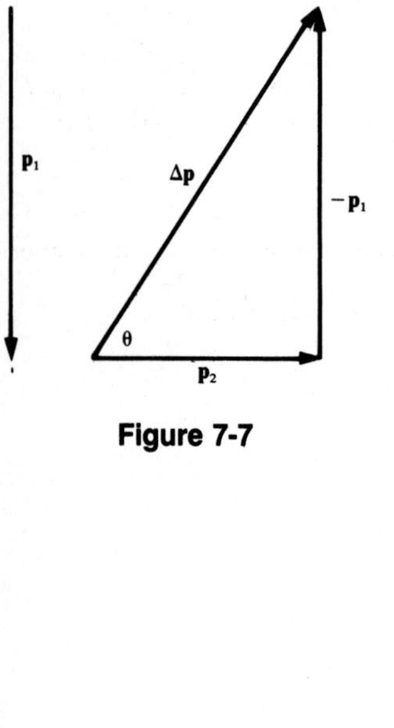

Figure 7-7

Now, draw the two momentum vectors and calculate the change in momentum by vector subtraction (see Fig. 7-7). **[recall:** To solve a problem in vector subtraction, change it into an addition problem. (Sec. 1-4)]

$$\Delta\mathbf{p} = \mathbf{p}_2 - \mathbf{p}_1 = \mathbf{p}_2 + (-\mathbf{p}_1)$$

Since the vectors to be added, $\mathbf{p}_2$ and $-\mathbf{p}_1$, are at right angles, you can find the magnitude of $\Delta\mathbf{p}$ using the Pythagorean theorem:

$$\Delta p = \sqrt{p_1^2 + p_2^2}$$

$$= \sqrt{(2.4 \text{ kg m/s})^2 + (2.0 \text{ kg m/s})^2} = 3.12 \text{ kg m/s}$$

and the angle of the change in momentum from the tangent function:

$$\tan\theta = \frac{p_1}{p_2} = \frac{2.4 \text{ kg m/s}}{2.0 \text{ kg m/s}} = 1.2$$

$$\theta = \arctan(1.2) = 50.2°$$

PROBLEM 7-2 A ballistic pendulum experiment is illustrated in Fig. 7-8. A steel ball of mass 40 g traveling at an unknown speed collides inelastically with a clay block of mass 160 g. The ball is trapped inside the block and the two swing together to a maximum height h. If $h = 5$ cm, determine the speed of the ball before it strikes the block.

Solution You have to divide this problem into two phases—first, the inelastic collision during which momentum is conserved, but energy is not; and second, the block swinging along the arc of a circle to a height h during which energy is conserved, but momentum is not. Start with the inelastic collision:

$$\begin{array}{ccc} \text{Before} & & \text{After} \\ mv + 0 & = & (m + M)v' \end{array}$$

$$v = \frac{m + M}{m} v' = \frac{40\text{ g} + 160\text{ g}}{40\text{ g}} v' = 5v'$$

Immediately after the collision, the block with the embedded ball has kinetic energy. When the block and ball come to rest at height h, their energy is entirely potential. Applying the principle of energy conservation,

$$\frac{1}{2}(m + M)(v')^2 = (m + M)gh$$

$$(v')^2 = 2gh$$

$$v' = \sqrt{2gh}$$

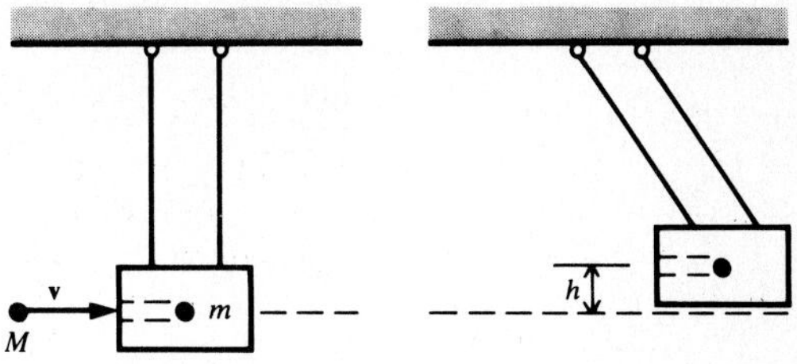

(a) (b)

Figure 7-8

Solve algebraically for the initial speed of the ball:

$$v = 5\sqrt{2gh}$$

Now, substitute the known data to obtain the initial speed of the ball:

$$v = 5\sqrt{2(9.8\text{ m/s}^2)(.05\text{ m})} = 4.95\text{ m/s}$$

PROBLEM 7-3 Two hockey pucks of the same mass (0.5 kg) move without friction on a horizontal surface. One puck moves at a speed of 5.70 m/s along the x axis before colliding with the second puck, which is initially at rest. Fig. 7-9 shows the situation before and after the collision. Because the hockey pucks are not perfectly elastic, there is an energy loss as a result of the collision. Find (**a**) the speed of each puck after the collision and (**b**) the energy loss.

Solution

(**a**) Since the total momentum is a vector quantity that does not change during the collision, you know that the x and y components of **p** are unchanged by the collision. The masses are equal, so they cancel out, thus

	Before collision	After collision
x component:	v_1	$= v_1' \cos 41.8° + v_2' \cos 30°$
y component:	0	$= v_1' \sin 41.8° - v_2' \sin 30°$

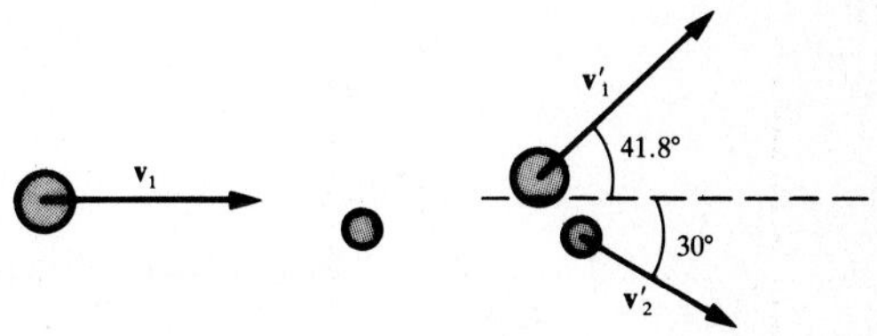

(a) Before collision (b) After collision

Figure 7-9

You can use the second equation to find the relationship between v_2' and v_1':

$$v_1' \sin 41.8° = v_2' \sin 30° \qquad v_2' = \frac{\sin 41.8°}{\sin 30°} v_1' = \frac{0.666}{0.5} v_1' = 1.333 v_1'$$

Now, use this result to eliminate v_2' from the equation for the x components:

$$v_1 = v_1' \cos 41.8° + 1.333 v_1' \cos 30° = v_1'(0.745 + 1.155)$$

The initial speed v_1 is given, so

$$v_1' = \frac{v_1}{1.90} = \frac{5.70\text{ m/s}}{1.90} = 3.00\text{ m/s}$$

Now, you can find the speed of the other puck:

$$v_2' = 1.333 v_1' = (1.333)(3\text{ m/s}) = 4\text{ m/s}$$

(**b**) The energy loss is

$$\Delta KE = KE\text{ (before)} - KE'\text{ (after)}$$

$$= \frac{1}{2} m_1 v_1^2 - \frac{1}{2} m_1 (v_1')^2 - \frac{1}{2} m_2 (v_2')^2$$

$$= \frac{1}{2}(0.5\text{ kg})(5.70\text{ m/s})^2 - \frac{1}{2}(0.5\text{ kg})(3\text{ m/s})^2 - \frac{1}{2}(0.5\text{ kg})(4\text{ m/s})^2$$

$$= (8.12 - 2.25 - 4.00)\text{ J} = 1.87\text{ J}$$

PROBLEM 7-4 While a truck weighing 36 000 N is parked on a hill, its brakes fail and it crashes into a car weighing 9000 N parked at the bottom of the hill. The two vehicles lock together during the collision and move a distance of 48 m before being brought to rest by the locked wheels of the car. The truck travels 46.4 m down the hill, which is elevated 15° above the horizontal (see Fig. 7-10). What is the coefficient of kinetic friction between the tires of the car and the road?

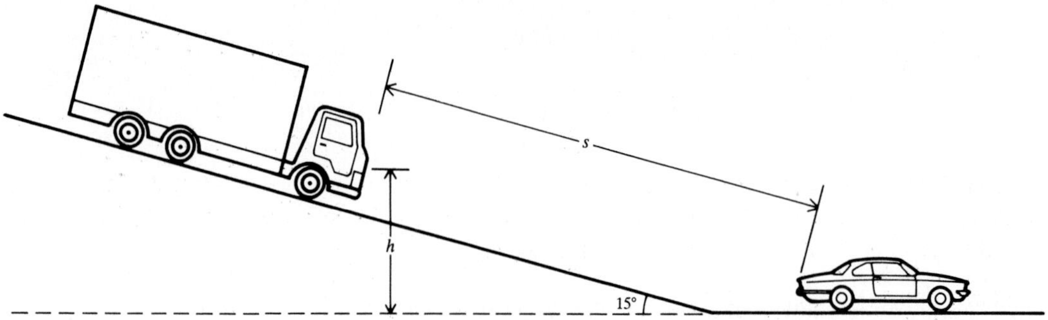

Figure 7-10

Solution You want to find the force of kinetic friction, $\mathbf{f}_k = \mu_k \mathbf{N}$ (Eq. 3-6). To do this, you'll have to figure the energy associated with 1) the truck before the collision; 2) the car and truck during the collision; and 3) the car and truck after the collision. Then, you can find the frictional force and finally the coefficient of kinetic friction. First, find the speed of the truck (v_t) just prior to the collision. Energy is conserved during this phase of the drama, so

$$\text{E (at bottom)} = \text{E (at top)}$$

$$\frac{1}{2}mv_t^2 = mgh$$

You can find the vertical distance h by means of trigonometry:

$$\sin\theta = \frac{h}{s}$$

$$h = s\sin\theta = (46.4 \text{ m})\sin 15° = 12.0 \text{ m}$$

Now, find the speed of the truck before the collision:

$$v_t^2 = 2gh$$

$$v_t = \sqrt{2gh} = \sqrt{2(9.8 \text{ m/s}^2)(12.0 \text{ m})} = 15.34 \text{ m/s}$$

The second phase of this situation is the inelastic collision in which momentum is conserved:

Before After

$$m_t v_t = (m_t + m_c)v'$$

$$v' = \left(\frac{m_t g}{m_t g + m_c g}\right)v_t$$

$$= \left(\frac{36\,000 \text{ N}}{36\,000 \text{ N} + 9000 \text{ N}}\right)(15.34 \text{ m/s}) = 12.27 \text{ m/s}$$

And now, you'll need to calculate the kinetic energy of the truck and car immediately after the collision:

$$\text{KE} = \frac{1}{2}(m_t + m_c)(v')^2 = \frac{1}{2}\left(\frac{45\,000 \text{ N}}{9.8 \text{ m/s}^2}\right)(12.27 \text{ m/s})^2 = 3.46 \times 10^5 \text{ J}$$

This kinetic energy is transformed into heat as the car slides distance s' along the road with its wheels locked. The work done against the frictional force is equal to the kinetic energy of the car and truck immediately after the collision, so

$$W = f_k s' = \text{KE}$$

The frictional force is

$$f_k = \mu_k N = \mu_k w_c$$

where w_c is the weight of the car. Substituting this into the previous equation, you find that

$$\mu_k w_c s' = \text{KE}$$

Now, solve for the coefficient of friction:

$$\mu_k = \frac{\text{KE}}{w_c s'} = \frac{3.46 \times 10^5 \text{ J}}{(9 \times 10^3 \text{ N})(48 \text{ m})} = 0.801$$

PROBLEM 7-5 A block of mass m and speed v_1 makes an elastic collision with an identical block, initially at rest. The collision is one-dimensional and takes place on a frictionless air track. Find the velocities of the two blocks after the collision.

Solution Because the momentum is conserved, you have

$$mv_1 + 0 = mv_1' + mv_2'$$

The collision is elastic, so energy is also conserved:

$$\frac{1}{2}mv_1^2 + 0 = \frac{1}{2}m(v_1')^2 + \frac{1}{2}m(v_2')^2$$

You can solve the momentum equation for v_1':

$$v_1' = v_1 - v_2'$$

Now, eliminate v_1' in the energy equation:

$$v_1^2 = (v_1 - v_2')^2 + (v_2')^2$$

$$v_1^2 = v_1^2 - 2v_1v_2' + (v_2')^2 + (v_2')^2$$

$$0 = -2v_1v_2' + 2(v_2')^2$$

Now divide by $2v_2'$ to obtain $v_2' = v_1$. You can find v_1' from the momentum equation:

$$v_1' = v_1 - v_2' = v_1 - v_1 = 0$$

You've found that the moving block is brought to rest by the collision, while the block of equal mass, which was originally at rest, moves toward the right with the same speed as the block which struck it. In this special situation, the energy of the moving block is transferred completely to the second block.

PROBLEM 7-6 A satellite goes around the earth in a circular orbit. (**a**) Is its linear momentum conserved? (**b**) Is its angular momentum conserved?

Solution

(**a**) In a circular orbit, the satellite's speed is constant, but the gravitational attraction of the earth, and outside force, is constantly changing the *direction* of the satellite's velocity. Therefore, linear momentum, a vector quantity, is not conserved, even though the *magnitude* of the satellite's linear momentum remains the same.

(**b**) The satellite's angular momentum **L** is the cross product of its position and linear momentum vectors. Since the satellite remains at a constant distance from the center of its orbit and moves at constant speed, the *magnitude* of **L** remains constant. Since the position vector and the velocity vector lie in the same plane, the *direction* of **L** is constant: always perpendicular to the plane of the orbit. And since the magnitude and direction of **L** both remain constant, angular momentum is conserved.

PROBLEM 7-7 At rest, a uranium atom ^{238}U decays into an alpha particle and a thorium atom ^{234}Th. In the process 8.64×10^{-13} J of energy is liberated. The mass of an alpha particle is 6.68×10^{-27} kg, and the mass of thorium is 3.91×10^{-25} kg. What is the momentum of the alpha particle after the decay?

Solution Radioactive decay is like an inelastic collision run backward; the particles begin stuck together and end by flying apart. The uranium starts at rest, so the initial momentum is zero. The particles fly apart with equal and opposite momenta. From Eq. (7-1),

$$p_{\text{total}} = m_\alpha v_\alpha + m_{\text{Th}} v_{\text{Th}} \quad \text{or} \quad m_\alpha v_\alpha = -m_{\text{Th}} v_{\text{Th}} = p_{\text{f}}$$

The initial kinetic energy is zero. The liberated energy is potential energy before the decay, so from Eq. (6-6),

$$\text{KE} + \text{PE} = \text{KE}_\alpha + \text{KE}_{\text{Th}}$$

$$0 + 8.64 \times 10^{-13}\ \text{J} = \frac{1}{2} m_\alpha v_\alpha^2 + \frac{1}{2} m_{\text{Th}} v_{\text{Th}}^2$$

$$= \frac{1}{2}\left(\frac{(m_\alpha v_\alpha)^2}{m_\alpha}\right) + \frac{1}{2}\left(\frac{(m_{\text{Th}} v_{\text{Th}})^2}{m_{\text{Th}}}\right)$$

$$= \frac{1}{2}\left(\frac{(p_{\text{f}})^2}{m_\alpha}\right) + \frac{1}{2}\left(\frac{(-p_{\text{f}})^2}{m_{\text{Th}}}\right)$$

$$= \frac{(p_{\text{f}})^2}{2}\left(\frac{1}{m_\alpha} + \frac{1}{m_{\text{Th}}}\right)$$

$$(p_{\text{f}})^2 = \frac{2(8.64 \times 10^{-13}\ \text{J})}{\left(\dfrac{1}{6.68 \times 10^{-27}\ \text{kg}} + \dfrac{1}{3.91 \times 10^{-25}\ \text{kg}}\right)}$$

$$= 1.13 \times 10^{-38}\ \text{kg}^2\ \text{m}^2/\text{s}^2$$

$$p_{\text{f}} = \sqrt{1.13 \times 10^{-38}\ \text{kg}^2\ \text{m}^2/\text{s}^2} = 1.07 \times 10^{-19}\ \text{kg m/s}$$

PROBLEM 7-8 A 300-g mass is attached to a string that goes through a small hole in a table top, as shown in Fig. 7-11. The distance from the mass to the hole is initially 40.0 cm and the mass rolls around the table in a circle with an angular velocity $\omega = 20.0$ rad/s. A physics lab assistant pulls the string down through the hole until the radius is reduced to 16.0 cm. How do (**a**) the angular velocity and (**b**) the energy of the mass change?

Figure 7-11

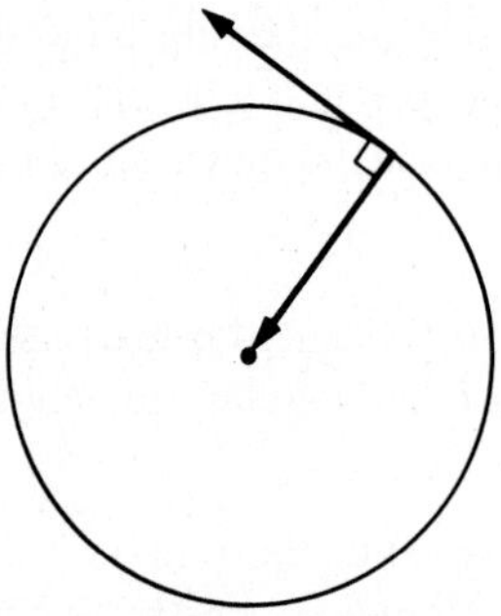

Figure 7-12

Solution

(**a**) The string exerts a force on the mass to keep it going in a circle and another force is needed to pull the mass in. This second force does *not* exert a torque on the mass because its direction is along the radius of the circle, at right angles to the momentum, which is tangent to the circle (see Fig. 7-12).

$$\tau = \mathbf{r} \times \mathbf{F} = rF \sin 0° = 0$$

Since there is no torque, angular momentum is conserved. From Eq. (7-3):

$$\mathbf{L} = I_1\omega_1 = I_2\omega_2 = mr_1^2\omega_1 = mr_2^2\omega_2$$

$$\omega_2 = \left(\frac{r_1}{r_2}\right)^2 \omega_1 = \left(\frac{0.400 \text{ m}}{0.160 \text{ m}}\right)^2 20 \text{ rad/s} = 125 \text{ rad/s}$$

So, the angular velocity increases.
(**b**) From Eq. (6-5b),

$$\text{KE}_1 = \frac{1}{2}I_1\omega_1^2 = \frac{1}{2}mr_1^2\omega_1^2 = \frac{1}{2}(0.300 \text{ kg})(0.400 \text{ m})^2(20 \text{ rad/s})^2 = 9.60 \text{ J}$$

$$\text{KE}_2 = \frac{1}{2}mr_2\omega_2^2 = \frac{1}{2}(0.300 \text{ kg})(0.160 \text{ m})^2(125 \text{ rad/s})^2 = 60.0 \text{ J}$$

As you can see, the kinetic energy is not conserved. The additional 50.4 J of energy comes from the work done in pulling the mass toward the center of the circle.

Supplementary Exercises

PROBLEM 7-9 What is the magnitude of the momentum of a 30-g baseball that has a speed of 40 m/s?

PROBLEM 7-10 A hard rubber ball of mass 50 g is dropped on a sidewalk from a height of 4 m. It bounces back to a height of 3 m. Calculate the magnitude of the change in momentum that results from the collision with the sidewalk.

PROBLEM 7-11 An ice skater starts spinning on one foot with her arms extended horizontally. When she pulls her arms in close to her body, the rate at which she is spinning increases from 0.5 rev/s to 0.9 rev/s. By what factor did the skater's moment of inertia change?

PROBLEM 7-12 A 100-g block moving along a frictionless horizontal surface collides with a 400-g block, initially at rest. The collision is inelastic so that the two blocks stick together and move as a unit after the collision. Assuming that the collision is one dimensional and that the 100-g block had a speed of 0.2 m/s before the collision, calculate the common speed of the two blocks after the collision.

Answers to Supplementary Exercises

7-9: 1.20 kg m/s

7-10: 0.826 kg m/s

7-11: *I* decreased by 5/9

7-12: 0.04 m/s

MIDSEMESTER EXAM (Chapters 1–7)

1. Two horizontal forces are being applied to a 6-kg object. One force has a magnitude of 10 N and is directed toward the north. The second force has a magnitude of 15 N and is directed toward the northwest. The angle between the two forces is 45°.

(**a**) Calculate the magnitude of the resultant force, R. **[Ch. 1]**
(**b**) Calculate the magnitude of the acceleration of the 6-kg object. **[Ch. 3]**
(**c**) Find the angle between the acceleration and the north–south axis. **[Ch. 1]**

2. An automobile is traveling due east at a speed of 90 km/h when it enters a slight curve. Two seconds later it is moving 10° north of east at the same speed.

(**a**) Calculate the magnitude of $\Delta \mathbf{v}$, the change of velocity, in km/h. **[Ch. 1]**
(**b**) Calculate the magnitude of the average acceleration that has taken place during the 2-second interval in m/s^2. **[Ch. 2]**
(**c**) Find the direction of the acceleration with respect to the north–south axis. **[Ch. 1]**

3. An automobile traveling at 20 m/s is approaching a traffic signal. When the driver is 150 m from the intersection, the light changes from green to yellow. Since the light will remain yellow for 5 s, the driver accelerates just enough to cross the intersection while the light is still yellow. **[Ch. 2]**

(**a**) Calculate the minimum acceleration, assumed to be constant.
(**b**) What is the speed of the automobile as its crosses the intersection?

4. The speed of a racing car increases from 30 m/s to 80 m/s while it travels 400 m. Calculate the acceleration, assumed to be constant. **[Ch. 2]**

5. A stone is thrown vertically upward. It reaches a maximum height of 18 m. **[Ch. 2]**

(**a**) What is its initial speed?
(**b**) How long does the stone take to reach its maximum height?

6. An arrow is shot vertically upward at an initial speed of 24 m/s. How many seconds will the arrow take to reach a speed of 12 m/s downward? **[Ch. 2]**

7. An arrow is shot into the air at an angle of 40° with respect to the horizontal. It reaches a maximum height of 8.432 m. **[Ch. 2]**

(**a**) What is the initial speed of the arrow?
(**b**) What is the speed of the arrow when it reaches its maximum height?

8. An arrow is shot upward at an angle of 50° with respect to the horizontal x axis. It leaves the bow at a height of 2 m above the level ground. Its initial speed is 16 m/s. **[Ch. 2]**

(**a**) What is the maximum height of the arrow measured from the ground?
(**b**) How long does it take for the arrow to return to the ground? [*Hint:* the quadratic formula is $x = (-b \pm \sqrt{b^2 - 4ac})/2a$].
(**c**) What is the horizontal distance traveled by the arrow when it reaches the ground?

9. An object with a mass of 1.5 kg is moving around a circle at a constant speed. It completes 5 revolutions in 2.5 s. The radius of the circle is 0.70 m. **[Ch. 4]**

(**a**) What is the angular speed of this object?
(**b**) What is the magnitude of the acceleration of this object?

10. When a 900-N box is dragged up a 20° ramp without the use of rollers under the box, a force of 500 N parallel to the ramp is required to keep the box moving at a constant speed.

(**a**) What is the coefficient of sliding friction between the box and the ramp? **[Ch. 5]**
(**b**) What is the efficiency of this inclined plane when friction is present? **[Ch. 6]**

11. Three masses are connected by two strings that pass over frictionless pulleys as illustrated in Fig. E-1. The coefficient of friction between the 400-g block and the horizontal surface is 0.30. **[Ch. 3]**

(a) What is the frictional force that opposes the movement of the 400-g block?
(b) What is the acceleration of the system?
(c) Calculate the tension in the string that is fastened to the 600-g block.

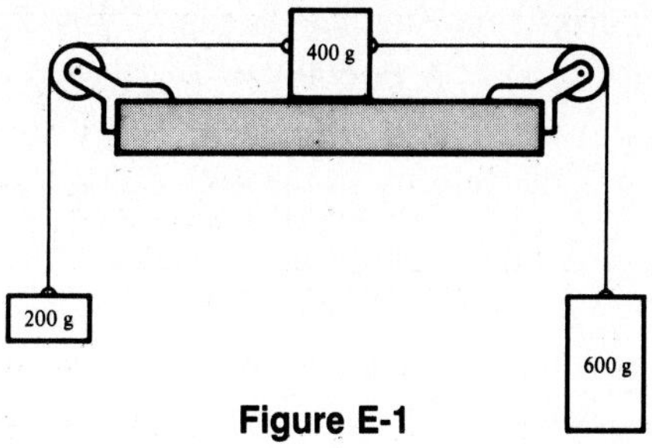

Figure E-1

12. An Atwood machine consists of two slightly different masses connected by a string that passes over a frictionless pulley. The masses are 208 g and 200 g.

(a) What is the acceleration of this system? **[Ch. 3]**
(b) How long does the 208-g mass take to move downward 150 cm if it starts from rest? **[Ch. 2]**

13. When the turntable of a record player is turned on, it reaches its correct speed of 33.3 rev/min in 1.5 s. **[Ch. 4]**

(a) What is the average angular acceleration of the turntable during this 1.5-s interval?
(b) If the angular acceleration is constant, what is the turntable's angular speed 1.0 s after the record player is turned on?
(c) What is the tangential acceleration of a point on the turntable that is 12 cm from the axis of rotation?
(d) What is the magnitude of the total acceleration of this point 1.0 s after the turntable has been turned on?

14. The curve of a race track has a radius of 600 m. If most racing cars take this curve at 240 km/h (approximately 150 mi/h), at what angle should the curve be banked? **[Ch. 4]**

15. A 950-N traffic signal is suspended from a cable that crosses an intersection. The cable on the right makes an angle of 15° with respect to the horizontal. The cable on the left makes an angle of 20° with respect to the horizontal. **[Ch. 5]**

(a) Calculate the ratio of the tensions in the two parts of the cable, T_1/T_2. The tension in the cable on the right is T_1.
(b) Calculate the tension T_2 in the cable on the left.

16. A steel beam weighing 400 N is held in a horizontal position by means of a pivot at the right end and a cable fastened to the left end. The cable makes an angle of 10° with respect to the horizontal beam. What is the tension in the cable? **[Ch. 5]**

17. A 400-g block of steel is being pulled up a 20° inclined plane. The coefficient of friction between the block and the inclined surface is 0.25. The force pulling the block is parallel to the inclined surface. What force will cause the block to move up the inclined plane with an acceleration of 2.0 m/s^2? **[Ch. 3]**

18. Suppose that the force found in the preceding problem is provided by a mass connected to a string passing over a frictionless pulley as illustrated in Fig. E-2. How large must the mass be to cause the acceleration of 2.0 m/s^2? **[Ch. 3]**

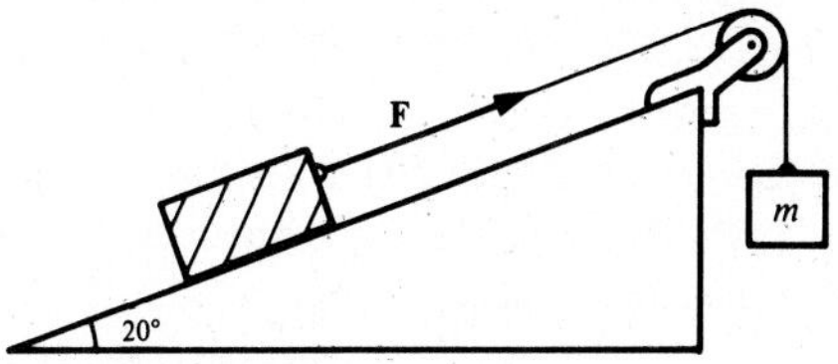

Figure E-2

19. A box weighing 900 N is being pulled up a ramp that makes an angle of 20° with the horizontal. Rollers under the box reduce friction practically to zero.

(a) How much work is required to move the box 4 m along the ramp? **[Ch. 6]**
(b) The box is pushed up the ramp at a constant speed by a force parallel to the ramp. What is the magnitude of this force? **[Ch. 5]**

20. A coal hopper weighing 2.0×10^4 N is being pulled up a track that makes an angle of 25° with respect to the horizontal. What power is required to pull the coal hopper at a constant speed of 5.30 m/s in the absence of friction? **[Ch. 6]**

21. A 400-g disk is moving due east on a frictionless air table at a speed of 4 m/s. This disk collides head-on with an 800-g disk that is at rest. The collision is perfectly elastic. **[Ch. 7]**

(**a**) What is the velocity of the 800-g disk after the collision?
(**b**) What is the velocity of the 400-g disk after the collision?

22. A 250-g disk moving north at a speed of 8 m/s on a frictionless, horizontal air table collides with a 400-g disk moving east at a speed of 5 m/s. **[Ch. 7]**

(**a**) What is the magnitude of the total momentum after the collision?
(**b**) What is the total kinetic energy after the collision if the collision is perfectly elastic?

23. A crane lifts a ball weighing 2×10^4 N by 40 m in 2 s. **[Ch. 6]**

(**a**) How much work against the force of gravity does the crane do on the ball?
(**b**) How much power is necessary to perform this work?

24. Imagine that you are one of a crew of Egyptian slaves laboring to put the last stone (which weighs 1.8×10^5 N) on top of a pyramid 30 m tall. The crew moves the stone by pushing it up a plane inclined at an angle of 10°, and must apply a force of 1.55×10^5 N to keep the stone moving at a constant speed. The coefficient of kinetic friction between the stone and the plane is 0.7. **[Ch. 6]**

(**a**) What is the ideal mechanical advantage?
(**b**) What is the actual mechanical advantage?
(**c**) What is the efficiency of the inclined plane?

25. A carpenter who weight 800 N leans a 20-m ladder that weighs 180 N against a wall at an angle of 60° with respect to the ground. The contact between the ladder and the wall is frictionless, so only the force of static friction between the ladder and the ground prevents the ladder from slipping. The carpenter, while carrying a 20-N hammer, can climb exactly 15 m along the ladder before the ladder starts to slip.

(**a**) What is the coefficient of static friction between the ladder and the ground? **[Ch. 5]**
(**b**) Suppose that the carpenter travels only 12 m along the ladder before he drops the hammer. What is the carpenter's potential energy at that height? **[Ch. 6]**
(**c**) How long does the hammer take to fall to the ground? **[Ch. 2]**
(**d**) What is the hammer's speed just before it strikes the ground? **[Ch. 2]**
(**e**) What is the hammer's momentum just before it strikes the ground? **[Ch. 7]**
(**f**) The hammer falls into the mud under the ladder in a perfectly inelastic collision. What is the hammer's change of momentum? **[Ch. 7]**
(**g**) If the collision between the hammer and the mud lasts for 0.01 s, what average force does the hammer experience? **[Ch. 7]**
(**h**) What average acceleration does the hammer experience? **[Ch. 3]**

Solutions to Midsemester Exam

1. (**a**) Use the law of cosines. The angle opposite the unknown side of the triangle whose two known sides are the two vectors placed tail-to-head is 180° − 45° = 135°.

$$R = \sqrt{(10\text{ N})^2 + (15\text{ N})^2 - 2(10\text{ N})(15\text{ N})\cos 135^\circ} = 23.18\text{ N}$$

(**b**) Use Newton's second law, $F = ma$.

$$a = \frac{F}{m} = \frac{23.18\text{ N}}{6\text{ kg}} = 3.86\text{ m/s}^2$$

(**c**) Use the law of sines.

$$\frac{\sin\alpha}{a} = \frac{\sin\beta}{b} \qquad \alpha = \text{arc sin}\left(\frac{a\sin\beta}{b}\right) = \text{arc sin}\left(\frac{15\text{ N}\sin 135^\circ}{23.18\text{ N}}\right) = 27.2^\circ$$

2. (**a**) Use the vector component method and $\Delta\mathbf{v} = \mathbf{v}_f - \mathbf{v}_i = \mathbf{v}_f + (-\mathbf{v}_i)$.

vector	east component	north component
$\mathbf{v}_f$ $-\mathbf{v}_i$	$\mathrm{v}_f \cos 10° = 88.63$ km/h -90 km/h	$\mathrm{v}_f \sin 10° = 15.63$ km/h 0
$\Delta\mathbf{v}$	-1.37 km/h	15.63 km/h

Now use the Pythagorean theorem to find the magnitude of $\Delta\mathbf{v}$:

$$\Delta v = \sqrt{(\Delta v_e)^2 + (\Delta v_n)^2} = \sqrt{1.37^2 + 15.63^2} = 15.7 \text{ km/h}$$

(**b**) The average acceleration is given by

$$a_{\text{ave}} = \frac{\Delta v}{\Delta t} = \left(\frac{15.7 \text{ km/h}}{2 \text{ s}}\right)\left(\frac{10^3 \text{ m}}{\text{km}}\right)\left(\frac{\text{h}}{3.6 \times 10^3 \text{ s}}\right) = 2.18 \text{ m/s}^2$$

(**c**) Use v_e, v_n, and the tangent function to find the angle θ between **a** and the north–south axis.

$$\tan\theta = \frac{v_e}{v_n} \qquad \theta = \arctan\left(\frac{v_e}{v_n}\right) = \arctan\left(\frac{-1.37}{15.63}\right) = -5.0° = 5.0° \text{ west of north}$$

3. (**a**) Use the equation for constant acceleration in a straight line.

$$s = v_0 t + \tfrac{1}{2}at^2$$

$$a = \frac{2(s - v_0 t)}{t^2} = \frac{2[150 \text{ m} - (20 \text{ m/s})(5 \text{ s})]}{(5 \text{ s})^2} = 4.00 \text{ m/s}^2$$

(**b**)
$$v = v_0 + at = 20 \text{ m/s} + (4 \text{ m/s}^2)(5 \text{ s}) = 40.0 \text{ m/s}$$

4. Use one of the equations for straight-line motion with constant acceleration.

$$v^2 = v_0^2 + 2as \qquad a = \frac{v^2 - v_0^2}{2s} = \frac{(80 \text{ m/s})^2 - (30 \text{ m/s})^2}{2(400 \text{ m})} = 6.88 \text{ m/s}^2$$

5. Choose "up" as the positive direction, so the acceleration of gravity is -9.8 m/s^2. Then use equations for straight-line motion with constant acceleration (and remember that $v = 0$ when the stone is at its maximum height).

(**a**)
$$v^2 = v_0^2 + 2as \qquad v_0 = \sqrt{v^2 - 2as} = \sqrt{0^2 - 2(-9.8 \text{ m/s}^2)(18 \text{ m})} = 18.78 \text{ m/s}$$

(**b**)
$$s = \tfrac{1}{2}(v_o + v)t \qquad t = \frac{2s}{v_0 + v} = \frac{2(18 \text{ m})}{18.78 \text{ m/s}} = 19.2 \text{ s}$$

6. Choose "up" as the positive direction, and use an equation for straight-line motion with constant acceleration.

$$v = v_0 + at \qquad t = \frac{v - v_0}{a} = \frac{-12 \text{ m/s} - 24 \text{ m/s}}{-9.8 \text{ m/s}^2} = 3.67 \text{ s}$$

7. Resolve the motion of the arrow into its vertical and horizontal components.

(**a**) First find the vertical component of v_0. Choose "up" as the positive direction (remember that $v_y = 0$ when the arrow is at its maximum height).

$$v_y^2 = v_{0y}^2 + 2a_y s_y \qquad v_{0y} = \sqrt{v_y^2 - 2a_y s_y} = \sqrt{0^2 - 2(-9.8 \text{ m/s}^2)(8.432 \text{ m})} = 12.86$$

Now use the sine function to find v_0.

$$\sin\theta = \frac{v_{0y}}{v_0} \qquad v_0 = \frac{12.86}{\sin 40°} = 20.0 \text{ m/s}$$

(**b**) When the arrow reaches its maximum height, $v_y = 0$, so $v = v_x$. Because there is no acceleration in the horizontal direction, $v_x = v_{0x}$.

$$v_x = v_{0x} \cos\theta = (20 \text{ m/s})(\cos 40°) = 15.3 \text{ m/s}$$

8. **(a)** First find the y component of the arrow's initial velocity: $v_{0y} = v_0 \sin 50°$. Choose "up" as the positive direction. At its maximum height, the arrow has a velocity of 0.

$$v_y^2 = v_{0y}^2 + 2as \qquad s = \frac{v_y^2 - v_{0y}^2}{2a} = \frac{0 - [(16\text{ m/s})(\sin 50°)]^2}{2(-9.8\text{ m/s}^2)} = 7.66\text{ m}$$

Don't forget to add the arrow's initial height of 2 m; $s_{max} = s_0 + s = 9.66$ m

(b) The arrow reaches the ground when $s_y = -2$ m. You need only the y component of the arrow's motion to answer this problem. Solve for t by using the quadratic formula.

$$s = v_{0y}t + \tfrac{1}{2}at^2 \qquad \tfrac{1}{2}at^2 + v_{0y}t + (-s) = 0 \qquad t = \frac{-v_{0y} \pm \sqrt{v_{0y}^2 - 4(\frac{1}{2}a)(-s)}}{2(\frac{1}{2}a)}$$

$$t = \frac{-(16\text{ m/s})(\sin 50°) \pm \sqrt{[(16\text{ m/s})(\sin 50°)]^2 - 2(-9.8\text{ m/s}^2)(2\text{ m})}}{9.8\text{ m/s}^2} = -0.154 \text{ and } 2.655\text{ s}$$

The positive number is the only solution appropriate, so $t = 2.655$ s.

(c) Because there is no acceleration in the x direction, the horizontal distance the arrow travels is simply

$$s_x = v_{0x}t = (16\text{ m/s})(\cos 50°)(2.655\text{ s}) = 27.3\text{ m}$$

9. **(a)**

$$\omega = \frac{\Delta\theta}{\Delta t} = \left(\frac{5\text{ rev}}{2.5\text{ s}}\right)\left(\frac{2\pi\text{ rad}}{\text{rev}}\right) = 12.6\text{ rad/s}$$

(b)

$$a = r\omega^2 = (0.7\text{ m})(12.6\text{ rad/s})^2 = 111\text{ m/s}^2$$

10. **(a)** Because the box is moving at a constant velocity, the sum of the forces acting on it is zero. Define the x axis as parallel to the ramp and write the equilibrium conditions for both the x and y directions. The frictional force f_k acts in the direction opposite the applied force F.

$$x\text{ direction:} \quad \Sigma F_x = F - f_k - w\sin 20° = 0 \qquad f_k = 500\text{ N} - (900\text{ N})(\sin 20°) = 192\text{ N}$$

$$y\text{ direction:} \quad \Sigma F_y = N - w\cos 20° = 0 \qquad N = (900\text{ N})(\cos 20°) = 846\text{ N}$$

Now use $\mu_k = f_k/N$ to get the coefficient of sliding friction.

$$\mu_k = \frac{f_k}{N} = \frac{192\text{ N}}{846\text{ N}} = 0.227$$

(b) Efficiency e is work output/work input. The work input $W_{in} = Fs\cos\theta$; here, $\theta = 0$ because the displacement is parallel to the force. The work output is the increase in the box's potential energy, so $W_{out} = mg\Delta h = mgs \sin 20°$.

$$e = \frac{mgs\sin 20°}{Fs} = \frac{\left(\frac{900\text{ N}}{9.8\text{ m/s}^2}\right)(9.8\text{ m/s}^2)(\sin 20°)}{500\text{ N}} = 0.616 = 61.6\%$$

11. **(a)** Because the 400-g block cannot move vertically and because the surface supporting it is horizontal, the normal force **N** acting on it is equal in magnitude to its weight mg.

$$f_s = \mu_s N = \mu_s mg = (0.30)(0.4\text{ kg})(9.8\text{ m/s}^2) = 1.176\text{ N}$$

(b) First find all the external forces acting on the system (the forces within the system do not contribute to its acceleration).

$$\text{net } F = (0.6\text{ kg})g - f - (0.2\text{ kg})g$$
$$= (0.6\text{ kg})(9.8\text{ m/s}^2) - (1.176\text{ N}) - (0.2\text{ kg})(9.8\text{ m/s}^2) = 2.744\text{ N}$$

Now use Newton's second law, $F = ma$, to find the acceleration a. The mass m is the system's total mass.

$$a = \frac{F}{m} = \frac{2.744\text{ N}}{1.2\text{ kg}} = 2.29\text{ m/s}^2$$

(c) Use your result from **(b)** and Newton's second law to find the net force acting on the 600-g block.

$$F = ma = (0.6\text{ kg})(2.29\text{ m/s}^2) = 1.374\text{ N}$$

The block has two opposite forces acting on it: its weight mg and the tension T from the string fastened to it.

$$F = mg - T \qquad T = mg - F = (0.6\text{ kg})(9.8\text{ m/s}^2) - 1.374\text{ N} = 4.51\text{ N}$$

12. (**a**) Find the system's acceleration by applying Newton's second law, $F = ma$. The total force is the difference between the weights of the two masses; the total mass is the sum of the two masses.

$$a = \frac{F}{m} = \frac{m_2 g - m_1 g}{m_1 + m_2} = \frac{(9.8\ \text{m/s}^2)(0.208 - 0.200)\ \text{kg}}{(0.208 + 0.200)\ \text{kg}} = 0.192\ \text{m/s}^2$$

(**b**) Use an equation for linear motion with constant acceleration, $s = v_0 t + \frac{1}{2}at^2$. Since the mass starts from rest, $v_0 = 0$.

$$t = \sqrt{\frac{2s}{a}} = \sqrt{\frac{2(1.5\ \text{m})}{0.192\ \text{m/s}^2}} = 35.9\ \text{s}$$

13. (**a**)
$$\alpha_{\text{ave}} = \frac{\Delta\omega}{\Delta t} = \frac{(33.3\ \text{rev/min})(2\pi\ \text{rad/rev})(1\ \text{min}/60\ \text{s})}{1.5\ \text{s}} = 2.325\ \text{rad/s}^2$$

(**b**)
$$\alpha_{\text{ave}} = \frac{\omega_2 - \omega_1}{t_2 - t_1} \qquad \omega_2 = \alpha_{\text{ave}}(t_2 - t_1) + \omega_1 = (2.325\ \text{rad/s}^2)(1\ \text{s} - 0) + 0 = 2.325\ \text{rad/s}$$

(**c**)
$$a_t = r\alpha_{\text{ave}} = (0.12\ \text{m})(2.325\ \text{rad/s}^2) = 0.279\ \text{m/s}^2$$

(**d**) The total acceleration is the vector sum of the tangential acceleration and the centripetal acceleration. The two vectors are at right angles to each other, so the Pythagorean theorem gives the magnitude of their sum.

$$a_c = r\omega^2 = (0.12\ \text{m})(2.325\ \text{rad/s})^2 = 0.649\ \text{m/s}^2$$

$$a_{\text{tot}} = \sqrt{a_t^2 + a_c^2} = \sqrt{(0.279\ \text{m/s}^2)^2 + (0.649\ \text{m/s}^2)^2} = 0.706\ \text{m/s}^2$$

14.

$$\tan\theta = \frac{v^2}{rg} = \frac{[(240\ \text{km/h})(10^3\ \text{m/km})(1\ \text{h}/3.6 \times 10^3\ \text{s})]^2}{(600\ \text{m})(9.8\ \text{m/s}^2)} = 0.756$$

$$\theta = \arctan 0.756 = 37.1°$$

15. The traffic signal is in translational equilibrium, so the sum of the forces acting on it in each direction is zero.

(**a**)
$$\Sigma F_x = 0 = T_1 \cos 15° - T_2 \cos 20° \qquad \frac{T_1}{T_2} = \frac{\cos 20°}{\cos 15°} = 0.973$$

(**b**)
$$\Sigma F_y = 0 = T_2 \sin 20° + T_1 \sin 15° - 950\ \text{N} = T_2 \sin 20° + 0.973 T_2 \sin 15° - 950$$

$$T_2 = \frac{950\ \text{N}}{\sin 20° + 0.973 \sin 15°} = 1600\ \text{N}$$

16. The beam is in rotational equilibrium, so the sum of the torques acting on it is zero. There are two torques acting on the beam: one from the cable τ_c and one from the beam's weight τ_w. Let ℓ represent the length of the beam, find the torques with τ = moment arm × force, and use $\Sigma\tau = 0 = \tau_w - \tau_c$.

$$\tau_w = (400\ \text{N})\left(\frac{\ell}{2}\right) \qquad \tau_c = T(\ell \sin 10°) \qquad T(\ell \sin 10°) = (400\ \text{N})\left(\frac{\ell}{2}\right)$$

$$T = \frac{200\ \text{N}}{\sin 10°} = 1.15 \times 10^3\ \text{N}$$

17. Choose the coordinate system where the x axis is parallel to the inclined plane. The net force that accelerates the block is equal to the applied force F minus the force of kinetic friction $f_k = \mu_k N$ and the x component of the block's weight. Because the block is not moving in the y direction, the normal force N is equal to the y component of the block's weight. Now apply Newton's second law.

$$\text{net } F = ma = F - \mu_k mg \cos 20° - mg \sin 20° \qquad F = ma + mg(\mu_k \cos 20° + \sin 20°)$$

$$F = (0.4\ \text{kg})(2.0\ \text{m/s}^2) + (0.4\ \text{kg})(9.8\ \text{m/s}^2)(0.25 \cos 20° + \sin 20°) = 3.06\ \text{N}$$

18. The falling mass has two opposite forces acting on it: its weight and the tension in the string. The sum of these two forces is the net force that accelerates the mass. Newton's third law states that the force of tension on the falling mass is equal to the force on the block.

$$\text{net } F = ma = mg - F \qquad m = \frac{F}{g - a} = \frac{3.062\ \text{N}}{(9.8\ \text{m/s}^2) - (2.0\ \text{m/s}^2)} = 0.393\ \text{kg}$$

19. (**a**) The work done in rolling the box up the ramp is converted to a change in the box's potential energy, so

$$W = mg\,\Delta h = (900\text{ N})(4\text{ m})(\sin 20°) = 1.23 \times 10^3\text{ J}$$

(**b**) Define the x axis as parallel to the ramp. Because there is no friction, there are only two forces acting in the x direction: the force F pushing the box and the x component of the box's weight. Because the box's velocity is constant

$$\Sigma F_x = 0 = F - w\sin 20° \qquad F = (900\text{ N})(\sin 20°) = 308\text{ N}$$

20. Define the x axis as parallel to the track. In the absence of friction, the force needed to pull the hopper at a constant speed is the force that equals the x component of the hopper's weight. Now apply the equation for power of a constant force.

$$P = Fv = (w\sin 25°)v = (2 \times 10^4\text{ N})(\sin 25°)(5.3\text{ m/s}) = 4.48 \times 10^4\text{ W}$$

21. In all collisions, linear momentum is conserved. And in elastic collisions such as this one, kinetic energy, too, is conserved. So set up two equations, one for conservation of kinetic energy, one for conservation of momentum, to solve for the two unknowns, v_1' and v_2'. Because m_2, the 800-g disk, starts from rest, it has no momentum and no kinetic energy before the collision.

momentum	**kinetic energy**
$m_1v_1 = m_1v_1' + m_2v_2'$	$\frac{1}{2}m_1v_1^2 = \frac{1}{2}m_1(v_1')^2 + \frac{1}{2}m_2(v_2')^2$
$(0.4\text{ kg})(4\text{ m/s}) = (0.4\text{ kg})v_1' + (0.8\text{ kg})v_2'$	$\frac{1}{2}(0.4\text{ kg})(4\text{ m/s})^2 = \frac{1}{2}(0.4\text{ kg})(v_1')^2 + \frac{1}{2}(0.8\text{ kg})(v_2')^2$
$v_1' = 4 - 2v_2'$	$v_1' = \sqrt{16 - 2(v_2')^2}$

(**a**) Now solve for v_2', the velocity of the 800-g disk after the collision.

$$4 - 2v_2' = \sqrt{16 - 2(v_2')^2} \qquad v_2' = \tfrac{8}{3}\text{ m/s}$$

The positive result indicates that the direction of the disk's motion is to the east.

(**b**) Solve for v_1'.

$$v_1' = 4 - 2v_2' = 4 - 2(\tfrac{8}{3}\text{ m/s}) = -\tfrac{4}{3}\text{ m/s}$$

The negative result indicates that this disk's motion is to the west.

22. (**a**) Because linear momentum, a vector quantity, is conserved, the total momentum after the collision equals the vector sum of the disks' individual momenta before the collision. The momenta are at right angles to each other, so find their vector sum by the Pythagorean theorem.

$$|\mathbf{p}_1| = m_1v_1 = (0.25\text{ kg})(8\text{ m/s}) = 2\text{ kg m/s} \qquad |\mathbf{p}_2| = m_2v_2 = (0.4\text{ kg})(5\text{ m/s}) = 2\text{ kg m/s}$$

$$p_{\text{tot}} = \sqrt{p_1^2 + p_2^2} = \sqrt{(2\text{ kg m/s})^2 + (2\text{ kg m/s})^2} = 2.83\text{ kg m/s}$$

(**b**) If the collision is perfectly elastic, kinetic energy, a scalar quantity, is conserved. So the total kinetic energy after the collision equals the sum of the disks' individual kinetic energies before the collision.

$$\text{KE} = \tfrac{1}{2}m_1v_1^2 + \tfrac{1}{2}m_2v_2^2 = \tfrac{1}{2}(0.25\text{ kg})(8\text{ m/s})^2 + \tfrac{1}{2}(0.4\text{ kg})(5\text{ m/s})^2 = 13.0\text{ J}$$

23. (**a**)

$$W = Fs\cos\theta = (2 \times 10^4\text{ N})(40\text{ m})(\cos 0°) = 8.00 \times 10^5\text{ J}$$

(**b**)

$$P = \frac{W}{t} = \frac{8 \times 10^5\text{ J}}{2\text{ s}} = 4 \times 10^5\text{ W}$$

24. (**a**)

$$\text{IMA} = \frac{\text{input distance}}{\text{output distance}} = \frac{d}{d\sin 10°} = 5.759$$

(**b**)

$$\text{AMA} = \frac{\text{output force}}{\text{input force}} = \frac{1.8 \times 10^5\text{ N}}{1.55 \times 10^5\text{ N}} = 1.161$$

(**c**)

$$e = \frac{\text{AMA}}{\text{IMA}} \times 100\% = \frac{1.161}{5.759} \times 100\% = 20.2\%$$

25. Figure E-3 depicts the forces in this system. Let w_l represent the weight of the ladder, w_c the weight of the carpenter, w_h the weight of the hammer, R the force of the wall on the ladder, N the force of the ground on the ladder, f_s the force of static friction between the ground and the ladder, and μ_s the coefficient of static friction.

(a) You must use the equations for both translational and rotational equilibrium to solve this problem. First, translational equilibrium:

$$\Sigma F_y = 0 = N - w_l - w_c - w_h \qquad N = 180\text{ N} + 800\text{ N} + 20\text{ N} = 1000\text{ N}$$

$$\Sigma F_x = 0 = f_s - R$$

Because $f_s = \mu_s N$, you can express ΣF_x as

$$R = f_s = \mu_s N = 1000\mu_s\text{ N}$$

Now, rotational equilibrium. Use the bottom of the ladder as the reference point for calculating the torques. Remember that torque τ = moment arm × force.

$$\Sigma\tau = 0 = (180\text{ N})(10\text{ m})(\cos 60°) + (820\text{ N})(15\text{ m})(\cos 60°) - R(20\text{ m})(\sin 60°)$$

Substitute 1000 μ_s N for R and solve for μ_s.

$$\mu_s = \frac{7050\text{ N m}}{(1000\text{ N})(20\text{ m})(\sin 60°)} = 0.407$$

(b)

$$\text{PE} = w_c h = (800\text{ N})(12\text{ m})(\sin 60°) = 8.31 \times 10^3\text{ J}$$

(c) Use an equation for straight-line motion with constant acceleration, $s = v_0 t + \frac{1}{2}at^2$. Because the hammer is at rest before the carpenter drops it, $v_0 = 0$ and $s = \frac{1}{2}at^2$.

$$t = \sqrt{\frac{2s}{a}} = \sqrt{\frac{2(12\text{ m})(\sin 60°)}{9.8\text{ m/s}^2}} = 1.456\text{ s}$$

(d)

$$v = v_0 + at = 0 + (9.8\text{ m/s}^2)(1.456\text{ s}) = 14.27\text{ m/s}$$

(e) Use the definition of linear momentum $p = mv$ and the definition of weight $w = mg$.

$$p = \frac{wv}{g} = \frac{(20\text{ N})(14.27\text{ m/s})}{9.8\text{ m/s}^2} = 29.1\text{ kg m/s}$$

(f) When the hammer is stuck in the mud, its momentum is zero, so its loss of momentum is 29.1 kg m/s.

(g) Use Newton's second law expressed as a function of momentum and time.

$$F = \frac{\Delta p}{\Delta t} = \frac{29.1\text{ kg m/s}}{0.01\text{ s}} = 2.91 \times 10^3\text{ N}$$

(h) Use Newton's second law expressed as a function of mass and acceleration

$$F = ma \qquad a = \frac{F}{m} = \frac{Fg}{w} = \frac{(2.91 \times 10^3\text{ N})(9.8\text{ m/s}^2)}{20\text{ N}} = 1.43 \times 10^3\text{ m/s}^2$$

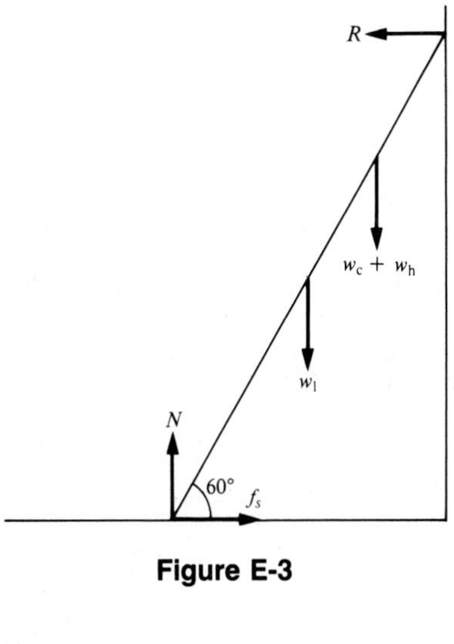

Figure E-3

8 GRAVITATION

THIS CHAPTER IS ABOUT

- ☑ Newton's Law of Universal Gravitation
- ☑ The Gravitational Field
- ☑ Kepler's Laws of Planetary Motion
- ☑ Inertial and Gravitational Mass
- ☑ Gravitational Potential Energy

8-1. Newton's Law of Universal Gravitation

Newton discovered that every object in the universe—regardless of its mass or composition—attracts every other object with a force that is proportional to the masses of the two objects and inversely proportional to the square of the distance between them:

NEWTON'S LAW OF UNIVERSAL GRAVITATION

$$F = G\frac{m_1 m_2}{r^2} \tag{8-1}$$

where G is a universal constant (often called "Newton's constant"). Its value is the same everywhere in the universe and does not change with time. In SI, $G = 6.67 \times 10^{-11}\ \text{N m}^2/\text{kg}^2$.

The distance r is measured from the center of mass of one object to the center of mass of the other, as illustrated in Figure 8-1.

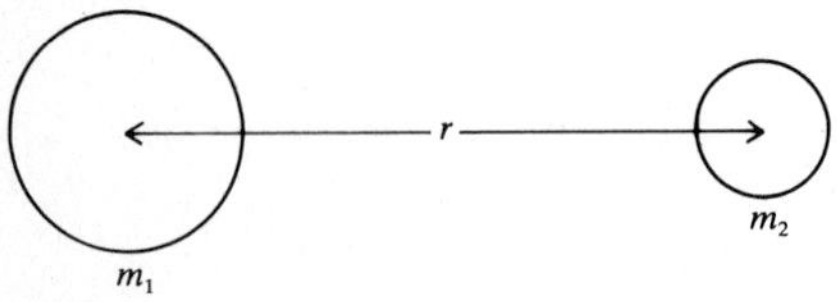

Figure 8-1. Uniform spheres separated by a distance r.

EXAMPLE 8-1: The mass of the sun is 1.99×10^{30} kg, and the mass of the moon is 7.36×10^{22} kg. The distance from the center of the sun to the center of the moon has an average value of 1.49×10^{11} m.
Calculate the magnitude of the gravitational force that the sun exerts on the moon.

Solution: You can calculate this gravitational force directly from Newton's law of universal gravitation (Eq. 8-1):

$$F = G\frac{m_1 m_2}{r^2}$$

$$= (6.67 \times 10^{-11}\ \text{N m}^2/\text{kg}^2)\frac{(1.99 \times 10^{30}\ \text{kg})(7.36 \times 10^{22}\ \text{kg})}{(1.49 \times 10^{11}\ \text{m})^2}$$

$$= 4.40 \times 10^{20}\ \text{N}$$

EXAMPLE 8-2: A small steel sphere located on the earth's surface at a distance of 6.38×10^6 m from the center of the earth experiences a gravitational acceleration of 9.8 m/s^2.
Use this information and the law of universal gravitation to calculate the mass of the earth.

Solution: The gravitational attraction of a body on the earth's surface is the weight of the body, $\mathbf{w} = m_b\mathbf{g}$ (Eq. 3-4). You can also calculate this gravitational force from Eq. (8-1):

$$F = G\frac{m_b m_e}{r_e^2}$$

where m_e represents the mass of the earth, r_e represents the radius of the earth, and m_b represents the mass of the body. Since the weight is the same as the gravitational force, you can write

$$m_b g = G\frac{m_b m_e}{r_e^2}$$

Now, cancel m_b and solve for the unknown mass of the earth m_e:

$$m_e = \frac{g r_e^2}{G} = \frac{(9.8\ \text{m/s}^2)(6.38 \times 10^6\ \text{m})^2}{6.67 \times 10^{-11}\ \text{N m}^2/\text{kg}^2} = 5.98 \times 10^{24}\ \text{kg}$$

EXAMPLE 8-3: The earth moves about the sun in a nearly circular orbit with an average radius of 1.49×10^{11} m.
Use the data of Examples 8-1 and 8-2 to calculate the magnitude of the centripetal acceleration of the earth.

Solution: The acceleration of an object is the net force on the object divided by its mass, $\mathbf{a} = \mathbf{F}/m$ (from Eq. 3-2), so begin by using Eq. (8-1) to calculate the force that the sun exerts on the earth:

$$F = G\frac{m_1 m_2}{r^2} = (6.67 \times 10^{-11}\ \text{N m}^2/\text{kg}^2)\frac{(1.99 \times 10^{30}\ \text{kg})(5.98 \times 10^{24}\ \text{kg})}{(1.49 \times 10^{11}\ \text{m})^2}$$

$$= 3.58 \times 10^{22}\ \text{N}$$

Now, substitute the magnitude of the net force into the equation to find the centripetal acceleration of the earth:

$$a_c = \frac{F}{m_2} = \frac{3.58 \times 10^{22}\ \text{N}}{5.98 \times 10^{24}\ \text{kg}} = 5.99 \times 10^{-3}\ \text{m/s}^2$$

8-2. The Gravitational Field

You may find it difficult to conceive of a force that acts on an object at a distance—even the "physicists" of Newton's day had a problem with the concept of gravitational force. And, it was another 150 years before Faraday developed the concept of a field. The general idea is that an object modifies the space surrounding it by establishing a gravitational field which extends outward in all directions, falling to zero at infinity. Any other mass located within this field experiences a force because of its location. So, it is the strength of the gravitational field at that location that produces the force—not the distant object. Of course, the situation is symmetrical—each object experiences a gravitational force because of the field set up by any other object. We'll use this concept a lot more when we deal with electricity and magnetism.

- **Gravitational field** is a vector quantity equal to the gravitational force acting on a particle divided by the mass of the particle:

GRAVITATIONAL FIELD

$$\mathbf{g} = \frac{\mathbf{F}_{\text{grav}}\ (\text{on } m)}{m} \qquad \textbf{(8-2)}$$

EXAMPLE 8-4: Find an expression for the gravitational field at a distance r from the center of a uniform sphere of mass M (see Figure 8-2).

Solution: Place a small test mass m at point P a distance r from the center of the sphere so that the gravitational field at P is

$$\mathbf{g} = \frac{\mathbf{F}_{\text{grav}}}{m}$$

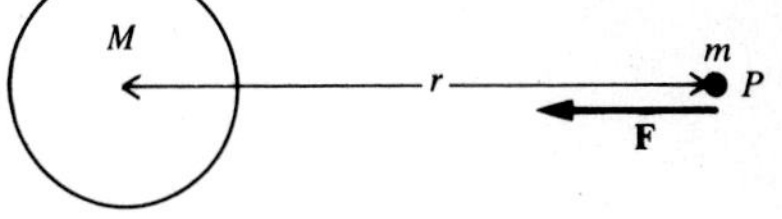

Figure 8-2

Then, obtain the magnitude of $\mathbf{F}_{grav}$ from Eq. (8-1):

$$F_{grav} = G\frac{mM}{r^2}$$

$$mg = G\frac{mM}{r^2}$$

Now, divide by the test mass m to obtain the magnitude of the gravitational field:

GRAVITATIONAL FIELD (UNIFORM SPHERE OR POINT MASS)

$$g = G\frac{M}{r^2} \tag{8-3}$$

Since the gravitational field is a vector quantity, it has direction as well as magnitude. The direction of this field is toward the center of mass M, which produces the field.

EXAMPLE 8-5: The magnitude of the earth's gravitational field is 9.8 m/s² at sea level.
How much less is the earth's field at an altitude of 100 km?

Solution: Write down expressions for the magnitude of **g** at the earth's surface and for the magnitude of **g**′ at altitude h:

$$g = \frac{GM}{r_e^2} \qquad g' = \frac{GM}{(r_e + h)^2}$$

Now, calculate the ratio g/g':

$$\frac{g}{g'} = \frac{(r_e + h)^2}{r_e^2} = \frac{r_e^2 + 2r_eh + h^2}{r_e^2} = 1 + 2\left(\frac{h}{r_e}\right) + \left(\frac{h}{r_e}\right)^2$$

And, calculate the value of h/r_e. (The distance from the earth's surface to its center is 6.38×10^6 m = 6380 km.)

$$\frac{h}{r_e} = \frac{100 \text{ km}}{6380 \text{ km}} = 1.567 \times 10^{-2}$$

Finally,

$$\frac{g}{g'} = 1 + 2(1.567 \times 10^{-2}) + (1.567 \times 10^{-2})^2 = 1.032$$

$$g' = \frac{g}{1.032} = 0.969\, g$$

You've found that the magnitude of the earth's gravitational field at an altitude of 100 km is 3% less than its magnitude at sea level.

EXAMPLE 8-6: A spaceship on its way to a distant planet passes through point P, as shown in Figure 8-3. The spaceship is 3.64×10^5 km from the earth and 6.02×10^4 km from the moon.
Calculate the magnitude of the gravitational field at P.

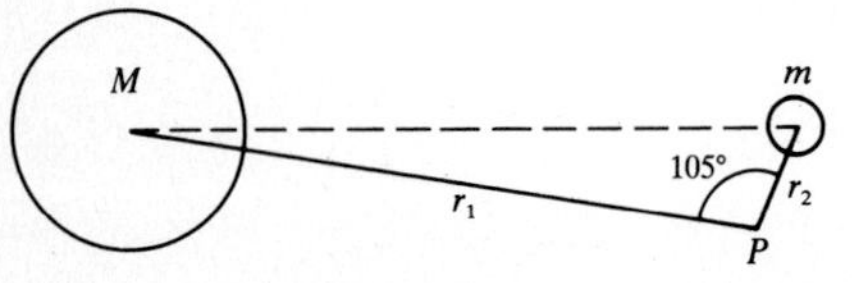

Figure 8-3

Solution: Let M be the mass of the earth and m be the mass of the moon, and use Eq. (8-3) to calculate the magnitude of the earth's gravitational field (g_1). To keep the units consistent, express the distances in meters:

$$g_1 = G\frac{M}{r_1^2} = (6.67 \times 10^{-11} \text{ N m}^2/\text{kg}^2)\frac{(5.98 \times 10^{24} \text{ kg})}{(3.64 \times 10^8 \text{ m})^2}$$

$$= 3.01 \times 10^{-3} \text{ m/s}^2$$

Now, calculate the magnitude of the moon's gravitational field (g_2) at P:

$$g_2 = G\frac{M}{r_2^2} = (6.67 \times 10^{-11}\ \text{N m}^2/\text{kg}^2)\frac{(7.36 \times 10^{22}\ \text{kg})}{(6.02 \times 10^7\ \text{m})^2}$$
$$= 1.35 \times 10^{-3}\ \text{m/s}^2$$

Now, draw vectors $\mathbf{g}_1$ and $\mathbf{g}_2$ head-to-tail (see Figure 8-4) and find their resultant from the law of cosines (Eq. 1-2). [**recall:** If you know the angle θ opposite the unknown side of a triangle, you can find the magnitude of the unknown side by means of the law of cosines. (Sec. 1-4)]

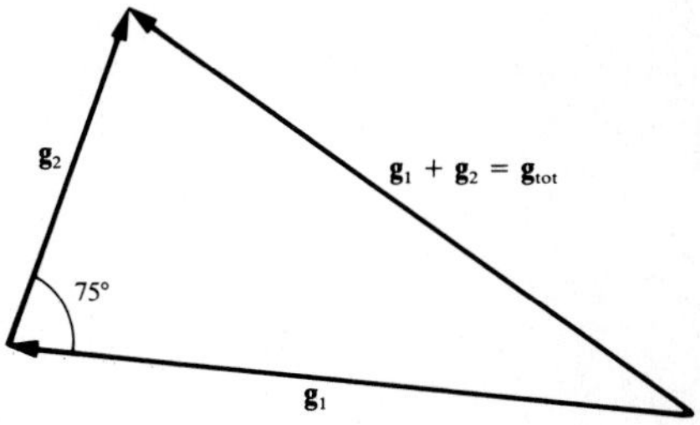

Figure 8-4

$$(g_{tot})^2 = g_1^2 + g_2^2 - 2g_1g_2 \cos\theta$$
$$= (3.01 \times 10^{-3}\ \text{m/s}^2)^2 + (1.35 \times 10^{-3}\ \text{m/s}^2)^2 - 2(3.01 \times 10^{-3}\ \text{m/s}^2)(1.35 \times 10^{-3}\ \text{m/s}^2)\cos 75°$$
$$g_{tot} = \sqrt{8.78 \times 10^{-6}\ \text{m}^2/\text{s}^4} = 2.96 \times 10^{-3}\ \text{m/s}^2$$

8-3. Kepler's Laws of Planetary Motion

Kepler formulated three kinematic laws to describe the motion of planets about the sun:

(1) The path of any planet about the sun is an ellipse with the sun at one focus of the ellipse.
(2) The position vector **r** from the sun to any planet sweeps out equal areas in equal times (see Figure 8-5, where the shaded areas are equal.). The time it takes for a planetary body to move from A to B is equal to the time it takes to move from A' to B'.
(3) The square of the period of a planet's revolution in its orbit about the sun is proportional to the cube of its mean distance from the sun. (See Example 8-8 for a derivation of this law.)

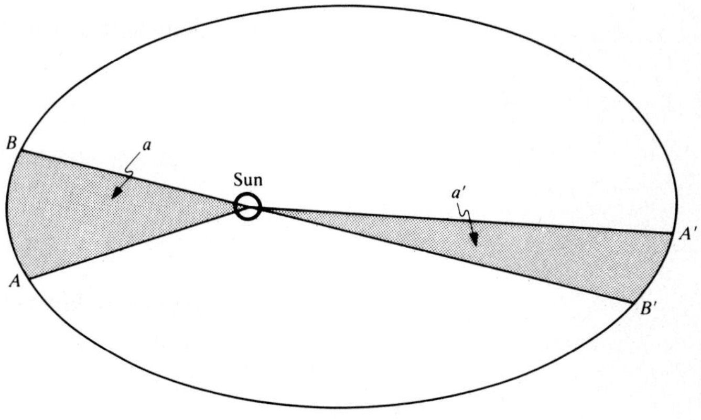

Figure 8-5. Kepler's law of areas: $a = a'$.

Kepler's laws apply to any body that orbits the sun, man-made spaceships as well as planets, comets, and other natural objects. The mass of the orbiting body does not enter into the calculation.

- The point of an orbit's closest approach to the sun is the **perihelion**.
- An orbit's farthest point from the sun is the **aphelion**.

(For an object orbiting the earth, the point closest to the earth is the *perigee* and the point farthest from the earth is the *apogee*.)

EXAMPLE 8-7: The perihelion distance r_1 of a comet orbiting the sun is two-fifths of its aphelion distance r_2.
Find the ratio of the speed at perihelion v_1 to the speed at aphelion v_2.

Solution: Your sketch should look like Figure 8-6. Choose some small time interval Δt during which the planet moves a distance $s = v\,\Delta t$. The area swept out by the line joining the planet to the sun at perihelion nearly has the shape of a thin triangle (see Figure 8-5), so its area is

$$A_1 = \frac{1}{2}r_1\,\Delta s_1 = \frac{1}{2}r_1v_1\,\Delta t$$

The planet at aphelion moves a distance $\Delta s_2 = v_2\,\Delta t$ during this same time interval. So, the area swept out by the position vector at aphelion is

$$A_2 = \frac{1}{2}r_2\,\Delta s_2 = \frac{1}{2}r_2v_2\,\Delta t$$

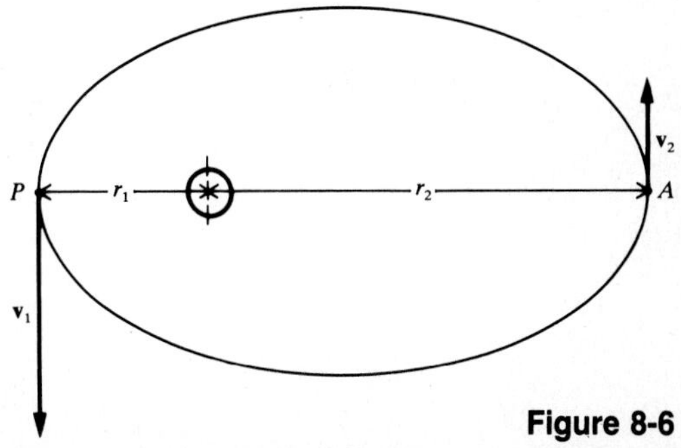

Figure 8-6

Kepler's second law states that these two areas are equal:

$$\frac{1}{2}r_1v_1\,\Delta t = \frac{1}{2}r_2v_2\,\Delta t \qquad r_1v_1 = r_2v_2 \qquad \frac{v_1}{v_2} = \frac{r_2}{\frac{2}{5}r_2} = \frac{5}{2} = 2.5$$

EXAMPLE 8-8: A planet is in a circular orbit about the sun.
Find the algebraic expression for the square of the period of revolution divided by the cube of the orbit's radius.

Solution: You know that the planet has a centripetal acceleration given by Eq. (4-12), $a_c = v^2/r$, so the corresponding centripetal force, as given by Newton's second law, is

$$F_c = m\frac{v^2}{r}$$

This centripetal force is supplied by the gravitational attraction of the sun on the planet; that is,

$$F_{\text{grav}} = G\frac{mM}{r^2}$$

Equate these two forces and cancel the common factor m, the mass of the planet:

$$\frac{GM}{r^2} = \frac{v^2}{r}$$

You can express the orbital speed v as the circumference divided by the period T:

$$v = \frac{2\pi r}{T}$$

Substitute this expression for v in the preceding equation and multiply by r:

$$\frac{GM}{r} = \frac{4\pi^2r^2}{T^2} \qquad \frac{T^2}{r^3} = \frac{4\pi^2}{GM}$$

And so, for any planet in the solar system, you have the relationship

KEPLER'S THIRD LAW

$$T^2 = \frac{4\pi^2}{GM}r^3 \tag{8-4}$$

This relationship is also valid for an elliptical orbit—provided r represents the *average distance* from the sun. This average distance is defined as one-half the sum of the aphelion distance and the perihelion distance.

EXAMPLE 8-9: The planet Mars has a mass of 6.418×10^{23} kg and completes one revolution about the sun in 687 days. The sun has a mass of 1.99×10^{30} kg. What is the average radius of the orbit of Mars?

Solution: You can solve this problem using Eq. (8-4). Of course, to keep the units straight, you'll have to express the period in seconds:

$$T = (687\ \text{days})\left(\frac{24\ \text{h}}{1\ \text{day}}\right)\left(\frac{3600\ \text{s}}{1\ \text{h}}\right) = 5.94 \times 10^7\ \text{s}$$

Now, solve Eq. (8-4) for r. Don't be misled by the data given—you need only the sun's mass to solve the problem:

$$r^3 = \frac{GMT^2}{4\pi^2} = \frac{(6.67 \times 10^{-11}\ \text{N m}^2/\text{kg}^2)(1.99 \times 10^{30}\ \text{kg})(5.94 \times 10^7\ \text{s})^2}{4\pi^2}$$

$$= 1.186 \times 10^{34}\ \text{m}^3$$

$$r = (1.186 \times 10^{34}\ \text{m}^3)^{1/3} = 2.28 \times 10^{11}\ \text{m}$$

8-4. Inertial and Gravitational Mass

We can measure the mass of an object in two ways:

The first method involves Newton's second law of motion. We apply a known force, measure the acceleration produced, and then calculate the mass from the equation $m = F/a$. The result is called inertial mass.

- The **inertial mass** of an object is a measure of the resistance of the object to a change in its state of motion.

The second method involves Newton's law of universal gravitation. We place the object in a known gravitational field, measure the gravitational force that acts on it, and then calculate the mass from the equation $m = F_{grav}/g$. The result is called gravitational mass.

- The **gravitational mass** of an object is a measure of the gravitational force exerted on it by some other body, such as the earth.

Empirical studies show that inertial mass and gravitational mass are strictly equivalent to within 3 parts in 10^{11}.

EXAMPLE 8-10: In an experiment performed on the surface of the moon, a horizontal force of 4 N is applied to a small lead sphere, causing it to accelerate 8 m/s^2. The lead sphere is then suspended vertically in the moon's gravitational field, and the gravitational force exerted on it is 0.817 N.
Use these data and the equivalence of inertial and gravitational mass to calculate the magnitude of the moon's gravitational field.

Solution: You'll need to know the gravitational mass of the lead sphere before you can calculate the moon's gravitational field, so first calculate the inertial mass of the lead sphere, which is also its gravitational mass:

$$m = \frac{F}{a} = \frac{4\ \text{N}}{8\ \text{m/s}^2} = 0.5\ \text{kg}$$

Now, you can calculate the magnitude of the gravitational field from Eq. (8-2):

$$g = \frac{F_{grav}\ (\text{on } m)}{m} = \frac{0.817\ \text{N}}{0.5\ \text{kg}} = 1.63\ \text{m/s}^2$$

8-5. Gravitational Potential Energy

- The **gravitational potential energy** of a body is the energy needed to move the body to an infinite distance from a second body.

Gravitational potential energy is negative because work must be done to separate the two bodies. Figure 8-7 shows a small mass m on the surface of the earth. In order to pull m out to a distance r_m from the center of the earth, an external force in opposition to the gravitational force must be applied to the mass.

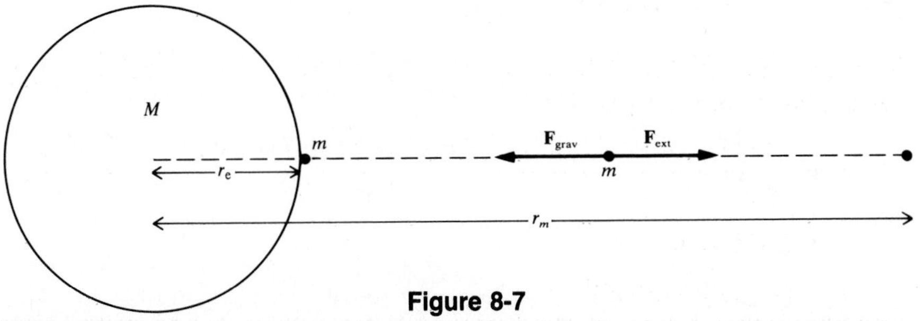

Figure 8-7

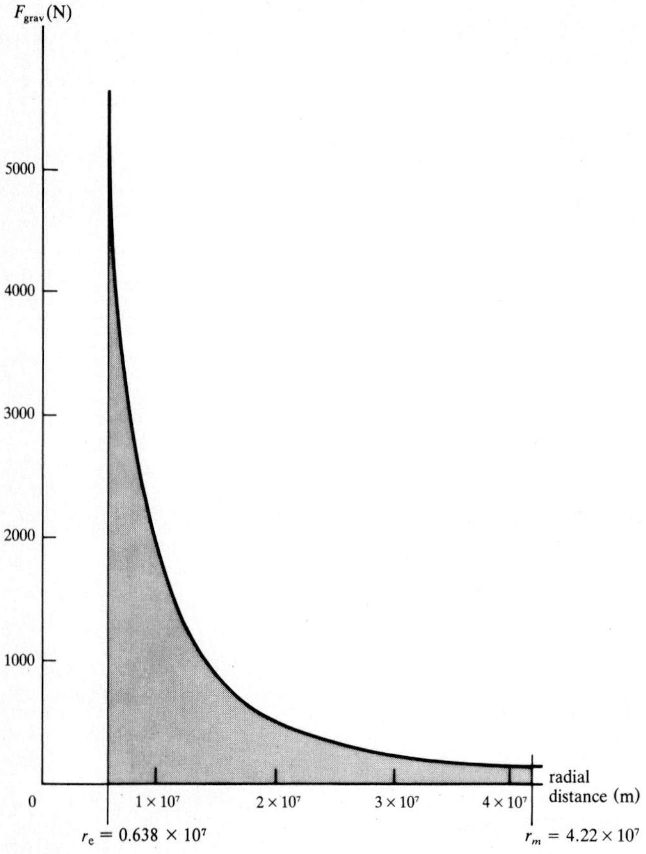

Figure 8-8

EXAMPLE 8-11: The mass of the earth is 5.98×10^{24} kg.
Calculate the energy required to move a 500-kg satellite from the earth's surface ($r_e = 6.38 \times 10^6$ m) to a distance of 4.22×10^7 m from the earth's center. At this distance the satellite will take 24 hours to circle the earth. If it is over the equator, it will appear to be stationary in the sky.

Solution: The force involved here is variable, so the work done is the area under the graph of the force plotted as a function of the distance (see Sec. 6-1B). A graph of the gravitational force versus the radial distance is shown in Figure 8-8. You can determine the work done against the gravitational force by estimating the shaded area under this curve between the two vertical lines at r_e and r_m. Or, integral calculus will give the area more accurately. This method gives the work done in launching the mass to a distance of r_m as

WORK REQUIRED TO SEPARATE TWO MASSES

$$W = GMm\left(\frac{1}{r_e} - \frac{1}{r_m}\right) \qquad \textbf{(8-5)}$$

Now, you can calculate the work required to move the satellite:

$$W = (6.67 \times 10^{-11}\ \mathrm{N\ m^2/kg^2})(5.98 \times 10^{24}\ \mathrm{kg})(500\ \mathrm{kg}) \cdot \left(\frac{1}{6.38 \times 10^6\ \mathrm{m}} - \frac{1}{4.22 \times 10^7\ \mathrm{m}}\right)$$
$$= 2.65 \times 10^{10}\ \mathrm{J}$$

The amount of work done by an external force in moving a mass m to an infinite distance from the earth is

$$W = GMm\left(\frac{1}{r_e}\right)$$

This system of two masses separated by an infinite distance has a potential energy defined as zero. This means that the potential energy of m on the earth's surface, PE_{grav}, plus the work W done to move it to infinity is also equal to zero:

$$PE_{grav} + \frac{GMm}{r_e} = 0$$

So that

GRAVITATIONAL POTENTIAL ENERGY OF TWO MASSES SEPARATED BY A DISTANCE *r*

$$PE_{grav} = -\frac{GMm}{r} \qquad \textbf{(8-6)}$$

Suppose you want to launch a rocket vertically and give it just enough kinetic energy to escape the earth and never return.

- The minimum initial velocity of an object at the earth's surface that would allow the object to escape from the earth, never to return, is the **escape velocity.**

The minimum KE is that which will make the total energy equal to zero. Neglecting air resistance,

$$KE + PE = 0$$

$$\frac{1}{2}mv_{esc}^2 - \frac{GMm}{r_e} = 0$$

Cancel the mass of the rocket and solve for the escape velocity:

ESCAPE VELOCITY FROM EARTH

$$v_{esc} = \sqrt{\frac{2GM}{r_e}} \qquad \textbf{(8-7)}$$

EXAMPLE 8-12: What initial speed must a rocket have to escape the gravitational attraction of the earth? Neglect air resistance and the rotational acceleration at the launch point.

Solution: Substitute the known values directly into Eq. (8-7):

$$v_{esc} = \sqrt{\frac{2(6.67 \times 10^{-11}\ \text{N m}^2/\text{kg}^2)(5.98 \times 10^{24}\ \text{kg})}{6.38 \times 10^6\ \text{m}}} = 1.12 \times 10^4\ \text{m/s}$$

note: You can also use Eq. (8-7) to determine the escape velocity from any body in space—provided that you know the mass of the body and its radius.

EXAMPLE 8-13: What speed would a satellite need to stay in a circular orbit at the earth's surface? Neglect the friction of air resistance and assume that no mountains get in the satellite's way.

Solution: Since the satellite is traveling in a circle, the centripetal force that it experiences is equal to the gravitational force on it.

$$F = \frac{mv^2}{r} = \frac{GmM}{r^2} \qquad v^2 = \frac{GM}{r}$$

So the speed of an object in any circular orbit is

SPEED IN A CIRCULAR ORBIT

$$v = \sqrt{\frac{GM}{r}} \qquad \textbf{(8-8)}$$

In this case, $r = r_e$, so

$$v = \sqrt{\frac{(6.67 \times 10^{-11}\ \text{N m}^2/\text{kg}^2)(5.98 \times 10^{24}\ \text{kg})}{6.38 \times 10^6\ \text{m}}} = 7.91 \times 10^3\ \text{m/s}$$

(At this speed the satellite would circle the earth in 1.41 h.)

EXAMPLE 8-14: How does the speed of a satellite in a circular orbit compare to the speed it needs to escape the earth's gravitational attraction?

Solution: Derive the ratio of the escape velocity to the speed of the satellite by dividing Eq. (8-7) by Eq. (8-8).

$$\frac{v_{esc}}{v} = \frac{\sqrt{\dfrac{2GM}{r_e}}}{\sqrt{\dfrac{GM}{r_e}}}$$

$$v_{esc} = \sqrt{2}\, v = 1.41\, v$$

A satellite needs 1.41 times its orbital speed to escape its orbit.

SUMMARY

Newton's law of universal gravitation	$F = G\dfrac{m_1 m_2}{r^2}$	the mutual attraction between two particles is proportional to the masses of the particles and inversely proportional to the square of the distance between them

Gravitational field	$\mathbf{g} = \dfrac{\mathbf{F}_{grav}\text{ (on } m)}{m}$	describes a vector quantity equal to the gravitational force acting on a particle divided by the mass of the particle
Gravitational field (uniform sphere or point mass)	$g = G\dfrac{M}{r^2}$	gives the magnitude of a gravitational field (its direction is always toward the mass that produces the field)
Kepler's third law	$T^2 = \dfrac{4\pi^2}{GM}r^3$	states that the square of the period of a planet's revolution in its orbit about the sun is proportional to the cube of its mean distance from the sun
Work required to separate two masses	$W = GMm\left(\dfrac{1}{r_e} - \dfrac{1}{r_m}\right)$	gives the work that must be done by an external force to separate two masses
Gravitational potential energy of two masses separated by a distance r	$PE_{grav} = -\dfrac{GMm}{r}$	describes the potential energy of a mass as equal to the work needed to move the mass to an infinite distance
Escape velocity from earth	$v_{esc} = \sqrt{\dfrac{2GM}{r_e}}$	gives the minimum initial velocity necessary to allow an object to escape from the earth and never return
Speed in a circular orbit	$v = \sqrt{\dfrac{GM}{r}}$	gives the tangential speed of an object in a circular orbit

RAISE YOUR GRADES

Do you know . . . ?

- ☑ Newton's law of universal gravitation
- ☑ the definition of gravitational field
- ☑ how to calculate the gravitational field of a homogeneous sphere or a point mass
- ☑ the shape of the orbit of a planet
- ☑ Kepler's law of areas
- ☑ the relationship between the period of a planet and the average radial distance from the sun
- ☑ the distinction between inertial and gravitational mass

Can you . . . ?

- ☑ calculate the force of gravitational attraction between two known masses
- ☑ calculate the gravitational field from the force on a test mass
- ☑ calculate the gravitational field at a known distance from the earth
- ☑ calculate the mass of the sun from the period and orbital radius of a planet
- ☑ calculate the gravitational potential energy of two masses separated by a known distance
- ☑ calculate the escape velocity from the earth or the moon

SOLVED PROBLEMS

PROBLEM 8-1 A number of communications satellites occupy *geosynchronous* orbits—that is, they appear to be in a fixed position in the sky directly above the equator. Since the earth makes one revolution about its axis every 24 hours, to appear fixed, these satellites must also have a period of 24 hours. Calculate the radius of the circular orbit of a geosynchronous satellite.

Solution Use the earth's mass (5.98×10^{24} kg) in Eq. (8-4), Kepler's third law, and solve for r:

$$T^2 = \frac{4\pi^2}{GM} r^3$$

$$r^3 = \frac{GMT^2}{4\pi^2} = \frac{(6.67 \times 10^{-11}\ \text{N m}^2/\text{kg}^2)(5.98 \times 10^{24}\ \text{kg})(24\ \text{h} \times 3600\ \text{s/h})^2}{4\pi^2}$$

$$r = (7.54 \times 10^{22}\ \text{m}^3)^{1/3} = 4.23 \times 10^7\ \text{m}$$

PROBLEM 8-2 The planet Venus has a period of 0.615 earth-years. Its average distance from the center of the sun is 1.082×10^{11} m. Use these data to calculate the mass of the sun.

Solution Use Eq. (8-4) to solve for the unknown mass:

$$M = \frac{4\pi^2 r^3}{GT^2}$$

Express the period in seconds:

$$T = 0.615\ \text{y}\left(\frac{365.3\ \text{d}}{1\ \text{y}}\right)\left(\frac{24\ \text{h}}{1\ \text{d}}\right)\left(\frac{3600\ \text{s}}{1\ \text{h}}\right) = 1.941 \times 10^7\ \text{s}$$

Then,

$$M = \frac{4\pi^2(1.082 \times 10^{11}\ \text{m})^3}{(6.67 \times 10^{-11}\ \text{N m}^2/\text{kg}^2)(1.941 \times 10^7\ \text{s})^2} = 1.99 \times 10^{30}\ \text{kg}$$

PROBLEM 8-3 The planet Saturn makes one revolution about the sun in 29.46 earth-years. Find the average radius of Saturn's elliptical orbit.

Solution First, calculate the period in seconds:

$$T = 29.46\ \text{y}\left(\frac{365.3\ \text{d}}{1\ \text{y}}\right)\left(\frac{24\ \text{h}}{1\ \text{d}}\right)\left(\frac{3600\ \text{s}}{1\ \text{h}}\right) = 9.30 \times 10^8\ \text{s}$$

Then, from Eq. (8-4), solve for r:

$$r^3 = \frac{GMT^2}{4\pi^2} = \frac{(6.67 \times 10^{-11}\ \text{N m}^2/\text{kg}^2)(1.99 \times 10^{30}\ \text{kg})(9.30 \times 10^8\ \text{s})^2}{4\pi^2} = 2.91 \times 10^{36}\ \text{m}^3$$

$$r = (2.91 \times 10^{36})^{1/3} = 1.43 \times 10^{12}\ \text{m}$$

PROBLEM 8-4 A lunar orbiter completes one revolution of the moon in 6848 s. Its altitude above the moon's surface is 60 km, making its distance from the center of the moon 1.80×10^6 m. Calculate the mass of the moon from these data.

Solution Calculate the mass M of the moon using Eq. (8-4):

$$M = \frac{4\pi^2 r^3}{GT^2} = \frac{4\pi^2(1.80 \times 10^6\ \text{m})^3}{(6.67 \times 10^{-11}\ \text{N m}^2/\text{kg}^2)(6.85 \times 10^3\ \text{s})^2} = 7.36 \times 10^{22}\ \text{kg}$$

PROBLEM 8-5 Two homogeneous spherical bodies are separated by a distance r, measured from the center of one sphere to the center of the other. The ratio of the masses is $m_1/m_2 = 3$. In the absence of external forces, these two objects will be accelerated toward each other. Find the ratio of the accelerations, $\mathbf{a}_1/\mathbf{a}_2$.

Solution The acceleration of an object is produced by the net force acting on it, so begin by writing down the relationship between the force that m_1 exerts on m_2 and the force that m_2 exerts on m_1. You know from Newton's third law (Eq. 3-3) that these two forces have opposite directions but equal magnitudes; that is, $F_{1\text{ on }2} = F_{2\text{ on }1}$. Use Newton's second law (Eq. 3-2) to relate these forces to the accelerations that they produce:

$$F_{1\text{ on }2} = m_2 a_2 \qquad F_{2\text{ on }1} = m_1 a_1$$

So that you get

$$m_1 a_1 = m_2 a_2 \qquad \frac{a_1}{a_2} = \frac{m_2}{m_1} = \frac{1}{3}$$

PROBLEM 8-6 A weather satellite is in a circular orbit about the earth with an average altitude of 120 km. Calculate (**a**) the speed of the satellite and (**b**) its period.

Solution

(**a**) Use Eq. (8-8) to find the satellite's speed in a circular orbit about the earth:

$$v = \sqrt{\frac{GM}{r}}$$

The radial distance from the center of the earth is

$$r = r_e + h = (6.38 \times 10^3 + 120)\text{ km} = 6.50 \times 10^3\text{ km} = 6.50 \times 10^6\text{ m}$$

You can now calculate the required speed for the circular orbit:

$$v = \sqrt{\frac{(6.67 \times 10^{-11}\text{ N m}^2/\text{kg}^2)(5.98 \times 10^{24}\text{ kg})}{6.50 \times 10^6\text{ m}}} = 7.83 \times 10^3\text{ m/s}$$

(**b**) Since this satellite is in a circular orbit, its speed is constant. [**recall:** The speed of an object is the displacement divided by the time it takes for the displacement to take place. (Sec. 2-12)] So, you can express the speed as follows:

$$v = \frac{2\pi r}{T}$$

Now, solve for the period:

$$T = \frac{2\pi r}{v} = \frac{2\pi(6.50 \times 10^6\text{ m})}{7.83 \times 10^3\text{ m/s}} = 5.22 \times 10^3\text{ s}$$

or

$$T = 5.22 \times 10^3\text{ s}\left(\frac{1\text{ h}}{3600\text{ s}}\right) = 1.45\text{ hours} = 1\text{ h }27\text{ min}$$

PROBLEM 8-7 After the satellite of Problem 8-6 has been in orbit for a few months, the weather service decides to increase the speed so that the satellite will have an elliptical orbit with an apogee that is twice its perigee. The perigee will be equal to the radius of the original circular orbit, as shown in Fig. 8-9. Determine the fraction by which the original speed should be multiplied to achieve the elliptical orbit.

Solution You can solve this problem by using the principle of conservation of energy and Kepler's law of areas (his second law). Because the orbit of the satellite is above the earth's atmosphere, there is no air resistance. This means that the total energy of the satellite—potential plus kinetic—is the same at every point in its elliptical orbit. The energy will be higher than the energy in the circular orbit because the average radius is larger. The total energy at perigee is

$$\text{E}_p = \text{PE}_{\text{grav}} + \text{KE} = -\frac{GMm}{r_p} + \frac{1}{2}mv_p^2$$

The speed at perigee is the original speed in the circular orbit multiplied by the unknown fraction f:

$$v_p = f\sqrt{\frac{GM}{r_p}}$$

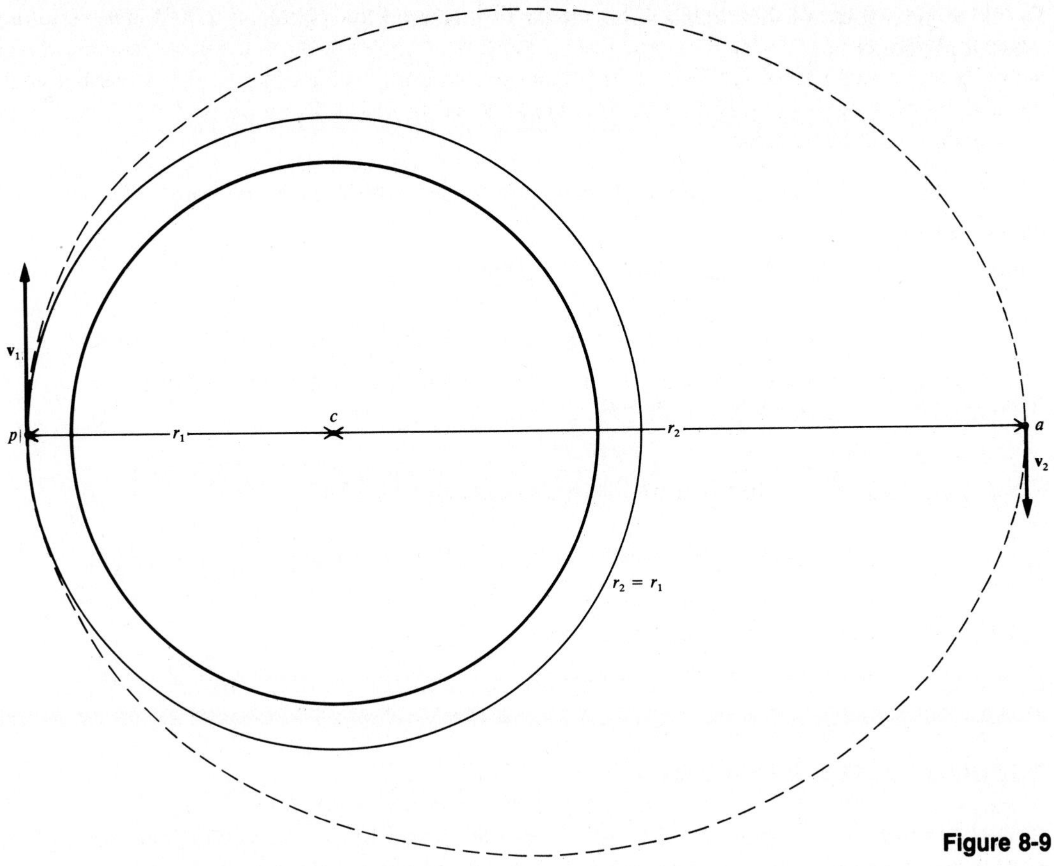

Figure 8-9

Substitute this into the energy equation to obtain

$$E_p = -\frac{GMm}{r_p} + \frac{1}{2}mf^2\left(\frac{GM}{r_p}\right) = \frac{GMm}{r_p}\left(\frac{f^2}{2} - 1\right)$$

When you used Kepler's second law in Example 8-7, you found that the speeds at apogee and perigee are related to the corresponding distances as follows:

$$r_a v_a = r_p v_p \qquad \frac{v_p}{v_a} = \frac{r_a}{r_p}$$

Since the apogee distance is twice the perigee distance, $r_a = 2r_p$ and $v_a = \frac{1}{2}v_p$. Now, write the expression for the total energy at apogee:

$$E_a = PE_{grav} + KE = \frac{GMm}{r_a} + \frac{1}{2}mv_a^2 = -\frac{GMm}{2r_p} + \frac{1}{2}m\left(\frac{1}{2}f\sqrt{\frac{GM}{r_p}}\right)^2$$

$$= \frac{GMm}{r_p}\left(\frac{f^2}{8} - \frac{1}{2}\right)$$

Now, you can equate these two expressions for the energy (E_p and E_a) and cancel out the common factor GMm/r_p:

$$\frac{f^2}{2} - 1 = \frac{f^2}{8} - \frac{1}{2}$$

$$\frac{f^2 - 2}{2} = \frac{f^2 - 4}{8}$$

$$4f^2 - 8 = f^2 - 4$$

$$3f^2 = 4$$

$$f = \frac{2}{\sqrt{3}}$$

So, the original speed of the circular orbit should be increased by a factor of $2/\sqrt{3}$. The resulting speed at perigee is

$$v_p = \frac{2}{\sqrt{3}}\sqrt{\frac{GM}{r_p}} = (1.155)(7.83 \times 10^3 \text{ m/s}) = 9.04 \times 10^3 \text{ m/s}$$

PROBLEM 8-8 Calculate the period of revolution for the satellite in the elliptical orbit described in Problem 8-7.

Solution First, calculate the mean radius of the elliptical orbit:

$$r_{av} = \frac{1}{2}(r_1 + r_2) = \frac{1}{2}(r_1 + 2r_1) = 1.5r_1 = (1.5)(6.50 \times 10^6 \text{ m}) = 9.75 \times 10^6 \text{ m}$$

Now, use Eq. (8-4) to calculate the period:

$$T^2 = \frac{4\pi^2}{GM}r^3 = \frac{4\pi^2(9.75 \times 10^6 \text{ m})^3}{(6.67 \times 10^{-11} \text{ N m}^2/\text{kg}^2)(5.98 \times 10^{24} \text{ kg})} = 9.174 \times 10^7 \text{ s}^2$$

$$T = (9.174 \times 10^7 \text{ s}^2)^{1/2} = 9.58 \times 10^3 \text{ s} = (9.58 \times 10^3 \text{ s})\left(\frac{1 \text{ h}}{3600 \text{ s}}\right) = 2.66 \text{ h}$$

Supplementary Exercises

PROBLEM 8-9 A spaceship is 1.2×10^7 m from the planet Mars. The gravitational force of attraction that the planet exerts on the spaceship is 300 N. What will this gravitational force become when the spaceship is only 5.0×10^6 m from the center of the planet?

PROBLEM 8-10 The planet Mars has a mass of 6.45×10^{23} kg and a radius of 3.38×10^6 m. Use these data to calculate the magnitude of the gravitational field on the surface of Mars.

PROBLEM 8-11 The acceleration of gravity on the earth at sea level is 9.8 m/s^2. At what altitude will the gravitational acceleration have one-half this magnitude?

PROBLEM 8-12 The mean orbital radius of the earth is 1.496×10^{11} m. The mean orbital radius of the planet Jupiter is 7.78×10^{11} m. Use these data with Kepler's third law to calculate the number of years for Jupiter to orbit the sun.

PROBLEM 8-13 A small object is subject to two horizontal forces: one force, toward the east, has a magnitude of 1.5 N; the second force, toward the north, has a magnitude of 2.0 N. The magnitude of the acceleration produced by these two forces is 5 m/s^2. What is the inertial mass of the object?

PROBLEM 8-14 The escape velocity from the surface of the earth is 7.91×10^3 m/s. The mass of the moon is 0.0123 times that of the earth, and the radius of the moon is 0.272 times that of the earth. Use these data to calculate the escape velocity from the surface of the moon.

PROBLEM 8-15 Two equal masses, 8.5×10^2 kg each, are separated by a distance of 0.80 m (measured from center to center). What is the gravitational potential energy of these two masses?

Answers to Supplementary Exercises

8-9: 1730 N

8-10: 3.77 m/s^2

8-11: 2.64×10^6 m

8-12: 11.9 years

8-13: 0.5 kg

8-14: 1.68×10^2 m/s

8-15: -6.02×10^{-5} J

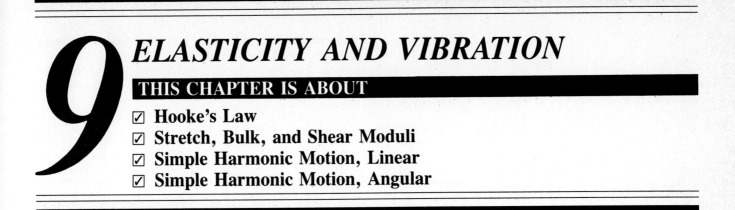

9 ELASTICITY AND VIBRATION

THIS CHAPTER IS ABOUT

☑ Hooke's Law
☑ Stretch, Bulk, and Shear Moduli
☑ Simple Harmonic Motion, Linear
☑ Simple Harmonic Motion, Angular

9-1. Hooke's Law

If a force of tension is applied to a wire, the length of the wire will increase slightly. When the force is removed, the length of the wire will usually return to its original length.

- **Elasticity** is the ability of a material to recover its original length—or shape or volume—after stress is removed.

Most solid materials, regardless of their size or shape, are elastic. Hooke's law describes elastic behavior in all kinds of distortion.

- According to **Hooke's law,** the stretch ΔL—the change in length—of an elastic material is proportional to the applied force F.

HOOKE'S LAW

$$F = -k\Delta L \tag{9-1}$$

The negative sign on the proportionality constant k is important, because the force exerted by the elastic material is in the direction opposite to the change in length. The constant k is always positive.

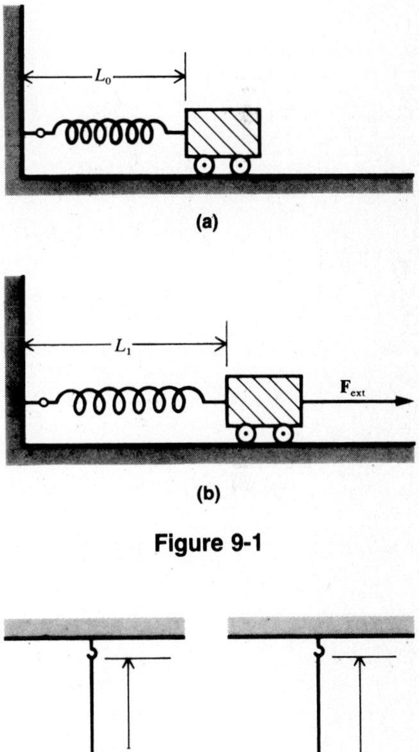

Figure 9-1

EXAMPLE 9-1: Figure 9-1a shows a coiled 10-cm spring attached to a cart that moves without friction along a horizontal surface. When no force is applied to the spring, its length is L_0. When an external force of 2 N is applied to the cart, the length of the spring increases from 10 cm to 12 cm (see Figure 9-1b), so that the elongation of the spring is $\Delta L = L_1 - L_0 = 2$ cm. Because the spring is being stretched toward the right, it exerts a force toward the left on the cart. The force $\mathbf{F}_s$ exerted by the spring and the elongation ΔL are in opposite directions. $\mathbf{F}_s$ is a **restoring force** because it tries to bring the cart back to its original position. Calculate the spring constant.

Solution: Use Hooke's law (Eq. 9-1) and solve for the spring constant k:

$$k = \frac{-F}{\Delta L} = \frac{-(-2\text{ N})}{0.02\text{ m}} = 100\text{ N/m}$$

9-2. Stretch, Bulk, and Shear Moduli

A. Stretch modulus

A straight wire doesn't stretch as easily as a coiled wire, but its *length* does increase when a force is applied. Figure 9-2a shows a weight hanger attached to a vertical steel wire. When an additional load Δw is placed on the hanger (see Figure 9-2b), the length of the wire increases from L_0 to L_1. How far the wire stretches depends on the stress and the strain.

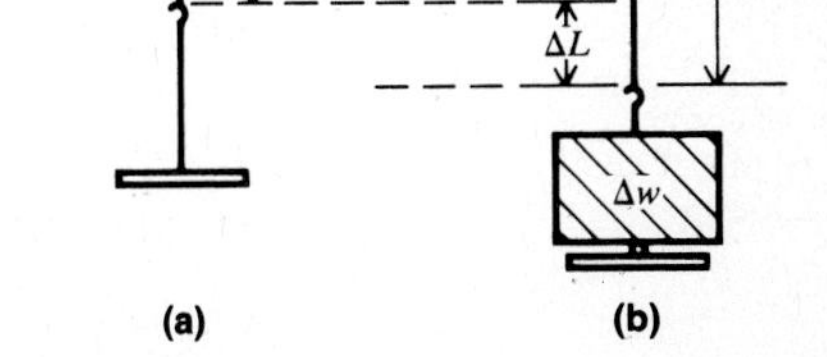

Figure 9-2. Weight and hanger system.

- **Stress** F/A is the applied force per cross-sectional area of the wire. Stress has the dimensions of pressure (force/area), and we often measure it in *pascals,* a derived SI unit (1 N/m^2 = 1 pascal = 1 Pa).

- **Strain** $\Delta L/L$ is the stretch or compression per unit length of the wire. Strain, which is length/length, is dimensionless.
- The **stretch modulus** E of the wire is the ratio of stress to strain when the *length* of the wire changes. It is also called the elastic modulus or Young's modulus and denoted by Y.

STRETCH MODULUS

$$E = \frac{\text{stretching stress}}{\text{stretching strain}} = \frac{F/A}{\Delta L/L} \quad \textbf{(9-2)}$$

EXAMPLE 9-2: When we add a mass of 2 kg to the weight hanger shown in Figure 9-2, the length of the cylindrical wire increases from 60.00 cm to 60.06 cm. The diameter of the wire is 0.5 mm. Calculate (**a**) the stress on the wire, (**b**) the strain produced, and (**c**) the stretch modulus of the wire.

Solution:

(**a**) The stress is the applied force divided by the cross-sectional area of the cylinder, which is a circle, so $A = \pi r^2 = \frac{1}{4}\pi d^2$:

$$\text{stress} = \frac{F}{A} = \frac{mg}{(\pi/4)d^2} = \frac{(2\text{ kg})(9.8\text{ m/s}^2)}{(\pi/4)(5 \times 10^{-4}\text{ m})^2} = 9.98 \times 10^7\text{ N/m}^2$$

(**b**) Now, calculate the strain produced by this stress:

$$\text{strain} = \frac{\Delta L}{L} = \frac{0.06\text{ cm}}{60\text{ cm}} = 1.00 \times 10^{-3}$$

(**c**) And calculate the stretch modulus:

$$E = \frac{F/A}{\Delta L/L} = \frac{9.98 \times 10^7\text{ N/m}^2}{1.00 \times 10^{-3}} = 9.98 \times 10^{10}\text{ N/m}^2$$

EXAMPLE 9-3: A compressional stress of $5.09 \times 10^5\text{ N/m}^2$ is applied to a steel rod 10 cm long and 1.0 cm in diameter. The elastic modulus of steel is $20 \times 10^{10}\text{ N/m}^2$. What strain does this stress produce?

Solution: Find the strain from Eq. (9-2):

$$\text{strain} = \frac{\text{stress}}{E} = \frac{5.09 \times 10^5\text{ N/m}^2}{20 \times 10^{10}\text{ N/m}^2} = 2.55 \times 10^{-6}$$

B. Bulk modulus

A solid object, such as a steel cylinder or a copper block, subjected to a high pressure decreases slightly in *volume*. For a given substance, the relative change in volume is proportional to the applied pressure.

- The **bulk modulus** B of a solid object is the ratio of stress to strain when the *volume* of the object changes.

BULK MODULUS

$$B = \frac{\text{volume stress}}{\text{volume strain}} = \frac{\Delta P}{-\Delta V/V} \quad \textbf{(9-3)}$$

The minus sign indicates that the increase in pressure causes a decrease in volume.

EXAMPLE 9-4: An aluminum cylinder 25 cm long and 8 cm in diameter is lowered approximately one kilometer into the ocean where the pressure is $1 \times 10^7\text{ N/m}^2$ greater than the atmospheric pressure. The bulk modulus of aluminum is $7.5 \times 10^{10}\text{ N/m}^2$.
Calculate the decrease in volume ΔV produced by this extreme pressure.

Solution: The original volume of the cylinder is

$$V = \pi r^2 h = \pi(0.04\ \text{m})^2(0.25\ \text{m}) = 1.26 \times 10^{-3}\ \text{m}^3$$

Now, you can use Eq. (9-3) to find the decrease in volume:

$$\Delta V = \frac{(\Delta P)V}{-B} = \frac{(1 \times 10^7\ \text{N/m}^2)(1.26 \times 10^{-3}\ \text{m}^3)}{-7.5 \times 10^{10}\ \text{N/m}^2} = -1.68 \times 10^{-7}\ \text{m}^3$$

C. Shear modulus

Under certain conditions, a force applied to a solid object can change the *shape* of the object. Figure 9-3 illustrates the result of applying a large horizontal force to the top of a rectangular block welded to a horizontal steel plane.

- A **shearing strain** $\Delta x/h$ is the displacement divided by the height.

If the force is not great enough to produce a permanent distortion of the block, the block will return to its original shape when the force is removed.

- The **shearing stress** F/A is the applied force divided by the area.
- The **shear modulus** n (sometimes denoted by S) of an object is the ratio of stress to strain when the *shape* of the object changes.

SHEAR MODULUS

$$n = \frac{\text{shearing stress}}{\text{shearing strain}} = \frac{F/A}{\Delta x/h} \qquad \textbf{(9-4)}$$

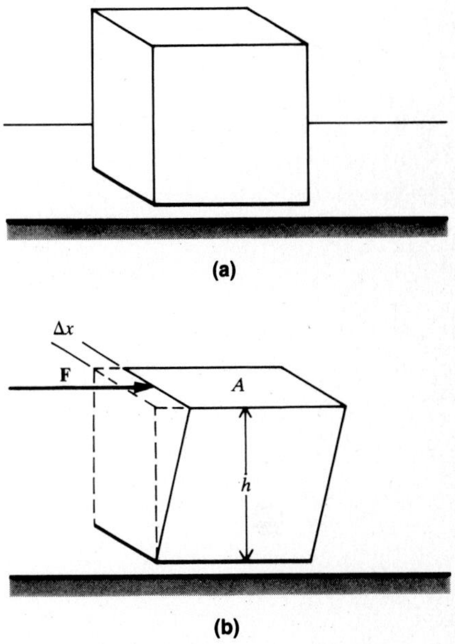

Figure 9-3. Shearing strain.

EXAMPLE 9-5: The shear modulus of cast iron is $4.6 \times 10^{10}\ \text{N/m}^2$. A 10-cm cube of iron is distorted by a horizontal force of 250 N.
What horizontal displacement will be produced?

Solution: First, solve Eq. (9-4) for the unknown displacement:

$$n = \frac{F/A}{\Delta x/h} \qquad \Delta x = \frac{Fh}{An}$$

Then, substitute the known quantities:

$$\Delta x = \frac{(250\ \text{N})(0.1\ \text{m})}{(0.1\ \text{m})^2(4.6 \times 10^{10}\ \text{N/m}^2)} = 5.43 \times 10^{-8}\ \text{m}$$

9-3. Simple Harmonic Motion, Linear

Harmonic motion is a to-and-fro motion, usually along a straight line (linear), or along the arc of a circle (angular). A body in harmonic motion must pass through an equilibrium position, slow down, eventually stop, and then return toward the equilibrium position—otherwise the motion would be lost and could not repeat itself.

- **Simple harmonic motion** (SHM) is a special type of *periodic motion*—motion that repeats itself regularly in time—resulting from the application of an elastic restoring force, $F = -kx$.

[**recall**: A restoring force is one that tries to bring a body back to its original position. (Sec. 9-1)]

In simple harmonic motion, the restoring force is proportional to the displacement of the body, so any body that obeys Hooke's law oscillates in simple harmonic motion. [**recall**: According to Hooke's law, the restoring force is proportional to the displacement. (Sec. 9-1)] Figure 9-4 illustrates a physical situation that produces the simple harmonic motion of a mass m along the x axis. Since the resulting motion is periodic, it can be illustrated by the displacement/time graph of Figure 9-5. The position of the moving mass as a function of time is

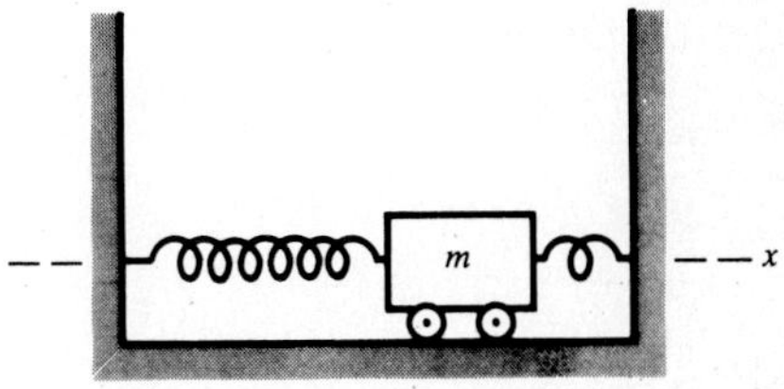

Figure 9-4. Simple harmonic motion.

DISPLACEMENT (SHM)

$$x = A \sin\left(\frac{2\pi}{T}t\right) \qquad \text{(9-5)}$$

where the **amplitude** A is the maximum value of the displacement and T is the period—the time for one complete oscillation.

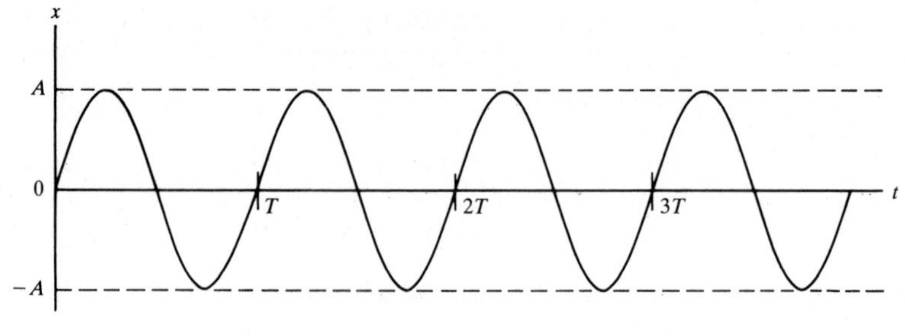

Figure 9-5. Periodicity of SHM.

A. SHM and the reference circle

An important characteristic of simple harmonic motion is its relationship to uniform circular motion. The key to solving most problems involving SHM is the fact that simple harmonic motion is the projection of uniform circular motion. By projection we mean a plot of the component along one axis as a function of time. A reference circle makes clear this relationship between SHM and uniform circular motion. [**recall**: An object moving along a circular path at constant speed is in uniform circular motion. (Sec. 4-3F)]

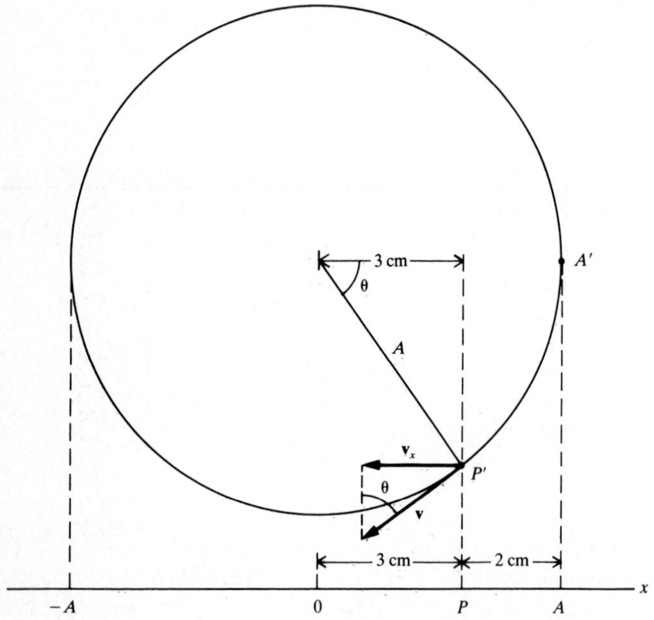

Figure 9-6

EXAMPLE 9-6: The frictionless cart illustrated in Figure 9-4 is pulled 5 cm from its equilibrium position and then released. As a result of the restoring force exerted by the springs, the cart oscillates horizontally with a period of 2 s. How long does it take for the cart to move 2 cm from its starting position?

Solution: Draw a circle with a radius representing the amplitude of the oscillation (see Figure 9-6). The cart's center of mass oscillates along the x axis between the points labeled A and $-A$. The SHM of the cart is the projection or the x component of motion of a point moving around the circle at a constant speed v. You can calculate the speed of this uniform circular motion as follows:

$$v = \frac{2\pi A}{T} = \frac{2\pi(5\text{ cm})}{2\text{ s}} = 15.71\text{ cm/s}$$

While the cart moves from point A to point P—a distance of 2 cm—the point in the reference circle moves through the angle θ from A' to P'. You can find θ from the cosine function:

$$\cos\theta = \frac{3\text{ cm}}{5\text{ cm}} = 0.60 \qquad \theta = \arccos(0.60) = 53.1^\circ$$

The angular speed in the reference circle is constant, so the time t it takes to move through θ is t/T, where the period T is the same as one complete revolution of the reference point.

$$\frac{t}{T} = \frac{53.1^\circ}{360^\circ} \qquad t = \frac{T(53.1^\circ)}{360^\circ} = \frac{(2\text{ s})(53.1^\circ)}{360^\circ} = 0.295\text{ s}$$

EXAMPLE 9-7: If the oscillating cart described in Example 9-6 has an initial speed of zero, what is the cart's speed as it passes through point P, 3 cm from the midpoint of the oscillation?

Solution: You can solve this problem by using the reference circle shown in Figure 9-6. Because the SHM along the x axis is the projection of uniform circular motion, the speed of the oscillating cart at P is the horizontal component of the velocity of P in the reference circle as it passes through P'. Use the sine function to find the horizontal component of v:

$$\sin\theta = \frac{v_x}{v} \qquad v_x = v\sin\theta = (15.71\text{ cm/s})(\sin 53.1°) = 12.6\text{ cm/s}$$

B. Frequency and period

- A **cycle** of simple harmonic motion is one complete oscillation. The **frequency** f of this periodic motion is the number of cycles per second, which is the reciprocal of the period: $f = 1/T$.
- The **angular frequency** ω of a body oscillating in simple harmonic motion is given by

ANGULAR FREQUENCY (SHM)

$$\omega = \sqrt{\frac{k}{m}} \tag{9-6}$$

where k is the spring constant and m is the mass of the oscillating object.

The period of oscillation is $2\pi/\omega$, so

PERIOD (SHM)

$$T = 2\pi\sqrt{\frac{m}{k}} \tag{9-7}$$

note: Don't confuse angular frequency ω, which is always a constant, with angular speed ω, which is variable.

EXAMPLE 9-8: Figure 9-7 shows a spring hanging vertically with a 100-g weight hanger attached to it. When an additional 200-g load is placed on the weight hanger, the length of the spring increases from 9 cm to 11 cm. When the 300-g mass is pulled down 8 cm from its equilibrium position and released, the mass vibrates vertically with an amplitude of 8 cm.
What is the period of this simple harmonic motion?

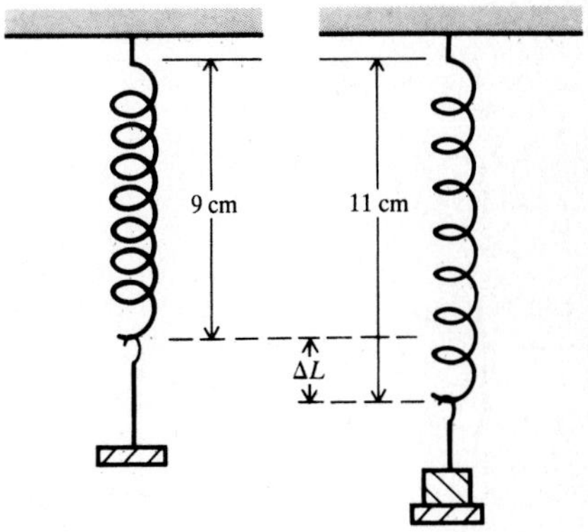

Figure 9-7

Solution: To find T, you need both m and the spring constant k. You can calculate k from Hooke's law (Eq. 9-1):

$$F = -k\,\Delta L$$

The weight of the 200-g load is the force that causes the 2-cm elongation of the spring. The restoring force exerted by the spring has the same magnitude as this force, but in the opposite direction, so

$$F = -mg = -(0.20\text{ kg})(9.8\text{ m/s}^2) = -1.96\text{ N}$$

Now, you can calculate k:

$$k = -\frac{F}{\Delta L} = \frac{-(-1.96\text{ N})}{0.02\text{ m}} = 98\text{ N/m}$$

And, from Eq. (9-7):

$$T = 2\pi\sqrt{\frac{m}{k}} = 2\pi\sqrt{\frac{0.3\text{ kg}}{98\text{ N/m}}} = 0.348\text{ s}$$

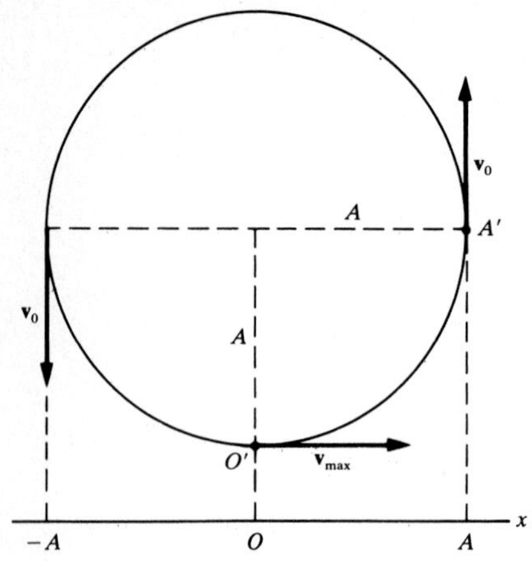

Figure 9-8

EXAMPLE 9-9: An object vibrates horizontally in simple harmonic motion with an amplitude of 4 cm and a frequency of 4 cycles/second.
What are the maximum and minimum speeds of this vibrating object?

Solution: Draw a reference circle and a horizontal line to represent the simple harmonic motion. Add velocity vectors to the reference circle to show the minimum velocity v_0 and the maximum velocity v_{max} of the reference point. Your drawing should resemble Figure 9-8. When the point in the reference circle passes through A', the direction of its velocity vector v is vertical. The horizontal component of v—the speed of the vibrating point at A—is zero.

- The *minimum speed* of an object moving in simple harmonic motion is zero; this occurs when the object is at the *maximum displacement* of its oscillation—that is, when it is farthest from its equilibrium point.

When the vibrating object passes through the midpoint of its path—point O in Figure 9-8—the direction of its velocity vector v_{max} is fully horizontal.

- A vibrating object reaches its *maximum speed* as it passes through the *midpoint* of its oscillation.

You can calculate the speed in the reference circle as follows:

MAXIMUM SPEED (SHM)

$$v_{max} = \pm \frac{2\pi A}{T} \tag{9-8}$$

So, for this example,

$$v_{max} = \pm \frac{2\pi A}{\frac{1}{f}} = \pm \frac{2\pi(4 \text{ cm})}{\frac{1}{4}} = 101 \text{ cm/s}$$

Note that the sign depends on the direction of the displacement from O—if the object is moving toward A, the sign is positive; if it is moving toward $-A$, the sign is negative.

C. Acceleration

You can find the acceleration of an object oscillating in simple harmonic motion from the acceleration of the point moving around the reference circle. [**recall**: An object moving in a circular path *always* has a centripetal acceleration directed toward the center of the circular path. (Sec. 4-3C)] The magnitude of centripetal acceleration is given by Eq. (4-12): $a_c = v^2/r$. Figure 9-9 shows the acceleration of the point in the reference circle as it passes through points O', P', and A'. You can see from the diagram that the horizontal component of a_c is zero when the vibrating object is at O.

- The acceleration of a vibrating mass is *zero* at the *midpoint* of its oscillation and reaches its *maximum* value of v^2/A at the two *end points*, A and $-A$.

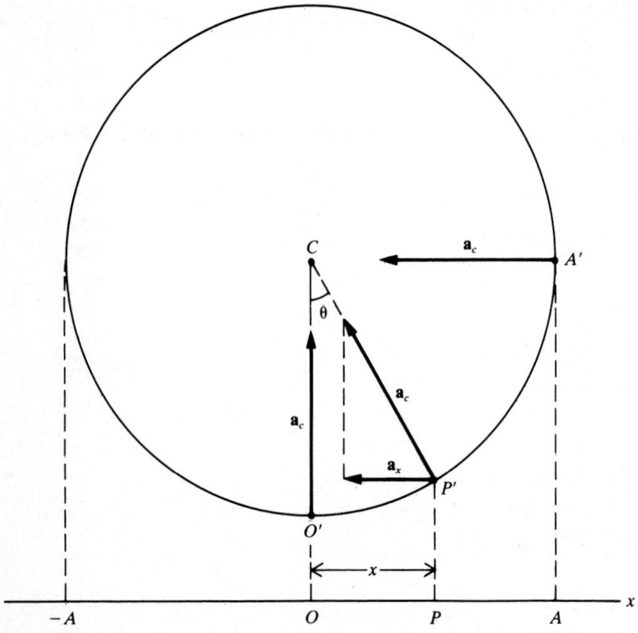

Figure 9-9. Acceleration in SHM.

EXAMPLE 9-10: A 50-g object vibrates horizontally with SHM. The amplitude of vibration is 4 cm and the frequency is 4 cycles/second, as in Example 9-9.

Find the displacement at which the acceleration of the vibrating mass is equal to one-half its maximum value.

Solution: Use the reference circle of Figure 9-9. You can find θ from the sine function:

$$\sin\theta = \frac{a_x}{a_c} = \frac{\left(\frac{1}{2}\right)a_c}{a_c} = 0.5$$

Now, use the larger triangle (hypotenuse $= |CP'| = A$) to find the unknown displacement x:

$$\sin\theta = \frac{x}{A} \qquad x = A\sin\theta = (4\ \text{cm})(0.5) = 2\ \text{cm}$$

EXAMPLE 9-11: Find the spring constant of the simple harmonic motion described in Example 9-10.

Solution: The spring constant appears in two equations associated with SHM: Eqs. (9-7) and (9-1). These equations will give you identical answers:

Equation (9-7)

$$T = 2\pi\sqrt{\frac{m}{k}}$$

$$k = \frac{4\pi^2 m}{T^2}$$

$$= \frac{4\pi^2(0.05\ \text{kg})}{(0.25\ \text{s})^2}\left(\frac{1\ \text{N}}{1\ \text{kg m/s}^2}\right)$$

$$= 31.6\ \text{N/m}$$

Equation (9-1)

$$k = \frac{F}{\Delta L} = \frac{F_{\text{max}}}{A}$$

$$= \frac{ma_c}{A} = \frac{mv^2/A}{A} = \frac{mv^2_{\text{max}}}{A^2}$$

$$= \frac{(0.05\ \text{kg})(1.005\ \text{m/s})^2}{(0.04\ \text{m})^2}$$

$$= 31.6\ \text{N/m}$$

D. The small-angle pendulum

If you pull a pendulum bob to one side and then release it, the bob oscillates back and forth. If the bob's maximum angular displacement is 10° or less, it oscillates with simple harmonic motion. Figure 9-10 shows a pendulum bob of mass m tied to a string of length ℓ. Two forces act on the pendulum bob: the gravitational force **w** and the tension **T** of the supporting string.

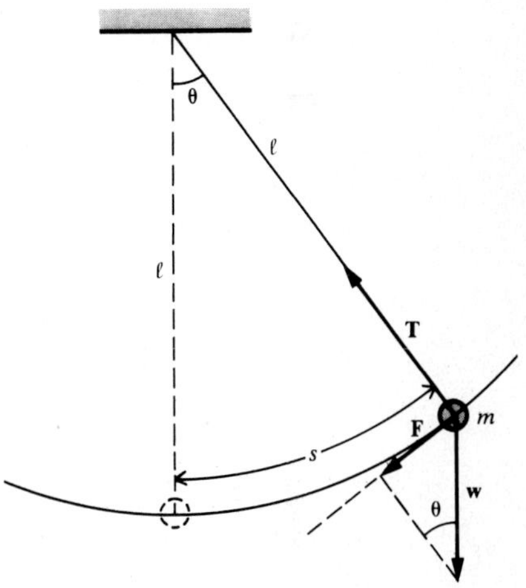

Figure 9-10. Small-angle pendulum.

EXAMPLE 9-12: Find an expression for a restoring force that acts on a pendulum bob when the angular displacement is 10° or less.

Solution: Refer again to Figure 9-10. The pendulum bob moves along the arc of a circle of radius ℓ. The tangential component of mg is the force **F** that causes the mass to accelerate along the arc:

$$F = -mg\sin\theta$$

Since this is a restoring force, its direction is opposite to the displacement s, indicated by the negative sign. If the angular displacement θ is 10° or less, you can approximate $\sin\theta$ by expressing θ in radians, as illustrated in the following table:

θ (in degrees)	θ (in rad)	$\sin\theta$	% difference
10°	0.1745	0.1736	0.5%
8°	0.1396	0.1392	0.3%
6°	0.1047	0.1045	0.2%

Even when the angle is 10°, the error made when θ (in radians) replaces $\sin\theta$ is only $\frac{1}{2}$%, and we can write

$$F = -mg\theta = -mg\left(\frac{s}{\ell}\right)$$

So, the force is proportional to the displacement. Using k to represent the constant of proportionality,

$$F = -ks = -\left(\frac{mg}{\ell}\right)s$$

Since the angular frequency $\omega = \sqrt{k/m}$, the angular frequency of the small-angle pendulum is

$$\omega = \sqrt{\frac{mg/\ell}{m}}$$

The mass of the pendulum bob cancels out, and

ANGULAR FREQUENCY (SMALL-ANGLE PENDULUM)

$$\omega = \sqrt{\frac{g}{\ell}} \tag{9-9}$$

Since the angular frequency is $\omega = 2\pi/T$, the period of the pendulum is

PERIOD (SMALL-ANGLE PENDULUM)

$$T = 2\pi\sqrt{\frac{\ell}{g}} \tag{9-10}$$

EXAMPLE 9-13: What is the length of a simple pendulum operating at sea level with a period of 1 s?

Solution: Use Eq. (9-10) and solve for the length ℓ:

$$T = 2\pi\sqrt{\frac{\ell}{g}} \qquad T^2 = 4\pi^2\left(\frac{\ell}{g}\right)$$

$$\ell = \frac{T^2 g}{4\pi^2} = \frac{(1.0\ \text{s})^2(9.8\ \text{m/s}^2)}{4\pi^2} = 0.248\ \text{m}$$

9-4. Simple Harmonic Motion, Angular

Figure 9-11 shows two views of a **torsional pendulum,** which consists of a mass at either end of a light rod suspended from a vertical wire. When the rod is rotated away from its equilibrium position through an angle β, the supporting wire exerts a torque on the rod that tends to twist it back to its equilibrium position. As a result of this restoring torque, the system oscillates about its equilibrium position. The magnitude of the restoring torque is proportional to the angle through which the system has rotated, so that it obeys Hooke's law:

HOOKE'S LAW (ANGULAR FORM)

$$\tau = -\kappa\beta \tag{9-11}$$

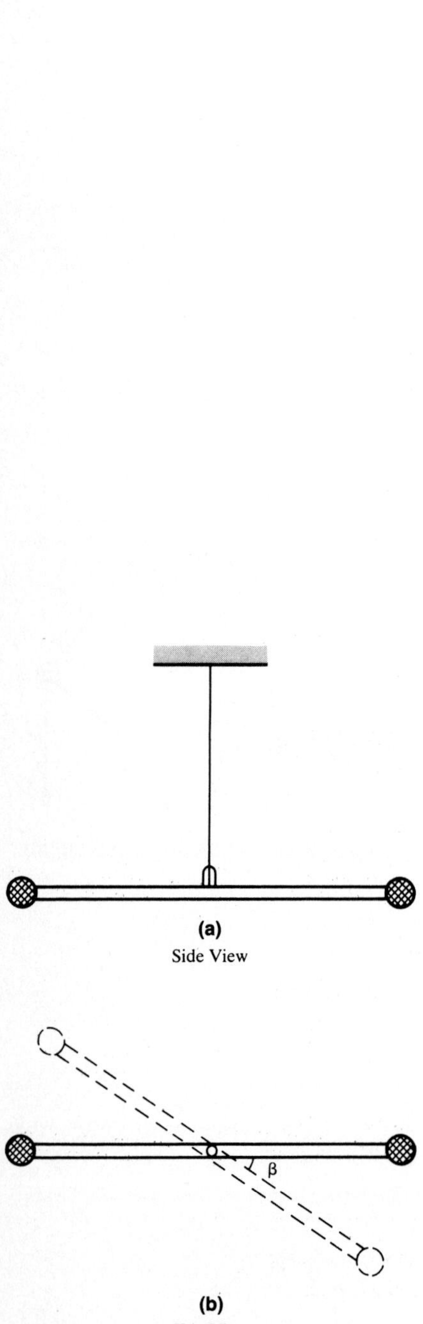

Figure 9-11. Torsional pendulum.

There is a close analogy between angular motion ($\tau = -\kappa\beta$) and linear motion ($F = -k\Delta L$) that we can use to transform the equations for linear harmonic motion into equations for angular harmonic motion. For example, the angular frequency of linear simple harmonic motion is given by Eq. (9-6): $\omega = \sqrt{k/m}$ or $\omega = \sqrt{(F/\Delta L)/m}$. Since torque is analogous to force, and the angular displacement β is analogous to linear displacement, you can replace $F/\Delta L$ with τ/β. In angular motion, the moment of inertia I plays the same role as mass does in linear motion, so we can give the angular frequency of a torsional pendulum as

ANGULAR FREQUENCY (TORSIONAL PENDULUM)

$$\omega = \sqrt{\frac{\tau/\beta}{I}} \quad \text{or} \quad \omega = \sqrt{\frac{\kappa}{I}} \tag{9-12}$$

where κ is the torsional constant. Since the period is $2\pi/\omega$,

PERIOD (TORSIONAL PENDULUM)

$$T = 2\pi\sqrt{\frac{I}{\kappa}} \qquad \textbf{(9-13)}$$

EXAMPLE 9-14: When a torsional pendulum is rotated 10° from its equilibrium position, the supporting wire exerts a restoring torque of 5×10^{-3} N m. The horizontal rod is 14 cm long and has a mass of 10 g. The spheres at the ends of the rod each have a mass of 35 g and a radius of 1.0 cm.
Calculate the period of this torsional pendulum.

Solution: The total moment of inertia of the pendulum is the sum of the moment of inertia of the rod and the moment of inertia of the two brass spheres, which you may consider to be point masses. Note that the center of rotation about the support wire is 8 cm from the center of either sphere.

$$I_{\text{tot}} = I(\text{spheres}) + I(\text{rod})$$

And, from Eqs. (4-17a) and (4-17d):

$$I_{\text{tot}} = 2mr^2 + \frac{1}{12}m\ell^2 = 2(0.035 \text{ kg})(0.08 \text{ m})^2 + \frac{1}{12}(0.01 \text{ kg})(0.14 \text{ m})^2$$
$$= 4.64 \times 10^{-4} \text{ kg m}^2$$

Now, calculate the torsional constant from Eq. (9-11):

$$\kappa = \frac{\tau}{\beta} = \frac{5 \times 10^{-3} \text{ N m}}{10° (\pi \text{ rad}/180°)} = 2.86 \times 10^{-2} \text{ N m/rad}$$

And, you can calculate the period from Eq. (9-13):

$$T = 2\pi\sqrt{\frac{I}{\kappa}} = 2\pi\sqrt{\frac{4.64 \times 10^{-4} \text{ kg m}^2}{2.86 \times 10^{-2} \text{ N m}}} = 0.800 \text{ s}$$

SUMMARY

Hooke's law	$F = -k\Delta L$	states that the stretch of an elastic material is proportional to the applied force—i.e., stress is proportional to strain
Stretch modulus	$E = \dfrac{\text{stretching stress}}{\text{stretching strain}} = \dfrac{F/A}{\Delta L/L}$	gives the ratio of stress to strain when the length of an object changes
Bulk modulus	$B = \dfrac{\text{volume stress}}{\text{volume strain}} = \dfrac{\Delta P}{-\Delta V/V}$	gives the ratio of stress to strain when the volume of an object changes
Shear modulus	$n = \dfrac{\text{shearing stress}}{\text{shearing strain}} = \dfrac{F/A}{\Delta x/h}$	gives the ratio of stress to strain when the shape of an object changes
Displacement (SHM)	$x = A \sin\left(\dfrac{2\pi}{T}t\right)$	gives the position of a mass moving with simple harmonic motion as a function of time

Angular frequency (SHM)	$\omega = \sqrt{\frac{k}{m}}$	gives the number of radians per second through which a body oscillating in simple harmonic motion moves
Period (SHM)	$T = 2\pi\sqrt{\frac{m}{k}}$	gives the time it takes a body moving in simple harmonic motion to complete one oscillation
Maximum speed (SHM)	$v_{max} = \pm\sqrt{\frac{2\pi A}{T}}$	gives the maximum speed of a vibrating object calculated from its reference circle
Angular frequency (small-angle pendulum)	$\omega = \sqrt{\frac{g}{\ell}}$	gives angular frequency for a small-angle pendulum
Period (small-angle pendulum)	$T = 2\pi\sqrt{\frac{\ell}{g}}$	gives the time it takes a small-angle pendulum to complete one oscillation
Hooke's law (angular form)	$\tau = -\kappa\beta$	states that the magnitude of a restoring torque is proportional to the angle through which a system rotates
Angular frequency (torsional pendulum)	$\omega = \sqrt{\frac{\tau/\beta}{I}}$ or $\omega = \sqrt{\frac{\kappa}{I}}$	gives the angular frequency for a torsional pendulum
Period (torsional pendulum)	$T = 2\pi\sqrt{\frac{I}{\kappa}}$	gives the time it takes a torsional pendulum to complete one oscillation

RAISE YOUR GRADES

Can you define . . . ?

- ☑ stress and strain
- ☑ stretch modulus, bulk modulus, and shear modulus
- ☑ simple harmonic motion

Can you . . . ?

- ☑ express Hooke's law for an elastic material
- ☑ describe the geometrical relationship between SHM and uniform circular motion
- ☑ calculate the maximum speed of an object vibrating with SHM, given the amplitude and period of vibration
- ☑ find the period of an oscillating mass, given its angular frequency
- ☑ write the relationship between period and frequency
- ☑ calculate the period of a vibrating mass from a knowledge of its mass and the spring constant
- ☑ calculate the time required for a vibrating mass to move a given distance
- ☑ calculate the period of a simple pendulum
- ☑ calculate the period of a torsional pendulum

SOLVED PROBLEMS

PROBLEM 9-1 Two steel wires of the same length have diameters of 0.2 mm and 0.3 mm. When you apply the same tensile force to the two wires, the thinner one stretches more than the thicker one. Calculate the ratio of the elongations, $\Delta L_1/\Delta L_2$.

Solution Solve Eq. (9-2) for the elongation, ΔL:

$$E = \frac{F/A}{\Delta L/L} = \frac{FL}{A\,\Delta L} \qquad \Delta L = \frac{FL}{AE}$$

The area of a circular cross section is $A = \frac{\pi}{4}d^2$. So, you can express the elongation of each wire as follows:

$$\Delta L_1 = \frac{FL}{(\pi/4)d_1^2 E} \qquad \Delta L_2 = \frac{FL}{(\pi/4)\,d_2^2 E}$$

Divide the first equation by the second to obtain the ratio:

$$\frac{\Delta L_1}{\Delta L_2} = \frac{d_2^2}{d_1^2} = \left(\frac{0.3\text{ mm}}{0.2\text{ mm}}\right)^2 = 2.25$$

PROBLEM 9-2 When a load of 150 g is hung on a spring, the spring's length is 20 cm. When the load is increased to 375 g, the length increases to 23 cm. When an unknown load is hung from the spring, the length is 22.2 cm. Calculate the unknown load.

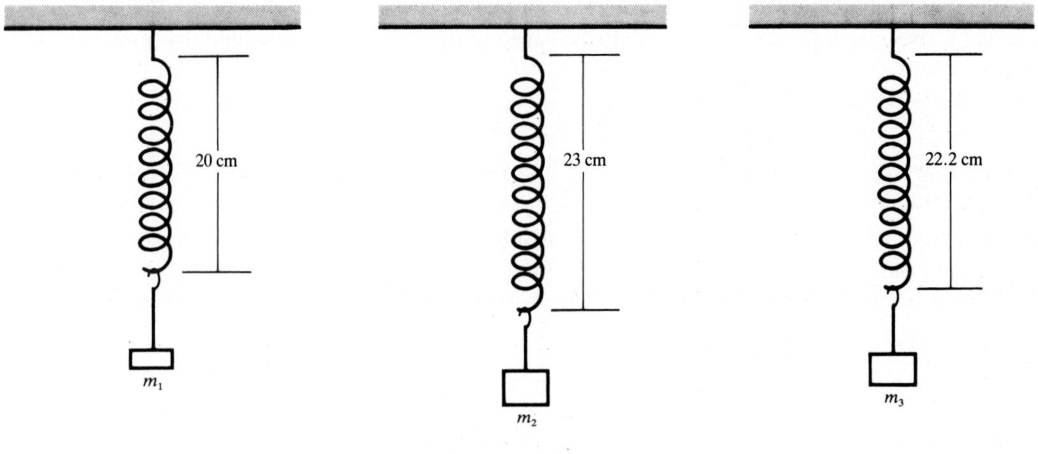

Figure 9-12

Solution The force exerted on the spring is the weight of the load (see Figure 9-12). The magnitude of this force is mg and its direction is opposite to the force exerted by the spring:

$$F = -mg$$

The added mass, Δm, causes a change ΔL in the length of the spring. When you substitute this force into Eq. (9-1), you get:

$$\Delta mg = k\,\Delta L \qquad \Delta m = \left(\frac{k}{g}\right)\Delta L$$

The change in the load which produces the elongation of 3.0 cm is

$$\Delta m = m_2 - m_1 = 375\text{ g} - 150\text{ g} = 225\text{ g}$$

Now, you can calculate the constant k/g:

$$\frac{k}{g} = \frac{\Delta m}{\Delta L} = \frac{225\text{ g}}{3.0\text{ cm}} = 75\text{ g/cm}$$

And, find the elongation produced by the unknown load:

$$\Delta L = 22.2 \text{ cm} - 20.0 \text{ cm} = 2.2 \text{ cm}$$

Calculate the change in the load which produced this elongation:

$$\Delta m = (k/g)\,\Delta L = (75 \text{ g/cm})(2.2 \text{ cm}) = 165 \text{ g}$$

So, the total unknown load is

$$m_3 = m_1 + \Delta m = 150 \text{ g} + 165 \text{ g} = 315 \text{ g}$$

PROBLEM 9-3 A rectangular brass block is firmly fastened to a horizontal surface. The block is 25 cm high, 8 cm long, and 5 cm wide. A horizontal force is applied to the upper edge of the block. As a result, the upper edge of the block is displaced a distance of 0.20 mm. Given that the shear modulus of brass is $3.4 \times 10^{10}\ \text{N/m}^2$, calculate the magnitude of the force that produced this distortion on the brass block.

Solution Solve the equation for shear modulus, Eq. (9-4), for the applied force. Watch the units carefully.

$$n = \frac{F/A}{\Delta x/h}$$

$$F = \frac{n\,\Delta x A}{h} = \frac{(3.4 \times 10^{10}\ \text{N/m}^2)(2 \times 10^{-4}\ \text{m})(0.08\ \text{m})(0.05\ \text{m})}{0.250\ \text{m}} = 1.09 \times 10^5\ \text{N}$$

PROBLEM 9-4 A 100-g mass attached to a vertical spring oscillates with an amplitude of 10 cm and a frequency of 0.5 cycles/second. The initial speed of the mass was zero and the initial displacement was −10 cm. How long does it take for the mass to reach a speed of 10 cm/s?

Solution Draw a reference circle and a vertical line to represent the path of the oscillating mass (see Fig. 9-13). Now, use Eq. (9-8) to calculate the speed of the point moving around the reference circle:

$$v_{\text{max}} = \pm\frac{2\pi A}{T} = \pm\frac{2\pi A}{1/f} = \pm\frac{2\pi(10\ \text{cm})}{(1/0.5\ \text{cycles/s})} = \pm 31.4\ \text{cm/s}$$

The moving mass is released from point $-A$ and reaches its maximum speed of 31.4 cm/s at the midpoint O of its oscillation; that is, in $\frac{1}{4}$ period. Since the period is 2 s, v_{max} is reached in 0.5 s, and the time required to reach a speed of 10 cm/s will be considerably less than 0.5 s. The speed of the oscillating mass as it passes through point P is the vertical projection of $\mathbf{v}_{\text{max}}$. Since the mass is moving toward point A, use the positive value of v_{max}, and the sine function to find the angle:

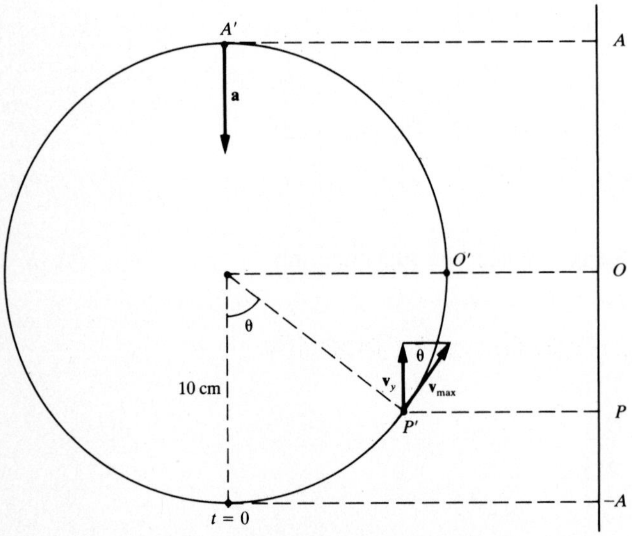

Figure 9-13

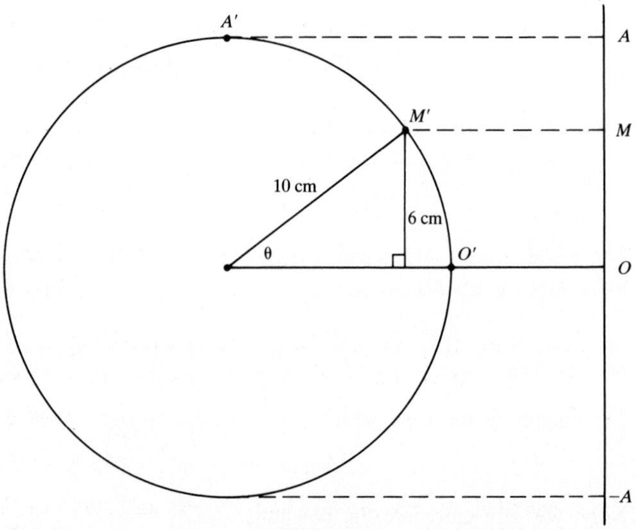

Figure 9-14

$$\sin\theta = \frac{v_y}{v_{\max}} = \frac{10\ \text{cm/s}}{31.4\ \text{cm/s}} = 0.318 \qquad \theta = \text{arc}\sin(0.318) = 18.6°$$

Because the angular speed in the reference circle is constant, you can find the time it takes to move through 18.6° by the following proportion:

$$\frac{t}{18.6°} = \frac{2\ \text{s}}{360°} \qquad t = (2\ \text{s})\left(\frac{18.6°}{360°}\right) = 0.103\ \text{s}$$

PROBLEM 9-5 Calculate the spring constant for the spring of Problem 9-4.

Solution You can find the spring constant k from Hooke's law, Eq. (9-1), or, since you already know that $m = 0.1$ kg and $T = 2$ s, more simply from Eq. (9-7):

$$T = 2\pi\sqrt{\frac{m}{k}}$$

$$k = \frac{4\pi^2 m}{T^2} = \frac{4\pi^2(0.1\ \text{kg})}{(2\ \text{s})^2} = 0.987\ \text{kg/s}^2 = 0.987\ \text{N/m}$$

PROBLEM 9-6 How long does it take for the oscillating mass in Problem 9-4 to move from $y = 0$ to $y = 6$ cm?

Solution Draw a reference circle and a vertical line, as shown in Fig. 9-14. You can find the angle θ in the reference circle from the sine function:

$$\sin\theta = \frac{y}{A} = \frac{6\ \text{cm}}{10\ \text{cm}} = 0.60 \qquad \theta = \text{arc}\sin(0.60) = 36.9°$$

and the time it takes to move through θ from the following proportion:

$$\frac{t}{36.9°} = \frac{T}{360°} \qquad t = (2\ \text{s})\left(\frac{36.9}{360}\right) = 0.205\ \text{s}$$

PROBLEM 9-7 A simple pendulum at sea level has a period of 1.0000 s. At what altitude will the period of this pendulum be 1.0002 s? (The mean radius of the earth at sea level is 6.38×10^6 m.)

Solution As you know from Eq. (9-10), the period of a simple pendulum depends on the local value of g. You can express the gravitational force the earth exerts on a mass m in two ways:

$$w = mg \quad \text{and} \quad F = \frac{GMm}{r^2}$$

Equate these two expressions for the gravitational force and cancel the common factor m:

$$mg = \frac{GMm}{r^2} \qquad g = \frac{GM}{r^2}$$

Substitute this expression for g in Eq. (9-10) for the period:

$$T = 2\pi\sqrt{\frac{\ell}{g}} = 2\pi\sqrt{\frac{\ell r^2}{GM}}$$

At sea level, the period is

$$T_e = 2\pi\left(\sqrt{\frac{\ell}{GM}}\right)r_e$$

where r_e is the radius of the earth. At an altitude of h, the distance from the center of the earth is $r = r_e + h$. The period at this greater distance from the earth is

$$T = 2\pi\left(\sqrt{\frac{\ell}{GM}}\right)r$$

Now, calculate the ratio of the two periods and solve for h:

$$\frac{T}{T_e} = \frac{r}{r_e} = \frac{r_e + h}{r_e} = 1 + \frac{h}{r_e} \qquad h = r_e\left(\frac{T}{T_e} - 1\right) = (6.38 \times 10^6\ \text{m})(2 \times 10^{-4}) = 1.28 \times 10^3\ \text{m}$$

PROBLEM 9-8 When a mass of 300 g is fastened to a spring, the frequency of its simple harmonic vibration is 2 cycles/s. When an additional mass Δm is added, the frequency of vibration becomes 1.5 cycles/s. Calculate how many grams were added.

Solution The period of a vibrating mass is given by Eq. (9-7):

$$T = 2\pi\sqrt{\frac{m}{k}}$$

The frequency is the reciprocal of the period, so

$$f = \frac{1}{2\pi}\sqrt{\frac{k}{m}}$$

Use f' to represent the new frequency associated with the increased mass, $m' = m + \Delta m$:

$$f' = \frac{1}{2\pi}\sqrt{\frac{k}{m'}}$$

Now, calculate the ratio of the two frequencies:

$$\frac{f}{f'} = \sqrt{\frac{m'}{m}}$$

And, solve for the new mass m':

$$m' = m\frac{f^2}{(f')^2} = (300\text{ g})\frac{(2)^2}{(1.5)^2} = 533\text{ g}$$

The mass that was added is

$$\Delta m = m' - m = 533\text{ g} - 300\text{ g} = 233\text{ g}$$

PROBLEM 9-9 A torsional pendulum is constructed by fastening a solid disk to a wire that hangs vertically. The wire is fastened to the center of the disk so that the disk will oscillate about its central axis. The torsional constant κ of the wire is 0.05 N m/rad. The disk has a mass of 600 g. Find the radius of the disk that will cause the period of the torsional pendulum to be 1.0 s.

Solution Write down the formula for the period of a torsional pendulum, Eq. (9-13):

$$T = 2\pi\sqrt{\frac{I}{\kappa}}$$

The moment of inertia of a solid disk about its central axis is $I = \frac{1}{2}mr^2$. Now, substitute this expression for I into Eq. (9-13):

$$T = 2\pi\sqrt{\frac{mr^2}{2\kappa}} = \pi\left(\sqrt{\frac{2m}{\kappa}}\right)r$$

And, solve for r:

$$r = \frac{T}{\pi}\sqrt{\frac{\kappa}{2m}} = \frac{(1.0\text{ s})}{\pi}\sqrt{\frac{0.05\text{ N m}}{2(0.60\text{ kg})}} = 6.50 \times 10^{-2}\text{ m} = 6.50\text{ cm}$$

Supplementary Exercises

PROBLEM 9-10 A coiled spring has a length of 14 cm when it is supporting a 150-g load. When the load is increased to 450 g, the length of the spring becomes 17 cm. What is the spring constant?

PROBLEM 9-11 A piece of wire 40 cm long is hanging vertically and supporting a weight of 0.5 N. When the weight is increased to 5.5 N, the length of the wire becomes 40.028 cm. The diameter of the wire is 0.3 mm. Calculate the stretch modulus of this wire.

PROBLEM 9-12 A plastic bag containing pure water is lowered to a depth of 1.5 km in the ocean. At this depth the pressure is 1.507×10^7 N/m^2 greater than the atmospheric pressure. If the original volume of the water was 500 cm^3, what will be the decrease in volume caused by this large increase in pressure? [The bulk modulus of water is 2.0×10^9 N/m^2.]

PROBLEM 9-13 A vertical steel shaft, having a diameter of 1.0 cm, is fastened to a horizontal plate. When a horizontal force of 150 N is applied to the top of the shaft, 0.50 m above the plate to which it is fastened, what horizontal displacement will be produced? [The shear modulus of steel is 8.0×10^{10} N/m^2.]

PROBLEM 9-14 A 50-g mass is vibrating vertically with simple harmonic motion. The amplitude of the vibration is 8 cm and the frequency is 0.398 cycles/s. Calculate the maximum speed of this vibrating mass.

PROBLEM 9-15 At what displacement will the vibrating mass in the preceding problem have a speed of 10.0 cm/s?

PROBLEM 9-16 An object vibrating in simple harmonic motion has an amplitude of 5 cm and a period of 1.2 s. How long does it take for this object to move 2 cm from its maximum displacement of 5 cm?

PROBLEM 9-17 A 100-g object hangs from a spring. The period of the simple harmonic vibration is 0.5 s. What is the spring constant?

Answers to Supplementary Exercises

9-10: 98 N/m

9-11: 10.1×10^{10} N/m^2

9-12: -3.77×10^{-6} m^3

9-13: 1.19×10^{-5} m

9-14: 20.0 cm/s

9-15: 6.93 cm

9-16: 0.177 s

9-17: 15.8 N/m

10 FLUIDS

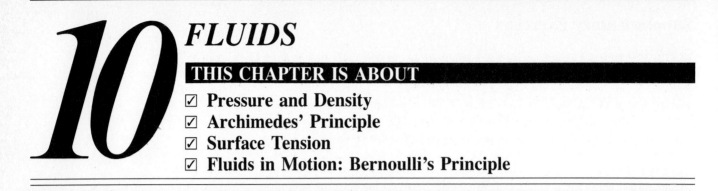

THIS CHAPTER IS ABOUT

☑ Pressure and Density
☑ Archimedes' Principle
☑ Surface Tension
☑ Fluids in Motion: Bernoulli's Principle

- **Fluids** are substances, such as liquids and gases, that have no rigidity. Consequently, a fluid lacks a fixed shape and assumes the shape of its container.

10-1. Pressure and Density

- The **pressure** P acting on a fluid is the force exerted perpendicularly per unit of the fluid's surface area.

PRESSURE $$P = \frac{F}{A} \qquad \textbf{(10-1)}$$

Since pressure has no particular direction, it is a scalar quantity measured in units of force per unit area. In SI, a pressure of one newton per square meter (N/m^2) is defined as one pascal (Pa). The atmospheric pressure at sea level (1 atm) is equivalent to 1.013×10^5 Pa or 101.3 kPa.

The shear modulus—the ratio of stress to strain when shape changes (see Sec. 9-2)—is zero for any fluid, since fluid has no fixed shape. For example, if a fluid were subjected to a tangential force, the "layers" of the liquid would slide past one another without friction. This means that a fluid can sustain only a perpendicular force and, conversely, can exert only a force perpendicular to the surface.

- The **density** d of a substance is its mass per unit volume.

DENSITY $$d = \frac{m}{V} \qquad \textbf{(10-2)}$$

Density is a characteristic property of a substance; objects composed of the same substance, whatever their size or mass, have the same density under the same conditions of temperature and pressure. Temperature and pressure affect the density of substances, appreciably for gases, but only slightly for liquids and solids.

EXAMPLE 10-1: A long cylindrical steel pipe with an inside diameter of 2 cm stands vertically, and is filled with water to a depth of 50 m. The density of water is 1×10^3 kg/m^3.
Calculate the pressure produced by the column of water at the bottom of the pipe.

Solution: The force the column of water exerts is the weight of the water, $F = w = mg$. Since density d is mass per unit volume, or $d = m/V$, you can write $m = dV$, where V is the volume of the water column. Volume is equal to the cross-sectional area times the height, so $V = Ah$. Now, substitute these terms into the definition of pressure (Eq. 10-1) and simplify:

$$P = \frac{F}{A} = \frac{mg}{A} = \frac{dVg}{A} = \frac{d(Ah)g}{A} = dhg$$

Since the cross-sectional area cancels out, the equation for the pressure exerted by a regular column of fluid of height h is

COLUMNAR FLUID PRESSURE

$$P = dgh \tag{10-3}$$

Now, you can substitute the numerical values:

$$P = dgh = (1 \times 10^3 \text{ kg/m}^3)(9.8 \text{ m/s}^2)(50 \text{ m})$$
$$= 4.9 \times 10^5 \text{ N/m}^2 = 4.9 \times 10^2 \text{ kPa}$$

note: You'll find that some textbooks use the symbol ρ for density.

EXAMPLE 10-2: What is the pressure due to water at a depth of 7.50 km below sea level?

Solution: The density of sea water is 1.025×10^3 kg/m^3, so from Eq. (10-3):

$$P = dgh = (1.025 \times 10^3 \text{ kg/m}^3)(9.8 \text{ m/s}^2)(7.5 \times 10^3 \text{ m}) = 7.53 \times 10^7 \text{ N/m}^2$$

note: The *total* pressure is greater because of the pressure of the atmosphere on the surface of the water.

- **Absolute** (or **total**) **pressure** is columnar fluid pressure (also called **gauge pressure**) plus atmospheric pressure.

A. Transmission of pressure: Pascal's principle

- **Pascal's principle** states that a change in pressure in a confined fluid is transmitted without change to all points in the fluid.

Consider the two connected cylindrical pipes shown in Figure 10-1, in which the taller pipe opens to the atmosphere. The absolute (or total) pressure at point M results from the pressure of the column of water plus the additional atmospheric pressure. Point N is at the same level, so the absolute pressure there is the same as at point M. The pressure at point O is somewhat less, because the relative depth at point O is less than at point M.

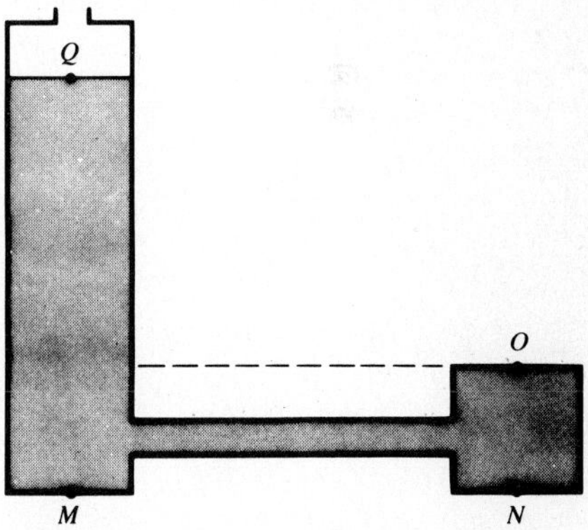

Figure 10-1. Two connected cylinders filled with water.

EXAMPLE 10-3: In the system shown in Figure 10-1, the depth of water in the two reservoirs is 4 m and 1 m, respectively, and point O is one meter above point N.
Calculate the absolute pressures at points M, N, and O when the applied pressure at point Q is (**a**) 1 atm and (**b**) 1.2 atm.

Solution:
(**a**) At one atmosphere,

$$P_M = P_N = 1 \text{ atm} + dgh$$
$$= 1.013 \times 10^2 \text{ kPa} + (1 \times 10^3 \text{ kg/m}^3)(9.8 \text{ m/s}^2)(4 \text{ m})$$
$$= (1.013 + 0.392) \times 10^2 \text{ kPa} = 1.405 \times 10^2 \text{ kPa}$$

Since the relative depth at point O is 3 m,

$$P_O = 1.013 \times 10^2 \text{ kPa} + (1 \times 10^3 \text{ kg/m}^3)(9.8 \text{ m/s}^2)(3 \text{ m})$$
$$= 1.307 \times 10^2 \text{ kPa}$$

(**b**) Calculate the change in pressure at Q:

$$P = 0.2 \text{ atm}\left(\frac{1.013 \times 10^5 \text{ Pa}}{1 \text{ atm}}\right) = 0.203 \times 10^2 \text{ Pa}$$

This pressure increase is transmitted to every point in the fluid, so all you have to do is add this pressure to the pressures already calculated. Use a prime to indicate the new pressure at each point.

$$P'_M = P_M + 0.203 \times 10^2 \text{ kPa} = 1.608 \times 10^2 \text{ kPa}$$

$$P'_N = P_N + 0.203 \times 10^2 \text{ kPa} = 1.608 \times 10^2 \text{ kPa}$$

$$P'_O = P_O + 0.203 \times 10^2 \text{ kPa} = 1.510 \times 10^2 \text{ kPa}$$

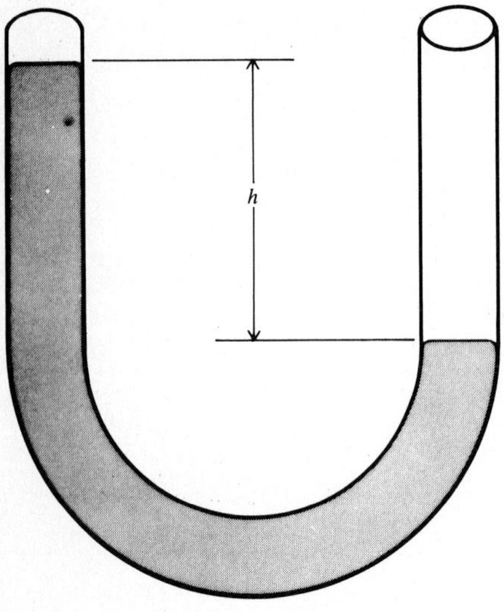

Figure 10-2. Mercury barometer.

EXAMPLE 10-4: The U-shaped glass tube shown in Figure 10-2 contains mercury (density = 13.6×10^3 kg/m^3). The pressure above the sealed, left-hand arm is zero; the pressure at the surface of the mercury in the right-hand arm is 1 atm. Calculate the height h of the mercury column.

Solution: The atmospheric pressure on the open surface of the mercury P_2 is equal to the pressure at the upper surface P_1 plus the additional pressure due to the height of the mercury. Since the left arm is sealed, $P_1 = 0$, and

$$P_2 = P_1 + dgh = 0 + dgh$$

Now, solve for the unknown height:

$$h = \frac{P_2}{dg} = \frac{1.013 \times 10^5 \text{ N/m}^2}{(13.6 \times 10^3 \text{ kg/m}^3)(9.8 \text{ m/s}^2)} = 0.760 \text{ m} = 760 \text{ mm}$$

This U-shaped tube is a mercury barometer that measures atmospheric pressure.

EXAMPLE 10-5: A small submarine operates at a depth of 6.50 km, taking scientific photographs of the ocean floor through a circular window 16 cm in diameter. The pressure inside the sub is 1 atm, and the density of sea water is 1.025×10^3 kg/m^3.
What is the total force acting on the window?

Solution: The pressure difference across the window is the total pressure minus the pressure inside the sub, $\Delta P = P_{tot} - 1$ atm. The total pressure at this depth is $P_{tot} = 1 \text{ atm} + dgh$. So, the pressure difference is

$$\begin{aligned}\Delta P &= dgh \\ &= (1.025 \times 10^3 \text{ kg/m}^3)(9.8 \text{ m/s}^2)(6.5 \times 10^3 \text{ m}) = 6.53 \times 10^7 \text{ N/m}^2\end{aligned}$$

To find the net force on the window, use Eq. (10-1):

$$F = PA = (6.53 \times 10^7 \text{ N/m}^2)\left(\frac{\pi}{4}\right)(0.16 \text{ m})^2 = 1.31 \times 10^6 \text{ N}$$

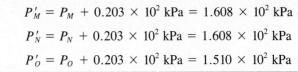

B. The hydraulic lift

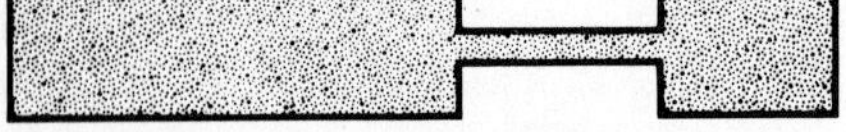

Figure 10-3. The hydraulic lift.

The hydraulic piston apparatus illustrated in Figure 10-3 uses an incompressible fluid to transmit pressure from the small cylinder to the large cylinder. According to Pascal's principle, the pressure in the small cylinder that results from applying the force $\mathbf{F}_1$ to the frictionless piston is transmitted undiminished to the larger piston, so $P_1 = P_2$. Since pressure is force per unit area, you can write this equation as $F_1/A_1 = F_2/A_2$, and by rearranging terms you can find the ratio of the forces:

HYDRAULIC FORCE MULTIPLICATION

$$\frac{F_2}{F_1} = \frac{A_2}{A_1} \qquad \textbf{(10-4)}$$

A_2 is much larger than A_1, so the force exerted *by* the large piston is much greater than the force exerted *on* the small piston.

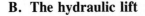

EXAMPLE 10-6: In Figure 10-3, the diameter of the large cylinder is 25 cm and the diameter of the small cylinder is 5 cm.
By what factor is the force multiplied?

Solution: From Eq. (10-3),

$$\frac{F_2}{F_1} = \frac{A_2}{A_1} = \frac{\left(\frac{\pi}{4}\right)d_2^2}{\left(\frac{\pi}{4}\right)d_1^2} = \left(\frac{d_2}{d_1}\right)^2 = \left(\frac{25\text{ cm}}{5\text{ cm}}\right)^2 = 25$$

Since the hydraulic lift produces a multiplication of the applied force, you can regard it as a machine with an actual advantage (AMA) equal to the ratio of the two forces (see Eq. 6-12):

AMA: HYDRAULIC LIFT

$$\text{AMA} = \frac{F_2}{F_1} \qquad \textbf{(10-5)}$$

The efficiency of a hydraulic lift with frictionless pistons is 100%.

10-2. Archimedes' Principle

When an object is immersed in a fluid, the object appears to weigh less.

- **Archimedes' principle** states that the buoyant force exerted on a body wholly or partly immersed in a fluid is equal to the weight of the fluid displaced by the body.

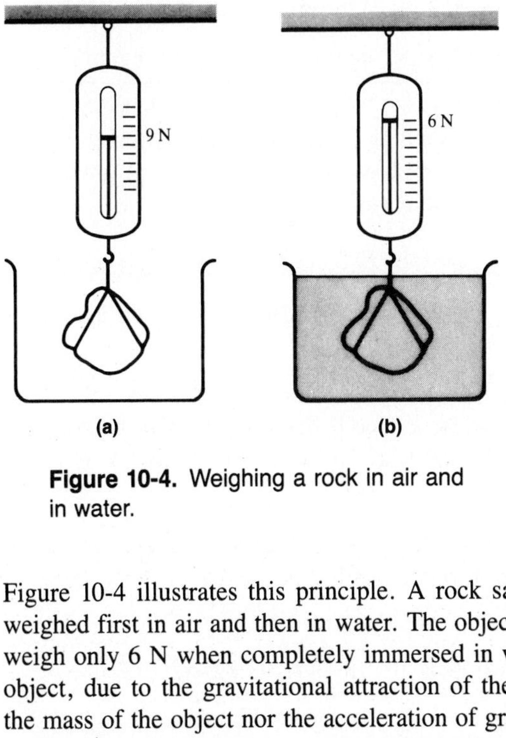

Figure 10-4. Weighing a rock in air and in water.

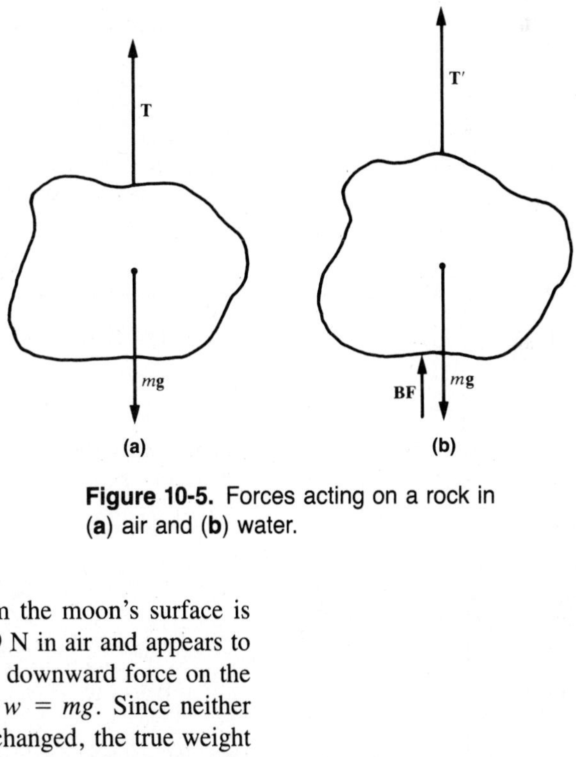

Figure 10-5. Forces acting on a rock in (**a**) air and (**b**) water.

Figure 10-4 illustrates this principle. A rock sample from the moon's surface is weighed first in air and then in water. The object weighs 9 N in air and appears to weigh only 6 N when completely immersed in water. The downward force on the object, due to the gravitational attraction of the earth, is $w = mg$. Since neither the mass of the object nor the acceleration of gravity has changed, the true weight of the object must be the same whether it is immersed in air or in water. The *apparent* loss of weight is due to an upward force of 3 N that acts on the object when it is immersed in water. The forces acting on the object in the two situations are illustrated in Figure 10-5.

- The upward force on the object when it is immersed in water is called the **buoyant force** (BF).
- The **specific gravity** (sp. gr.) of a substance is the ratio of the density of the substance d_x to the density of water d_w. By choosing equal volumes and a constant gravitational force **g**, you can also express specific gravity as a ratio of weights:

SPECIFIC GRAVITY

$$\text{sp. gr.} = \frac{d_x}{d_w} = \frac{w_x}{w_w} \qquad \textbf{(10-6)}$$

Density is mass per unit volume, so you can find the specific gravity of any substance as follows:

SPECIFIC GRAVITY (SUBSTANCE)

$$\text{sp. gr.} = \frac{\text{weight of substance}}{\text{weight of equal volume of water}} \qquad \textbf{(10-7)}$$

Specific gravity is dimensionless and has no units.

EXAMPLE 10-7: Find (**a**) the specific gravity and (**b**) the density of the moon rock pictured in Figures 10-4 and 10-5.

Solution:

(**a**) Use Eq. (10-6) to find the specific gravity. The weight of the moon rock is 9 N. The weight of an equal volume of water is the buoyant force on the rock when immersed in water:

$$w_w = \text{BF} = \text{true weight} - \text{apparent weight}$$
$$= 9\text{ N} - 6\text{ N} = 3\text{ N}$$

And now you can calculate the specific gravity:

$$\text{sp. gr.} = \frac{w_x}{w_w} = \frac{9\text{ N}}{3\text{ N}} = 3$$

(**b**) You can use the same definition (Eq. 10-6) to solve for the density of the rock:

$$d_x = (\text{sp. gr.})d_w = 3(1.0 \times 10^3\text{ kg/m}^3) = 3.0 \times 10^3\text{ kg/m}^3$$

There is a buoyant force present when an object is immersed in any fluid, including air. The density of air is so small, however, that its buoyant force is usually negligible when compared to the weight of a solid substance.

EXAMPLE 10-8: Given that the density of air is 1.293 kg/m^3, find the buoyant force of air on the moon rock.

Solution: The buoyant force on the rock when immersed in air is equal to the weight of the air displaced by the rock, so you'll first have to find the volume of the rock. From the definition of density (Eq. 10-2): $d = m/V$, and $V = m/d = (w/g)/d$, so

$$V = \frac{9\text{ N}}{(3.0 \times 10^3\text{ kg/m}^3)(9.8\text{ m/s}^2)} = 3.06 \times 10^{-4}\text{ m}^3$$

The buoyant force is the weight of an equal volume of air:

$$\text{BF} = w = mg = Vdg = (3.06 \times 10^{-4}\text{ m}^3)(1.293\text{ kg/m}^3)(9.8\text{ m/s}^2)$$
$$= 3.88 \times 10^{-3}\text{ N}$$

The percentage of error made by neglecting the buoyant force of the air is very small:

$$\frac{\text{BF}}{w_x} = \frac{3.88 \times 10^{-3}\text{ N}}{9\text{ N}} \times 100\% = 0.04\%$$

EXAMPLE 10-9: A helium-filled balloon has a volume of 2000 m^3. The density of helium is 0.178 kg/m^3, and the total weight of the balloon (fabric, ropes, and gondola) is 4000 N. Calculate the maximum load the balloon can lift.

Solution: First, calculate the total weight of the system (helium and balloon). The weight of the helium is

$$w = mg = Vdg = (2 \times 10^3 \text{ m}^3)(0.178 \text{ kg/m}^3)(9.8 \text{ m/s}^2) = 3.49 \times 10^3 \text{ N}$$

The total downward force is

$$w_{tot} = (3.49 + 4.00) \times 10^3 \text{ N} = 7.49 \times 10^3 \text{ N}$$

Next, calculate the buoyant force due to the displaced air:

$$\text{BF} = w = mg = Vdg$$
$$= (2 \times 10^3 \text{ m}^3)(1.293 \text{ kg/m}^3)(9.8 \text{ m/s}^2) = 2.53 \times 10^4 \text{ N}$$

Now, let L represent the maximum load that the balloon can lift, which is the buoyant force minus the total system weight:

$$L = \text{BF} - w_{tot} = 2.53 \times 10^4 \text{ N} - 7.49 \times 10^3 \text{ N} = 1.78 \times 10^4 \text{ N}$$

10-3. Surface Tension

Left to itself, a liquid tends to assume a shape that has a minimum surface area for a given volume, so that the potential energy (see Sec. 6-2) is also at a minimum.

- Cohesive forces attract the molecules of a liquid toward each other. In the interior of the liquid, where each molecule is surrounded by others, the forces are acting in all directions and so cancel each other out. But at the liquid's surface, there is a net force toward the interior, creating **surface tension**.

To increase the surface area of a liquid, we must supply energy to pull some of the molecules away from their neighbors. As a result, the surface PE of the liquid also increases.

Two of the phenomena that result from surface tension are shown in Figure 10-6.

(1) It is possible to overfill a container to a point where the surface of the liquid is higher than the edge of the container.
(2) A needle placed carefully on the surface of water will remain there indefinitely.

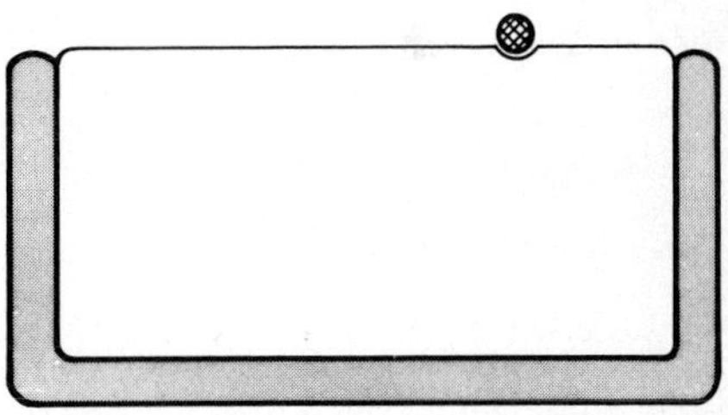

Figure 10-6. A needle supported by surface tension.

Figure 10-7 illustrates the tensile forces of the surface layer that hold the needle in equilibrium.

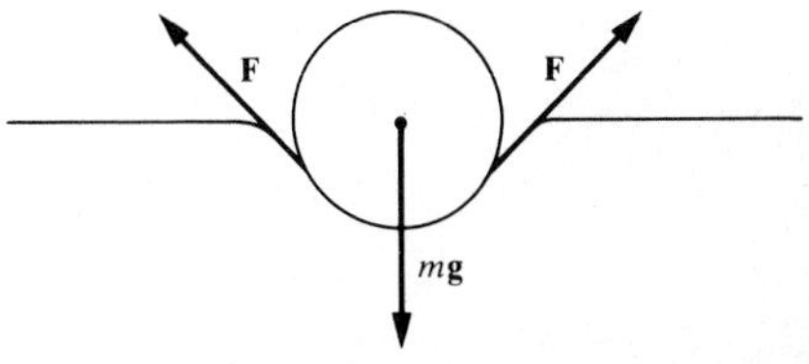

Figure 10-7. Forces acting on a floating needle.

- The potential energy per unit surface area—or the work necessary to increase the liquid's surface area per unit surface area—is the **coefficient of surface tension** γ:

COEFFICIENT OF SURFACE TENSION

$$\gamma = \frac{F\,\Delta s}{L\,\Delta s} = \frac{F}{L} \tag{10-8}$$

where $F\,\Delta s$ represents the work required to increase the surface area by $L\,\Delta s$. Since the distance Δs can be cancelled, γ can be interpreted as a *contractile force* per unit length.

The **Du Nouy torsion balance** is a standard method of measuring the coefficient of surface tension. A ring of fine platinum wire, 4.00 cm in circumference, is lowered into the liquid and then carefully pulled upward so that a vertical, thin film of liquid remains in contact with the ring, as shown in Figure 10-8.

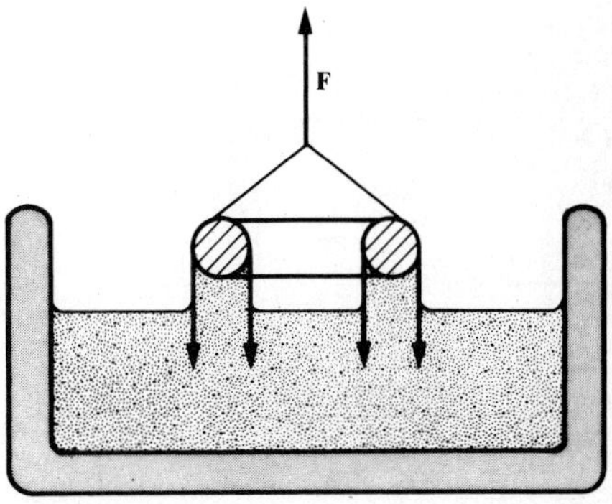

Figure 10-8. The Du Nouy torsion balance.

EXAMPLE 10-10: The upward force on the platinum ring in a Du Nouy torsion balance pulled above the surface of water is 5.82×10^{-3} N. The weight of the ring, 4.0 cm in circumference, has been subtracted.
Calculate the coefficient of surface tension of water from these data.

Solution: The total length of the surface film in contact with the ring is approximately twice the circumference of the ring—that is, one film layer on the outside of the ring and a second layer inside the ring. So, $L = 2C$, and you can calculate the coefficient of surface tension from Eq. (10-8):

$$\gamma = \frac{F}{L} = \frac{5.82 \times 10^{-3}\ \text{N}}{2(4\ \text{cm})(1\ \text{m}/100\ \text{cm})} = 7.28 \times 10^{-2}\ \text{N/m}$$

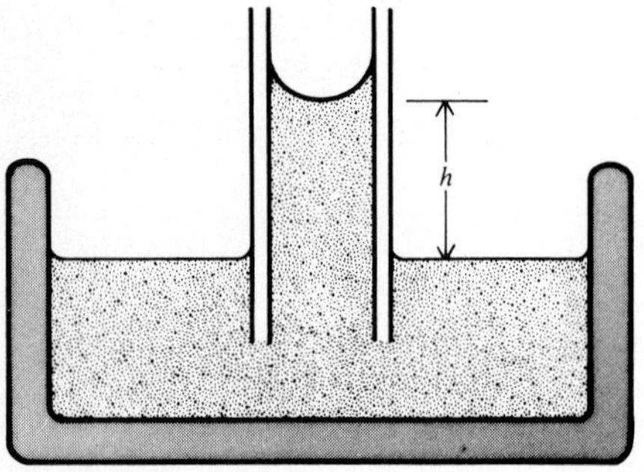

Figure 10-9. Rise of a liquid in a capillary tube.

A second method of measuring the coefficient of a liquid's surface tension is based on the vertical rise of liquid in a capillary tube—a tube with a very small bore. Figure 10-9 shows the rise of a liquid in a small glass tube in which the liquid "wets" the tube—that is, the adhesive forces between glass and liquid are greater than the cohesive forces within the liquid. The liquid hangs from the circle where it contacts the tube's inner wall, and surface tension produces the total upward force on the cylindrical column of liquid. The magnitude of this force is the force per unit length γ multiplied by the inside circumference of the capillary tube:

$$\text{upward force} = \gamma(2\pi r)$$

The downward force on the cylinder of liquid is its weight:

$$\text{downward force} = mg = Vdg = (\pi r^2 h)dg$$

After the liquid has risen to its maximum height, the column of liquid is in equilibrium, which means that the upward force and the downward force have the same magnitude:

$$\gamma(2\pi r) = (\pi r^2 h)dg$$

Now, we can find the height of the liquid:

CAPILLARY RISE $$h = \frac{2\gamma}{rdg} \qquad \textbf{(10-9)}$$

EXAMPLE 10-11: Ethyl alcohol has a coefficient of surface tension of 2.23×10^{-2} N/m, and a density of 8.0×10^2 kg/m^3.
To what height will ethyl alcohol rise in a capillary tube with an inside diameter of 0.2 mm?

Solution: You can calculate the height directly from Eq. (10-9):

$$h = \frac{2\gamma}{rdg} = \frac{2(2.23 \times 10^{-2}\ \text{N/m})}{(1.0 \times 10^{-4}\ \text{m})(8.0 \times 10^2\ \text{kg/m}^3)(9.8\ \text{m/s}^2)}$$
$$= 5.69 \times 10^{-2}\ \text{m} = 56.9\ \text{mm}$$

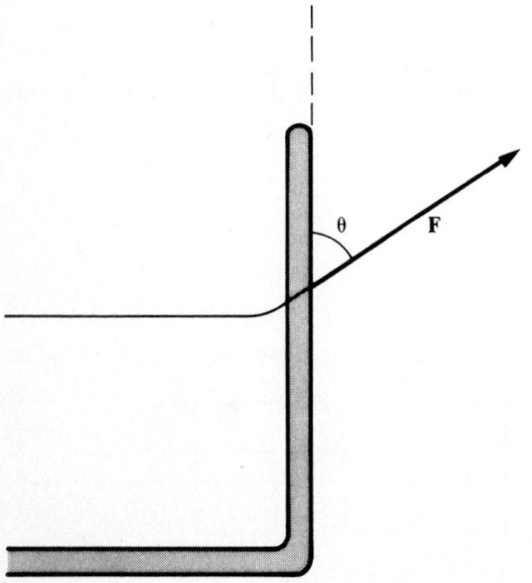

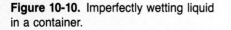

Figure 10-10. Imperfectly wetting liquid in a container.

If a liquid were to wet the surface of its container perfectly, the liquid would rise along the container wall so smoothly as to meet the wall at an angle of 0°. But for most combinations of liquid and solid, the liquid meets the solid wall at an angle greater than 0° (see Figure 10-10). If the cohesive forces within the liquid are less than the adhesive forces between the liquid and the container wall, then $\theta < 90°$ and we say that the contact is wetting. If the cohesive forces are stronger than the adhesive, then $\theta > 90°$ and we say that the contact is nonwetting. For example, the angle of contact of kerosene in a glass container is 26° (a wetting contact), while for mercury in a glass container the angle is 140° (a nonwetting contact). The vertical component of the force due to surface tension provides the upward force, so the equation for the height of the liquid in the capillary tube becomes

CAPILLARY RISE $$h = \frac{2\gamma \cos\theta}{rdg} \qquad \textbf{(10-10)}$$

10-4. Fluids in Motion: Bernoulli's Principle

Figure 10-11 shows a tank filled with fluid. The fluid enters the tank through a small inlet pipe and leaves the tank through a large outlet pipe. If the fluid is incompressible—that is, a change in pressure does not cause a change in volume as in water, for example— the volume of fluid entering the tank per second must be equal to the volume leaving the tank per second. Let ΔV_1 represent the volume of fluid entering the tank during a time interval Δt, and ΔV_2 the volume leaving the tank, so that we have the following relationship for an incompressible fluid:

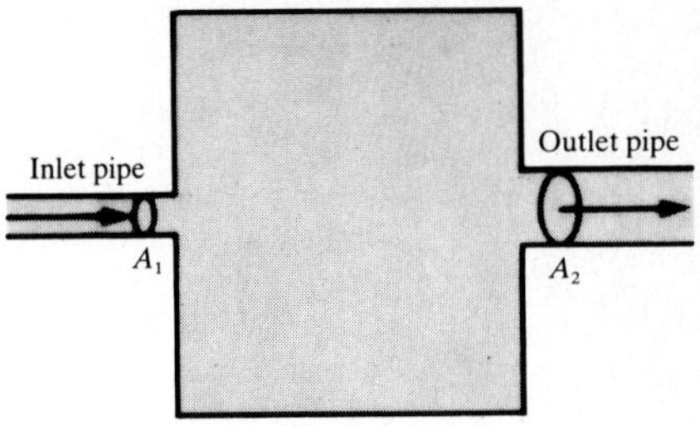

Figure 10-11. Volumetric flow rate.

VOLUMETRIC FLOW RATE (INCOMPRESSIBLE FLUID)

$$\frac{\Delta V_1}{\Delta t} = \frac{\Delta V_2}{\Delta t} \quad \textbf{(10-11)}$$

Let v represent the speed with which a liquid moves in a cylindrical pipe, so that during a time interval Δt, the liquid moves a distance $v\,\Delta t$. The volume of liquid that passes a cross-section A is then given by $\Delta V = Av\,\Delta t$. Dividing this by Δt gives the volumetric flow rate:

VOLUMETRIC FLOW RATE

$$\frac{\Delta V}{\Delta t} = Av \quad \textbf{(10-12)}$$

And, because the liquid is incompressible, the volumetric flow rate is the same in both pipes. So, we have

VOLUMETRIC FLOW RATE (SYSTEM)

$$A_1 v_1 = A_2 v_2 \quad \textbf{(10-13)}$$

EXAMPLE 10-12: The flow rate into the tank shown in Figure 10-11 is 8.0 cm^3/s. The inside diameters of the inlet pipe and outlet pipe are 1.0 cm and 2.0 cm, respectively.
Calculate the speed of the liquid in (**a**) the inlet pipe and (**b**) the outlet pipe in cm/s.

Solution:
(**a**) Solve Eq. (10-12) for the speed of liquid in the inlet pipe:

$$\frac{\Delta V}{\Delta t} = Av \qquad v_1 = \frac{\Delta V/\Delta t}{A_1} = \frac{8.0\ \text{cm}^3/\text{s}}{(\pi/4)(1.0\ \text{cm})^2} = 10.19\ \text{cm/s}$$

(**b**) Use Eq. (10-13) to find the speed of liquid in the outlet pipe:

$$A_1 v_1 = A_2 v_2 \qquad v_2 = v_1\frac{A_1}{A_2} = (10.19\ \text{cm/s})\frac{(\pi/4)(1.0\ \text{cm})^2}{(\pi/4)(2.0\ \text{cm})^2} = 2.55\ \text{cm/s}$$

- **Bernoulli's principle** states that the quantity $\frac{1}{2}dv^2 + hdg + P$ has the same value at every point in an incompressible fluid moving in streamline (nonturbulent) flow.

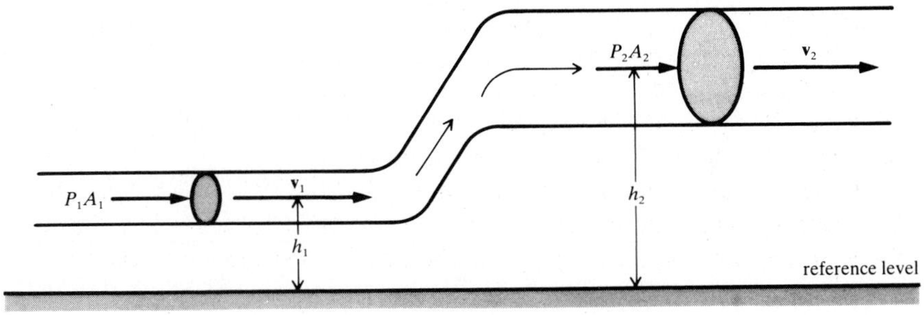

Figure 10-12. Bernoulli's principle.

For a tube like the one shown in Figure 10-12, we can write Bernoulli's principle in the form of an equation:

BERNOULLI'S EQUATION

$$\frac{1}{2}dv_1^2 + h_1dg + P_1 = \frac{1}{2}dv_2^2 + h_2dg + P_2 \qquad \textbf{(10-14)}$$

You may recognize Eq. (10-14) as a statement of the conservation of energy principle.

EXAMPLE 10-13: Water flows in the lower part of the tube shown in Figure 10-12 at a speed $v_1 = 2$ cm/s. The pressure there is $P_1 = 2 \times 10^4$ N/m^2. The diameters in the system are $x_1 = 2$ cm and $x_2 = 3$ cm; the heights are $h_1 = 4$ cm and $h_2 = 34$ cm.
Find (**a**) the speed of the flowing water and (**b**) the pressure in the upper portion of the tube.

Solution:
(**a**) You can find v_2 from Eq. (10-13):

$$v_2 = v_1\frac{A_1}{A_2} = (2\text{ cm/s})\frac{(\pi/4)(2\text{ cm})^2}{(\pi/4)(3\text{ cm})^2} = 0.889\text{ cm/s}$$

(**b**) Now, use Bernoulli's equation (10-14) to find the unknown pressure:

$$\begin{aligned} P_2 &= P_1 + (h_1 - h_2)dg + \frac{1}{2}d(v_1^2 - v_2^2) \\ &= 2 \times 10^4\text{ N/m}^2 + (-0.3\text{ m})(10^3\text{ kg/m}^3)(9.8\text{ m/s}^2) \\ &\quad + \frac{1}{2}(10^3\text{ kg/m}^3)[(0.02\text{ m/s})^2 - (0.889 \times 10^{-2}\text{ m/s})^2] \\ &= (2 \times 10^4 - 2.94 \times 10^3 + 1.60 \times 10^{-1})\text{ N/m}^2 = 1.71 \times 10^4\text{ N/m}^2 \end{aligned}$$

SUMMARY

Pressure	$P = \dfrac{F}{A}$	gives the pressure acting on a fluid as the force exerted perpendicularly on the area
Density	$d = \dfrac{m}{V}$	gives the density of an object as its mass divided by its volume
Columnar fluid pressure	$P = dgh$	gives the pressure exerted by a regular column of fluid of height h
Hydraulic force multiplication	$\dfrac{F_2}{F_1} = \dfrac{A_2}{A_1}$	gives the ratio of the forces transmitting pressure in a hydraulic piston apparatus
AMA: hydraulic lift	$\text{AMA} = \dfrac{F_2}{F_1}$	states that the actual advantage of a hydraulic lift is equal to the ratio of the two forces acting on it
Specific gravity	$\text{sp. gr.} = \dfrac{d_x}{d_w} = \dfrac{w_x}{w_w}$	gives the specific gravity of a substance as the ratio of the substance's density to the density of water

Specific gravity (substance)	$\text{sp. gr.} = \dfrac{\text{weight of substance}}{\text{weight of equal volume of water}}$	gives the specific gravity of any substance by weight
Coefficient of surface tension	$\gamma = \dfrac{F\,\Delta s}{L\,\Delta s} = \dfrac{F}{L}$	gives the work per unit area necessary to increase a liquid's surface area—or the potential energy per unit surface area
Capillary rise (perfectly wetting liquid)	$h = \dfrac{2\gamma}{rdg}$	gives the maximum height of a liquid in a capillary tube when the liquid meets the wall at 0°
Capillary rise	$h = \dfrac{2\gamma \cos\theta}{rdg}$	gives the maximum height of a liquid in a capillary tube in which the liquid does not "wet" the tube perfectly
Volumetric flow rate (incompressible fluid)	$\dfrac{V_1}{\Delta t} = \dfrac{V_2}{\Delta t}$	gives the relationship for the volume of an incompressible fluid entering a tank during a time interval and the volume of the fluid leaving the tank
Volumetric flow rate	$\dfrac{\Delta V}{\Delta t} = Av$	is the volume of liquid passing a cross-section A during a time interval
Volumetric flow rate (system)	$A_1v_1 = A_2v_2$	states that the volumetric flow rate of an incompressible liquid is the same in both pipes of a system
Bernoulli's equation	$\frac{1}{2}dv_1^2 + h_1dg + P_1 = \frac{1}{2}dv_2^2 + h_2dg + P_2$	is Bernoulli's principle as a statement of the conservation of energy

RAISE YOUR GRADES

Can you define . . . ?

- ☑ fluids
- ☑ pressure
- ☑ buoyant force
- ☑ specific gravity
- ☑ surface tension

Can you . . . ?

- ☑ write an expression for the pressure exerted by a fluid
- ☑ write the equation for the pressure increase at a depth h in a fluid of constant density
- ☑ state Pascal's principle
- ☑ state the height of the mercury column in a barometer at sea level
- ☑ write the equation for the force multiplication of a hydraulic lift
- ☑ state Archimedes' principle
- ☑ write the definition of specific gravity
- ☑ calculate the specific gravity of an object, given its weight and the weight of an equal volume of water

☑ write the definition of surface tension
☑ write the equation for the height of a liquid in a capillary tube if the liquid wets the inner surface of the tube
☑ write the equation for the height of a liquid in a capillary tube if the angle of contact is not zero
☑ write the definition of volumetric flow rate of a fluid
☑ express the volumetric flow rate of a fluid in terms of the speed with which it is moving
☑ write the relationship between the volumetric flow rate at different locations within an incompressible fluid
☑ write Bernoulli's equation for two points in an incompressible fluid in streamline flow

SOLVED PROBLEMS

PROBLEM 10-1 An open tank contains a liquid with a density of 0.7×10^3 kg/m^3. The *gauge pressure* (total pressure minus atmospheric pressure) at the bottom of the tank is 7.55×10^4 N/m^2. Find the height of liquid in the tank.

Solution The total pressure at the bottom of the tank is $P = 1 \text{ atm} + dgh$. The gauge pressure P' is the excess above one atmosphere: $P' = dgh$. Use Eq. (10-3) to solve for the height of the liquid:

$$h = \frac{P'}{dg} = \frac{7.55 \times 10^4 \text{ N/m}^2}{(0.7 \times 10^3 \text{ kg/m}^3)(9.8 \text{ m/s}^2)} = 11.0 \text{ m}$$

PROBLEM 10-2 Each rear tire of a certain car supports a total load of 898 lb (4000 N). The recommended gauge pressure for the tires is 32 lb/in^2 (2.21×10^5 N/m^2). Calculate the approximate area of the tire's contact with the road. (Express your answer in cm^2.)

Solution The gauge pressure is the amount by which the pressure within the tire exceeds the external pressure of one atmosphere. Use Eq. (10-1) to solve for the area:

$$P = \frac{F}{A}$$

$$A = \frac{F}{P} = \frac{4 \times 10^3 \text{ N}}{2.21 \times 10^5 \text{ N/m}^2}$$

$$= 1.81 \times 10^{-2} \text{ m}^2 \frac{10^4 \text{ cm}^2}{1 \text{ m}^2} = 1.81 \times 10^2 \text{ cm}^2$$

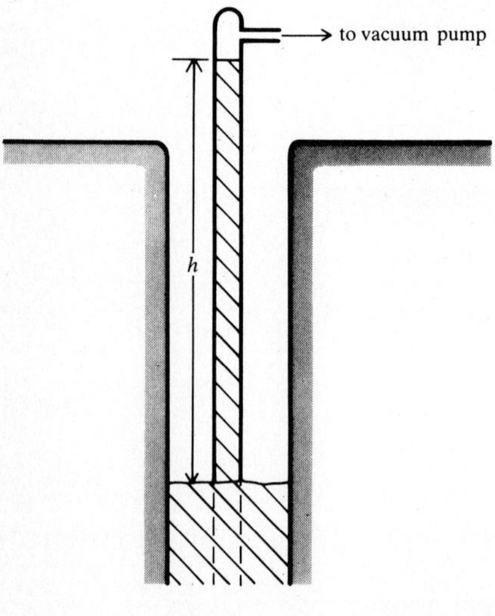

Figure 10-13

PROBLEM 10-3 Water is pumped out of a well by pumping out the air inside a long pipe, as illustrated in Fig. 10-13. What is the maximum height to which the water can be lifted by this technique?

Solution Since the pressure above the column of water in the pipe is nearly zero, the pressure at depth h is $P = dgh$. The pressure at the surface of the water in the well is 1 atm or 1.013×10^5 N/m^2. Solving for the unknown height,

$$h = \frac{P}{gd} = \frac{1.013 \times 10^5 \text{ N/m}^2}{(9.8 \text{ m/s}^2)(1.0 \times 10^3 \text{ kg/m}^3)} = 10.3 \text{ m}$$

PROBLEM 10-4 A smooth-fitting frictionless piston is placed in a fluid-filled vertical pipe, which is connected to a sealed tank, as shown in Fig. 10-14. The inside diameter of the pipe containing the piston is 1.2 cm. How much will the pressure reading at the top of the tank change when a 60 N load is placed on the piston?

Solution The added weight on the piston produces a pressure change within the pipe which will be transmitted to all points within the confined liquid. This change in pressure is

$$\Delta P = \frac{\Delta F}{A} = \frac{60 \text{ N}}{(\pi/4)(1.2 \times 10^{-2} \text{ m})^2} = 5.31 \times 10^5 \text{ N/m}^2$$

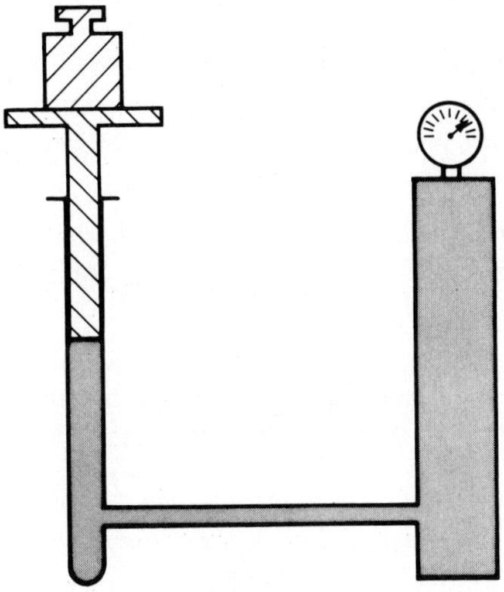

Figure 10-14

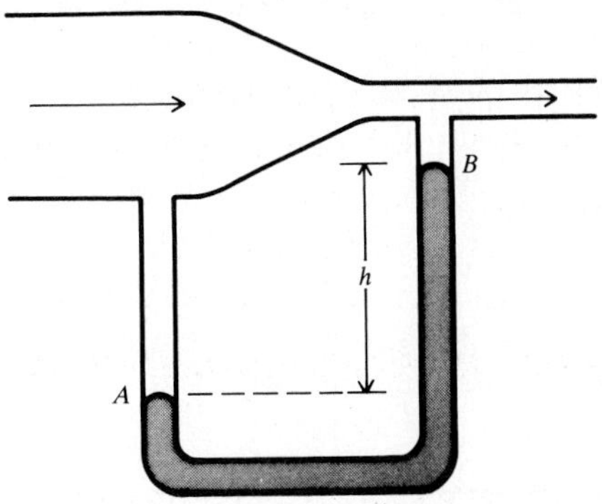

Figure 10-15

PROBLEM 10-5 A U-tube containing a colored liquid measures the pressure difference between the two points to which the tube is connected. The U-tube manometer, illustrated in Fig. 10-15, contains a reference liquid with a density of 0.8×10^3 kg/m^3. The difference in the heights of this liquid on the two sides of the U-tube is 14 cm. What is the pressure difference between points *A* and *B*?

Solution The pressure difference between these two points is given directly by Eq. (10-3):

$$P = dgh = (0.8 \times 10^3 \text{ kg/m}^3)(9.8 \text{ m/s}^2)(0.14 \text{ m}) = 1.10 \times 10^3 \text{ N/m}^2$$

PROBLEM 10-6 A miniature submarine has been designed to withstand an external pressure of 8 atm. What is the maximum depth to which this submarine can descend?

Solution The external pressure on the hull of the submarine is given by $P_{\text{tot}} = 1 \text{ atm} + dgh$. Since the maximum external pressure is 8 atm, the quantity dgh equals $(8 - 1)$ atm $= 7$ atm. Then,

$$h = \frac{7 \text{ atm}}{dg} = \frac{(7 \text{ atm})(1.013 \times 10^5 \text{ (N/m}^2\text{)/atm})}{(1.025 \times 10^3 \text{ kg/m}^3)(9.8 \text{ m/s}^2)} = 70.6 \text{ m}$$

PROBLEM 10-7 A block of African hardwood weighs 8.5 N in air and 2.2 N when completely immersed in ethyl alcohol (sp. gr. = 0.79). Calculate the specific gravity of the wood block.

Solution Use Eq. (10-7) to calculate the weight of a quantity of water having the same volume as the wood block:

$$\text{sp. gr. (alcohol)} = \frac{w \text{ (alcohol)}}{w \text{ (H}_2\text{O)}} \qquad w \text{ (H}_2\text{O)} = \frac{w \text{ (alcohol)}}{\text{sp. gr. (alcohol)}} = \frac{(8.5 - 2.2) \text{ N}}{0.79} = 7.97 \text{ N}$$

Now, use Eq. (10-6) again to find the specific gravity of the wood block:

$$\text{sp. gr. (wood)} = \frac{w \text{ (wood)}}{w \text{ (H}_2\text{O)}} = \frac{8.5 \text{ N}}{7.97 \text{ N}} = 1.07$$

PROBLEM 10-8 A small glass capillary tube has an inside diameter of 0.16 mm. A clear liquid with a specific gravity of 1.0 wets the glass perfectly and rises to a height of 16.6 cm. What is the coefficient of surface tension of this liquid?

Solution Since the liquid wets the glass perfectly, use Eq. (10-9) and solve for γ:

$$h = \frac{2\gamma}{rdg} \qquad \gamma = \frac{hrdg}{2}$$

The density of the liquid is not given, but its specific gravity is 1.0, so it must have the same density as water (10^3 kg/m^3). Thus,

$$\gamma = \frac{(18.6 \times 10^{-2}\ \text{m})(8.0 \times 10^{-5}\ \text{m})(10^3\ \text{kg/m}^3)(9.8\ \text{m/s}^2)}{2} = 7.29 \times 10^{-2}\ \text{N/m}$$

PROBLEM 10-9 A cylindrical water-tank has a small hole 60 cm above the floor on which the tank stands. The depth of water in the tank is 1.8 m. Find the horizontal distance R from the side of the tank to the point on the floor where the stream of water lands. This situation is illustrated in Fig. 10-16.

Solution Eventually you will use the equation for speed (Eq. 2-1), $v = r/t$, to find the horizontal distance. So, first calculate the speed of the jet of water as it leaves the hole in the tank by applying Bernoulli's equation (10-14) to a point A at the surface of the water and to a point B inside the jet of water:

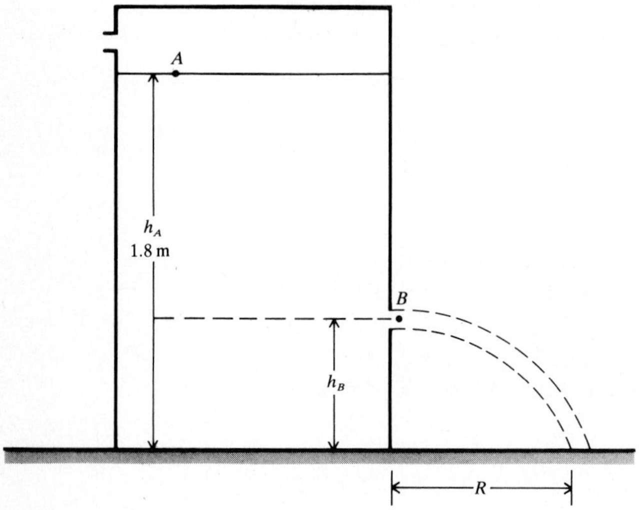

Figure 10-16

$$\frac{1}{2}dv_A^2 + h_A dg + P_A = \frac{1}{2}dv_B^2 + h_B dg + P_B$$

If the diameter of the tank is much greater than that of the hole the speed of the water at A (v_A) is essentially zero. Point B is outside the tank where the pressure is 1 atm, the same as at A.

$$0 + h_A dg + 1\ \text{atm} = \frac{1}{2}dv_B^2 + h_B dg + 1\ \text{atm}$$

Now, you can solve for the speed of the stream of water leaving the tank:

$$v_B = \sqrt{2g(h_A - h_B)} = \sqrt{2(9.8\ \text{m/s}^2)(1.2\ \text{m})}$$
$$= 4.85\ \text{m/s}$$

The water streaming out of the tank is in free fall, and its initial velocity is horizontal. You'll need to find the time for the water to fall a distance of 0.60 m (Eq. 2-8):

$$s = v_0 t + \frac{1}{2}gt^2$$

Since the initial speed in the vertical direction is zero, the time is given by

$$t = \sqrt{\frac{2s}{g}} = \sqrt{\frac{2(0.60\ \text{m})}{9.8\ \text{m/s}^2}} = 0.350\ \text{s}$$

The horizontal component of the velocity is constant, so now you can find the horizontal distance R from Eq. (2-1):

$$v_B = \frac{R}{t} \qquad R = v_B t = (4.85\ \text{m/s})(0.350\ \text{s}) = 1.70\ \text{m}$$

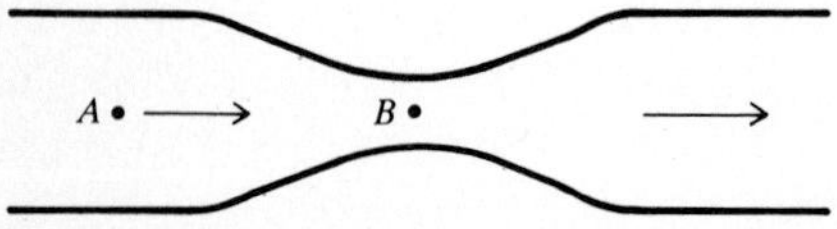

Figure 10-17

PROBLEM 10-10 An incompressible fluid (density = 0.8 g/cm^3) moves at a rate of 12 cm^3/s through a smooth, horizontal glass tube which contains a constriction, as illustrated in Figure 10-17. The inside diameter of the tube is 1.4 cm, while the inside diameter of the constriction is 0.7 cm. Calculate the pressure difference between points A and B.

Solution You'll need to find the speed of the moving fluid at points A and B from Eq. (10-12):

$$\frac{\Delta V}{\Delta t} = Av \qquad v = \frac{\Delta V/\Delta t}{A}$$

$$v_A = \frac{12\ \text{cm}^3/\text{s}}{(\pi/4)(1.4\ \text{cm})^2} = 7.80\ \text{cm/s} \qquad v_B = \frac{12\ \text{cm}^3/\text{s}}{(\pi/4)(0.7\ \text{cm})^2} = 31.2\ \text{cm/s}$$

Then, use Bernoulli's equation (10-14). Since the tube is horizontal, the dgh terms cancel:

$$\frac{1}{2}dv_A^2 + P_A = \frac{1}{2}dv_B^2 + P_B$$

$$\Delta P = P_A - P_B = \frac{1}{2}d(v_B^2 - v_A^2)$$

$$= \frac{1}{2}(0.8\ \text{g/cm}^3)[(31.2\ \text{cm/s})^2 - (7.8\ \text{cm/s})^2]$$

$$= \left(3.65 \times 10^2 \frac{\text{g}}{\text{cm s}^2}\right)\left(\frac{1\ \text{kg}}{10^3\ \text{g}}\right)\left(\frac{10^2\ \text{cm}}{1\ \text{m}}\right) = 36.5\frac{\text{kg}}{\text{m s}^2} = 36.5\ \text{N/m}^2$$

Supplementary Exercises

PROBLEM 10-11 The upper surface of the water in a water tower is 35 m above ground level. What is the gauge pressure at ground level in a pipe connected to the reservoir?

PROBLEM 10-12 A liquid with a specific gravity of 0.80 is pumped out of a tank by removing the air from a vertical pipe containing the liquid. What is the maximum height to which the liquid can be lifted by this technique?

PROBLEM 10-13 A hydraulic lift designed to lift an automobile has a piston that is 20 cm in diameter. What pressure is required to lift a load of 11 600 N?

PROBLEM 10-14 What is the actual mechanical advantage of a hydraulic lift if the large piston has a diameter of 20 cm and the small piston has a diameter of 4 cm?

PROBLEM 10-15 A sample of ore weighs 7.5 N in air and 3.2 N in water. What is the density of this ore?

PROBLEM 10-16 A rectangular block of wood is floating on the surface of water. Three-fifths of the wood is above the surface. What is the density of this wood?

PROBLEM 10-17 The coefficient of surface tension of water at 20°C is 7.28×10^{-2} N/m. To what height will water rise in a capillary tube that has an inside diameter of 1.0 mm?

PROBLEM 10-18 Water flows through a pipe at the rate of 6.0 gallons per minute. The inside diameter of the pipe is 2.5 cm. Calculate the speed at which the water is moving within the pipe. [1.00 gallon = 3.79 liters, and 1.00 liter = 0.001 m^3.]

Answers to Supplementary Exercises

10-11: 3.43×10^5 N/m^2

10-12: 12.9 m

10-13: 3.69×10^5 N/m^2

10-14: 25

10-15: 1.74 g/cm^3

10-16: 0.4 g/cm^3

10-17: 2.97 cm

10-18: 77.2 cm/s

11 WAVE MOTION

THIS CHAPTER IS ABOUT

☑ Waves and Pulses
☑ Superposition of Waves
☑ Doppler Effect

11.1 Waves and Pulses

A mechanical **wave** is set in motion in an elastic medium by a **disturbance**; that is, a rapid displacement of a small section of the medium. Only the disturbance is propagated. The particles of the medium do not travel with the wave; they merely oscillate about some equilibrium position. In other words, a wave is a disturbance's way of getting from one place to another. A one-time disturbance is a **pulse**; repetitive disturbances generate a **wave train.**

A. Transverse waves

- A **transverse wave** is generated when the particles in the medium oscillate at right angles to the direction of propagation of the wave.

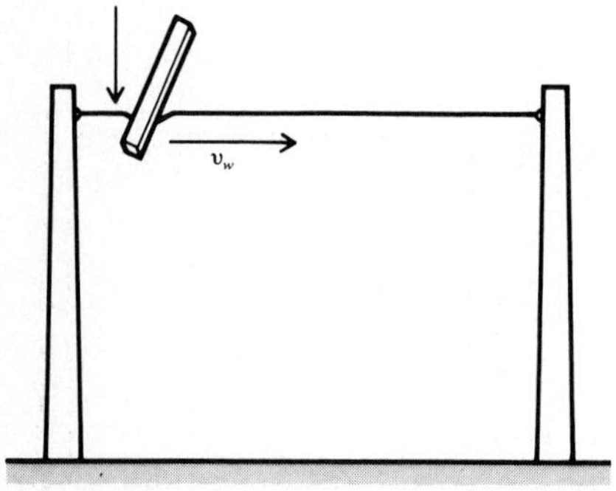

Figure 11-1. Transverse pulse.

Consider the tightly stretched rope shown in Figure 11-1. Striking the rope sharply with a stick near the left end generates a transverse pulse that travels rapidly toward the right. When the pulse passes a point on the rope, that portion of the rope moves downward, then upward, perpendicular to the direction of propagation. The **speed of propagation** of the pulse along the rope depends on the magnitude of the force of tension **F** in the rope and on its mass per unit length m/L:

SPEED OF TRANSVERSE WAVES (STRETCHED STRING)

$$v_w = \sqrt{\frac{|\mathbf{F}|}{m/L}} \tag{11-1}$$

EXAMPLE 11-1: A wire is stretched tightly between two posts 15 m apart. The mass of the wire is 1.2 kg and the tension is 60 N.
How long will it take for a transverse pulse to travel from one end of the wire to the other?

Solution: From Eq. (11-1), calculate the speed of propagation:

$$v_w = \sqrt{\frac{|\mathbf{F}|}{m/L}} = \sqrt{\frac{60\ \text{kg m/s}^2}{(1.2\ \text{kg})/(15\ \text{m})}} = 27.4\ \text{m/s}$$

The speed is constant, so you can express it as distance divided by time:

$$v_w = \frac{L}{t}$$

$$t = \frac{L}{v_w} = \frac{15\ \text{m}}{27.4\ \text{m/s}} = 0.548\ \text{s}$$

If a wire is subjected to repeated disturbances at regular intervals, a continuous transverse wave train is propagated, as illustrated in Figure 11-2. The speed of propagation of this **sinusoidal wave** is the same as that of a single transverse pulse.

note: A sinusoidal wave occurs when each particle in a medium undergoes SHM oscillations (see Sec. 9-3).

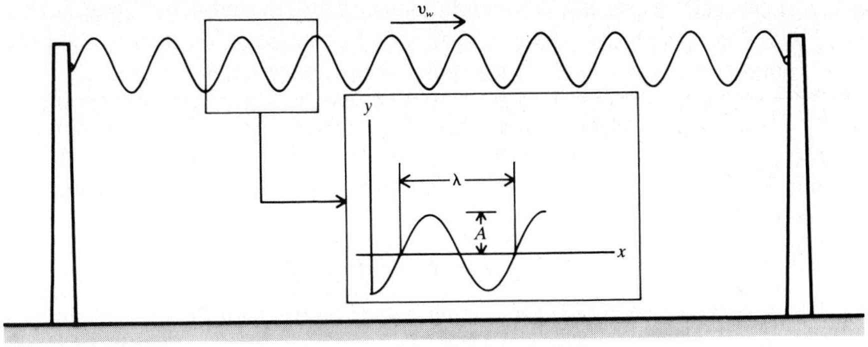

Figure 11-2. Transverse wave.

We can describe wave motion with the same conventions we use for simple harmonic motion. The insert in Figure 11-2 shows the relationships:

- **Frequency** f is the number of oscillations per unit time occurring at a given point.
- **Amplitude** A is the magnitude of maximum displacement in the y direction.
- **Wavelength** λ is the distance between two points that occupy the same relative position on the wave.
- **Period** T is the time required for one complete oscillation to pass a given point.

B. Longitudinal waves

- A **longitudinal wave** is generated when particle oscillation is parallel to the direction of propagation of the wave.

Striking the support pole of a tautly stretched rope causes the rope fibers to move horizontally in the same direction in which the pulse is propagated (see Figure 11-3). Because the rope is not displaced transversely, you can't see the longitudinal pulse from a distance. However, you'd be able to feel the longitudinal pulse if you gripped the rope with your hand, particularly near the left end.

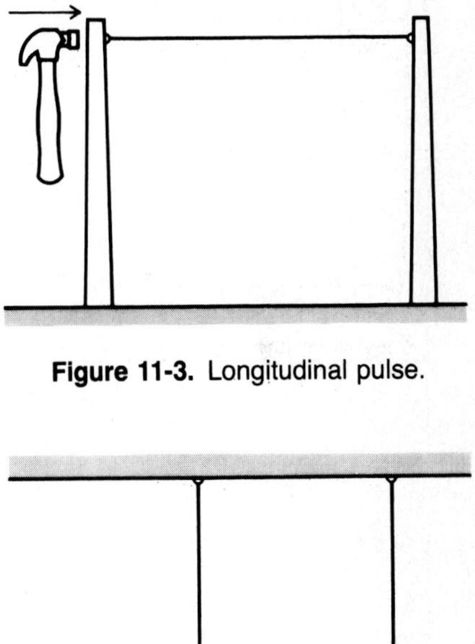

Figure 11-3. Longitudinal pulse.

The friction between rope fibers causes a longitudinal pulse to "fade out" after traveling only a few feet, so a rigid rod is a better medium for the propagation of a longitudinal pulse. Figure 11-4 shows a metal rod suspended horizontally from the ceiling of a room. If you strike the rod on the left end with a hammer, a longitudinal pulse will travel rapidly to the right. The hammer blow causes the rod's molecules to move closer together, so this longitudinal pulse is called a **compression pulse.** The **speed of propagation** of the compression pulse is given by the formula

SPEED OF COMPRESSION WAVES (LIQUID OR SOLID ROD)

$$v_w = \sqrt{\frac{E}{d}} \tag{11-2}$$

where E is the rod's elastic or stretch modulus (see Eq. 9-2) and d is its density.

Figure 11-4

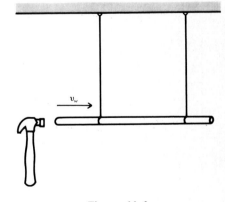

EXAMPLE 11-2: A steel rod has a density of 7.8×10^3 kg/m^3 and an elastic modulus of 20×10^{10} N/m^2.
Calculate the speed of propagation of a compression pulse in this rod.

Solution: You can calculate the speed directly from Eq. (11-2):

$$v_w = \sqrt{\frac{E}{d}} = \sqrt{\frac{20 \times 10^{10}\ \text{N/m}^2}{7.8 \times 10^3\ \text{kg/m}^3}} = 5.06 \times 10^3\ \text{m/s}$$

note: If the end of the rod is vibrated sinusoidally, a continuous longitudinal wave is produced in the rod. The speed of propagation of this compressional wave is the same as the speed of the compressional pulse.

Although a stretched wire or a rigid rod can propagate *both* transverse and longitudinal waves, a fluid such as air can propagate *only* longitudinal waves.

- **Sound waves** are longitudinal waves propagated in a liquid or a gas.

The **speed of propagation** of sound waves depends on the bulk modulus B (see Eq. 9-3) and the density of the fluid:

SPEED OF SOUND WAVES

$$v_w = \sqrt{\frac{B}{d}} \tag{11-3}$$

note: Waves on the *surface* of water involve both transverse and longitudinal motions simultaneously.

EXAMPLE 11-3: The bulk modulus of air is 1.418×10^5 N/m^2 and the density of air at 20°C is 1.205 kg/m^3.
Calculate the speed of sound in air at sea level.

Solution: From Eq. (11-3):

$$v_w = \sqrt{\frac{B}{d}} = \sqrt{\frac{1.418 \times 10^5 \text{ N/m}^2}{1.205 \text{ kg/m}^3}} = 343 \text{ m/s}$$

The relationship between the speed of a wave and its frequency f is given by

SPEED OF A WAVE

$$v_w = f\lambda \tag{11-4}$$

The SI unit for frequency of a wave is a hertz (Hz), which is equivalent to 1 second^{-1} or 1 cycle per second.

EXAMPLE 11-4: A horn produces a sound with a frequency of 440 Hz.
What is the wavelength of the sound wave propagated at sea level at a temperature of 20°C? [Hint: You calculated the speed of sound in Example 11-3.]

Solution: Use Eq. (11-4) and solve for λ:

$$\lambda = \frac{v_w}{f} = \frac{343 \text{ m/s}}{440 \text{ Hz}} = 0.780 \text{ m}$$

11-2. Superposition of Waves

- The **principle of superposition** states that the resultant disturbance due to the presence of two or more waves at a point in space is the algebraic sum of the disturbances produced by each wave.

If we let y_1 and y_2 represent the disturbances caused by two waves, then the resultant disturbance is

SUPERPOSITION OF TWO WAVES

$$y = y_1 + y_2 \tag{11-5}$$

The superposition principle is illustrated in Figure 11-5.

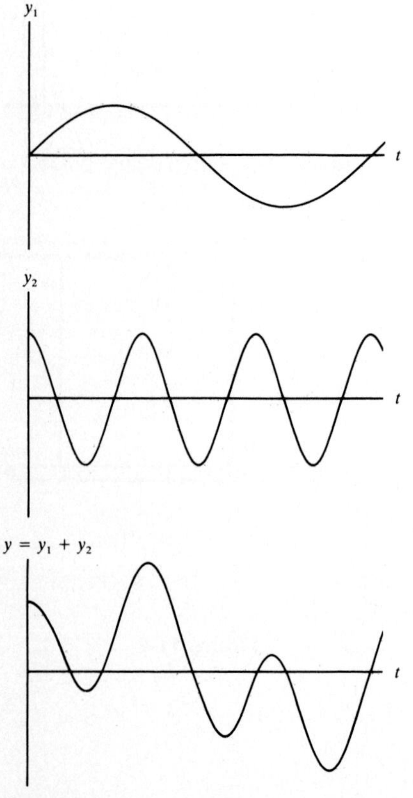

Figure 11-5. The superposition of two waves.

- When two waves reach their maximum displacement at the same time they are **in phase**. If their maximum displacements do not coincide, the waves are **out of phase.** The waves are 180° out of phase when one wave reaches its maximum at the same time the other reaches its minimum (think of rotating the wave 180° around the axis of its line of propagation).

Consider two waves of equal amplitudes but slightly different frequencies. At some point the waves are in phase, so their resultant amplitude is twice the amplitude of the individual waves. A little later, the waves will be 180° out of phase so their resultant amplitude will be zero. This means that the superposition of two waves of equal amplitudes and slightly different frequencies causes a regular fluctuation in resultant amplitude (Figure 11-6).

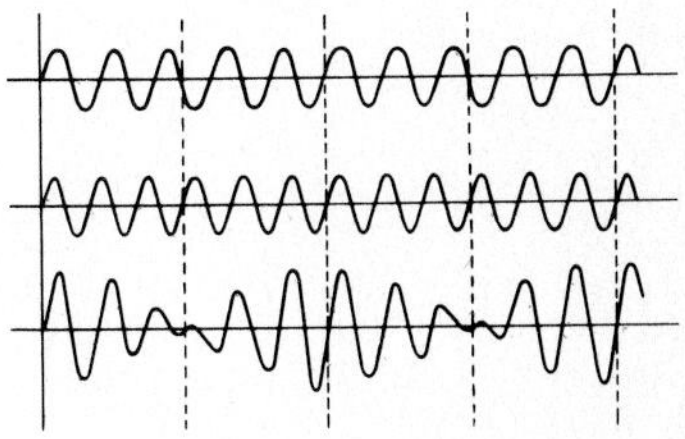

Figure 11-6. Beat frequency.

- The frequency of the regular fluctuation of two superimposed waves of equal amplitudes is the **beat frequency,** which is equal to the difference of the frequencies of the individual waves.

BEAT FREQUENCY

$$f_{\text{beat}} = f_1 - f_2 \tag{11-6}$$

EXAMPLE 11-5: Two sources of sound produce pure tones (i.e., single frequencies) of the same amplitude. The wavelengths of the two sources are 0.79 m and 0.78 m.
Determine the beat frequency that results from the superposition of these two sound waves.

Solution: You'll first have to calculate the frequencies of the two sources from Eq. (11-4). [**recall**: The speed of sound is approximately 343 m/s. (Example 11-3)]

$$v_w = f\lambda \qquad f = \frac{v_w}{\lambda}$$

$$f_1 = \frac{v_w}{\lambda_1} = \frac{343 \text{ m/s}}{0.79 \text{ m}} = 434 \text{ Hz} \qquad f_2 = \frac{v_w}{\lambda_2} = \frac{343 \text{ m/s}}{0.78 \text{ m}} = 440 \text{ Hz}$$

From Eq. (11-6), the beat frequency is

$$f_{\text{beat}} = f_2 - f_1 = (440 - 434) \text{ Hz} = 6 \text{ Hz}$$

note: In the examples that follow, we will use 340 m/s as an approximation of the speed of sound in air (this is its value at 15°C).

11-3. Doppler Effect

- The **Doppler effect** describes the apparent change in frequency of a wave due to the relative motion of the source and the observer.

If a source of sound is moving toward you, the frequency you hear is *higher* than the frequency actually emitted by the source. On the other hand, if the source is moving away from you, the frequency you hear is *lower* than the true frequency of the source.

If the observer is at rest and the source is moving at a speed v, the apparent frequency detected by the observer is

APPARENT FREQUENCY (SOURCE MOVING TOWARD OBSERVER)

$$f' = \frac{f}{1 - v/v_w} \tag{11-7}$$

APPARENT FREQUENCY (SOURCE MOVING AWAY FROM OBSERVER)

$$f' = \frac{f}{1 + v/v_w} \tag{11-8}$$

where v_w is the speed of sound (approximately 340 m/s).

If the source is at rest and the observer is moving at a speed v, the frequency the observer hears differs from the actual frequency of the sound that the source emits:

APPARENT FREQUENCY (OBSERVER MOVING TOWARD SOURCE)
$$f' = f\left(1 + \frac{v}{v_w}\right) \quad \textbf{(11-9)}$$

APPARENT FREQUENCY (OBSERVER MOVING AWAY FROM SOURCE)
$$f' = f\left(1 - \frac{v}{v_w}\right) \quad \textbf{(11-10)}$$

Keep in mind that the apparent frequency is *greater* than the actual frequency when the source and observer are moving closer together. Conversely, the frequency detected by the observer is *less* than the frequency actually emitted by the source when the source and observer are moving farther apart.

EXAMPLE 11-6: A train approaching you sounds its horn as you stand at a railroad crossing. The actual frequency of the horn is 1200 Hz and the frequency you hear is 1300 Hz.
At what speed is the train approaching you?

Solution: A moving source is approaching a stationary observer, so use Eq. (11-7):

$$f' = \frac{f}{1 - v/v_w} \qquad 1 - \frac{v}{v_w} = \frac{f}{f'} \qquad \frac{v}{v_w} = 1 - \frac{f}{f'}$$

$$v = \frac{v_w(f' - f)}{f'} = \frac{(340 \text{ m/s})(1300 - 1200) \text{ Hz}}{1300 \text{ Hz}} = 26.2 \text{ m/s}$$

EXAMPLE 11-7: When the train in Example 11-6 has passed the crossing, the engineer sounds the horn again.
What frequency will you hear now?

Solution: Use Eq. (11-8), because the source is moving away from the stationary observer:

$$f' = \frac{f}{1 + v/v_w} = \frac{1200 \text{ Hz}}{1 + (26.2 \text{ m/s})/(340 \text{ m/s})} = 1114 \text{ Hz}$$

The Doppler effect is not restricted to sound; it also applies to light, a wave phenomenon that obeys $f\lambda = v_w$. The speed of a moving source or moving observer is generally a very small fraction of the speed of light c, which is 3×10^8 m/s. When v is very small compared to c, Eqs. (11-7) through (11-10) can be expressed as

DOPPLER EFFECT FOR LIGHT WAVES
$$\frac{v}{c} = \frac{\Delta f}{f} \quad \textbf{(11-11)}$$

note: The rule for the direction of the change in observed frequency is the same as for sound waves—if the source and observer approach each other, the observer detects a higher frequency.

EXAMPLE 11-8: The observed wavelength of a hydrogen emission from the atmosphere of a distant star is 6.5681×10^{-7} m. The same emission line observed in a laboratory on earth is 6.5693×10^{-7} m.
Calculate the relative speed of the star with respect to the earth.

Solution: First, calculate the actual frequency and the observed frequency. You can do this by substituting c for v_w in Eq. (11-4):

$$f = \frac{c}{\lambda} = \frac{3.00 \times 10^8 \text{ m/s}}{6.5693 \times 10^{-7} \text{ m}} = 4.5667 \times 10^{14} \text{ Hz}$$

$$f' = \frac{3.00 \times 10^8 \text{ m/s}}{6.5681 \times 10^{-7} \text{ m}} = 4.5675 \times 10^{14} \text{ Hz}$$

Because the observed frequency is higher than the actual frequency of the source, the star is approaching the earth, so use Eq. (11-11) to solve for the relative speed of the star:

$$v = \frac{\Delta f}{f} c$$

$$= \frac{(4.5675 - 4.5667) \times 10^{14} \text{ Hz}}{4.5667 \times 10^{14} \text{ Hz}} (3.00 \times 10^8 \text{ m/s}) = 5.26 \times 10^4 \text{ m/s}$$

SUMMARY

Speed of transverse waves (stretched string)	$v_w = \sqrt{\dfrac{\lvert\mathbf{F}\rvert}{m/L}}$	the speed of propagation of the pulse on a stretched string depends on the force of tension and the mass per unit length
Speed of compression waves (liquid or solid rod)	$v_w = \sqrt{\dfrac{E}{d}}$	the speed of propagation of a compression pulse depends on the stretch modulus and the density of the medium
Speed of sound waves	$v_w = \sqrt{\dfrac{B}{d}}$	the speed of propagation of sound waves depends on the bulk modulus and the density of the fluid
Speed of a wave	$v_w = f\lambda$	gives the relationship between the speed of a wave and its frequency
Superposition of two waves	$y = y_1 + y_2$	states that the resultant disturbance due to the presence of two or more waves at a point in space is the algebraic sum of the disturbances produced by each wave
Beat frequency	$f_{beat} = f_1 - f_2$	states that the frequency of the regular fluctuation of two waves is equal to the difference of the frequencies of the individual waves
Apparent frequency (source moving toward observer)	$f' = \dfrac{f}{1 - v/v_w}$	gives observed frequency when source is moving at speed v toward a stationary observer
Apparent frequency (source moving away from observer)	$f' = \dfrac{f}{1 + v/v_w}$	gives observed frequency when source is moving at speed v away from a stationary observer

Apparent frequency (observer moving toward source)	$f' = f\left(1 + \frac{v}{v_w}\right)$	gives observed frequency when observer is moving at speed v toward source
Apparent frequency (observer moving away from source)	$f' = f\left(1 - \frac{v}{v_w}\right)$	gives observed frequency when observer is moving at speed v away from source
Doppler effect for light waves	$\frac{v}{c} = \frac{\Delta f}{f}$	gives the change in the frequency of light when v is much less than c

RAISE YOUR GRADES

Can you define . . . ?

☑ a wave ☑ wavelength ☑ a pulse ☑ a sound wave

Can you . . . ?

☑ write the formula for the speed of propagation of a transverse wave in a stretched string
☑ write the formula for the speed of a longitudinal wave in a solid rod
☑ write the formula for the speed of a sound wave in a fluid
☑ calculate the wavelength of a wave if you know the speed of the wave and its frequency
☑ use the superposition principle to find the resultant displacement when two waves are present at a point in space
☑ find the beat frequency that results when two waves of slightly different frequencies are superimposed
☑ calculate the apparent frequency of a source of sound moving toward or away from a stationary observer
☑ calculate the apparent frequency that results when an observer is moving toward or away from a stationary source of sound
☑ write the equation for the Doppler effect due to relative motion between a light source and an observer

SOLVED PROBLEMS

PROBLEM 11-1 A thin wire is stretched tightly by means of a frictionless pulley and a weight, as illustrated in Fig. 11-7. The wire has a mass of 50 g and a length of 2.0 m. The observed speed of propagation of transverse waves in the wire is 34.3 m/s. Determine (**a**) the tension in the wire and (**b**) the total mass hanging from the wire.

Solution Use Eq. (11-1) for the speed of propagation of a transverse wave in a stretched wire and solve algebraically for the unknown tension F:

$$v = \sqrt{\frac{F}{m/L}} \qquad v^2 = FL/m$$

$$F = \frac{v^2 m}{L} = \frac{(34.3 \text{ m/s})^2(.05 \text{ kg})}{2.0 \text{ m}} = 29.4 \text{ N}$$

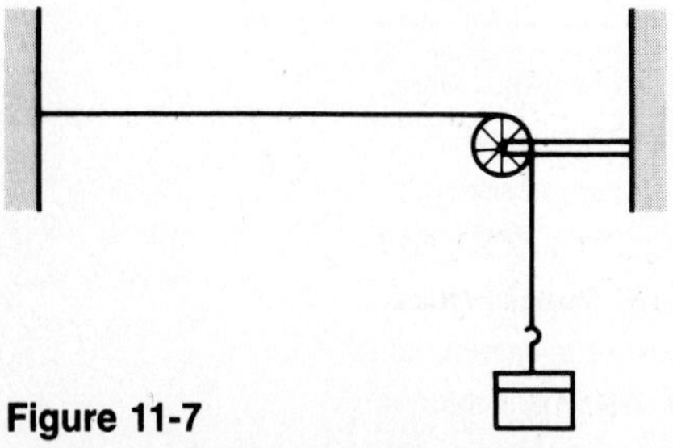

Figure 11-7

Because the pulley is frictionless, the tension is equal to the hanging weight:

$$m = \frac{w}{g} = \frac{29.4 \text{ N}}{9.8 \text{ m/s}^2} = 3.00 \text{ kg}$$

PROBLEM 11-2 Two steel rods made of slightly different alloys are used to transmit sound waves. The two rods have the same density, but one has an elastic modulus 1% higher than the other. Determine the ratio of the speeds of propagation in the two rods.

Solution Begin by writing Eq. (11-2) for the speed of sound in the first rod:

$$v_1 = \sqrt{\frac{E_1}{d}}$$

For the second rod, $E_2 = 1.01\,E_1$, so the speed of propagation of sound in the second rod is slightly greater than in the first:

$$v_2 = \sqrt{\frac{1.01\,E_1}{d}}$$

Now, calculate the ratio of the two speeds:

$$\frac{v_2}{v_1} = \sqrt{1.01} = 1.005$$

Sound travels $\frac{1}{2}$% faster in the rod with the greater elastic modulus.

PROBLEM 11-3 The bulk modulus of sea water is 2.10×10^9 N/m^2 and its density is 1.025×10^3 kg/m^3. What is the speed of propagation of sound in sea water?

Solution Use Eq. (11-3):

$$v_w = \sqrt{B/d} = \sqrt{\frac{2.10 \times 10^9 \text{ N/m}^2}{1.025 \times 10^3 \text{ kg/m}^3}} = 1.43 \times 10^3 \text{ m/s}$$

PROBLEM 11-4 A metal rod has a diameter of 0.5 cm and a length of 20 cm; its mass is 10.6 g. The time required for a sound wave to travel from one end of the rod to the other is 3.91×10^{-5} s. Calculate the stretch modulus of this metal.

Solution First, calculate the speed of sound propagation in the rod:

$$v = \frac{L}{t} = \frac{0.20 \text{ m}}{3.91 \times 10^{-5} \text{ s}} = 5.12 \times 10^3 \text{ m/s}$$

Now, calculate the density of the metal:

$$d = \frac{m}{V} = \frac{m}{\pi r^2 L} = \frac{10.6 \text{ g}}{\pi(0.25 \text{ cm})^2(20 \text{ cm})} = \left(\frac{2.70 \text{ g}}{\text{cm}^3}\right)\left(\frac{1 \text{ kg}}{10^3 \text{ g}}\right)\left(\frac{10^6 \text{ cm}^3}{1 \text{ m}^3}\right) = 2.70 \times 10^3 \text{ kg/m}^3$$

Next, use Eq. (11-2) for the speed of sound in a solid and solve for the stretch modulus:

$$v_w = \sqrt{E/d}$$

$$E = v_w^2\, d = (5.12 \times 10^3 \text{ m/s})^2(2.70 \times 10^3 \text{ kg/m}^3) = 7.08 \times 10^{10} \text{ N/m}^2$$

PROBLEM 11-5 The E-string of a piano can be tuned with the aid of a tuning fork that produces a pure tone with a frequency of 660 Hz. When the piano string and the tuning fork are sounded simultaneously, a beat frequency of 2 Hz is produced. What is the frequency of the piano string?

Solution The beat frequency is the difference between the frequencies of the piano string and the tuning fork. Since you don't know which of the two frequencies is higher, the piano string may be vibrating at a frequency of 658 Hz or 662 Hz. Either of these two frequencies will produce a beat frequency of 2 Hz when the 660 Hz tuning fork is struck.

PROBLEM 11-6 A stationary observer at a railroad crossing hears the horn of a distant train at an apparent frequency of 860 Hz. The train's horn produces an actual frequency of 880 Hz. How fast is the train moving and in what direction relative to the observer?

Solution The fact that the apparent frequency (860 Hz) is less than the actual frequency (880 Hz) tells you that the train is moving away from the observer. So, use Eq. (11-8) to find the train's speed:

$$f' = \frac{f}{1 + v/v_w} \qquad v = v_w\left(\frac{f - f'}{f'}\right) = (340 \text{ m/s})\left(\frac{880 \text{ Hz} - 860 \text{ Hz}}{860 \text{ Hz}}\right) = 7.91 \text{ m/s}$$

PROBLEM 11-7 Some very distant galaxies are known to be moving away from our solar system at a speed equal to $\frac{1}{3}c$. If the actual wavelength of an emission line of hydrogen is 6.5693×10^{-7} m, find the apparent wavelength recorded by a laboratory on earth that is observing the distant galaxy. [Hint: You may neglect the motion of the earth in this situation.]

Solution Calculate the actual frequency of the hydrogen emission line from Eq. (11-4):

$$f = \frac{c}{\lambda} = \frac{3.00 \times 10^8 \text{ m/s}}{6.5693 \times 10^{-7} \text{ m}} = 4.57 \times 10^{14} \text{ Hz}$$

Then, use Eq. (11-11) to find the change in frequency:

$$\Delta f = f\left(\frac{v}{c}\right) = (4.57 \times 10^{14} \text{ Hz})\left(\frac{\frac{1}{3}c}{c}\right) = 1.52 \times 10^{14} \text{ Hz}$$

Since the galaxy is moving away from the earth, the observed frequency will be less than the actual frequency:

$$f' = f - \Delta f = (4.57 - 1.52) \times 10^{14} \text{ Hz} = 3.05 \times 10^{14} \text{ Hz}$$

Supplementary Exercises

PROBLEM 11-8 A steel wire having a mass of 2.0 kg is stretched tightly between two buildings. The length of the wire is 22 m and the tension is 50 N. What is the speed of propagation of transverse waves in this wire?

PROBLEM 11-9 An aluminum rod has a stretch modulus E of 7.0×10^{10} N/m^2 and a density of 2.70 g/cm^3. What is the speed of a sound wave in this rod?

PROBLEM 11-10 Sound waves propagate through ethyl alcohol at a speed of 1.125×10^3 m/s. The density of ethyl alcohol is 0.791 g/cm^3. Calculate the bulk modulus of ethyl alcohol.

PROBLEM 11-11 An organ pipe produces a tone with a frequency of 1200 Hz. What is the wavelength of this sound wave when it travels at a speed of 340 m/s?

PROBLEM 11-12 Two waves having the same frequency, the same wavelength, and the same amplitude pass through a point in space. The phase difference is 180° (or π rad). What is the amplitude of the wave that results from the superposition of these two sound waves?

PROBLEM 11-13 Two guitar strings have the same length and the same mass. The tension in one string, however, is 2% greater than that in the other string. The string with the lower tension propagates transverse waves at a speed of 450 m/s. What is the speed of propagation in the string that has a higher tension?

PROBLEM 11-14 Two very similar tuning forks produce frequencies of 440 Hz and 445 Hz. When these two sources are vibrating simultaneously, what beat frequency is produced?

PROBLEM 11-15 You are moving along a street at 30 mi/h (13.41 m/s) when a parked car ahead of you sounds its horn. If you hear a frequency of 1200 Hz, what is the frequency actually produced by the horn? (Use 340 m/s for the speed of sound.)

PROBLEM 11-16 The speed of a transverse wave in a taut wire is 200 m/s. The tension in this wire is 150 N and its length is 1.20 m. Find the mass of the wire in grams.

Answers to Supplementary Exercises

11-8: 23.5 m/s **11-9:** 5.09 km/s **11-10:** 1.00×10^9 N/m^2 **11-11:** 28.3 cm **11-12:** zero **11-13:** 454 m/s **11-14:** 5 Hz **11-15:** 1154 Hz **11-16:** 4.50 g

12 SOUND

THIS CHAPTER IS ABOUT

☑ Standing Waves and Interference

☑ Intensity and Loudness

12-1. Standing Waves and Interference

A. Vibrating strings

The superposition (see Sec. 11-2) of two waves of the same frequency traveling in opposite directions on a stretched string creates **standing waves.**

- A **node** (N) is a point on the string at which there is *minimum* disturbance.
- An **antinode** (A) is a point on the string at which there is *maximum* disturbance.

Nodes are separated from one another by half a wavelength, as are antinodes. An antinode is separated from a node by one-quarter wavelength, so that nodes are found at 0, $\lambda/2$, λ, $3\lambda/2, \ldots$, and antinodes are found at $\lambda/4$, $3\lambda/4$, $5\lambda/4$, and so on.

At the antinodes, the superposition of the two waves in a standing wave produces *increased* amplitude called **constructive interference**. At the nodes, the superposition of the two waves produces *decreased* amplitude called **destructive interference.** Since the two ends of the string in Figure 12-1 are fixed, they are necessarily nodes. This means that the amplitude of vibration of the string is zero at the nodes, and greatest at the midpoint between two nodes—at the antinodes.

Figure 12-1 illustrates three possible *modes* of vibration for a string fastened at each end.

- A **mode** is a way in which it is physically possible for a string to vibrate.

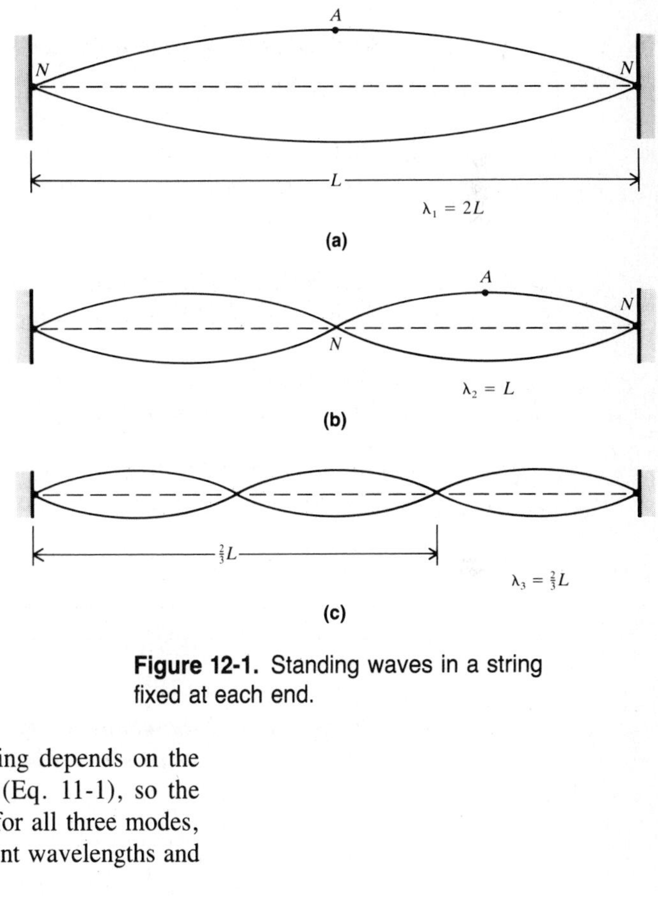

Figure 12-1. Standing waves in a string fixed at each end.

The velocity of propagation of a transverse wave in a string depends on the tension and the mass per unit length, $v_w = \sqrt{F/(m/L)}$ (Eq. 11-1), so the *speed of propagation* of the transverse waves is the same for all three modes, but the standing waves in the three modes will have different wavelengths and different frequencies.

- The **frequency of a standing wave** is equal to the speed of propagation divided by the wavelength.

FREQUENCY OF STANDING WAVES

$$f = \frac{v_w}{\lambda} \tag{12-1}$$

The standing wave with the longest possible wavelength has the lowest frequency, which is called its **fundamental mode** of vibration (Figure 12-1a). The **fundamental frequency** (also called the **first harmonic**) that corresponds to the fundamental mode is

FUNDAMENTAL FREQUENCY (FIXED STRING)

$$f_1 = \frac{v_w}{\lambda_1} = \frac{v_w}{2L} \tag{12-2}$$

• **Harmonic frequencies** are integer multiples of the fundamental frequencies.

HARMONIC FREQUENCIES (FIXED STRING) $$f_n = \frac{v_w}{\lambda_n} = \frac{nv_w}{2L} \qquad n = 1, 2, 3, \ldots \qquad \textbf{(12-2a)}$$

For example, the mode of vibration following the fundamental frequency, called the **second harmonic** or the **first overtone,** has a frequency of $f_2 = v_w/\lambda_2 = v_w/L = 2(v_w/2L) = 2f_1$. The third mode of vibration, called the third harmonic or the second overtone, has a frequency of $f_3 = v_w/\lambda_3 = v_w/\frac{2}{3}L = 3(v_w/2L) = 3f_1$. These harmonic frequencies are the basis for almost all musical sound.

EXAMPLE 12-1: A string fixed at each end is set into vibration so that there are a total of five nodes, including one at each end.
What is the relation of the vibration frequency of this mode to the fundamental frequency?

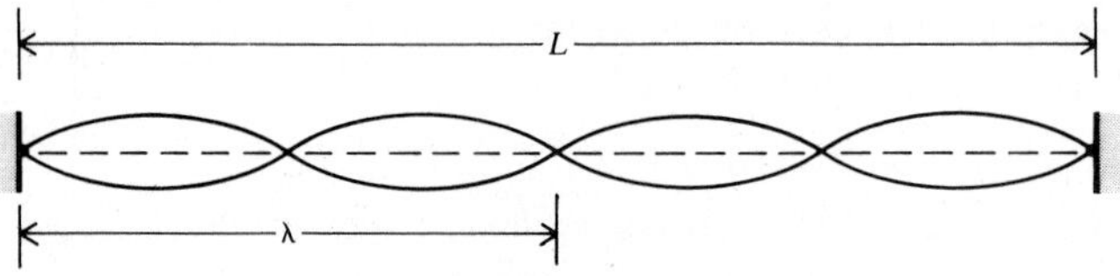

Figure 12-2

Solution: A sketch of the standing wave in this string is shown in Figure 12-2. You can see from the sketch that the wavelength is $\lambda = \frac{1}{2}L$. So, use Eq. (12-2) to find the frequency of this mode of vibration as a multiple of the fundamental frequency:

$$f = \frac{v_w}{\lambda} = \frac{v_w}{\frac{1}{2}L} = 2\left(\frac{v_w}{L}\right) = 4\left(\frac{v_w}{2L}\right)$$

This frequency is four times the fundamental frequency, and is called the fourth harmonic or third overtone.

EXAMPLE 12-2: A guitar string has a length of 40 cm and a mass of 5 g.
What string tension is needed for a fundamental frequency of 440 Hz?

Solution: First, solve Eq. (12-2) for v_w, the speed of propagation of transverse waves in a string:

$$f_1 = \frac{v_w}{2L}$$

$$v_w = f_1(2L)$$

Now, set the result equal to the tension term in Eq. (11-1):

$$v_w = f_1(2L) = \sqrt{\frac{F}{m/L}}$$

And, solve for the tension:

$$f_1(2L) = \sqrt{\frac{F}{m/L}} \qquad f_1^2(2L)^2 = \frac{F}{m/L}$$

$$F = \frac{f_1^2(2L)^2m}{L} = 4Lf_1^2m = 4(0.40 \text{ m})(440/\text{s})^2(5 \times 10^{-3} \text{ kg}) = 1.55 \times 10^3 \text{ N}$$

B. Vibrating air columns

Vibrating air columns in open or closed pipes produce longitudinal standing waves. Figure 12-3 shows three modes of vibration in a horizontal pipe closed at one end. Since the air molecules cannot move longitudinally at the closed end, a minimum disturbance occurs there. This means that there must always be a node at the closed end. The maximum disturbance occurs at the open end, indicating an antinode. Because the frequency of vibration is the speed of sound in air divided by the wavelength (Eq. 12-1), the fundamental frequency corresponds to the longest wavelength:

FUNDAMENTAL FREQUENCY (CLOSED PIPE)

$$f_1 = \frac{v_w}{\lambda_1} = \frac{v_w}{4L} \qquad \textbf{(12-3)}$$

The first overtone (Figure 12-3b) has a frequency of

$$f_2 = \frac{v_w}{\lambda_2} = \frac{v_w}{\frac{4}{3}L} = 3\left(\frac{v_w}{4L}\right)$$

which is three times the fundamental frequency.

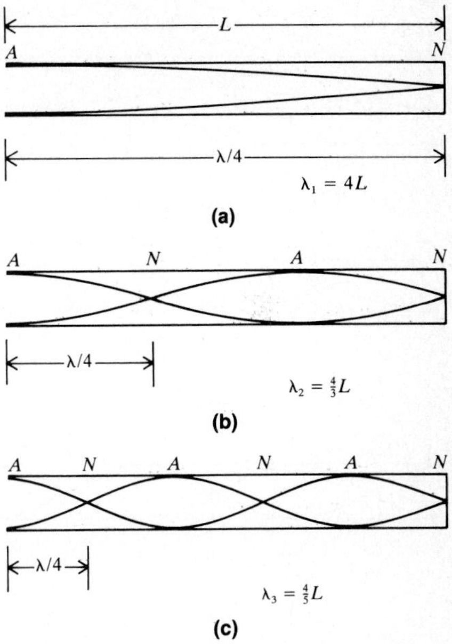

Figure 12-3. The three lowest modes of vibration in a closed-end pipe.

EXAMPLE 12-3: A 1.5-m organ pipe is closed at one end. The speed of sound within the column of air enclosed by the pipe is 340 m/s.
What is the fundamental frequency?

Solution: Calculate the fundamental frequency directly from Eq. (12-3):

$$f_1 = \frac{v_w}{4L} = \frac{340 \text{ m/s}}{4(1.5 \text{ m})} = 56.7 \text{ Hz}$$

The second overtone of a closed organ pipe is illustrated in Figure 12-3c. The distance from a node to the nearest antinode is $\lambda/4$, so the length of the pipe is equal to five quarter-wavelengths: $L = \frac{5}{4}\lambda_3$ or $\lambda_3 = \frac{4}{5}L$. The frequency of vibration of these standing waves is

$$f_3 = \frac{v_w}{\lambda_3} = \frac{v_w}{\frac{4}{5}L} = 5\left(\frac{v_w}{4L}\right)$$

The frequency of the second overtone is five times the fundamental frequency. Table 12-1 lists the fundamental frequencies and the first three overtones in a vibrating string and a pipe closed at one end. A vibrating string can produce all harmonics of the fundamental frequency. The closed-end pipe, however, produces only odd harmonics, so the first overtone is the third harmonic, the second overtone is the fifth harmonic, and so on. Your ear distinguishes musical sounds by the number and amplitude of all the harmonics present.

HARMONIC FREQUENCIES (CLOSED PIPE)

$$f_n = \frac{v_w}{\lambda_n} = \frac{nv}{4L} \qquad n = 1, 3, 5, 7, \ldots \qquad \textbf{(12-3a)}$$

TABLE 12-1: Frequencies of Standing Waves

	Vibrating string or open-end pipe	Closed-end pipe
1st harmonic (Fundamental frequency)	$\frac{v_w}{2L}$	$\frac{v_w}{4L}$
2nd harmonic	$\frac{2v_w}{2L}$	
3rd harmonic	$\frac{3v_w}{2L}$	$\frac{3v_w}{4L}$
4th harmonic	$\frac{4v_w}{2L}$	
5th harmonic	$\frac{5v_w}{2L}$	$\frac{5v_w}{4L}$
6th harmonic	$\frac{6v_w}{2L}$	
7th harmonic	$\frac{7v_w}{2L}$	$\frac{7v_w}{4L}$

EXAMPLE 12-4: A closed-end organ pipe vibrates in its second overtone at a frequency of 2000 Hz. The speed of sound within the pipe is 340 m/s. Calculate the length of the pipe.

Solution: The frequency of the second overtone is listed in Table 12-1. Solve the expression for L:

$$f_3 = 5\left(\frac{v_w}{4L}\right) \qquad L = \frac{5v_w}{4f_3} = \frac{5(340 \text{ m/s})}{4(2000/\text{s})} = 0.213 \text{ m}$$

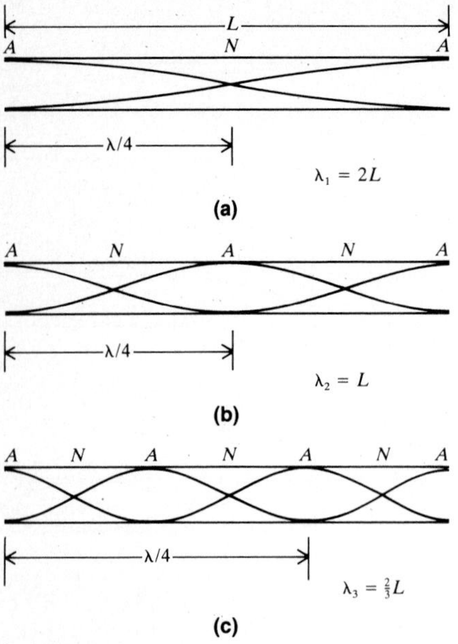

Figure 12-4. Standing waves in an open-end pipe.

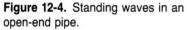

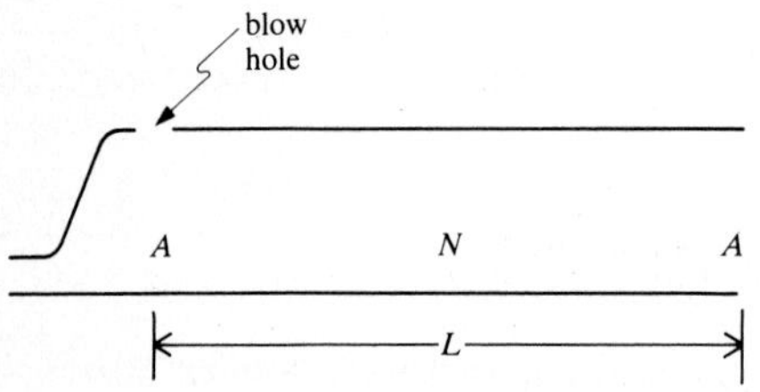

Figure 12-5

An open-end pipe can also produce standing waves. Figure 12-4 illustrates the three lowest frequencies in such a pipe. Compare Figure 12-4 with Figure 12-1 and observe that the wavelengths of the standing waves in an open-end pipe are the same as the wavelengths in a string fixed at both ends, so Eq. (12-2) applies also to open-end pipes. Like the vibrating string, the open-end pipe can produce both odd and even harmonics.

EXAMPLE 12-5: A whistle constructed from a small open tube with a blow-hole at one end has a fundamental frequency of 2000 Hz. (See Figure 12-5.) What is the length of the whistle from the blow-hole to the open end?

Solution: Compare the whistle to the open-end pipe in Figure 12-4a: The wavelength of the fundamental frequency is equal to twice the length of the whistle; that is, $\lambda = 2L$. Start with Eq. (12-1) and solve for L:

$$f = \frac{v_w}{\lambda} = \frac{v_w}{2L} \qquad L = \frac{v_w}{2f} = \frac{340 \text{ m/s}}{2(2000 \text{ Hz})} = 8.5 \times 10^{-2} \text{ m} = 8.5 \text{ cm}$$

12-2. Intensity and Loudness

Sound waves propagating through air or through any other medium carry energy outward from a source to a detector—for example, the human ear.

- The **intensity level** β of a sound wave is the rate at which energy is transported by the wave; that is, the power per unit cross-sectional area.

Although the SI unit for intensity is a watt (J/s) per square meter (W/m^2), relative intensity levels of sound waves are often measured in decibels (dB), expressed on a logarithmic scale:

RELATIVE INTENSITY (dB)

$$\beta = 10 \log \frac{I}{I_0} \qquad \textbf{(12-4)}$$

where I represents the actual intensity and I_0 represents an arbitrary reference level.

EXAMPLE 12-6: If we choose the reference level of sound intensity to be the minimum sound intensity that can be detected by the human ear (10^{-12} W/m^2), what is the relative intensity in decibels of a sound wave with an intensity of 10^{-9} W/m^2?

Solution: You can calculate the relative intensity directly from Eq. (12-4):

$$\beta = 10 \log \frac{I}{I_0} = 10 \log \frac{10^{-9} \text{ W/m}^2}{10^{-12} \text{ W/m}^2} = 10 \log(10^3) = 30 \text{ dB}$$

Since intensity is power divided by area (P/A), you can write the ratio of two intensities detected by the same detector as follows:

$$\frac{I_2}{I_1} = \frac{P_2/A}{P_1/A} = \frac{P_2}{P_1}$$

If the two intensities are compared with the same detector, such as a human ear or a microphone, the area cancels out and the intensity level in decibels can also be expressed as

RELATIVE INTENSITY IN dB (SINGLE DETECTOR)

$$\beta = 10 \log \frac{P}{P_0} \qquad \textbf{(12-5)}$$

The difference between the intensity levels of two sound waves is

$$\Delta\beta = \beta_2 - \beta_1 = 10 \log \frac{I_2}{I_0} - 10 \log \frac{I_1}{I_0}$$

When you combine the log terms, the result is

INTENSITY DIFFERENCE (dB)

$$\Delta\beta = 10 \log \frac{I_2}{I_1} \qquad \textbf{(12-6)}$$

If the comparison is made with the same detector, you can compare the two sounds by using the power ratios:

INTENSITY DIFFERENCE IN dB (SINGLE DETECTOR)

$$\Delta\beta = 10 \log \frac{P_2}{P_1} \qquad \textbf{(12-7)}$$

EXAMPLE 12-7: What is the difference in intensity level between the sound from one violin and the sound from twelve violins playing together?

Solution: Begin by writing down the power ratio: $P_2/P_1 = 12$. Now, use Eq. (12-7) to calculate the difference in the two sound levels:

$$\Delta\beta = 10 \log \frac{P_2}{P_1} = 10 \log(12) = 10.8 \text{ dB}$$

EXAMPLE 12-8: The maximum intensity the human ear can tolerate without pain is approximately 1.0 W/m^2.
What is the relative intensity level of this sound wave in dB as compared to the minimum detectable intensity of 10^{-12} W/m^2?

Solution: Directly from Eq. (12-4):

$$\beta = 10 \log \frac{I}{I_0} = 10 \log \frac{1 \text{ W/m}^2}{10^{-12} \text{ W/m}^2} = 10 \log(10^{12}) = 120 \text{ dB}$$

SUMMARY

Frequency of standing waves	$f = \dfrac{v_w}{\lambda}$	gives the frequency as the speed of propagation of a standing wave divided by the wavelength
Fundamental frequency (fixed string)	$f_1 = \dfrac{v_w}{\lambda_1} = \dfrac{v_w}{2L}$	gives the lowest natural frequency of a standing wave for a string fixed at both ends
Fundamental frequency (closed pipe)	$f_1 = \dfrac{v_w}{\lambda_1} = \dfrac{v_w}{4L}$	gives the lowest natural frequency of a standing wave in a pipe closed at one end
Relative intensity (dB)	$\beta = 10 \log \dfrac{I}{I_0}$	gives the relative rate at which energy is transported by a sound wave, measured in decibels
Relative intensity in dB (single detector)	$\beta = 10 \log \dfrac{P}{P_0}$	gives the ratio of two intensities compared with the same detector
Intensity difference (dB)	$\Delta\beta = 10 \log \dfrac{I_2}{I_1}$	gives the difference between the intensity levels of two sound waves
Intensity difference in dB (single detector)	$\Delta\beta = 10 \log \dfrac{P_2}{P_1}$	gives the ratio of the difference between two intensity levels compared with the same detector

RAISE YOUR GRADES

Can you define . . . ?

- ☑ a standing wave
- ☑ a node
- ☑ constructive interference
- ☑ an overtone

Can you . . . ?

- ☑ determine the location of nodes and antinodes on a standing wave pattern
- ☑ calculate the frequency of a standing wave when you know the wavelength and the speed of propagation
- ☑ calculate the fundamental frequency of a stretched wire
- ☑ calculate the frequency of the second overtone in a guitar string
- ☑ determine the tension in a guitar string that will produce a given fundamental frequency
- ☑ calculate the fundamental frequency of vibration in a closed-end organ pipe of a given length
- ☑ determine the number of nodes in an organ pipe vibrating in its second overtone
- ☑ calculate the length of an organ pipe that has a given fundamental frequency
- ☑ write the equation for the relative intensity in decibels of a sound wave
- ☑ calculate the difference in the intensity levels of two sound waves when one has 1000 times as much power as the other

Do you know . . . ?

- ☑ which harmonics are missing in a pipe closed at one end
- ☑ how many nodes are present in a stretched string vibrating in its third harmonic
- ☑ how the frequency of vibration of a guitar string changes when the tension is doubled
- ☑ the wavelength of a standing wave when the distance between adjacent nodes is given

SOLVED PROBLEMS

PROBLEM 12-1 An 80-cm steel wire is vibrating in its first overtone at a frequency of 220 Hz. The mass of the wire is 8 g. Calculate the wave velocity and tension in this wire.

Solution Begin with Table (12-1) for the frequency of the first overtone in a vibrating wire and solve for the wave velocity:

$$f_2 = \frac{v_w}{L} \qquad v_w = f_2 L = (220\ \text{Hz})(0.80\ \text{m}) = 176\ \text{m/s}$$

Now, use Eq. (11-1) for the speed of propagation of a transverse wave in a stretched string and solve for the tension:

$$v_w = \sqrt{\frac{F}{m/L}}$$

$$F = v_w^2\left(\frac{m}{L}\right) = \frac{(176\ \text{m/s})^2(8 \times 10^{-3}\ \text{kg})}{0.80\ \text{m}} = 3.10 \times 10^2\ \text{N}$$

PROBLEM 12-2 A closed-end organ pipe has a fundamental frequency of 44 Hz. If the speed of sound within the pipe is 340 m/s, determine the length of the pipe.

Solution You can calculate the length of the pipe from Eq. (12-3):

$$f_1 = \frac{v_w}{4L} \qquad L = \frac{v_w}{4f_1} = \frac{340\ \text{m/s}}{4(44/\text{s})} = 1.93\ \text{m}$$

PROBLEM 12-3 What are the frequencies of the first and second overtones of the organ pipe described in Problem 12-2?

Solution The overtones of a closed-end pipe consist only of the odd harmonics, so the first and second overtones correspond to the third and fifth harmonic frequencies:

$$f_2 = 3f_1 = 3(44\text{ Hz}) = 132\text{ Hz} \qquad f_3 = 5f_1 = 5(44\text{ Hz}) = 220\text{ Hz}$$

PROBLEM 12-4 The apparatus illustrated in Fig. 12-6 measures the speed of propagation of sound in a gas. When the 60-cm tube is filled with hydrogen gas at 1 atm, its fundamental frequency is 535 Hz. Calculate the speed of propagation of sound in hydrogen at atmospheric pressure.

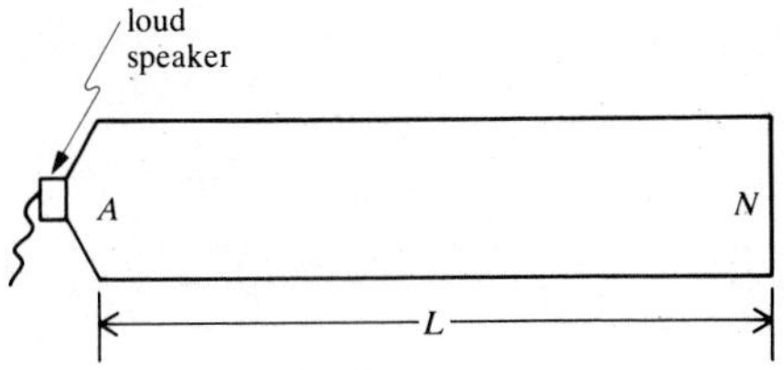

Figure 12-6

Solution You can calculate the speed of sound from Eq. (12-3):

$$f_1 = \frac{v_w}{4L}$$

$$v_w = 4Lf_1 = 4(0.60\text{ m})(535\text{ Hz}) = 1284\text{ m/s}$$

PROBLEM 12-5 A closed-end organ pipe is 70 cm long. The speed of sound in the pipe is 340 m/s, and it has one overtone with a frequency of 850 Hz. (**a**) Which harmonic of the fundamental frequency corresponds to the overtone? (**b**) Make a diagram showing the nodes and antinodes.

Solution

(**a**) First, calculate the fundamental frequency of the pipe (Eq. 12-3):

$$f_1 = \frac{v_w}{4L} = \frac{340\text{ m/s}}{4(0.70\text{ m})} = 121.4\text{ Hz}$$

Then, divide the frequency of the overtone by the fundamental frequency to determine which harmonic corresponds to the overtone:

$$\frac{850\text{ Hz}}{121.4\text{ Hz}} = 7$$

This overtone is the seventh harmonic.

(**b**) In order to make a diagram of the nodes and antinodes, you need to know the wavelength. From Table 12-1, you can see that the seventh harmonic corresponds to the third overtone, and you can write the frequency as

$$f_4 = 7\left(\frac{v_w}{4L}\right) = \frac{v_w}{\frac{4}{7}L} = \frac{v_w}{\lambda}$$

so, $\lambda = \frac{4}{7}L$. This means that the length of the pipe is equal to seven quarter-wavelengths, $L = 7(\lambda/4)$. Now you can make your drawing, which should look like Fig. 12-7.

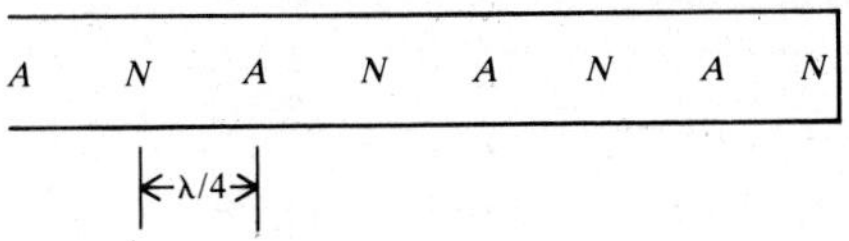

Figure 12-7

PROBLEM 12-6 A certain high-fidelity amplifier produces 20 W of power to drive the speakers over the frequency range of 200 Hz to 12 kHz. At 18 kHz, however, the output decreases to only 5 W. Calculate the loss in power output (in dB) at 18 kHz.

Solution Use Eq. (12-7) to find the difference between these two power levels:

$$\Delta\beta = 10 \log \frac{P_2}{P_1} = 10 \log \frac{20 \text{ W}}{5 \text{ W}} = 10(0.602) = 6.02 \text{ dB}$$

PROBLEM 12-7 Let I_0 represent the average intensity level of sound produced by twelve people conversing with each other in a room. Approximately how many people engaged in conversation in the same room would cause the average intensity level in decibels to double?

Solution Use $12I_0$ to represent the average sound intensity produced by twelve people. Then, let nI_0 represent the average intensity level due to n people. And from Eq. (12-6) find n:

$$\Delta\beta = 10 \log\left(\frac{nI_0}{12I_0}\right)$$

$$2 = 10 \log\left(\frac{n}{12}\right)$$

$$\log\left(\frac{n}{12}\right) = 0.2$$

Now, take the antilog of each side of the equation:

$$\frac{n}{12} = \text{antilog}(0.2) = 1.585$$

$$n = 12(1.585) = 19$$

Supplementary Exercises

PROBLEM 12-8 By what factor does the speed of propagation of a transverse wave in a wire increase when the tension is doubled?

PROBLEM 12-9 A wire 60 cm long has a fundamental frequency of 250 Hz. What is the frequency and the wavelength of the third harmonic?

PROBLEM 12-10 A closed-end organ pipe has a fundamental frequency of 440 Hz. The speed of sound is 340 m/s. What is the length of this pipe?

PROBLEM 12-11 A closed-end pipe contains a gas other than air. The pipe is 120 cm long and the frequency of the first overtone is 250 Hz. What is the speed of sound in this gas?

PROBLEM 12-12 If the reference level of sound has an intensity of 2×10^{-10} W, what is the relative intensity in decibels of sound wave whose intensity is 8×10^{-6} W?

PROBLEM 12-13 Two sound waves measured by the same detector have a difference in sound level of 25 dB. What is the ratio of the intensities associated with these two waves?

Answers to Supplementary Exercises

12-8: $\sqrt{2}$

12-9: 750 Hz; 40 cm

12-10: 19.3 cm

12-11: 400 m/s

12-12: 46.0 dB

12-13: 316

13 TEMPERATURE

THIS CHAPTER IS ABOUT

☑ Temperature and Temperature Scales
☑ Thermal Expansion

13-1. Temperature and Temperature Scales

- **Temperature** is a measure of kinetic energy (see Sec. 6-2B).

The SI unit for temperature is the **kelvin** (K). [Note that you do not use a degree sign with this unit.]

- On the Kelvin or absolute scale, **absolute zero** (0 K) is the point at which all molecules possess zero kinetic energy.

Water freezes at 273.15 K and boils at 373.15 K. The Celsius scale (°C) defines the freezing point of water as 0 °C and its boiling point as 100 °C. Since both the Kelvin and Celsius scales contain 100 degrees between the freezing and boiling points, you can interconvert them as follows:

TEMPERATURE CONVERSION

$$\begin{aligned} K &= {}^\circ C + 273.15 \\ {}^\circ C &= K - 273.15 \end{aligned} \qquad \textbf{(13-1)}$$

You'll still see the Fahrenheit scale (°F) in everyday use in cooking, weather forecasting, and so on. Its relationship to the Celsius scale is

$$^\circ F = \frac{9}{5}(^\circ C) + 32 \qquad \textbf{(13-1a)}$$

One way to measure temperature is with a *thermometer*. Common thermometers use the expansion of a liquid—mercury, alcohol, kerosene—as the basis for measuring temperature. You can also use other physical properties to measure temperature: the expansion of a gas, the generation of a voltage at a junction of dissimilar metals (a *thermocouple*), or the resistance of a wire to the passage of an electric current.

EXAMPLE 13-1: What temperature on the Fahrenheit scale corresponds to 50 °C?

Solution: You can find the Fahrenheit temperature directly from Eq. (13-1a):

$$^\circ F = \frac{9}{5}(^\circ C) + 32^\circ = \frac{9}{5}(50^\circ) + 32 = 122\ ^\circ F$$

EXAMPLE 13-2: At what Celsius temperature are the Celsius and Fahrenheit temperatures equal?

Solution: Use x to represent the numerical value. Since x has the same value on both scales, you can write Eq. (13-1a) as

$$x = \frac{9}{5}x + 32$$

and solve for x:

$$x\left(\frac{9}{5} - 1\right) = -32^\circ \qquad x = \frac{5}{4}(-32^\circ) = -40^\circ$$

EXAMPLE 13-3: Determine the Kelvin temperature that corresponds to −100 °F.

Solution: First, convert the Fahrenheit temperature to the Celsius scale (Eq. 13-1a):

$$°F = \frac{9}{5}(°C) + 32$$

$$°C = \frac{5}{9}(°F - 32°) = \frac{5}{9}(-100° - 32°) = -73.3\ °C$$

Then, from Eq. (13-1):

$$K = °C + 273.15 = -73.3 + 273.15 = 199.85\ K$$

13-2. Thermal Expansion

A. Solids

All of the internal and external dimensions of a solid increase when it is heated. When the temperature of a piece of steel is increased by ΔT, the change in its length is

LINEAR EXPANSION (SOLID)

$$\Delta L = L\alpha\Delta T \tag{13-2}$$

where α is the **coefficient of linear expansion** for a particular solid. The units of α are reciprocal Celsius degrees, 1/°C, commonly expressed as $(°C)^{-1}$. Table 13-1 lists the coefficients of linear expansion of several solids. The **coefficient of volume expansion** β for a solid is three times the coefficient of linear expansion:

COEFFICIENT OF VOLUME EXPANSION (SOLID)

$$\beta = 3\alpha \tag{13-3}$$

TABLE 13-1: Coefficients of Linear Expansion

Material	$(°C)^{-1}$
aluminum	2.6×10^{-5}
brass	1.9×10^{-5}
concrete	1.2×10^{-5}
glass (soft)	8.5×10^{-6}
iron or steel	1.1×10^{-5}
lead	2.9×10^{-5}
platinum	9.0×10^{-6}
Pyrex	3.3×10^{-6}
quartz (fused)	4.0×10^{-7}

EXAMPLE 13-4: At 20 °C, a thin sheet of steel contains a rectangular slot 3.0 cm × 5.0 cm. The temperature of the steel is increased to 220 °C.
What is the fractional increase in the area of the rectangular opening? (Assume that the steel's crystalline structure prevents distortion of the hole or buckling of the sheet during the expansion).

Solution: Find the coefficient of linear expansion for steel from Table 13-1, and then calculate the increase in the length and height of the slot from Eq. (13-2):

$$\Delta\ell = \ell\alpha\Delta T = (5.0\ \text{cm})(1.1 \times 10^{-5}\ °C)(200\ °C) = 1.1 \times 10^{-2}\ \text{cm}$$

$$\Delta h = h\alpha\Delta T = (3.0\ \text{cm})(1.1 \times 10^{-5}\ °C)(200\ °C) = 0.66 \times 10^{-2}\ \text{cm}$$

At 220 °C, the length and height of the slot are

$$\ell' = (5.0 + 1.1 \times 10^{-2})\ \text{cm} = 5.011\ \text{cm}$$

$$h' = (3.0 + 0.66 \times 10^{-2})\ \text{cm} = 3.0066\ \text{cm}$$

which means that the new area at 220 °C is

$$A' = (5.011 \times 3.0066)\ \text{cm}^2 = 15.066\ \text{cm}^2$$

Now, you can subtract the original area of the slot from the new area:

$$\Delta A' = (15.066 - 15)\ \text{cm}^2 = 0.066\ \text{cm}^2$$

And the fractional increase in area is

$$\frac{\Delta A'}{A} = \frac{0.066\ \text{cm}^2}{15.0\ \text{cm}^2} \times 100\% = 0.44\%$$

EXAMPLE 13-5: A cylindrical steel bar, 4 cm in diameter and 80 cm in length, is supported between two fixed blocks, as illustrated in Figure 13-1. The stretch modulus for steel is 20×10^{10} N/m^2 (see Sec. 9-2A.) The bar is heated from 20 °C to 100 °C.
What is the total force the bar exerts against the rigid support blocks?

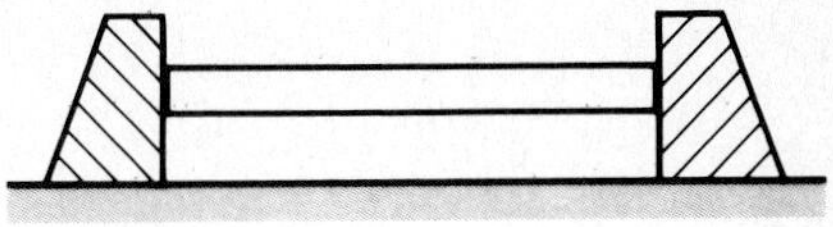

Figure 13-1

Solution: First, use Eq. (13-2) to calculate the amount by which the bar would expand if it were free to do so. Since the bar cannot expand because of its rigid supports, ΔL in this equation represents the amount by which the bar is *compressed* at 100 °C. Now, find the force required to compress the rod by substituting the expression for ΔL into the equation for the stretch modulus (Eq. 9-2):

$$E = \frac{F/A}{\Delta L/L}$$

$$F = EA\frac{\Delta L}{L} = EA\frac{L\alpha\,\Delta T}{L} = EA\alpha\,\Delta T = E\pi r^2\alpha\,\Delta T$$

$$= (20 \times 10^{10}\ \text{N/m}^2)\pi(0.02\ \text{m})^2(1.1 \times 10^{-5}\ (°\text{C})^{-1})(80\ °\text{C}) = 2.21 \times 10^5\ \text{N}$$

B. Fluids

The volume of a fluid increases when it is heated. When the temperature of a fluid is increased by ΔT, the change in its volume is

VOLUME EXPANSION OF A FLUID $$\Delta V = V\beta\,\Delta T \qquad \textbf{(13-4)}$$

where β is the coefficient of volume expansion. The coefficients of volume expansion of several fluids are listed in Table 13-2.

TABLE 13-2: Coefficients of Volume Expansion

Material	$(°\text{C})^{-1}$
Liquids	
glycerin	4.85×10^{-4}
mercury	1.82×10^{-4}
methyl alcohol	1.134×10^{-3}
*Gases**	
air	3.67×10^{-3}
carbon dioxide	3.74×10^{-3}
helium	3.665×10^{-3}
hydrogen	3.660×10^{-3}

*At constant pressure

EXAMPLE 13-6: A certain liquid has a volume of 20.000 cm^3 at 30 °C. When the liquid is heated to 180 °C, its volume increases to 20.546 cm^3.
Determine the probable identity of this liquid.

Solution: Solve Eq. (13-3) for β and compare your result with the values in Table 13-2:

$$\Delta V = V\beta\,\Delta T$$

$$\beta = \frac{\Delta V}{V\,\Delta T} = \frac{0.546\ \text{cm}^3}{(20.0\ \text{cm}^3)(180\ °\text{C} - 30\ °\text{C})} = 1.82 \times 10^{-4}\ (°\text{C})^{-1}$$

This is the coefficient of volume expansion of mercury.

C. The ideal gas

The pressure, volume, Kelvin temperature, and mass of a gas specify its **state**.

- The **ideal gas law** gives the state of a gas in the form of an equation:

IDEAL GAS LAW $$PV = nRT \qquad \textbf{(13-5)}$$

where n represents the **number of moles** of the gas, which is proportional to the mass, T is the temperature of the gas in kelvins, and R is the **universal gas constant.** In SI, $R = 8.314$ J/(mol K).

note: The **mole** (mol) is the fundamental SI unit of amount of substance. It is defined as the amount of substance that contains **Avogadro's number** of atoms or molecules, where Avogadro's number is defined as the number of atoms in exactly 0.012 kg (exactly 12 g) of ^{12}C, approximately 6.02×10^{23}.

An **ideal gas** obeys the ideal gas law *exactly;* and at ordinary temperatures and pressures, most gases obey it *rather closely.* At extremely low temperatures and high pressures, however, the behavior of real gases deviates significantly from that predicted by the ideal gas law.

Consider a given mass of an ideal gas held at constant pressure. From Eq. (13-5), the volume of this ideal gas is directly proportional to its Kelvin temperature:

VOLUME (IDEAL GAS)

$$V = \left(\frac{nR}{P}\right)T \tag{13-5a}$$

If you let T_2 represent an absolute temperature slightly greater than T_1, the volumes corresponding to the two temperatures are

$$V_2 = \left(\frac{nR}{P}\right)T_2 \quad \text{and} \quad V_1 = \left(\frac{nR}{P}\right)T_1$$

Now, let ΔV represent the change in volume and ΔT the change in temperature. Subtract the second equation from the first to obtain a relationship for the change in volume:

CHANGE IN VOLUME (IDEAL GAS AT CONSTANT PRESSURE)

$$\Delta V = \left(\frac{nR}{P}\right)\Delta T \tag{13-6}$$

EXAMPLE 13-7: Calculate the coefficient of volume expansion for an ideal gas at 0 °C.

Solution: Use Eq. (13-4) to get an expression for β, the coefficient of volume expansion:

$$\Delta V = V\beta\Delta T$$

$$\beta = \frac{\Delta V}{V\Delta T}$$

Now, divide Eq. (13-6) by Eq. (13-5) to obtain an expression for $\Delta V/V$:

$$\frac{\Delta V}{V} = \frac{(nR/P)\,\Delta T}{(nR/P)T} = \frac{\Delta T}{T}$$

Then, use this result to eliminate $\Delta V/V$ (remember that T is in kelvins):

$$\beta = \frac{\Delta T}{T\Delta T} = \frac{1}{T} = \frac{1}{273\ \text{K}} = 3.66 \times 10^{-3}\ (°\text{C})^{-1}$$

Note that the coefficients of volume expansion for all the gases listed in Table 13-2 are close to this value.

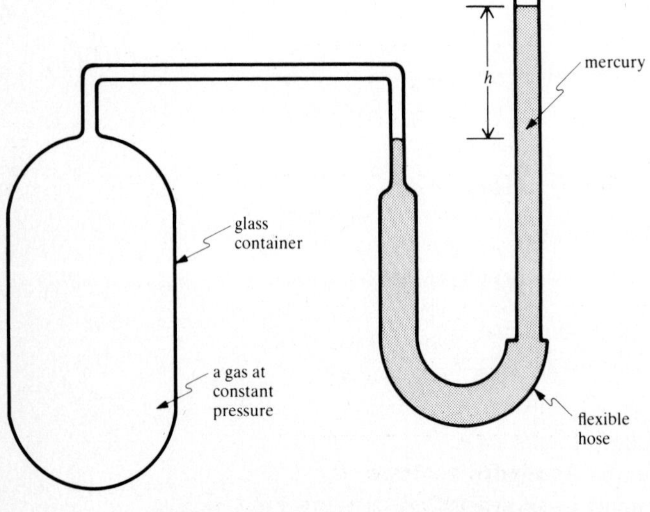

Figure 13-2. The constant-pressure gas thermometer.

The **constant-pressure gas thermometer** is one of the most reliable methods of measuring temperature (see Figure 13-2). Although the volume of the gas increases as the temperature rises, the mercury reservoir in the flexible hose can be adjusted so that the height, which determines the pressure, stays constant.

EXAMPLE 13-8: A constant-pressure gas thermometer contains 40 cm^3 of helium. The temperature increases from 0 °C to 1.5 °C.

How much does the volume of the helium increase?

Solution: Find the coefficient of volume expansion for helium from Table 13-2. Then calculate the volume increase with Eq. (13-4):

$$\Delta V = V\beta\,\Delta T = (40\ \text{cm}^3)\,(3.665 \times 10^{-3}\ (^\circ\text{C})^{-1})\,(1.5\ ^\circ\text{C}) = 0.220\ \text{cm}^3$$

EXAMPLE 13-9: You determine the change in the volume of the helium in the thermometer of Example 13-8 by measuring how much the mercury is lowered in the narrow-bore capillary tube. The inside diameter of this tube is 0.5 cm. Determine the change in height that corresponds to a temperature change of 1.5 °C.

Solution: Write the formula for the volume of a cylinder, $\Delta V = \pi r^2 h$, and solve for h:

$$h = \frac{V}{\pi r^2} = \frac{0.220\ \text{cm}^3}{\pi (0.25\ \text{cm})^2} = 1.12\ \text{cm}$$

EXAMPLE 13-10: The hydrogen gas in a constant-pressure gas thermometer has a volume of 620 cm³ at 20 °C. When the temperature is increased to 100 °C, the gas occupies 789 cm³.
Determine the temperature of the hydrogen gas when its volume is 700 cm³.

Solution: Use Eq. (13-5a) to find the absolute temperature that corresponds to a volume of 700 cm³. First, evaluate the constant, nR/P:

$$V = \left(\frac{nR}{P}\right)T \qquad \frac{nR}{P} = \frac{V}{T} = \frac{620\ \text{cm}^3}{(20^\circ + 273^\circ)\ \text{K}} = 2.116\ \text{cm}^3/\text{K}$$

Then, calculate the new absolute temperature:

$$T = \frac{V}{nR/P} = \frac{700\ \text{cm}^3}{2.116\ \text{cm}^3/\text{K}} = 331\ \text{K}$$

And finally, the Celsius temperature:

$$^\circ\text{C} = \text{K} - 273 = 331 - 273 = 58\ ^\circ\text{C}$$

SUMMARY

Temperature conversion	$\text{K} = {}^\circ\text{C} + 273.15$ ${}^\circ\text{C} = \text{K} - 273.15$	gives the interconversion of the Kelvin and Celsius scales
	${}^\circ\text{F} = \frac{9}{5}({}^\circ\text{C}) + 32$	gives the relationship of the Fahrenheit scale to the Celsius scale
Linear expansion (solid)	$\Delta L = L\alpha\,\Delta T$	gives the change of length of a solid when its temperature increases by ΔT
Coefficient of volume expansion (solid)	$\beta = 3\alpha$	gives the coefficient of volume expansion as three times the coefficient of linear expansion
Volume expansion of a fluid	$\Delta V = V\beta\,\Delta T$	gives the change in volume of a fluid when its temperature increases by ΔT
Ideal gas law	$PV = nRT$	gives the relationship between the pressure, volume, Kelvin temperature, and mass of a gas

Volume (ideal gas)	$V = \left(\frac{nR}{P}\right)T$	the mass of an ideal gas held at constant pressure is directly proportional to its Kelvin temperature
Change in volume (ideal gas at constant pressure)	$\Delta V = \left(\frac{nR}{P}\right)\Delta T$	gives the change in volume for an ideal gas at constant pressure when its temperature changes by ΔT

RAISE YOUR GRADES

Can you . . . ?

- ☑ interconvert Celsius, Fahrenheit, and Kelvin temperatures
- ☑ determine the change in length of a rod, given the composition and change in temperature of the rod
- ☑ calculate the change in volume of a fluid, given the change in temperature
- ☑ write the relationship between the coefficient of volume expansion and the coefficient of linear expansion
- ☑ calculate the change in volume of an ideal gas due to a temperature increase

SOLVED PROBLEMS

PROBLEM 13-1 Calculate the temperature on the Celsius scale that corresponds to 59 °F.

Solution Use Eq. (13-1a) to convert Fahrenheit to Celsius temperature:

$$°F = \frac{9}{5}(°C) + 32 \qquad \frac{9}{5}(°C) = °F - 32°$$

$$°C = (F - 32°)\frac{5}{9} = (59° - 32°)\frac{5}{9} = 15\ °C$$

PROBLEM 13-2 What is the Fahrenheit temperature that corresponds to 293 K?

Solution First, find the corresponding Celsius temperature from Eq. (13-1):

$$°C = 293\ K - 273\ K = 20\ °C$$

Then, Eq. (13-1a):

$$°F = \frac{9}{5}(°C) + 32° = \frac{9}{5}(20°) + 32° = 68\ °F$$

PROBLEM 13-3 A piece of pure metal is 30.0000 mm long at 0 °C and 30.0324 mm long at 120 °C. (**a**) Calculate the coefficient of linear expansion of this metal, and then (**b**) identify it from Table 13-1.

Solution

(**a**) Use Eq. (13-2) and solve for α:

$$\Delta L = L\alpha\,\Delta T \qquad \alpha = \frac{\Delta L}{L\,\Delta T} = \frac{0.0324\ \text{mm}}{(30\ \text{mm})(120\ °C)} = 9.00 \times 10^{-6}\ (°C)^{-1}$$

(**b**) This is the coefficient of linear expansion of platinum.

PROBLEM 13-4 A thermometer made of Pyrex glass has a capillary tube 0.2 mm in diameter and a bulb whose volume is 0.5 cm^3. The mercury inside the thermometer just fills the bulb initially. When the temperature increases 5 °C, what is the increase in height of the mercury in the capillary tube?

Solution Begin by calculating the increase in the volume of the glass and the increase in the volume of the mercury. You can obtain the coefficient of volume expansion of glass from Eq. (13-4):

$$\Delta V = V\beta\,\Delta T$$

$$\beta = 3\alpha = 3(3.3 \times 10^{-6}\ 1/°\text{C}) = 9.9 \times 10^{-6}\ (°\text{C})^{-1}$$

Then, calculate the increase in the volume of the glass bulb. You can ignore the change in the volume of the capillary bore since it is so small.

$$\Delta V_1 = V\beta\,\Delta T = (0.5\ \text{cm}^3)\left(9.9 \times 10^{-6}\ (°\text{C})^{-1}\right)(5\ °\text{C}) = 2.475 \times 10^{-5}\ \text{cm}^3$$

And the increase in the volume of the mercury:

$$\Delta V_2 = V\beta\,\Delta T = (0.5\ \text{cm}^3)\left(1.82 \times 10^{-4}\ (°\text{C})^{-1}\right)(5\ °\text{C}) = 4.55 \times 10^{-4}\ \text{cm}^3$$

The difference in these two changes in volume causes the mercury to rise:

$$\Delta V = \Delta V_2 - \Delta V_1 = (4.55 - 0.25) \times 10^{-4}\ \text{cm}^3 = 4.30 \times 10^{-4}\ \text{cm}^3$$

Now, let h represent the increase in height of the column of mercury, and from the equation for volume:

$$\Delta V = \pi r^2 h$$

$$h = \frac{\Delta V}{\pi r^2} = \frac{4.30 \times 10^{-4}\ \text{cm}^3}{\pi(0.01\ \text{cm})^2} = 1.37\ \text{cm}$$

PROBLEM 13-5 A sealed glass bulb is filled with pure nitrogen gas at a pressure of 2.0 atm. The volume of the container is 500 cm^3 and the temperature is 22 °C. Calculate the number of moles and the mass of nitrogen gas within the bulb. [The mass of 1 mole of N_2 is 28 g and 1 atm = 1.013 × 10^5 N/m^2 (Sec. 10-1).]

Solution Solve Eq. (13-5) for the number of moles:

$$PV = nRT$$

$$n = \frac{PV}{RT} = \frac{\left(2.0\ \text{atm}\left(\dfrac{1.013 \times 10^5\ \text{N/m}^2}{1\ \text{atm}}\right)\right)\left(500\ \text{cm}^3\left(\dfrac{(1\ \text{m})^3}{(10^2\ \text{cm})^3}\right)\right)}{(8.314\ \text{J/mol K})(22 + 273)\ \text{K}} = 0.0413\ \text{moles}$$

Then, calculate the mass of the gas:

$$m = (0.0413\ \text{moles})(28\ \text{g/mole}) = 1.16\ \text{g}$$

PROBLEM 13-6 If the nitrogen gas in Problem 13-5 is heated to 100 °C, what will the pressure become?

Solution Solve Eq. (13-5) for the pressure:

$$P = \frac{nRT}{V} = \frac{(0.0413\ \text{mol})(8.314\ \text{J/mol K})(100 + 273)\ \text{K}}{500\ \text{cm}^3\left(\dfrac{(1\ \text{m})^3}{(10^2\ \text{cm})^3}\right)} = 2.56 \times 10^5\ \text{N/m}^2$$

$$= (2.56 \times 10^5\ \text{N/m}^2)\frac{(1\ \text{atm})}{(1.013 \times 10^5\ \text{N/m}^2)} = 2.53\ \text{atm}$$

PROBLEM 13-7 A constant-pressure gas thermometer contains 0.7 g of nitrogen gas at a pressure of 1 atm. If the absolute temperature of this gas changes by 20 K, what is the change in volume?

[***note:*** You can express the gas constant as R = 0.0821 (L atm)/(mol K) where 1 liter (L) is a volume equivalent to 1000 cm^3 or 0.001 m^2.]

Solution Use Eq. (13-7) to find ΔV. A mole of N_2 has a mass of 28 g.

$$\Delta V = \left(\frac{nR}{P}\right)\Delta T$$

$$= \frac{(0.025 \text{ mol})(0.0821 \text{ L atm/mol K})}{(1 \text{ atm})}(20 \text{ K})$$

$$= (.0411 \text{ L})\frac{(10^3 \text{ cm}^3)}{(1 \text{ L})} = 41.1 \text{ cm}^3$$

PROBLEM 13-8 Calculate the volume occupied by 1 mole of an ideal gas at a pressure of 1 atm and a temperature of 0 °C.

Solution From Eq. (13-5):

$$V = \frac{nRT}{P} = \frac{(1 \text{ mol})(0.0821 \text{ L atm/mol K})(273 \text{ K})}{(1 \text{ atm})} = 22.4 \text{ L}$$

This is called the **molar volume** of an ideal gas.

Supplementary Exercises

PROBLEM 13-9 What is the absolute temperature that corresponds to −58 °F?

PROBLEM 13-10 If the Fahrenheit temperature of a liquid increases by 27°, what is the corresponding change in temperature on the Celsius scale?

PROBLEM 13-11 A sheet of aluminum contains a circular hole whose diameter is 2.5400 cm at 20 °C. What is the diameter of this hole when the aluminum sheet is heated to 220 °C?

PROBLEM 13-12 A rectangular piece of lead has dimensions 8 cm × 16 cm × 20 cm. Calculate the increase in the volume of this lead block when it is heated from 10 °C to 160 °C.

PROBLEM 13-13 An unknown liquid has a volume of 400 cm^3 at −10 °C. When the liquid is heated to 110 °C, its volume increases by 23.28 cm^3. Calculate the coefficient of volume expansion of this liquid.

PROBLEM 13-14 A sealed glass bulb contains 800 cm^3 of helium gas at a pressure of 1.5 atm and a temperature of 25 °C. What is the pressure when the gas is heated to 225 °C? [The expansion of the glass container is small enough to be neglected.]

PROBLEM 13-15 A constant-pressure gas thermometer contains 250 cm^3 of helium gas at a pressure of 1.4 atm and a temperature of 0 °C. When the gas thermometer is placed in a warm liquid, its volume increases by 25 cm^3. What is the temperature of the liquid?

Answers to Supplementary Exercises

13-9: 223 K

13-10: 15 °C

13-11: 2.5532 cm

13-12: 33.4 cm^3

13-13: 4.85×10^{-4} 1/°C

13-14: 2.51 atm

13-15: 27.3 °C

14 HEAT

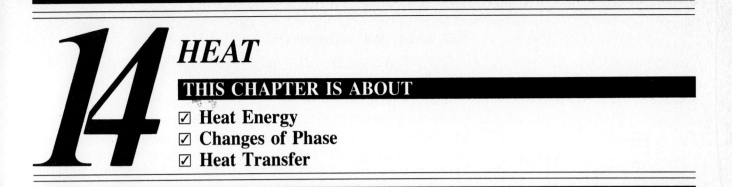

THIS CHAPTER IS ABOUT

☑ Heat Energy
☑ Changes of Phase
☑ Heat Transfer

14.1 Heat Energy

- **Heat energy** is the kinetic energy of the particles that make up matter.

In SI, heat energy is measured in joules. In the centimeter–gram–second system, heat energy is measured in **calories**, where 1 calorie (cal) equals exactly 4.184 J, which is often called the **mechanical equivalent** of heat. You need 1 cal to increase the temperature of exactly 1 g of water by one degree Kelvin or Celsius at ordinary room temperature (20 °C or 293 K).

- **Heat capacity** is the amount of heat energy needed to raise the temperature of a certain amount of material by 1 K or 1 °C.

The **amount of heat** Q required to increase the temperature of a substance depends on the mass of the substance, the temperature increase, and the specific heat capacity c of the material, defined as the heat capacity per unit mass:

HEAT REQUIRED TO INCREASE THE TEMPERATURE OF A SUBSTANCE

$$Q = mc\,\Delta T \tag{14-1}$$

- The term **specific heat** refers to the amount of heat energy in *calories* required to raise the temperature of 1 g of a substance 1 °C. Table 14-1 lists the specific heats of some common substances.

note: The calorie that diet-watchers count equals 1000 cal or 1 kilocalorie (kcal), sometimes abbreviated Cal.

TABLE 14-1: Specific heat of some common substances

Substance	Temperature range (°C)	Specific heat (cal/g °C)
Solids		
aluminum	15–100	0.22
copper	15–100	0.093
glass	20–100	0.20
ice	−10–0	0.50
iron	10–160	0.11
lead	10–160	0.0306
silver	10–100	0.056
steel	10–106	0.11
Liquids		
alcohol (methyl)	10–40	0.60
glycerin	40–60	0.60
mercury	10–40	0.033
water	0–100	1.00
Gases (at constant pressure)		
air	0–100	0.25
helium	0–100	1.24
steam	100–500	0.48

EXAMPLE 14-1: You need 30 cal of heat to increase the temperature of 10 g of glycerine from 20 °C to 25 °C.
Calculate the specific heat of glycerine.

Solution: Solve Eq. (14-1) for c:

$$Q = mc\,\Delta T \qquad c = \frac{Q}{m\,\Delta T} = \frac{30\ \text{cal}}{(10\ \text{g})(5\ °\text{C})} = 0.60\ \text{cal/g °C}$$

EXAMPLE 14-2: The specific heat of water is one calorie per gram per Celsius degree ($c = 1$ cal/g °C).
Calculate the amount of energy in joules required to raise the temperature of 2 liters of water from 20 °C to 90 °C.

Solution: You'll need to calculate the mass of the water being heated. [**recall**: Mass is equal to volume times density. (Eq. 10-2)]

$$m = Vd = (2\ \text{L})\frac{(10^3\ \text{cm}^3)}{(1\ \text{L})}\left(\frac{1\ \text{g}}{1\ \text{cm}^3}\right) = 2 \times 10^3\ \text{g}$$

Now, use Eq. (14-1) to calculate Q:

$$Q = mc\,\Delta T = (2 \times 10^3 \text{ g})(1 \text{ cal/g °C})(90 - 20) \text{ °C}$$

$$= (1.40 \times 10^5 \text{ cal})\frac{(4.184 \text{ J})}{(1 \text{ cal})} = 5.86 \times 10^5 \text{ J}$$

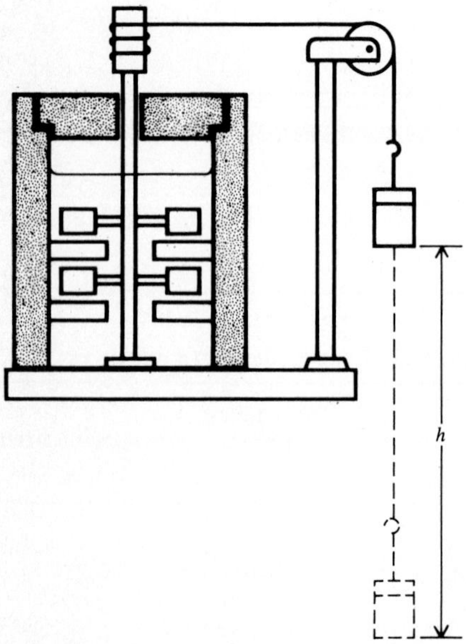

Figure 14-1. Mechanical apparatus for heating a liquid.

EXAMPLE 14-3: It is also possible to use friction to transform mechanical energy into heat energy. Figure 14-1 illustrates a mechanical device for heating a liquid by stirring it vigorously. A 50-N weight falls a distance of 1.2 m, stirring 400 cm^3 of methyl alcohol (density 0.810 g/cm^3).
Calculate the methyl alcohol's increase in temperature.

Solution: Calculate the mass of the liquid being heated. (You may neglect the heat absorbed by the insulated container and the stirring paddles.)

$$m = Vd = (400 \text{ cm}^3)(0.810 \text{ g/cm}^3) = 324 \text{ g}$$

You'll also need to know the potential energy of the 50-N weight. [**recall**: Potential energy equals weight times height. (Eq. 6-2)]

$$\text{PE} = (mg)h = (50 \text{ N})(1.2 \text{ m}) = 60 \text{ J}$$

As the paddles stir the liquid, this mechanical energy is converted into heat energy, so

$$Q = \text{PE} = (mg)h = (60 \text{ J})\left(\frac{1 \text{ cal}}{4.184 \text{ J}}\right) = 14.34 \text{ cal}$$

Then, from Eq. (14-1),

$$\Delta T = \frac{Q}{mc} = \frac{14.34 \text{ cal}}{(324 \text{ g})(0.60 \text{ cal/g °C})} = 7.38 \times 10^{-2} \text{ °C}$$

EXAMPLE 14-4: When a high-speed lead bullet strikes a heavy steel plate, most of the kinetic energy of the bullet is transformed into heat energy. A 5-g lead bullet moving at a speed of 200 m/s strikes a steel plate and absorbs 75% of the heat produced on impact. Calculate the bullet's rise in temperature.

Solution: First, calculate the kinetic energy of the bullet. [**recall**: Kinetic energy equals one-half mass times velocity squared. (Eq. 6-5a)]

$$\text{KE} = \frac{1}{2}mv^2 = \frac{1}{2}(5 \times 10^{-3} \text{ kg})(200 \text{ m/s})^2 = 100 \text{ J}$$

Then, calculate the heat energy absorbed in calories:

$$Q = 0.75 \text{ KE} = \frac{(0.75)(100 \text{ J})}{4.184 \text{ J/cal}} = 17.93 \text{ cal}$$

Now, you can use Eq. (14-1) and Table 14-1 to find ΔT:

$$\Delta T = \frac{Q}{mc} = \frac{17.93 \text{ cal}}{(5 \text{ g})(0.0306 \text{ cal/g °C})} = 117.2 \text{ °C}$$

14-2. Changes of Phase

- Many substances can exist in three states or **phases**—solid, liquid, and gas. The phase of a substance depends on temperature and pressure.

Water, for example, is in the liquid phase if it has a temperature between 0 °C and 100 °C and is subject to a pressure of 1 atm.

A. Solid to liquid transformations

Solid ice can be transformed into liquid water at a temperature of 0 °C. This change of phase requires a considerable amount of energy: You need 80 cal of heat energy to transform 1 g of ice into 1 g of water. To transform water from a liquid to a solid phase, you must remove 80 cal of heat energy for each gram.

- The **heat of fusion** h_f, sometimes called the latent heat of fusion, is the quantity of heat needed to transform a substance from a solid to a liquid.

The heat that must be added to melt a substance, or must be removed to freeze it, is given by the following formula:

HEAT OF TRANSFORMATION (SOLID ↔ LIQUID)

$$Q = mh_f \qquad \textbf{(14-2)}$$

EXAMPLE 14-5: The specific heat of ice is 0.50 cal/g °C. How much heat is required to transform a 400-g block of ice at −20 °C into water at 60 °C?

Solution: This operation takes three steps: (1) heating the ice to 0 °C; (2) transforming the ice to water at 0 °C; and (3) heating the water from 0 °C to 60 °C. From Eq. (14-1), calculate the heat required to raise the ice to 0 °C:

$$Q_1 = mc\,\Delta T = (400\text{ g})(0.50\text{ cal/g °C})(20\text{ °C}) = 4 \times 10^3\text{ cal}$$

And, from Eq. (14-2), obtain the heat required to melt the ice:

$$Q_2 = mh_f = (400\text{ g})(80\text{ cal/g}) = 32 \times 10^3\text{ cal}$$

Again, from Eq. (14-1), calculate the heat required to heat the water:

$$Q_3 = mc\,\Delta T = (400\text{ g})(1\text{ cal/g °C})(60\text{ °C}) = 24 \times 10^3\text{ cal}$$

So, the total heat energy required is

$$\begin{aligned} Q &= Q_1 + Q_2 + Q_3 \\ &= (4 + 32 + 24) \times 10^3\text{ cal} = 60 \times 10^3\text{ cal} = 60\text{ kcal} \end{aligned}$$

B. Liquid to gas transformations

- The **heat of vaporization** h_v is the amount of heat required to transform a liquid into a gas.

For example, the heat of vaporization required to transform water to steam is 540 cal/g. The heat that must be added to boil a substance, or must be removed to condense it, is given by the following formula:

HEAT OF TRANSFORMATION (LIQUID ↔ GAS)

$$Q = mh_v \qquad \textbf{(14-3)}$$

EXAMPLE 14-6: The specific heat of steam is 0.48 cal/g °C. How much heat is required to change 1.5 L of water at 30 °C to steam at 150 °C?

Solution: The heat required to raise the water temperature to 100 °C is

$$Q_1 = mc\,\Delta T = (1.5 \times 10^3\text{ g})(1\text{ cal/g °C})(70\text{ °C}) = 1.05 \times 10^5\text{ cal}$$

The heat required to vaporize the water is

$$Q_2 = mh_v = (1.5 \times 10^3\text{ g})(540\text{ cal/g}) = 8.10 \times 10^5\text{ cal}$$

And, the heat required to increase the water temperature of the steam to 150 °C is

$$Q_3 = mc\,\Delta T = (1.5 \times 10^3\text{ g})(0.48\text{ cal/g °C})(50\text{ °C}) = 0.36 \times 10^5\text{ cal}$$

So, the total heat required is

$$Q = (1.05 + 8.10 + 0.36) \times 10^5\text{ cal} = 9.51 \times 10^5\text{ cal} = 951\text{ kcal}$$

14-3. Heat Transfer

A. Conduction

- **Conduction** is the transfer of heat due to physical contact within a substance or from one material to another.

The amount of heat conducted through a rectangular block of material is proportional to the time t allowed for the heat to flow through the block, the area A of the block, and the temperature difference $T_2 - T_1$ between the surfaces. The amount of heat is inversely proportional to the thickness d of the slab, as given by

HEAT CONDUCTION

$$Q = KAt\frac{T_2 - T_1}{d} \tag{14-4}$$

- The constant of proportionality K is called the **thermal conductivity** of the material. Table 14-2 lists the thermal conductivity of some common solids.

TABLE 14-2: Thermal Conductivity of Some Common Solids

Substance	Thermal conductivity [cal/(cm s °C)]
concrete	0.002
copper	0.92
glass	0.0025
steel	0.11
white pine	0.0002

EXAMPLE 14-7: The thermal conductivity of steel is 0.11 cal/(cm s °C). The bottom surface of a steel kettle has a thickness of 2 mm and an effective area of 175 cm^2. The lower surface of the kettle bottom has a temperature of 500 °C and the inner surface has a temperature of 100 °C.
Determine the rate of heat conduction through the bottom of the kettle.

Solution: Choose a time interval of 1 s, and use Eq. (14-4):

$$Q = KAt\frac{T_2 - T_1}{d}$$

$$= (0.11 \text{ cal/cm s °C})(175 \text{ cm}^2)(1 \text{ s})\left(\frac{(500 - 100) \text{ °C}}{0.2 \text{ cm}}\right) = 3.85 \times 10^4 \text{ cal}$$

The rate of heat conduction is 3.85×10^4 cal/s.

B. Convection

- **Convection** is the transfer of heat due to the flow of a liquid or gas.

The air surrounding a hot radiator is less dense than the cooler air in the rest of the room. As a result, the warm air rises toward the ceiling and the cooler air moves in toward the radiator. The process of circulating hot air from the radiator to the rest of the room is called **thermal convection.** When a blower circulates the hot air, say from a furnace throughout a house, the process is called **forced convection.**

EXAMPLE 14-8: A rectangular heat duct has interior dimensions of 10 cm × 18 cm. Heated air flows through the duct at a speed of 30 cm/s. The temperature of the air is 81 °F as it enters a room, where the temperature is 72 °F. [The specific heat of air is 0.25 cal/g °C, and the density of air is 1.293×10^{-3} g/cm^3.]
How many calories of heat are added to the room in 5 minutes if the air leaves the room via the exhaust duct at 72 °F?

Solution: First, you'll have to calculate the volumetric flow rate of the air through the duct. [**recall**: The volumetric flow rate is equal to the speed multiplied by the cross-sectional area. (Sec. 10-4)]

$$Av = (30 \text{ cm/s})(10 \text{ cm})(18 \text{ cm}) = 5400 \text{ cm}^3\text{/s}$$

Then, find how much air enters the room in 5 minutes:

$$\left(\frac{5400\ \text{cm}^3}{\text{s}}\right)(5\ \text{min})\left(\frac{60\ \text{s}}{1\ \text{min}}\right) = 1.62 \times 10^6\ \text{cm}^3$$

So, the mass of air added to the room in 5 minutes is

$$m = Vd = (1.62 \times 10^6\ \text{cm}^3)(1.293 \times 10^{-3}\ \text{g/cm}^3) = 2.095 \times 10^3\ \text{g}$$

Now, use Eq. (14-1) and Table 14-1 to calculate the quantity of heat this mass of air adds to the room:

$$Q = mc\,\Delta T = (2.095 \times 10^3\ \text{g})(0.25\ \text{cal/g °C})((81 - 72)\ \text{°F})\left(\frac{5\ \text{°C}}{9\ \text{°F}}\right)$$

$$= 2.62 \times 10^3\ \text{cal}$$

C. Radiation

- **Radiation** is the transfer of heat from one object to another by electromagnetic waves.

Neither conduction nor convection of heat can take place in empty space, so radiation is the only way heat energy from the sun can reach the earth. Both light and heat from the sun are electromagnetic waves that travel through empty space at a speed of 3.0×10^8 m/s.

Every object emits some radiation unless it is at absolute zero (-273.15 °C = 0 K). **Stefan's law** gives the amount of radiation per unit time that an object emits:

RATE OF RADIATION

$$\frac{Q}{t} = Ae\sigma T^4 \qquad \textbf{(14-5)}$$

A is the object's surface area, T is its temperature in kelvins, σ is a constant called the **Stefan–Boltzmann constant** with a value of 5.67×10^{-8} W/(m² K⁴), and e is **emissivity**, a property dependent on the nature of the object's substance. The value of e may be between 0, for a perfect mirror, and 1 for a perfect radiation absorber, or **blackbody**. Objects at room temperature emit only a little invisible infrared radiation.

EXAMPLE 14-9: A sphere of carbon, 2.00 cm in diameter, is heated in a vacuum from 300 K to 600 K and then to 900 K. Calculate the rate at which the sphere emits radiation at each temperature.

Solution: Since the sphere is in a vacuum, there can be no heat transfer due to conduction or convection. Therefore, radiation alone accounts for the heat loss, and we can apply Stefan's Law. The rate of energy radiation, Q/t, is measured in watts (W). Since carbon is black, we assume the emissivity e of the sphere is 1. The area of the sphere is

$$A = 4\pi r^2 = 4\pi(0.0100\ \text{m})^2 = 1.257 \times 10^{-3}\ \text{m}^2$$

So at 300 K

$$Q/t = (1.257 \times 10^{-3})(1)(5.67 \times 10^{-8}\ \text{W/(m}^2\ \text{K}^4))(300\ \text{K})^4$$

$$= (7.125 \times 10^{-11}\ \text{W/K}^4)(8.1 \times 10^9\ \text{K}^4) = 0.577\ \text{W}$$

at 600 K

$$Q/t = (7.125 \times 10^{-11}\ \text{W/K}^4)(600\ \text{K})^4 = 9.23\ \text{W}$$

and at 900 K

$$Q/t = (7.125 \times 10^{-11}\ \text{W/K}^4)(900\ \text{K})^4 = 46.7\ \text{W}$$

SUMMARY

Heat required to increase the temperature of a substance	$Q = mc\,\Delta T$	the amount of heat required to increase the temperature of a substance depends on its mass, the temperature increase, and the specific heat capacity of the material
Heat of transformation (solid ↔ liquid)	$Q = mh_f$	amount of heat required to transform *m* grams of a substance from a solid to a liquid
Heat of transformation (liquid ↔ gas)	$Q = mh_v$	amount of heat required to transform *m* grams of liquid into a gas or vapor
Heat conduction	$Q = KAt\dfrac{T_2 - T_1}{d}$	the amount of heat conducted through a rectangular block is proportional to the time, area, and temperature difference, and inversely proportional to the thickness of the slab
Rate of radiation (Stefan's law)	$\dfrac{Q}{t} = Ae\sigma T^4$	the rate of heat radiation is proportional to the radiating object's surface area, emissivity, the fourth power of its Kelvin temperature, and to the Stefan–Boltzmann constant

RAISE YOUR GRADES

Can you define . . . ?

- ☑ heat capacity
- ☑ heat of fusion
- ☑ conduction
- ☑ convection
- ☑ radiation

Do you know . . . ?

- ☑ the specific heat of water
- ☑ the mechanical equivalent of heat
- ☑ the difference between specific heat and specific heat capacity
- ☑ the amount of heat required to melt one gram of ice
- ☑ the amount of heat required to vaporize one gram of water
- ☑ the three methods by which heat can be transferred from a warm object to a cooler object

Can you . . . ?

- ☑ write the equation for the heat required to raise the temperature of a substance
- ☑ name all the steps involved in transforming a block of ice at a temperature below 0 °C into steam at a temperature above 100 °C
- ☑ calculate the amount of heat required to melt a given amount of ice
- ☑ calculate the amount of heat required to transform a given amount of water at 100 °C into steam at the same temperature
- ☑ calculate the rate of heat conduction through a solid material of given area and thickness

SOLVED PROBLEMS

PROBLEM 14-1 Three-hundred grams of lead shot is heated to 95 °C and then poured into an insulated beaker containing 100 g of water at 15 °C. The mass of the beaker is 90 g and its specific heat is 0.20 cal/g °C. The mixture of water and lead has an equilibrium temperature of 20.8 °C. Calculate the specific heat of the lead.

Solution The heat gained by the water and the glass beaker is equal to the heat lost by the lead shot. So, use Eq. (14-1) to calculate the heat gained and lost:

$$\text{Heat gained} = \text{Heat lost}$$

$$(Q = mc\,\Delta T) \qquad (Q = mc\,\Delta T)$$

$$(\text{water} + \text{beaker}) \qquad (\text{lead})$$

$$[(100\text{ g})(1\text{ cal/g °C}) + (90\text{ g})(0.20\text{ cal/g °C})](20.8 - 15)\text{ °C} = (300\text{ g})c(95 - 20.8)\text{ °C}$$

$$(100 + 18)(5.8)\text{ cal} = c(22\ 260\text{ g °C})$$

$$c = \frac{684.4\text{ cal}}{22\ 260\text{ g °C}} = 0.0307\text{ cal/g °C}$$

PROBLEM 14-2 A block of ice weighs 4.9 N and has a temperature of −6 °C. The specific heat of ice is 0.50 cal/g °C, and the specific heat of steam is 0.48 cal/g °C. How much heat energy is required to convert this block of ice into steam at a temperature of 140 °C?

Solution Break the process down into its individual steps: (1) heating the ice to 0 °C, (2) melting the ice, (3) heating the water to 100 °C, (4) vaporizing the water, and (5) heating the steam to 140 °C; and then calculate the heat required for each step. But, first you'll have to calculate the mass of the ice. [**recall**: Mass equals weight divided by gravitational acceleration. (From Eq. 3-4)]

$$m = \frac{w}{g} = \frac{4.9\text{ N}}{9.8\text{ m/s}^2} = 0.5\text{ kg}$$

Now, calculate the heat required for each of the five steps. From Eq. (14-1):

$$Q_1 = mc\,\Delta T = (500\text{ g})(0.50\text{ cal/g °C})(6\text{ °C}) = 1.5\text{ kcal}$$

Use Eq. (14-2) to find the heat needed to melt the ice:

$$Q_2 = mh_f = (500\text{ g})(80\text{ cal/g}) = 40\text{ kcal}$$

Again, from Eq. (14-1), the heat needed to heat the water:

$$Q_3 = mc\,\Delta T = (500\text{ g})(1\text{ cal/g °C})(100\text{ °C}) = 50\text{ kcal}$$

Use Eq. (14-3) to find the heat needed to transform the water into steam:

$$Q_4 = mh_v = (500\text{ g})(540\text{ cal/g}) = 270\text{ kcal}$$

And, Eq. (14-1) for the heat needed to heat the steam:

$$Q_5 = mc\,\Delta T = (500\text{ g})(0.48\text{ cal/g °C})(140 - 100)\text{ °C} = 9.6\text{ kcal}$$

Now, add the quantities of heat required:

$$Q_{tot} = Q_1 + Q_2 + Q_3 + Q_4 + Q_5 = (1.5 + 40 + 50 + 270 + 9.6)\text{ kcal} = 371\text{ kcal}$$

PROBLEM 14-3 The melting point of pure mercury is −39 °C. The amount of heat required to melt 120 cm^3 of solid mercury is 4.57 kcal. The density of frozen mercury is 13.6 g/cm^3. Calculate the heat of fusion of mercury.

Solution First, calculate the mass of 120 cm^3 of mercury. [**recall**: Mass is equal to volume times density. (Equation 10-2)]

$$m = Vd = (120\text{ cm}^3)(13.6\text{ g/cm}^3) = 1.632\text{ kg}$$

Now, use Eq. (14-2) to solve for the heat of fusion:

$$h_f = \frac{Q}{m} = \frac{4.57 \times 10^3 \text{ cal}}{1.632 \times 10^3 \text{ g}} = 2.80 \text{ cal/g}$$

PROBLEM 14-4 The density of lead is 11.3 g/cm^3 and its specific heat capacity is 0.0306 cal/g °C. A lead ball with a diameter of 1 cm is dropped onto a concrete surface from a height of 4 m. If 80% of the kinetic energy is transformed into thermal energy in the lead ball, calculate its temperature increase.

Solution The mass of the lead sphere is

$$m = Vd = \frac{4}{3}\pi r^3 d = \frac{4}{3}\pi(0.5 \text{ cm})^3(11.3 \text{ g/cm}^3) = 5.92 \text{ g}$$

The kinetic energy of the lead ball when it reaches the ground is equal to its initial potential energy (see Sec. 6-3), so

$$\text{KE} = \text{PE} = mgh = (5.92 \times 10^{-3} \text{ kg})(9.8 \text{ m/s}^2)(4 \text{ m}) = 0.232 \text{ J}$$

The heat energy produced is 80% of the KE, so

$$Q = 0.80 \text{ KE} = \frac{(0.80)(0.232 \text{ J})}{4.184 \text{ J/cal}} = 4.44 \times 10^{-2} \text{ cal}$$

Now, you can calculate the increase in temperature from Eq. (14-1):

$$\Delta T = \frac{Q}{mc} = \frac{4.44 \times 10^{-2} \text{ cal}}{(5.92 \text{ g})(0.0306 \text{ cal/g °C})} = 0.245 \text{ °C}$$

PROBLEM 14-5 A 100-g insulated glass beaker contains 300 g of water at 7 °C. The specific heat of the beaker is 0.20 cal/g °C. After 60 g of ice at −5 °C is added to the water, the final equilibrium is 0 °C. How many grams of ice remain in the beaker?

Solution The heat gained by the ice is equal to the heat lost by the water and the glass beaker. Let Δm represent the mass of ice that melts:

$$\underset{\text{(ice)}}{\text{heat gained}} = \underset{\text{(water + beaker)}}{\text{heat lost}}$$

$$(60 \text{ g})(0.50 \text{ cal/g °C})(5 \text{ °C}) + \Delta m(80 \text{ cal/g}) = [(300 \text{ g})(1 \text{ cal/g °C}) + (100 \text{ g})(0.20 \text{ cal/g °C})](7 \text{ °C})$$

$$150 \text{ cal} + \Delta m(80 \text{ cal/g}) = 2240 \text{ cal}$$

$$\Delta m = \frac{(2240 - 150) \text{ cal}}{80 \text{ cal/g}} = 26.1 \text{ g}$$

Now, you can calculate how much ice is left: (60 − 26.1) g = 33.9 g.

PROBLEM 14-6 The thermal conductivity of white pine is 2.0 × 10^{-4} cal/(cm s °C). For a piece of white pine with an area of 400 cm^2 and a thickness of 2 cm, determine how much heat will be conducted through the slab in one hour if the temperature difference is 40 °C.

Solution You can calculate the amount of heat conducted through the pine board directly from Eq. (14-4):

$$\frac{Q}{t} = KA\frac{T_2 - T_1}{d} = \left(\frac{2.0 \times 10^{-4} \text{ cal}}{\text{cm s °C}}\right)(400 \text{ cm}^2)\left(\frac{60 \text{ s}}{1 \text{ min}}\right)\left(\frac{60 \text{ min}}{1 \text{ hr}}\right)\frac{40 \text{ °C}}{2 \text{ cm}} = 5.76 \text{ kcal/hr}$$

PROBLEM 14-7 A 20-cm by 30-cm pane of glass has a thickness of 8 mm. When the temperature difference between the two surfaces is 50 °C, the rate of heat flow through the glass is 93.8 cal/s. Calculate the thermal conductivity of this glass.

Solution Solve Eq. (14-4) for K:

$$K = \frac{Qd}{At\Delta T} = \frac{(93.8 \text{ cal})(0.8 \text{ cm})}{(20 \text{ cm} \times 30 \text{ cm})(1 \text{ s})(50 \text{ °C})} = 2.5 \times 10^{-3} \text{ cal/cm s °C}$$

PROBLEM 14-8 A 30-g aluminum calorimeter cup contains 150 g of water at 20 °C. After 203.4 g of small steel pellets at 60 °C are poured into the calorimeter, the equilibrium temperature is 25 °C. The specific heat of aluminum is 0.22 cal/g °C. Calculate the specific heat of steel.

Solution Equate the heat lost by the steel pellets to the heat gained by the water and its aluminum container:

$$\underset{\text{(by steel)}}{\text{Heat lost}} = \underset{\text{(by water + cup)}}{\text{heat gained}}$$

$$(203.4\ \text{g})c(60 - 25)\ °\text{C} = [(150\ \text{g})(1\ \text{cal/g °C}) + (30\ \text{g})(0.22\ \text{cal/g °C})](25 - 20)\ °\text{C}$$

$$c = \frac{(156.6)(5)\ \text{cal}}{7.119 \times 10^3\ \text{g °C}} = 0.110\ \text{cal/g °C}$$

PROBLEM 14-9 Steam is flowing into a condenser at the rate of 5 m^3/min. The density of the steam is 1.50 kg/m^3 and its input temperature is 160 °C. The steam is cooled to 100 °C and then condensed, producing water at 100 °C. The specific heat capacity of steam is 0.48 cal/g °C. Calculate the amount of heat added to the condenser each minute.

Solution First, calculate the mass of steam entering the condenser per minute:

$$m = Vd = (5\ \text{m}^3)(1.50\ \text{kg/m}^3) = 7.5\ \text{kg}$$

Then, calculate the amount of heat produced when the steam is cooled to 100 °C:

$$Q_1 = mc\,\Delta T = (7.5 \times 10^3\ \text{g})(0.48\ \text{cal/g °C})(60\ °\text{C}) = 2.16 \times 10^5\ \text{cal}$$

Now, calculate the amount of heat that must be removed from the steam to condense it into water:

$$Q_2 = mh_v = (7.5 \times 10^3\ \text{g})(540\ \text{cal/g}) = 4.05 \times 10^6\ \text{cal}$$

Add these two figures to obtain the total amount of heat that enters the condenser each minute:

$$Q_{tot} = Q_1 + Q_2 = (2.16 + 40.5) \times 10^5\ \text{cal} = 42.7 \times 10^5\ \text{cal}$$

PROBLEM 14-10 A copper slab with a mass of 30 g is pulled across a steel surface at a constant speed of 5 cm/s by means of a pulley and a falling weight, as illustrated in Fig. 14-2. A heavy block is placed on top of the copper slab. A high-quality insulator on top of the copper slab prevents heat from leaving its upper surface. The normal force which pushes the piece of copper against the steel surface is 90 N. The coefficient of kinetic friction between the copper slab and the steel surface is 0.4. Assuming that 50% of the heat generated by friction between the two surfaces is absorbed by the piece of copper, how much will the temperature of the copper increase if it is pulled a horizontal distance of 80 cm.

Solution First, calculate the force of kinetic friction (Eq. 3-6):

$$f_k = \mu_k N = 0.4(90\ \text{N}) = 36\ \text{N}$$

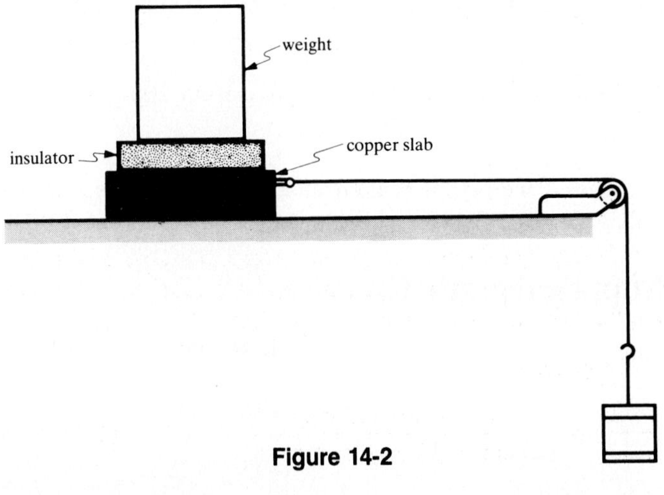

Figure 14-2

Then, calculate the work done by the falling weight. Since the apparatus moves at a constant speed, the hanging mass must weigh 36 N. From Eq. (6-2), the amount of potential energy lost by the hanging weight is

$$\text{PE} = wh = (36\text{ N})(0.80\text{ m}) = 28.8\text{ J}$$

Now, you can calculate the amount of heat added to the piece of copper:

$$Q = (0.5)\text{ PE} = (0.5)(28.8\text{ J})\frac{(1\text{ cal})}{(4.184\text{ J})} = 3.44\text{ cal}$$

Finally, calculate the increase in temperature from Eq. (14-1) and Table 14-1:

$$\Delta T = \frac{Q}{mc} = \frac{3.44\text{ cal}}{(30\text{ g})(0.093\text{ cal/g °C})} = 1.23\text{ °C}$$

PROBLEM 14-11 The filament in one of Edison's original light bulbs was a bamboo cylinder 10 cm long and 2 mm in diameter. The emissivity of carbonized bamboo is 0.950. Find the filament's Kelvin temperature when the filament radiated light at 50 W.

Solution First, find the surface area of the filament.

$$A = 2\pi rh + 2(\pi r^2) = 2\pi[(1\times 10^{-3}\text{ m})(1\times 10^{-1}\text{ m}) + (1\times 10^{-3}\text{ m})^2] = 6.346\times 10^{-4}\text{ m}^2$$

Then use Stefan's law, Eq. (14-5), and solve for temperature.

$$Q/t = Ae\sigma T^4$$

$$T^4 = \frac{Q/t}{Ae\sigma}$$

$$T = \left(\frac{Q/t}{Ae\sigma}\right)^{1/4} = \left(\frac{50\text{ W}}{(6.346\times 10^{-4}\text{ m}^2)(0.950)(5.67\times 10^{-8}\text{ W/m}^2\text{ K}^4)}\right)^{1/4}$$

$$= 1.10\times 10^3\text{ K} = 1100\text{ K}$$

Supplementary Exercises

PROBLEM 14-12 How many calories of heat does it take to increase the temperature of 300 g of silver from 20 °C to 80 °C?

PROBLEM 14-13 How much heat is required to transform 500 g of ice at 0 °C to water at the same temperature?

PROBLEM 14-14 If the apparatus illustrated in Fig. 14-1 is used to heat 150 g of water, what temperature increase will result from a weight of 100 N falling through a distance of 2.0 m?

PROBLEM 14-15 A steel pellet having a mass of 0.6 g strikes a concrete wall at a speed of 220 m/s. If 75% of the heat produced is absorbed by the steel pellet, what is its increase in temperature?

PROBLEM 14-16 How much heat is required to transform 200 g of ice at 0 °C into steam at 100 °C?

PROBLEM 14-17 What is the rate of heat conduction through a piece of white pine that has an area of 400 cm^2, a thickness of 2.5 cm, and a temperature difference of 40 °C?

Answers to Supplementary Exercises

14-12: 1.01×10^3 cal

14-13: 40 kcal

14-14: 0.319 °C

14-15: 39.4 °C

14-16: 144 kcal

14-17: 1.28 cal/s

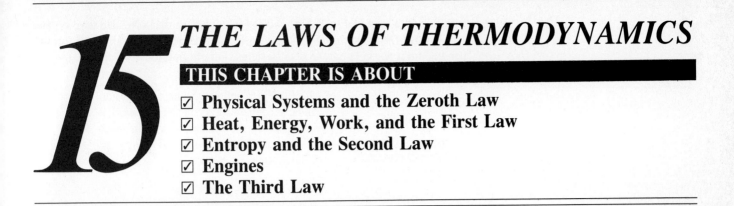

15 THE LAWS OF THERMODYNAMICS

THIS CHAPTER IS ABOUT

☑ Physical Systems and the Zeroth Law
☑ Heat, Energy, Work, and the First Law
☑ Entropy and the Second Law
☑ Engines
☑ The Third Law

15-1. Physical Systems and the Zeroth Law

- A **system** in thermodynamics is any part of the physical universe in which we have a special interest; everything else in the universe forms the **surroundings**.
- An **isolated** or **adiabatic** system is one in which no transfer of energy occurs between the system of interest and the surroundings.
- An **open** system is in thermal or mechanical contact with other systems or the surroundings. Open systems are usually designed so that we can measure the amount of heat or work transferred.
- The physical **state** of a system is described by its measurable physical parameters, such as temperature, pressure, volume, composition, etc.
- An **equation of state** relates a system's measurable parameters mathematically. For example, if a particular isolated system contains n moles of an ideal gas, the ideal gas law ($PV = nRT$) is its equation of state. Other systems, of course, will have different parameters represented by particular equations of state.
- When the measurable physical parameters are constant throughout a system and do not vary with time, the system is in a condition of **thermodynamic equilibrium.**
- A condition of **thermal equilibrium** exists if two systems are in thermal contact and no heat flow occurs.

According to the **zeroth law of thermodynamics,** systems at the same temperature are in thermal equilibrium. This means that we can use the temperature of a system as the physical parameter to measure conditions of thermal equilibrium between systems:

ZEROTH LAW OF THERMODYNAMICS If system A and system B are in thermal equilibrium with a third system (e.g., a thermometer), then they are in thermal equilibrium with each other.

15-2. Heat, Energy, Work, and the First Law

Energy is conserved! You've used the principle of energy conservation to solve numerous problems dealing with mechanical energy, including the *macroscopic* kinetic and potential energy of a mass due to its position, and energy transfers in the form of work.

- The **internal energy** U of a system includes energy in chemical, thermal, and nuclear forms, and is due to the system's *microscopic* atomic motion and molecular configurations.

A system's temperature reflects its internal energy. Note the word "reflects." We can't measure the internal energy directly, but we can measure changes in internal energy ΔU in terms of some referent.

The **first law of thermodynamics** is a statement of energy conservation. We can express it in a number of ways. For example, you've probably heard this expression of the first law: *The energy of the universe is constant*. The law in this form is easy to remember, but impractical to apply. We can also state the first law in this way: *The energy of an isolated system is constant.* This is an easier form to handle, but it doesn't apply to open systems. For our purposes, we can use the following statement of the first law:

FIRST LAW OF THERMODYNAMICS

All of the heat energy Q added to a system can be accounted for as mechanical work W performed by the system on the surroundings, as an increase in internal energy ΔU of the system, or as both. **(15-1)**

or

$$Q = W + \Delta U$$

where Q is positive when heat is added to the system, W is positive when the system does work on the surroundings, and ΔU is positive when the internal energy of the system increases.

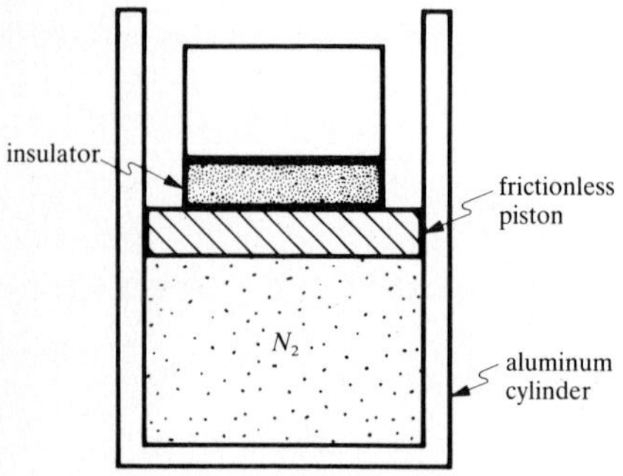

Figure 15-1

EXAMPLE 15-1: A system consisting of a frictionless aluminum piston in a leakproof aluminum cylinder containing 2.0 g of nitrogen gas at 27 °C is illustrated in Figure 15-1. The piston supports a load insulated so that there is no heat transfer between piston and load. The surface area of the piston in contact with the nitrogen is 200 cm^2. The total mass of the cylinder and piston is 140 g, and the total downward force on the gas is 1013 N. You may take as given that one mole of N_2 has a mass of 28 g, atmospheric pressure is 1 atm, the specific heat of aluminum is 0.22 cal/g °C, and that of N_2 is 0.25 cal/g °C. Also, for our purposes, you may consider that N_2 behaves ideally.
Calculate (**a**) the pressure exerted on the N_2 gas by the piston and load; (**b**) the initial volume of the gas; (**c**) the volume of the gas after the system is heated to 327 °C; (**d**) the displacement of the piston after heating; (**e**) the work done by the gas on the piston; (**f**) the increase in the internal energy of the system; and (**g**) the total amount of heat added to the system.

Solution: The problem may seem complicated at first glance, but don't panic. You'll find that each step in first-law problems such as this one forms a logical link to the next calculation.

(**a**) [**recall**: Pressure is force per unit area. (Equation 10-1)] The pressure in atm that the piston and load exert on the gas is

$$P = \frac{F}{A} = \left(\frac{1013\ \text{N}}{200\ \text{cm}^3}\right)\left(\frac{100\ \text{cm}}{1\ \text{m}}\right)^2\left(\frac{1\ \text{atm}}{1.013 \times 10^5\ \text{N/m}^2}\right) = 0.5\ \text{atm}$$

(**b**) Use the ideal gas law (Equation 13-5) to find the initial volume of gas V_i, with $R = 0.0821$ L atm/(mol K) and temperature in Kelvin degrees. [**recall**: K = °C + 273.15. (Equation 13-1)] The total pressure on the gas is the pressure of piston, load, and atmosphere:

$$V_i = \frac{nRT_i}{P} = \frac{(2/28\ \text{mol})(0.0821\ \text{L atm/mol K})(300\ \text{K})}{(0.5 + 1.0)\ \text{atm}} = 1.173\ \text{L}$$

(**c**) The initial temperature is 300 K, and the final temperature is 600 K, twice the initial temperature. Volume and temperature are directly proportional (Equation 13-5a), so the final volume V_f is twice the initial volume: $V_f = 2(1.173\ \text{L}) = 2.346$ L.

(**d**) You can find the displacement of the piston by using the formula for the change in volume of a cylinder:

$$\Delta V = \pi r^2 \Delta h = A\,\Delta h$$

$$\Delta h = \frac{\Delta V}{A} = \left(\frac{1.173\ \text{L}}{200\ \text{cm}^3}\right)\left(\frac{10^3\ \text{cm}^3}{1\ \text{L}}\right) = 5.865\ \text{cm}$$

(e) Before you can calculate the work done by the gas, you'll need to find the force that the gas exerts against the piston and load:

$$F = PA = (1.5\text{ atm})\left(\frac{1.013 \times 10^5\text{ N/m}^2}{1\text{ atm}}\right)(0.02\text{ m}^2) = 3039\text{ N}$$

The force exerted and the displacement are in the same direction, so from Equation (6-1):

$$W = Fh\cos\theta° = (3039\text{ N})(0.0587\text{ m}) = 178.4\text{ J}$$

$$= (178.4\text{ J})\left(\frac{1\text{ cal}}{4.184\text{ J}}\right) = 42.6\text{ cal}$$

(f) The increase in internal energy of the system is the quantity of heat absorbed by the aluminum cylinder and piston (Q_{Al}) and by the N_2 gas (Q_{N_2}). So, from Equation (14-1),

$$Q_{\text{Al}} = mc\,\Delta T = (140\text{ g})(0.22\text{ cal/g °C})(300\text{ °C}) = 9240\text{ cal}$$

$$Q_{N_2} = mc\,\Delta T = (2.0\text{ g})(0.25\text{ cal/g °C})(300\text{ °C}) = 150\text{ cal}$$

$$\Delta U = Q_{\text{Al}} + Q_{N_2} = (9240 + 150)\text{ cal} = 9390\text{ cal}$$

(g) Now, use the first law (Eq. 15-1) to calculate the total amount of heat added to the system; that is, the increase in internal energy plus the work done by the system on the load:

$$Q = W + \Delta U = (42.6 + 9390)\text{ cal} = 9433\text{ cal}$$

15-3. Entropy and the Second Law

A. Spontaneous changes

If you drop an ice cube into a glass of warm soda, you know the ice cube will melt and the soda will cool down. If you saw a movie scene in which the reverse happened—extra ice cubes appeared in the glass—you'd know you were watching a "special effect." However, there is nothing in the first law to suggest that there's anything odd about that movie scene. Energy is conserved no matter which way the change goes. However, you know from experience in which direction the equilibrium lies.

- If, given the starting conditions, a system progresses to the equilibrium state without an external source of energy or driving force, the process is **spontaneous**.

Processes such as the melting of an ice cube in a warm liquid or the explosive reaction of gaseous hydrogen and oxygen to give water are spontaneous.

- If, given the starting conditions, a system requires the addition of some external source of energy to reach the equilibrium state, the process is **nonspontaneous**.

You also know that heat flows spontaneously from a hotter region to a cooler region until thermal equilibrium is established. When change occurs only in one direction, as in the melting ice cube, the process is **irreversible**. This means, then, that all natural processes are irreversible.

B. Entropy

- **Entropy** S is a measure of the degree of disorder of a system.

Think of a new deck of playing cards arranged by suit in numerical order. This deck is a highly ordered system with low entropy. But, if you throw the deck out of a second-story window and then pick the cards up at random, the suits will be mixed up and the numerical order destroyed. The deck is now a highly disordered system and its entropy is much higher than before.

- The **entropy change** ΔS is a measure of the increase ($\Delta S > 0$) or decrease ($\Delta S < 0$) in disorder of a system that undergoes a change of state.

On a molecular level, the particles in all pure substances move more and more vigorously as temperature increases, so that the entropies of all pure substances increase as T increases. At 0 K, the entropy of any perfect crystalline material is zero. We can calculate entropy *change*. For example, heat flow from a hot object to a cooler object is a spontaneous process. When this process occurs, the change in entropy for the warmer object ΔS_H is given by

ENTROPY CHANGE (HOT OBJECT)

$$\Delta S_H = \frac{\Delta Q}{T_H} \tag{15-2}$$

The entropy of the warmer object will decrease, so $\Delta S_H < 0$. Heat flows out, so $\Delta Q < 0$. The change in entropy of the cooler object ΔS_C is given by

ENTROPY CHANGE (COOL OBJECT)

$$\Delta S_C = \frac{\Delta Q}{T_C} \tag{15-3}$$

The entropy of the cooler object will increase as heat flows in, so $\Delta S_C > 0$.

C. The second law

If you think of the hot and cold objects as part of a system undergoing a spontaneous process, you'll see that the entropy changes the two objects experience are not equal; in fact, $|\Delta S_C| > |\Delta S_H|$. ΔQ is the same, but the temperatures are not. In the spontaneous process of heat flow from a hot to a cold object, the system becomes more disordered. The **second law of thermodynamics** deals with this natural trend from order to disorder, which we can express in terms of entropy:

SECOND LAW OF THERMODYNAMICS

The entropy of the universe never decreases. During any natural process, the entropy either remains constant or else increases.

EXAMPLE 15-2: A 30-g ice cube is placed in a styrofoam cup containing 100 g of water. The ice and water are at the equilibrium temperature of 0 °C. During a 10-min period, heat enters the system from the surroundings, causing half of the ice cube to melt. Hydrogen and oxygen atoms in ice have a highly ordered arrangement; their arrangement in liquid water is less ordered.
Calculate the change in entropy of this system.

Solution: First, calculate the change in the heat content of the system (in joules) required to melt half the ice. From Eq. (14-2):

$$\Delta Q = mh_f = (15 \text{ g})(80 \text{ cal/g}) = 1200 \text{ cal} = (1200 \text{ cal})(4.184 \text{ J/cal}) = 5021 \text{ J}$$

Now, you can calculate the increase in entropy of the system from Eq. (15-3). Note that the temperature of the system is constant at 273 K.

$$\Delta S = \frac{\Delta Q}{T} = \frac{5021 \text{ J}}{273 \text{ K}} = 18.4 \text{ J/K}$$

The change in entropy of the system is positive, which means the entropy of the system has increased.

Suppose you placed two small pieces of metal at room temperature into an insulated container. Is it possible that you could remove them an hour later only to discover that one piece of metal has become a few degrees warmer while the other has become a few degrees cooler? Certainly not! Heat *never* flows from a cool object to a warmer one. This leads to another expression of the second law of thermodynamics:

SECOND LAW OF THERMODYNAMICS

Within any closed system it is impossible to have a process whose only effect is to transfer energy from a cold object to a warm object.

EXAMPLE 15-3: Two identical blocks of copper, each with a mass of 150 g, have temperatures of 20 °C and 24 °C, respectively. The two blocks are put in physical contact inside an insulated container. After approximately 30 min, they reach a common equilibrium temperature of 22 °C. (The specific heat capacity of copper is 0.093 cal/g °C.)
Calculate the change in entropy of the system.

Solution: First, calculate the heat energy ΔQ transferred from the warmer block to the cooler one:

$$\Delta Q = mc\,\Delta T = (150\text{ g})(0.093\text{ cal/g °C})(2\text{ °C}) = 27.9\text{ cal}$$

$$= (27.9\text{ cal})(4.184\text{ J/cal}) = 116.7\text{ J}$$

Now, from Eq. (15-2), calculate the loss of entropy of the warmer block. Because the block's temperature drops from 24 °C to 22 °C, use the average temperature of the warmer block (23 °C) in the calculation:

$$\Delta S_{\mathrm{H}} = \frac{\Delta Q}{T_{\mathrm{H}}} = \frac{-116.7\text{ J}}{(273 + 23)\text{ K}} = -0.3943\text{ J/K}$$

Note that the value of ΔQ for the warmer block is negative—it is losing heat energy.

Use Eq. (15-3) to calculate the gain in entropy of the cooler block, whose average temperature is 21 °C:

$$\Delta S_{\mathrm{C}} = \frac{\Delta Q}{T_{\mathrm{C}}} = \frac{116.7\text{ J}}{(273 + 21)\text{ K}} = 0.3969\text{ J/K}$$

And, combine the loss of entropy of the warmer block with the gain in entropy of the cooler block to find the net gain of the system:

$$\Delta S = (0.3969 + (-0.3943))\text{ J/K} = 2.6 \times 10^{-3}\text{ J/K}$$

15-4. Engines

A. Thermodynamic cycles

A fluid, such as a gas, becomes hotter as it is compressed, as when a hard-working cyclist pumps up his tires. A gas cools off when it expands, as in a turbine doing work to make electricity. The concept of a thermodynamic cycle helps our understanding of these processes, which are the basis of engines.

- In a **thermodynamic cycle** the state (pressure, volume, temperature) of a fluid changes, then returns to its original values.

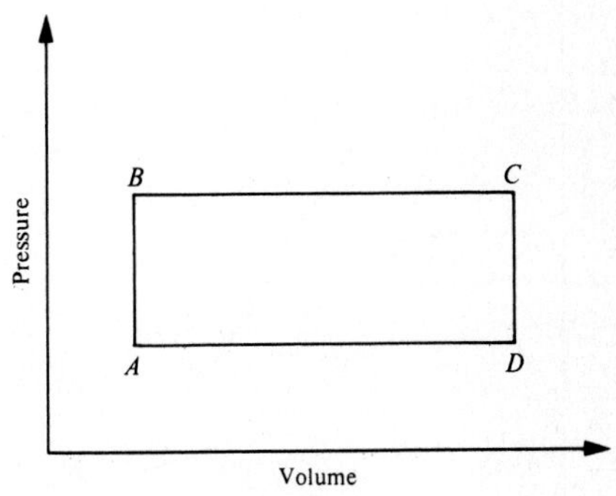

Figure 15-2

We usually use a diagram to follow a cycle. Figure 15-2 is a pressure–volume (P–V) diagram. If one mole of an ideal gas is in the state at A, we know all the quantities that describe the gas: pressure and volume from the graph, temperature from the ideal gas law, $PV = nRT$. We can create a cycle by changing the state of the gas to different points on the graph, then returning it to its original values at point A. In going around the cycle, the gas will absorb some heat and reject some, do some work and have some work done on it.

- The net work done in a thermodynamic cycle is proportional to the area inside the curve on a $P-V$ diagram.

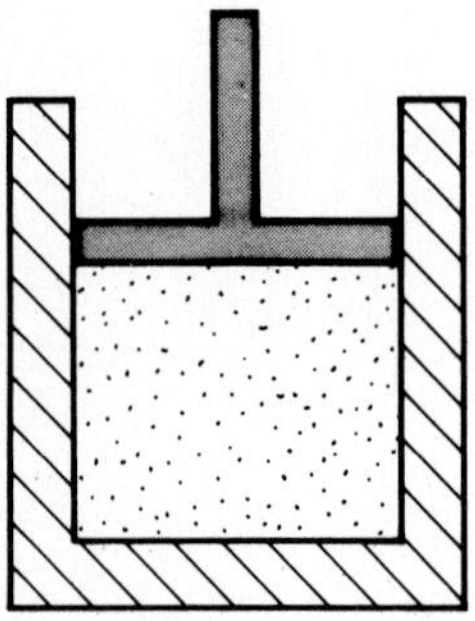

Figure 15-3

EXAMPLE 15-4: For each segment of the cycle in Figure 15-2, determine whether heat flows into or out of the system, whether work is done, and whether that work is done *by* the gas or *on* it.

Solution: Imagine that the gas is enclosed in a cylinder with a movable piston as in Figure 15-3.

In segment AB, the volume of the gas is constant, and, because the piston does not move, no work is done. From the ideal gas law, $PV = nRT$, you see that the temperature is higher at B and thus heat must be supplied to the gas.

In segment BC, the volume increases and so the piston does work on the surroundings.

$$W = F\Delta x = \frac{F}{A}(A\Delta x) = P\Delta V$$

Because $V_C > V_B$ and P is constant, $T_C > T_B$ [from Eq. (13-6)]. Therefore, heat is being supplied during BC as well.

In segment CD, as P decreases, so does T, and heat flows out. The volume doesn't change, so no work is done.

In segment DA, the work, $W = P\Delta V$, is negative because the volume decreases. Work must be done on the gas by some outside force to compress it. However, the amount of work is less than that done by the gas in going from B to C because the pressure is lower. The temperature also decreases and the system loses or rejects heat.

Certain changes in a thermodynamic system or cycle are easy to understand and calculate. Changes that keep the pressure or volume constant, like those above, are examples. Others are

- An **isothermal** change, which takes place at a constant temperature.
- An **adiabatic** changes, where no heat flows into or out of the system.

B. The Carnot cycle

- A **Carnot cycle** is a cycle in which the paths that the system follows are either isothermal or adiabatic and all the changes are reversible.

Figure 15-4 shows a Carnot cycle for an ideal gas. From A to B the change is adiabatic: the gas is compressed, and its pressure and temperature rise so that there is no heat exchange with the surroundings. From B to C to gas expands isothermally: its pressure drops but the temperature remains the same at T_H, so a high-temperature source in the surroundings must be supplying heat to the gas. From C to D the gas expands adiabatically: the temperature and pressure decreases compensate for the volume increase so that the gas loses no heat to the surroundings. From D to A the gas is compressed isothermally at a temperature T_C lower than in segment BC; as its pressure increases and volume decreases, it must lose heat to a low-temperature reservoir in the surroundings. Work is done along all the segments because the volume changes in each of them. In order for the changes to be reversible there must be no friction and only microscopically small temperature differences between any part of the system and the surroundings.

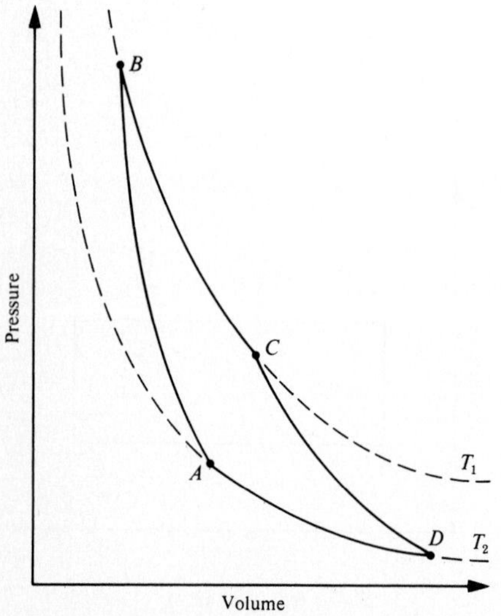

Figure 15-4. A Carnot cycle

- A **Carnot engine** is an engine that operates in a Carnot cycle.

All reversible engines have the same efficiency when they operate between the same T_H and T_C temperatures, while all nonreversible engines have a lower efficiency than reversible engines between these two temperatures. The maximum efficiency of any engine, called the **Carnot efficiency,** is given by

CARNOT EFFICIENCY

$$e_{max} = \frac{T_H - T_C}{T_H} \times 100\% = \frac{\Delta T}{T_H} \times 100\% \qquad \textbf{(15-4)}$$

In this section, we'll look at nonreversible engines with efficiencies significantly lower than this maximum efficiency. For engines, we state the second law of thermodynamics as follows:

SECOND LAW OF THERMODYNAMICS No engine or other cyclic process can have as its sole result the conversion of heat entirely into work; that is, no engine is 100% efficient.

C. Combustion engines

- A **combustion engine** burns a fuel to produce the heat input for a thermodynamic cycle.

The gasoline engine of a car is one example of a combustion engine. When gasoline is burned inside the cylinders, chemical energy is transformed into heat energy. The by-products of combustion have a very high temperature and produce a very high pressure. As a result, a piston is forced downward and a fraction of the combustion heat is tranformed into mechanical work. Of course, some of the heat energy is carried away by the high temperature exhaust gases, and there is also a loss of heat energy to the cylinder walls.

Let Q_H represent the input heat energy due to the combustion of the fuel and Q_C represent the heat energy that is lost. You can then write the first law of thermodynamics for a combustion engine as

FIRST LAW OF THERMODYNAMICS (COMBUSTION ENGINE)

$$Q_H = Q_C + W \qquad \textbf{(15-5)}$$

You can express the output work produced by the engine as

OUTPUT WORK (COMBUSTION ENGINE)

$$W = Q_H - Q_C \qquad \textbf{(15-6)}$$

The efficiency of the engine is then the output work divided by the heat input:

EFFICIENCY (COMBUSTION ENGINE)

$$e = \frac{W}{Q_H} \times 100\% = \frac{Q_H - Q_C}{Q_H} \times 100\% \qquad \textbf{(15-7)}$$

EXAMPLE 15-5: A gasoline engine has a combustion temperature of 1500 °C and an exhaust temperature of 400 °C. When the car climbs a slight hill at 55 mi/h, the heat input power is 100 kW and the exhaust heat output power is 62.7 kW. Calculate (**a**) the maximum possible efficiency (Carnot efficiency) of this combustion engine and (**b**) its actual efficiency.

Solution:

(**a**) Calculate the Carnot efficiency directly from Eq. (15-4). [Hint: Don't forget to use absolute temperatures.]

$$e_{max} = \frac{T_H - T_C}{T_H} = \frac{(1773 - 673)\ \text{K}}{1773\ \text{K}} \times 100\% = 62.0\%$$

(**b**) And, you can calculate the actual efficiency directly from Eq. (15-7):

$$e = \frac{Q_H - Q_C}{Q_H} = \frac{Q_H/t - Q_C/t}{Q_H/t} \frac{(100 - 62.7)\ \text{kW}}{100\ \text{kW}} \times 100\% = 37.3\%$$

EXAMPLE 15-6: Suppose that the gasoline engine described in Example 15-4 is tuned so that its efficiency is increased to 40%.
Calculate (**a**) the rate of heat input and (**b**) the rate of heat output when the engine is producing 40 hp of useful work.

Solution:
(**a**) First, you'll have to convert the engine's power output from hp to kW (see Example 6-15):

$$W = (40 \text{ hp})\left(\frac{0.746 \text{ kW}}{1 \text{ hp}}\right) = 29.8 \text{ kW}$$

Then, from Eq. (15-7), you can find the corresponding rate of heat input:

$$Q_H = \frac{W}{(e/100\%)} \qquad \text{so} \qquad \frac{Q_H}{t} = \frac{W/t}{(e/100\%)} = \frac{29.8 \text{ kW}}{0.40} = 74.5 \text{ kW}$$

(**b**) Use Eq. (15-6) and solve for the rate at which heat is exhausted:

$$\frac{W}{t} = \frac{Q_H}{t} - \frac{Q_C}{t} \qquad \frac{Q_C}{t} = \frac{Q_H}{t} - \frac{W}{t} = (74.5 - 29.8) \text{ kW} = 44.7 \text{ kW}$$

D. Heat pumps

- A **heat pump** is a device that uses input energy to transfer heat from a low temperature area to a warmer area.

A refrigerator is an example of a heat pump. Energy W_{in} allows the compressor to remove heat from the ice trays and the inside of the refrigerator and to transfer it to the kitchen. A heat pump can be described by a thermodynamic cycle just like that of an engine. In such a cycle, the system absorbs heat at a low temperature and rejects it at a higher temperature. This is equivalent to following path *ADCBA* in Figures 15-2 and 15-4.

Let Q_C represent the heat that is removed from the inside of the refrigerator and Q_H represent the heat that is pumped into the kitchen. Then, you can write the first law of thermodynamics for a heat pump as

FIRST LAW OF THERMODYNAMICS (HEAT PUMP)

$$Q_C + W_{in} = Q_H \qquad \textbf{(15-8)}$$

Heat pumps can be used for heating houses, especially in mild climates, since the amount of heat Q_H pumped into the house is greater than the work input W_{in} to the heat pump, as you can see from Eq. (15-8). The effectiveness of a heat pump is the heat Q_H moved to the higher temperature region divided by the work input. This ratio is called the **coefficient of performance** (COP) of the heat pump:

COEFFICIENT OF PERFORMANCE (HEAT PUMP)

$$\text{COP} = \frac{Q_H}{W_{in}} \qquad \textbf{(15-9)}$$

The maximum possible COP (or Carnot COP)—which is never achieved by a real heat pump—depends on the high and low temperatures between which the pump operates:

CARNOT COP

$$\text{COP}_{max} = \frac{T_H}{T_H - T_C} \qquad \textbf{(15-10)}$$

EXAMPLE 15-7: A refrigerator maintains an inside temperature of 34 °F and pumps heat to a radiator mounted on its back. The temperature of the radiator is 86 °F. The average power consumed by this refrigerator is 110 watts and heat is

pumped to the radiator at a rate of 2.15×10^2 kcal/h.
Calculate (**a**) the actual COP and (**b**) the Carnot COP of the refrigerator.

Solution:

(**a**) You can use Eq. (15-9) to find the actual COP, but first you'll have to calculate the work input *per hour*:

$$W_{in} = (110\ \text{J/s})(3600\ \text{s}) = 3.96 \times 10^5\ \text{J}$$

and the heat output in joules:

$$Q_H = (2.15 \times 10^5\ \text{cal})(4.184\ \text{J/cal}) = 9.00 \times 10^5\ \text{J}$$

Then, you can calculate the actual COP:

$$\text{COP} = \frac{Q_H}{W_{in}} = \frac{9.00 \times 10^5\ \text{J}}{3.96 \times 10^5\ \text{J}} = 2.27$$

Now, convert the Fahrenheit temperatures to the Kelvin scale (Eqs. 13-1 and 13-1a):

$$T_H = (86 - 32)\ ^\circ\text{F}\left(\frac{5\ ^\circ\text{C}}{9\ ^\circ\text{F}}\right) = 30\ ^\circ\text{C} = (273 + 30)\ \text{K} = 303\ \text{K}$$

$$T_C = (34 - 32)\ ^\circ\text{F}\left(\frac{5\ ^\circ\text{C}}{9\ ^\circ\text{F}}\right) = 1.1\ ^\circ\text{C} = (273 + 1)\ \text{K} = 274\ \text{K}$$

(**b**) Calculate the Carnot COP directly from Eq. (15-10):

$$\text{COP}_{max} = \frac{T_H}{T_H - T_C} = \frac{303\ \text{K}}{(303 - 274)\ \text{K}} = 10.4$$

EXAMPLE 15-8: A room air conditioner removes heat from the cooling coils that have a temperature of 9 °C. The heat is pumped to an exterior heat exchanger that has a temperature of 32 °C. The air conditioner pumps 1.4×10^7 J into the exterior heat exchanger each hour. Its actual COP is 2.0.
Calculate the power consumed by the motor of this air conditioner.

Solution: Solve Eq. (15-9) for W_{in}:

$$W_{in} = \frac{Q_H}{\text{COP}} = \frac{1.4 \times 10^7\ \text{J}}{2.0} = 7.0 \times 10^6\ \text{J}$$

Then, calculate the power of the motor. [**recall**: Power is the rate at which work is done. (Eq. 6-7)]

$$P = \frac{W}{t} = \frac{7.0 \times 10^6\ \text{J}}{3600\ \text{s}} = 1.94\ \text{kW}$$

15.5 The Third Law

There are both empirical and theoretical reasons for us to accept the **third law of thermodynamics,** which states

THIRD LAW OF THERMODYNAMICS It is impossible to reduce the temperature of a body to absolute zero in a finite number of steps.

We know from experiment that each succeeding step in an effort to reach colder and colder temperatures becomes more difficult to accomplish. And, theoretically, if we operated a Carnot cycle with the cold reservoir at absolute zero, no heat would be rejected to the cold reservoir, so $Q_C = 0$. Since work W is equal to $Q_H - Q_C$, this engine would transform a quantity of thermal energy Q_H entirely into work, which would violate the second law of thermodynamics. The third law, therefore, is an essential part of the basic foundation upon which the science of thermodynamics rests.

SUMMARY

First law of thermodynamics	$Q = W + \Delta U$	states that all the heat energy added to a system can be accounted for as the sum of the mechanical work performed by the system on the surroundings, and the increase in internal energy of the system
Entropy change (hot object)	$\Delta S_H = \dfrac{\Delta Q}{T_H}$	the change in entropy for a hot object is given by the ratio of the change in heat energy to the temperature of the hot reservoir
Entropy change (cool object)	$\Delta S_C = \dfrac{\Delta Q}{T_C}$	the change in entropy for a cool object is given by the ratio of the change in heat energy to the temperature of the cool reservoir
Carnot efficiency	$e_{max} = \dfrac{T_H - T_C}{T_H} \times 100\% = \dfrac{\Delta T}{T_H} \times 100\%$	the maximum efficiency of any engine is given by the ratio of the change in temperature to the temperature of the hot reservoir
First law of thermodynamics (combustion engine)	$Q_H = Q_C + W$	the input heat energy due to fuel combustion is equal to the heat energy lost *plus* the output work produced by the engine
Output work (combustion engine)	$W = Q_H - Q_C$	the output work of a combustion engine is equal to the input heat energy *minus* the heat lost
Efficiency (combustion engine)	$e = \dfrac{W}{Q_H} \times 100\% = \dfrac{Q_H - Q_C}{Q_H} \times 100\%$	the efficiency of an engine is the output work divided by the heat input
First law of thermodynamics (heat pump)	$Q_C + W_{in} = Q_H$	heat removed from a low temperature area plus the input energy is equal to the heat pumped into a warmer area
Coefficient of performance (heat pump)	$COP = \dfrac{Q_H}{W_{in}}$	the effectiveness of a heat pump is the heat moved to a higher temperature area divided by the work input
Carnot COP	$COP_{max} = \dfrac{T_H}{T_H - T_C}$	the maximum possible COP depends on the high and low temperatures between which a pump operates

RAISE YOUR GRADES

Can you define . . . ?

- ☑ a system in thermodynamics
- ☑ an isolated system
- ☑ an open system
- ☑ thermal equilibrium

Can you . . . ?

- ☑ calculate the work done by a gas when it expands
- ☑ state the relationship between entropy and disorder
- ☑ calculate the gain in entropy of an object when heat is added to it
- ☑ state what happens to the entropy of a system during a natural or spontaneous process
- ☑ write the first law of thermodynamics for a combustion engine
- ☑ calculate the efficiency of a heat engine, given the heat input and the work output
- ☑ write the equation for the Carnot efficiency of a combustion engine
- ☑ write the first law of thermodynamics as applied to a heat pump
- ☑ calculate the efficiency of a heat pump, given the work input and amount of heat transferred
- ☑ write the equation for the efficiency of a Carnot heat pump

SOLVED PROBLEMS

PROBLEM 15-1 Calculate the increase of the internal energy of a system to which 200 kcal is added, resulting in 5.37×10^5 J of work done on the surroundings.

Solution Solve Eq. (15-1) for ΔU:

$$Q = W + \Delta U$$

$$\Delta U = Q - W = 2.00 \times 10^5 \text{ cal} - (5.37 \times 10^5 \text{ J})\left(\frac{1 \text{ cal}}{4.184 \text{ J}}\right) = 7.17 \times 10^4 \text{ cal} = 71.7 \text{ kcal}$$

PROBLEM 15-2 A cylindrical aluminum rod (diameter = 1 cm; length = 50 cm; density = 2.7 g/cm^3), mounted vertically inside an insulated chamber, supports a 10 000-N load. The rod is heated from its original temperature of 22 °C to a final temperature of 330 °C. Because of the rod's expansion, work is done to lift the load. (The coefficient of linear expansion of aluminum is 26×10^{-6} (°C)$^{-1}$.) The internal energy of the rod also increases. (The specific heat of aluminum is 0.22 cal/g °C.) Calculate (**a**) the increase in internal energy of the rod, (**b**) the work done in lifting the load, and (**c**) the heat added to the system.

Solution

(**a**) The increase in the rod's internal energy is equal to the heat added, so you'll use Eq. (14-1). First, though, you'll have to calculate the volume and mass of the rod:

$$V = \pi r^2 h = \pi(0.5 \text{ cm})^2(50 \text{ cm}) = 39.3 \text{ cm}^3 \qquad m = Vd = (39.3 \text{ cm}^3)(2.7 \text{ g/cm}^3) = 106 \text{ g}$$

Then, from Eq. (14-1):

$$\Delta U = Q = mc\,\Delta T = (106 \text{ g})(0.22 \text{ cal/g °C})(308 \text{ °C}) = 7.18 \times 10^3 \text{ cal}$$

(**b**) Calculate the rod's elongation from Eq. (13-2):

$$\Delta L = L\alpha\,\Delta t = (50 \text{ cm})(26 \times 10^{-6} \text{ °C}^{-1})(308 \text{ °C}) = 0.400 \text{ cm} = 4.00 \times 10^{-3} \text{ m}$$

So, you can calculate the work done in lifting the load. [**recall**: Work is the product of a constant force acting on an object and the object's displacement (Sec. 6-1). In this case, the force is the weight of the load and the displacement is the length of the elongated rod.]

$$W = wh = (1 \times 10^4 \text{ N})(4.00 \times 10^{-3} \text{ m}) = 40 \text{ J}$$

(**c**) Use Eq. (15-1) to calculate the total heat added:

$$Q = W + \Delta U = \left(\frac{40 \text{ J}}{4.184 \text{ J/cal}}\right) + 7.18 \times 10^3 \text{ cal} = 7.19 \times 10^3 \text{ cal}$$

PROBLEM 15-3 A styrofoam cup contains 100 g of water at 21 °C. After 502 g of lead shot (specific heat = 0.0306 cal/g °C) at 51 °C is poured into the cup, the equilibrium temperature of the mixture is 25 °C (even though there is five times as much warm lead as cool water, the equilibrium temperature is closer to that of the water because the specific heat of water is 32.7 times that of lead). Calculate the mixture's net increase in entropy.

Solution You can calculate the loss of entropy of the lead shot from Eq. (15-2). [Hint: Don't forget to use the average temperature of the lead.]

$$\Delta S_{\text{H}} = \frac{\Delta Q}{T_{\text{H}}} = \frac{mc\,\Delta T}{T_{\text{H}}} = \frac{-(502 \text{ g})(0.0306 \text{ cal/g K})(26 \text{ K})}{(273 + 38) \text{ K}} = -1.28 \text{ cal/K}$$

Then, calculate the gain in entropy of the water from Eq. (15-3). Remember to use the average temperature of the water.

$$\Delta S_{\text{C}} = \frac{\Delta Q}{T_{\text{C}}} = \frac{mc\,\Delta T}{T_{\text{C}}} = \frac{(100 \text{ g})(1 \text{ cal/g K})(4 \text{ K})}{(273 + 23) \text{ K}} = 1.35 \text{ cal/K}$$

So, the net gain in entropy of the system is

$$\Delta S = \Delta S_{\text{C}} + \Delta S_{\text{H}} = 1.35 - 1.28 = 0.07 \text{ cal/K}$$

PROBLEM 15-4 What is the maximum possible efficiency of a steam turbine if the input steam has a temperature of 300 °C and the exhaust steam has a temperature of 100 °C?

Solution You can calculate the maximum efficiency of a heat engine directly from Eq. (15-4). You must use absolute temperatures.

$$T_{\text{H}} = (273 + 300) \text{ K} = 573 \text{ K} \qquad T_{\text{C}} = (273 + 100) \text{ K} = 373 \text{ K}$$

$$e_{\text{max}} = \frac{T_{\text{H}} - T_{\text{C}}}{T_{\text{H}}} \times 100\% = \frac{(573 - 373) \text{ K}}{573 \text{ K}} \times 100\% = 34.9\%$$

PROBLEM 15-5 The actual efficiency of the steam turbine described in Problem 15-4 is 25%. The work done by the turbine is 1.20×10^6 J per minute. Calculate the input heat and the exhaust heat for one minute of operation of the steam turbine.

Solution You can calculate the heat input directly from Eq. (15-7):

$$Q_{\text{H}} = \frac{W}{e/100\%} = \frac{1.20 \times 10^6 \text{ J}}{0.25}\left(\frac{1 \text{ cal}}{4.184 \text{ J}}\right) = 1.147 \times 10^6 \text{ cal}$$

Now, you can use Eq. (15-5) to find the heat exhausted per minute:

$$Q_{\text{C}} = Q_{\text{H}} - W = 1.147 \times 10^6 \text{ cal} - \left(\frac{1.20 \times 10^6 \text{ J}}{4.184 \text{ J/cal}}\right) = 8.60 \times 10^5 \text{ cal}$$

PROBLEM 15-6 A heat pump extracts heat from a deep well at a temperature of 59 °F. The heat exchanger into which the heat is pumped has a temperature of 86 °F. The power consumed by the heat pump is 1800 watts. During each minute 2.50×10^5 J of heat is pumped into a house. Calculate the (**a**) maximum and (**b**) actual COPs of this heat pump.

Solution

(**a**) You can calculate the maximum possible COP from Eq. (15-10). First, convert the Fahrenheit temperatures to the Kelvin scale:

$$T_{\text{C}} = (59 - 32) \text{ °F}\left(\frac{5 \text{ °C}}{9 \text{ °F}}\right) = 15 \text{ °C} \qquad T_{\text{C}} = (273 + 15) \text{ K} = 288 \text{ K}$$

$$T_{\text{H}} = (86 - 32) \text{ °F}\left(\frac{5 \text{ °C}}{9 \text{ °F}}\right) = 30 \text{ °C} \qquad T_{\text{H}} = (273 + 30) \text{ K} = 303 \text{ K}$$

$$\text{COP}_{\text{max}} = \frac{T_{\text{H}}}{T_{\text{H}} - T_{\text{C}}} = \frac{303 \text{ K}}{(303 - 288) \text{ K}} = 20.2$$

(b) Now, calculate the input work in joules during one minute of operation:

$$W_{in} = Pt = (1800 \text{ J/s})(60 \text{ s}) = 1.08 \times 10^5 \text{ J}$$

Use Eq. (15-9) to calculate the actual COP:

$$\text{COP} = \frac{Q_H}{W_{in}} = \frac{2.50 \times 10^5 \text{ J}}{1.08 \times 10^5 \text{ J}} = 2.3$$

Supplementary Exercises

PROBLEM 15-7 When 240 kcal of heat energy is added to a system the amount of work done by the system is 7.55×10^5 J. Calculate the increase in calories of the internal energy of this system.

PROBLEM 15-8 When a certain gas is heated it expands against atmospheric pressure (1 atm = 1.013×10^5 N/m^2). How much work is done by this expanding gas if its volume increases by 2.5 L?

PROBLEM 15-9 One-half mole of neon gas at standard temperature and pressure (273 K, 1 atm) occupies a volume of 11.2 L. At what temperature will the volume increase by 2.5 L if the pressure remains the same?

PROBLEM 15-10 When an ice cube with a mass of 35 g and a temperature of 0 °C melts completely, what is the increase in entropy?

PROBLEM 15-11 Two identical blocks of aluminum, each with a mass of 400 g, and with temperatures of 22 °C and 30 °C, are placed in contact inside an insulated container. After about 40 min they reach a common equilibrium temperature of 26 °C. Calculate the increase in entropy of this system. [The specific heat of aluminum is 0.22 cal/g °C.]

PROBLEM 15-12 The heat input to a certain heat engine is 1.911×10^5 cal/min. The work done in 1 min by this heat engine is 2.40×10^5 J. What is the efficiency of this heat engine?

PROBLEM 15-13 How much heat is exhausted during one minute of operation of the heat engine described in Problem 15-14?

PROBLEM 15-14 A certain Carnot heat engine has an efficiency of 54%. If the input temperature is 650 K, what is the exhaust temperature?

PROBLEM 15-15 What is the maximum possible efficiency (Carnot efficiency) of a diesel engine whose combustion temperature is 1650 °C and whose exhaust temperature is 450 °C?

PROBLEM 15-16 During a 10-min interval a certain heat pump removes 86 kcal of heat from a deep well. The input power to the heat pump is 600 watts. How much heat is pumped into the house by this heat pump during a 10-min interval?

PROBLEM 15-17 What is the coefficient of performance of the heat pump in Problem 15-18?

PROBLEM 15-18 A heat pump extracts heat from a reservoir at 55 °F and pumps heat into a radiator at 86 °F. What is the maximum COP of this heat engine?

Answers to Supplementary Exercises

15-7: 5.96×10^4 cal

15-8: 2.53×10^2 J

15-9: $T_2 = 334$ K

15-10: $\Delta S = 42.9$ J/K

15-11: 6.6×10^{-2} J/K

15-12: 30.0%

15-13: 1.34×10^5 cal

15-14: $T_C = 299$ K

15-15: 62.4%

15-16: $Q_H = 172$ kcal

15-17: COP = 2.0

15-18: 17.8

FIRST SEMESTER EXAM (Chapters 1–15)

1. An automobile is moving due north at 60 mi/h. Five seconds later it is moving 10° east of north at a speed of 55 mi/h. Calculate the change in velocity. **[Ch. 1]**

2. Calculate the average acceleration of the automobile in Problem 1 in mi/h^2. **[Ch. 2]**

3. A mass of 300 g is subjected to a force that causes it to accelerate by 4 m/s^2. What is the magnitude of the force acting on this mass? **[Ch. 3]**

4. A small object is moving along the arc of a circle at a constant speed. The radius of the circle is 12 cm and the object moves through an angle of 50° in 6 s. What is the speed of this object in cm/s? **[Ch. 4]**

5. How long does a small lead sphere take to fall 1.3 m if its initial speed is zero? **[Ch. 2]**

6. Calculate the speed of the lead sphere in Problem 5 after it has fallen 1.2 m. **[Ch. 2]**

7. What is the speed of a small lead sphere 0.6 s after it is dropped from rest? **[Ch. 2]**

8. How high will a small rock go if it is thrown vertically upward with an initial speed of 15 m/s? **[Ch. 2]**

9. An object having a mass of 400 g is subjected to two horizontal forces. One force is directed toward the east and has a magnitude of 6 N. The other force is directed 25° north of east and has a magnitude of 8 N. Calculate the resultant force. **[Ch. 1]**

10. Calculate the magnitude of the acceleration of the 400-g object in Problem 9. **[Ch. 3]**

11. A horizontal force of 3.5 N is applied to a 2-kg steel block resting on a horizontal surface. The coefficient of static friction is 0.30. What is the magnitude of the frictional force that is acting on this block? **[Ch. 3]**

12. A 2-kg block is on a frictionless horizontal surface. It is connected to a second 2-kg block by means of a string passing over a frictionless pulley. The second 2-kg block is hanging vertically. What is the acceleration of this system? **[Ch. 3]**

13. A small object moves 3 cm along the arc of a circle whose radius is 10 cm. Through what angle in radians does it move? **[Ch. 4]**

14. If the object described in Problem 13 moves 3 cm along the arc of the circle in 2 s, what is its angular speed in radians per second? **[Ch. 4]**

15. A rotating flywheel has a speed of 5 rad/s. Three seconds later, its speed becomes 8 rad/s. What is the average angular acceleration of this flywheel? **[Ch. 4]**

16. A small object is moving around a circle at a constant angular speed of 2.0 rad/s. The radius of the circle is 8.0 cm. What is the tangential speed of this object? **[Ch. 4]**

17. What is the centripetal acceleration of the object described in Problem 16? **[Ch. 4]**

18. A highway curve has a radius of 500 m. At what angle should the curve be banked if automobiles round the curve at 100 km/h? **[Ch. 4]**

19. A small object is subjected to two forces, a force of 4.0 N toward the north and a force of 6.0 N toward the east. Find the magnitude and direction of a third force that will put the object in equilibrium. **[Ch. 5]**

20. A horizontal steel beam weighing 250 N is supported by a hinge at the right end and a cable fastened to the left end. The cable makes an angle of 15° with respect to the beam. Calculate the force of tension in the cable. **[Ch. 5]**

21. Three spherical masses are connected by a wooden rod whose weight is negligible. The masses are 2 kg, 4 kg, and 3 kg, and their distances from the left end of the rod are 10 cm, 25 cm, and 50 cm. Find the center of mass of these three objects as measured from the left end of the rod. **[Ch. 5]**

22. A frictionless cart weighing 6 N is being pulled up an inclined plane at a constant speed. The angle of the inclined plane is 20° and the string that is pulling the cart is parallel to the inclined surface. What is the tension in the string? **[Ch. 5]**

23. A force of 20 N pulls a barge 2 km through a canal. The angle between the force and the displacement is 25°. Calculate the work done. **[Ch. 6]**

24. How much potential energy does an object weighing 16 N gain when it is lifted up from the floor and placed on a table 0.9 m high? **[Ch. 6]**

25. How much potential energy does a spring with a spring constant of 0.5 N/m gain when it is stretched a distance of 5 cm? **[Ch. 6]**

26. What is the kinetic energy of a mass of 400 g moving at a speed of 2.5 m/s? **[Ch. 6]**

27. What is the power output of a force of 50 N pulling a cart up an inclined plane at a speed of 2.0 m/s? **[Ch. 6]**

28. What is the linear momentum of a mass of 600 g moving at a speed of 3.0 m/s? **[Ch. 7]**

29. A 200-g block moving along a frictionless horizontal surface at a speed of 0.8 m/s strikes a 300-g block which is at rest. Clay on the sides of the two blocks causes them to stick together and move as a unit. Calculate the speed of the two blocks after their collision. **[Ch. 7]**

30. Calculate the energy lost in the inelastic collision described in Problem 29. **[Ch. 6]**

31. Two lead spheres whose masses are 3 kg and 5 kg are placed with their centers 12 cm apart. Calculate the gravitational force that one sphere exerts on the other sphere. [*Hint:* Newton's universal constant of gravitation is $G = 6.67 \times 10^{-11}$ N m^2/kg.] **[Ch. 8]**

32. Calculate the magnitude of the gravitational field when a mass of 2 kg is subjected to a gravitational force of 11.0 N. **[Ch. 8]**

33. Calculate the magnitude of the gravitational field 20 cm from the center of a lead sphere whose mass is 8 kg. **[Ch. 8]**

34. The average distance of the planet Venus from the center of the sun is 1.0821×10^{11} m. The sun's mass is 1.99×10^{30} kg. Calculate the period in days of the orbit of Venus about the sun. **[Ch. 8]**

35. When a certain spring is stretched with a force of 2 N, its length increases by 4 cm. Calculate the spring constant k. **[Ch. 9]**

36. A steel wire is 50 cm long and has a diameter of 0.5 mm. The stretch modulus of steel is 20×10^{10} N/m^2. What force is required to increase the length of this wire by 1.0 mm? **[Ch. 9]**

37. The bulk modulus of water is 0.20×10^{10} N/m^2. What pressure is required to reduce the volume of 0.5 m^3 of water by 2×10^{-3} m^3? **[Ch. 9]**

38. A mass fastened to a spring is vibrating vertically in simple harmonic motion. The amplitude of the vibration is 5 cm and the period is 2.0 s. How long does the mass take to go 1.0 cm upward from its lowest position? **[Ch. 9]**

39. Calculate the pressure at the bottom of a tank filled with water to a depth of 12 m. (Remember that the density of water is 1×10^3 kg/m^3 and atmospheric pressure is 1.013×10^5 N/m^2.) **[Ch. 10]**

40. How high is the column of mercury in a vertical glass tube if the pressure at the top of the mercury is 1 atm and the pressure at the bottom of the column is 3 atm? (The density of mercury is 1.36×10^4 kg/m^3.) **[Ch. 10]**

41. A brass cylinder with a diameter of 2 cm and a height of 5 cm is suspended in a beaker of water. Calculate the buoyant force that the water exerts on the cylinder. **[Ch. 10]**

42. Water is flowing at a speed of 2.5 cm/s through a horizontal tube whose inside diameter is 1.0 cm. Calculate the volumetric flow rate in cm^3/s. **[Ch. 10]**

43. Calculate the speed of propagation of a transverse wave in a piano wire with a mass of 2 g, a length of 1.0 m, and under a tension of 10 N. **[Ch. 11]**

44. The stretch modulus E of aluminum is 7.0×10^{10} N/m^2. The density of aluminum is 2.70 g/cm^3. Calculate the speed of a compression pulse in an aluminum rod. **[Ch. 11]**

45. Calculate the wavelength of a sound wave traveling through air at a speed of 340 m/s at a frequency of 440 Hz. **[Ch. 11]**

46. A stationary source of sound is producing a pure tone whose frequency is 880 Hz. An observer is moving toward the source at a speed of 17 m/s. The speed of sound in air is 340 m/s. Calculate the frequency that the observer will hear. **[Ch. 11]**

47. The speed of propagation of a transverse wave in a stretched wire is 80 m/s. The length of the wire is 0.80 m. Calculate the frequency of the fundamental mode of vibration. **[Ch. 12]**

48. What is the frequency of the second overtone of the stretched wire in Problem 47? **[Ch. 12]**

49. An organ pipe 1.8 m long is closed at one end and open at the other end. The speed of propagation of sound within this pipe is 320 m/s. What is the fundamental frequency of this pipe? **[Ch. 12]**

50. The reference level I_0 of sound intensity is 1×10^{-12} W/m^2. What is the relative intensity level in decibels of a sound wave whose intensity is 2×10^{-8} W/m^2? **[Ch. 12]**

51. What is the Fahrenheit temperature that corresponds to a temperature on the Celsius scale of 18 °C? **[Ch. 13]**

52. Express the boiling point of liquid nitrogen, −195.8 °C, in kelvins. **[Ch. 13]**

53. An aluminum rod 1.2 m long is heated 35 °C. As a result of this temperature increase the length of the rod increases by 1.092 mm. Calculate the coefficient of linear expansion of aluminum. **[Ch. 13]**

54. Calculate the volume of 2 moles of an ideal gas heated to a temperature of 400 K. The pressure of this gas is 0.5 atm. The universal gas constant $R = 8.314$ J/(mol K). **[Ch. 13]**

55. How much heat (in calories) is required to raise the temperature of 400 g of lead from 22 °C to 102 °C? (The specific heat capacity of lead is 0.0306 cal/g °C.) **[Ch. 14]**

56. How much heat (in calories) is required to melt 200 g of ice at a temperature of 0 °C? (The heat of fusion of water is 80 cal/g.) **[Ch. 14]**

57. How much heat is required to transform 2.0 L of water at 25 °C into steam at 100 °C? (The heat of vaporization of water is 540 cal/g.) **[Ch. 14]**

58. Calculate the amount of heat transmitted per minute through a slab of steel 5 mm thick if the area of one side of the slab is 60 cm^2 and one side is 200 °C hotter than the other. The thermal conductivity of steel is 0.11 cal/(cm s °C). **[Ch. 14]**

59. Calculate the increase in volume of 3 moles of nitrogen gas when heated from 35 °C to 435 °C if the pressure remains constant at 1 atm. **[Ch. 13]**

60. The nitrogen gas described in Problem 59 is contained within a cylinder with a leakproof piston which has an area of 1200 cm^2. Calculate the vertical distance that the expanding gas moves the piston. **[Ch. 13]**

61. Calculate the force on this piston due to the pressure of 1 atm. **[Ch. 10]**

62. The weight of the piston in Problem 60 is 3 N and the force of friction acting on it is 5 N. Calculate the work done by the expanding gas. **[Ch. 6]**

63. A gasoline engine has a combustion temperature of 1800 °C and an exhaust temperature of 600 °C. Calculate the maximum possible efficiency (Carnot efficiency) of this heat engine. **[Ch. 15]**

64. In this Problem, you may need to use these characteristics of aluminum:

specific heat $c = 0.22$ cal/(g °C)
coefficient of linear expansion $\alpha = 2.6 \times 10^{-5}$ (°C)$^{-1}$
stretch modulus $E = 7.0 \times 10^{10}$ N/m^2
density $d = 2.70 \times 10^3$ kg/m^3

(a) How much work is required to raise a bar of aluminum with a mass of 1 kg and a length of 10 cm to a height of 100 km above the earth's surface? The earth's radius is 6.38×10^6 m and its mass is 5.98×10^{24} kg. **[Ch. 8]**
(b) The aluminum bar parachutes back to earth. When it strikes the ground, 1×10^4 cal of its kinetic energy is converted to heat energy. What is its temperature after it strikes the ground if its temperature was 23 °C before the collision? **[Ch. 14]**
(c) By how much does this temperature change increase the length of the aluminum bar? **[Ch. 13]**
(d) The aluminum bar is now dropped into an insulated container with 50 g of ice and 50 g of liquid water. What is this system's final temperature? **[Ch. 14]**
(e) The aluminum bar is removed from the water, dried off, and used to transmit compressional waves. What is the speed of these waves? **[Ch. 11]**
(f) Finally, a portion of the bar is made into a wire 0.5 m long. What is the wire's fundamental frequency? **[Ch. 12]**

Solutions to First Semester Exam

1. Use the vector component method and the Pythagorean theorem to find the magnitude of the car's change in velocity $\Delta\mathbf{v}$.

vector	north component	east component
$\mathbf{v}_f$ $-\mathbf{v}_i$	55(cos 10°) −60	55(sin 10°) 0
$\Delta\mathbf{v}$	−5.84	9.55

$$\Delta v = \sqrt{(\Delta v_n)^2 + (\Delta v_e)^2} = \sqrt{(-5.84 \text{ mi/h})^2 + (9.55 \text{ mi/h})^2} = 11.2 \text{ mi/h}$$

Find the direction of $\Delta\mathbf{v}$ with the tangent function.

$$\tan\theta = \frac{-5.84}{9.55} \qquad \theta = \arctan\left(\frac{-5.84}{9.55}\right) = -31.4°$$

The negative sign indicates clockwise rotation from the east axis, or 31.4° south of east.

2.
$$a_{ave} = \frac{\Delta v}{\Delta t} = \left(\frac{11.2 \text{ mi/h}}{5 \text{ s}}\right)\left(\frac{3600 \text{ s}}{\text{h}}\right) = 8.06 \times 10^3 \text{ mi/h}^2$$

The direction of $\mathbf{a}_{ave}$ is the same as that of $\Delta\mathbf{v}$, 31.4° south of east.

3. Use Newton's second law: $F = ma = (0.3 \text{ kg})(4 \text{ m/s}^2) = 1.2 \text{ N}$

4.
$$s = r\theta = (12 \text{ cm})(50°)\left(\frac{\pi \text{ rad}}{180°}\right) = 10.5 \text{ cm} \qquad v = \frac{\Delta s}{\Delta t} = \frac{10.5 \text{ cm}}{6 \text{ s}} = 1.75 \text{ cm/s}$$

5. Use one of the equations for linear motion with constant acceleration, $s = v_0 t + \frac{1}{2}at^2$. Because $v_0 = 0$

$$s = \tfrac{1}{2}at^2 \qquad t = \sqrt{2s/a} = \sqrt{2(1.3 \text{ m})/(9.8 \text{ m/s}^2)} = 0.515 \text{ s}$$

6. Use another equation for straight-line motion with constant acceleration, and set $v_0 = 0$.

$$v^2 = v_0^2 + 2as \qquad v = \sqrt{2as} = \sqrt{2(9.8 \text{ m/s}^2)(1.2 \text{ m})} = 4.85 \text{ m/s}$$

7. $$v = v_0 + at = 0 + (9.8\text{ m/s}^2)(0.6\text{ s}) = 5.88\text{ m/s}$$

8. Choose "up" as the positive direction, and remember that when the rock reaches its highest point, its velocity is zero.

$$v^2 = v_0^2 + 2as \qquad s = \frac{-v_0^2}{2a} = \frac{-(15\text{ m/s})^2}{2(-9.8\text{ m/s}^2)} = 11.5\text{ m}$$

9. Use the law of cosines to find the magnitude and the law of sines to find the direction of the resultant force.

$$c^2 = a^2 + b^2 - 2ab\cos\theta \qquad c = \sqrt{(6\text{ N})^2 + (8\text{ N})^2 - 2(6\text{ N})(8\text{ N})\cos(180 - 25)^\circ} = 13.67\text{ N}$$

$$\frac{\sin\beta}{b} = \frac{\sin\theta}{c} \qquad \beta = \arcsin\left(\frac{b\sin\theta}{c}\right) = \arcsin\left(\frac{(8\text{ N})\sin(180 - 25)^\circ}{13.67\text{ N}}\right) = 14.3^\circ$$

10. Use Newton's second law, $F = ma$: $$a = \frac{F}{m} = \frac{13.67\text{ N}}{0.4\text{ kg}} = 34.2\text{ m/s}^2$$

11. The real question here is whether or not the block is moving. If it is, the frictional force is given by $f_k = \mu_k N$. If not, the frictional force is equal but opposite to the applied force, up to a maximum of

$$f_s = \mu_s N = \mu_s mg = (0.3)(2\text{ kg})(9.8\text{ m/s}^2) = 5.88\text{ N}$$

Since the applied horizontal force is only 3.5 N, $f_s = 3.5$ N.

12. Use the equation for total acceleration of a system, $a = \Sigma F/m_{\text{tot}}$. The only force acting on the system is the weight $m_2 g$ of the hanging mass.

$$a = \frac{m_2 g}{m_1 + m_2} = \frac{(2\text{ kg})(9.8\text{ m/s}^2)}{(2\text{ kg}) + (2\text{ kg})} = 4.9\text{ m/s}^2$$

13. $$\theta = s/r = (3\text{ cm})/(10\text{ cm}) = 0.3\text{ rad}$$

14. $$\omega = \Delta\theta/\Delta t = (0.3\text{ rad})/(2\text{ s}) = 0.15\text{ rad/s}$$

15. $$\alpha_{\text{ave}} = \Delta\omega/\Delta t = (8\text{ rad/s} - 5\text{ rad/s})/(3\text{ s}) = 1\text{ rad/s}^2$$

16. $$v_t = r\omega = (8\text{ cm})(2.0\text{ rad/s}) = 16.0\text{ cm/s}$$

17. $$a_c = r\omega^2 = (8\text{ cm})(2.0\text{ rad/s})^2 = 32\text{ cm/s}^2$$

18. $$\tan\theta = \frac{v_2}{rg} \qquad \theta = \arctan\left(\frac{v^2}{rg}\right) = \arctan\left(\frac{\left[\left(\frac{1\times 10^2\text{ km}}{\text{h}}\right)\left(\frac{10^3\text{ m}}{\text{km}}\right)\left(\frac{\text{h}}{3.6\times 10^3\text{ s}}\right)\right]^2}{(500\text{ m})(9.8\text{ m/s}^2)}\right) = 8.95^\circ$$

19. For an object to be in equilibrium, the sum of the forces acting on it in all directions must be zero. Let **R** represent the force that will put the object in equilibrium.

$$\Sigma F_x = 0 = 4.0\text{ N} + R_x \qquad R_x = -4.0\text{ N}$$

$$\Sigma F_y = 0 = 6.0\text{ N} + R_y \qquad R_y = -6.0\text{ N}$$

Now use the Pythagorean theorem and the tangent function to find the magnitude and direction of **R**.

$$R = \sqrt{(-4.0\text{ N})^2 + (-6.0\text{ N})^2} = 7.21\text{ N}$$

$$\tan\theta = \frac{-4\text{ N}}{-6\text{ N}} \qquad \theta = \arctan\left(\frac{4}{6}\right) = 33.7^\circ\text{ south of west}$$

20. The beam is in rotational equilibrium, so the sum of the torques on it is zero. Choose the right end of the beam as the reference point and let ℓ represent the length of the beam and T the unknown tension—and remember that torque = force × moment arm.

$$\Sigma\tau = 0 = (250\text{ N})\left(\frac{\ell}{2}\right) - T(\ell\sin 15^\circ) \qquad T = \frac{(250\text{ N})(\ell/2)}{\ell\sin 15^\circ} = 483\text{ N}$$

21. Because the spheres lie on one line, you need only one equation to solve this problem.

$$x_{\text{cm}} = \frac{(10\text{ cm})(2\text{ kg}) + (25\text{ cm})(4\text{ kg}) + (50\text{ cm})(3\text{ kg})}{2\text{ kg} + 4\text{ kg} + 3\text{ kg}} = 30\text{ cm}$$

22. Because it is moving at a constant velocity, the cart is in translational equilibrium and the sum of the forces acting on it is zero. Define the x axis as parallel to the plane so that you need only the forces acting in the x direction to solve the problem.

$$\Sigma F_x = 0 = F - w \sin 20^\circ \qquad F = (6 \text{ N})(\sin 20^\circ) = 2.05 \text{ N}$$

23. $$W = Fs \cos \theta = (20 \text{ N})(2 \times 10^3 \text{ m})(\cos 25^\circ) = 3.63 \times 10^4 \text{ W}$$

24. $$\text{PE} = wh = (16 \text{ N})(0.9 \text{ m}) = 14.4 \text{ J}$$

25. $$\text{PE} = \tfrac{1}{2}kx^2 = \tfrac{1}{2}(0.5 \text{ N/m})(0.05 \text{ m})^2 = 6.25 \times 10^{-4} \text{ J}$$

26. $$\text{KE} = \tfrac{1}{2}mv^2 = \tfrac{1}{2}(0.4 \text{ kg})(2.5 \text{ m/s})^2 = 1.25 \text{ J}$$

27. $$P = Fv = (50 \text{ N})(2.0 \text{ m/s}) = 100 \text{ W}$$

28. $$p = mv = (0.6 \text{ kg})(30 \text{ m/s}) = 1.8 \text{ kg m/s}$$

29. Linear momentum is conserved in all collisions, elastic or inelastic, so the sum of the blocks' individual momenta before their collision equals their common momentum after the collision. Before the collision, the 300-g block was at rest, so its momentum was zero.

$$m_1 v = (m_1 + m_2)v' \qquad v' = \frac{m_1 v}{m_1 + m_2} = \frac{(0.2 \text{ kg})(0.8 \text{ m/s})}{0.2 \text{ kg} + 0.3 \text{ kg}} = 0.320 \text{ m/s}$$

30. Find the difference between the blocks' total kinetic energies before and after the collision.

$$\text{KE} = \tfrac{1}{2}m_1 v_1^2 + \tfrac{1}{2}m_2 v_2^2 = \tfrac{1}{2}(0.2 \text{ kg})(0.8 \text{ m/s})^2 + 0 = 0.0640 \text{ J}$$

$$\text{KE}' = \tfrac{1}{2}(m_1 + m_2)(v')^2 = \tfrac{1}{2}(0.2 \text{ kg} + 0.3 \text{ kg})(0.32 \text{ m/s})^2 = 0.0256 \text{ J}$$

$$\Delta\text{KE} = \text{KE}' - \text{KE} = 0.256 \text{ J} - 0.0640 \text{ J} = -0.0384 \text{ J}$$

31. Use Newton's law of universal gravitation.

$$F = \frac{Gm_1 m_2}{r^2} = \frac{(6.67 \times 10^{-11} \text{ N m}^2/\text{kg}^2)(3 \text{ kg})(5 \text{ kg})}{(0.12 \text{ m})^2} = 6.95 \times 10^{-8} \text{ N}$$

32. $$g = F_{\text{grav}}/m = (11.0 \text{ N})/(2 \text{ kg}) = 5.50 \text{ m/s}^2$$

33. $$g = \frac{Gm}{r^2} = \frac{(6.67 \times 10^{-11} \text{ N m}^2/\text{kg}^2)(8 \text{ kg})}{(0.2 \text{ m})^2} = 1.33 \times 10^{-8} \text{ m/s}^2$$

34. Use Kepler's third law, $T^2 = 4\pi^2 r^3/GM$.

$$T = \sqrt{\frac{4\pi^2 r^3}{GM}} = \sqrt{\frac{4\pi^2(1.0821 \times 10^{11} \text{ m})^3}{(6.67 \times 10^{-11} \text{ N m}^2/\text{kg}^2)(1.99 \times 10^{30} \text{ kg})}}\left(\frac{\text{h}}{3.6 \times 10^3 \text{ s}}\right)\left(\frac{\text{day}}{24 \text{ h}}\right) = 225 \text{ days}$$

35. $$k = F/\Delta L = (2 \text{ N})/(0.04 \text{ m}) = 50 \text{ N/m}$$

36. Use the equation relating stretch modulus, stretching stress, and stretching strain, and solve for the force.

$$E = \frac{F/A}{\Delta L/L} \qquad F = \frac{EA\,\Delta L}{L} = \frac{(20 \times 10^{10} \text{ N/m}^2)(\pi/4)(5 \times 10^{-4} \text{ m})^2(1 \times 10^{-3} \text{ m})}{0.5 \text{ m}} = 78.5 \text{ N}$$

37. Use the equation relating bulk modulus, volume stress, and volume strain, and solve for the pressure.

$$B = \frac{\Delta P}{-\Delta V/V} \qquad \Delta P = \frac{-B\,\Delta V}{V} = \frac{-(0.2 \times 10^{10} \text{ N/m}^2)(-2 \times 10^{-3} \text{ m}^3)}{0.5 \text{ m}^3} = 8.00 \times 10^6 \text{ N/m}^2$$

38. Use the equation for displacement in simple harmonic motion, $x = A \sin(2\pi t/T)$, and solve for t when $x = 5$ cm and $x = 4$ cm. The difference between the two results is the time the mass takes to move 1 cm up from its lowest position of 5 cm. (Remember to express angular quantities in radians.)

$$t_1 = \frac{T \arcsin(x_1/A)}{2\pi} = \frac{(2 \text{ s}) \arcsin(5 \text{ cm}/5 \text{ cm})}{2\pi} = 0.5 \text{ s}$$

$$t_2 = \frac{T \arcsin(x_2/A)}{2\pi} = \frac{(2 \text{ s}) \arcsin(4 \text{ cm}/5 \text{ cm})}{2\pi} = 0.295 \text{ s}$$

$$\Delta t = t_1 - t_2 = 0.500 \text{ s} - 0.295 \text{ s} = 0.205 \text{ s}$$

39. The total pressure at the bottom of the tank is the pressure due to the water, $P = dgh$, plus the pressure due to the atmosphere above the water.

$$P_{tot} = dgh + P_{atm} = (1 \times 10^3\ \text{kg/m}^3)(9.8\ \text{m/s}^2)(12\ \text{m}) + 1.013 \times 10^5\ \text{N/m}^3 = 2.19 \times 10^5\ \text{N/m}^2$$

40. Use the equation for columnar fluid pressure, $P = dgh$, and solve for the height h. The pressure due to the column of mercury is the pressure at the bottom of the column minus the atmospheric pressure above the column.

$$h = \frac{P}{dg} = \frac{(3\ \text{atm} - 1\ \text{atm})\left(\dfrac{1.013 \times 10^5\ \text{N/m}^2}{\text{atm}}\right)}{(1.36 \times 10^4\ \text{kg/m}^3)(9.8\ \text{m/s}^2)} = 1.52\ \text{m}$$

41. Archimedes' principle states that the buoyant force on an immersed object equals the weight of the fluid that the object displaces. Find the mass of the displaced water by multiplying the density of water and volume of the cylinder.

$$\text{BF} = \pi r^2 hdg = \pi(0.01\ \text{m})^2(0.05\ \text{m})(1 \times 10^3\ \text{kg/m}^3)(9.8\ \text{m/s}^2) = 0.154\ \text{N}$$

42. Use the equation relating volumetric flow rate to cross-sectional area.

$$\Delta V/\Delta t = Av = \pi r^2 v = \pi(0.5\ \text{cm})^2(2.5\ \text{cm/s}) = 1.96\ \text{cm}^3/\text{s}$$

43. $$v_w = \sqrt{\frac{F}{m/L}} = \sqrt{\frac{10\ \text{N}}{(0.002\ \text{kg})(1.0\ \text{m})}} = 70.7\ \text{m/s}$$

44. $$v_w = \sqrt{\frac{E}{d}} = \sqrt{\frac{7.0 \times 10^{10}\ \text{N/m}^2}{2.7 \times 10^3\ \text{kg/m}^3}} = 5.09 \times 10^3\ \text{m/s}$$

45. $$v_w = f\lambda \qquad \lambda = \frac{v_w}{f} = \frac{340\ \text{m/s}}{440/\text{s}} = 0.773\ \text{m} \qquad [\textit{note}\text{: } 440/\text{s} = 440\ \text{Hz}]$$

46. Use the equation for the Doppler effect with a stationary source and an observer moving toward the source.

$$f' = f\left(1 + \frac{v}{v_w}\right) = (880\ \text{Hz})\left(1 + \frac{17\ \text{m/s}}{340\ \text{m/s}}\right) = 924\ \text{Hz}$$

47. $$f_1 = \frac{v_w}{2L} = \frac{80\ \text{m/s}}{2(0.8\ \text{m})} = \frac{50}{\text{s}} = 50\ \text{Hz}$$

48. Harmonic frequencies are given by $f_n = nf_1$ for $n = 1, 2, 3, \ldots$. The second overtone is the third harmonic frequency, so

$$f_3 = 3f_1 = 3(50\ \text{Hz}) = 150\ \text{Hz}$$

49. $$f_1 = \frac{v_w}{4L} = \frac{320\ \text{m/s}}{4(1.8\ \text{m})} = \frac{44.4}{\text{s}} = 44.4\ \text{Hz}$$

50. $$\beta = 10\log\frac{I}{I_0} = 10\log\frac{2 \times 10^{-8}\ \text{W/m}^2}{1 \times 10^{-12}\ \text{W/m}^2} = 43.0\ \text{dB}$$

51. $$°\text{F} = \frac{9}{5}(°\text{C}) + 32° = \frac{9}{5}(18\ °\text{C}) + 32° = 64.4\ °\text{F}$$

52. $$\text{K} = °\text{C} + 273.15 = -195.8\ °\text{C} + 273.15 = 77.4\ \text{K}$$

53. Solve the equation for linear expansion of a solid, $\Delta L = L\alpha\Delta T$, for the coefficient α.

$$\alpha = \frac{\Delta L}{L\,\Delta T} = \frac{1.092 \times 10^{-3}\ \text{m}}{(1.2\ \text{m})(35\ °\text{C})} = 2.60 \times 10^{-5}\ (°\text{C})^{-1}$$

54. Use the ideal gas law, $PV = nRT$, and solve for volume V.

$$V = \frac{nRT}{P} = \frac{(2\ \text{mol})\left(8.314\dfrac{\text{J}}{\text{mol K}}\right)(400\ \text{K})}{(0.5\ \text{atm})\left(1.013 \times 10^5\dfrac{\text{N/m}^2}{\text{atm}}\right)} = 0.131\ \text{m}^3$$

55. Use the definition of specific heat capacity c.

$$Q = mc\,\Delta T = (400\text{ g})\left(0.0306\frac{\text{cal}}{\text{g °C}}\right)(102\text{ °C} - 22\text{ °C}) = 979\text{ cal}$$

56. Use the equation for heat of transformation between the solid and liquid phases.

$$Q = mh_f = (200\text{ g})(80\text{ cal/g}) = 1.60 \times 10^4\text{ cal}$$

57. Work this problem in two steps. First find the heat necessary to raise the temperature of the water as a liquid from 25 °C to 100 °C. Then find the heat needed to boil the water at a constant temperature of 100 °C. Recall that the specific heat of liquid water is 1 cal/(g °C) and that 1 L = 10^{-3} m^3, so the mass of 1 L of water is 10^3 g.

$$Q_1 = mc\,\Delta T = (2\text{ L})(10^3\text{ g/L})\left(\frac{1\text{ cal}}{\text{g °C}}\right)(100\text{ °C} - 25\text{ °C}) = 1.50 \times 10^5\text{ cal}$$

$$Q_2 = mh_v = (2\text{ L})(10^3\text{ g/L})(540\text{ cal/g}) = 1.08 \times 10^6\text{ cal}$$

$$Q_{tot} = Q_1 + Q_2 = 0.15 \times 10^6\text{ cal} + 1.08 \times 10^6\text{ cal} = 1.23 \times 10^6\text{ cal}$$

58.

$$Q = KAt\left(\frac{T_2 - T_1}{d}\right) = \left(\frac{0.11\text{ cal}}{\text{cm s °C}}\right)(60\text{ cm}^2)(60\text{ s})\left(\frac{200\text{ °C}}{0.5\text{ cm}}\right) = 1.58 \times 10^5\text{ cal}$$

59. Use the expression for change in volume of an ideal gas at constant pressure.

$$\Delta V = \frac{nR\,\Delta T}{P} = \frac{(3\text{ mol})\left(8.314\frac{\text{J}}{\text{mol K}}\right)(708 - 308)\text{ K}}{(1\text{ atm})\left(1.013 \times 10^5\frac{\text{N/m}^2}{\text{atm}}\right)} = 0.098\,49\text{ m}^3 = 98.49\text{ L}$$

60. The volume of a cylinder is $\pi r^2 h$, so the change in volume of the expanding gas is $\Delta V = \pi r^2\,\Delta h = A\,\Delta h$. Now solve for Δh.

$$\Delta h = \frac{\Delta V}{A} = \frac{0.098\,49\text{ m}^3}{(1200\text{ cm}^2)(10^{-4}\text{ cm}^2/\text{m}^2)} = 0.8207\text{ m}$$

61. Use the definition of pressure, $P = F/A$, and solve for F.

$$F = PA = (1.013 \times 10^5\text{ N/m}^2)(0.12\text{ m}^2) = 1.216 \times 10^4\text{ N}$$

62. Use the definition of work, $W = Fs\cos\theta$. You calculated the force F in Problem 61 and the displacement $s = \Delta h$ in Problem 60. The direction of the force is parallel to the displacement, so $\theta = 0$. Neither the weight of the piston nor the force of friction affects the work that the expanding gas does if its pressure remains constant.

$$W = Fs\cos\theta = (1.216 \times 10^4\text{ N})(0.8207\text{ m})(\cos 0) = 9.98 \times 10^3\text{ J}$$

63. Remember to convert Celsius temperatures to kelvins before solving this problem.

$$e_{max} = \frac{T_H - T_C}{T_H} \times 100\% = \frac{2073\text{ K} - 873\text{ K}}{2073\text{ K}} \times 100\% = 57.9\%$$

64. **(a)** Use the equation for the work required to separate two masses against the force of gravity.

$$W = GMm\left(\frac{1}{r_e} - \frac{1}{r_m}\right)$$

$$= \left(6.67 \times 10^{-11}\frac{\text{N m}^2}{\text{kg}}\right)(5.98 \times 10^{24}\text{ kg})(1\text{ kg})\left(\frac{1}{6.38 \times 10^6\text{ m}} - \frac{1}{(0.10 + 6.38) \times 10^6\text{ m}}\right) = 9.65 \times 10^5\text{ J}$$

(b) Solve the equation for the amount of heat required to increase the temperature of a substance, $Q = mc\,\Delta T$, for the change in temperature. Then add this amount of change to the initial temperature to get the final temperature.

$$\Delta T = \frac{Q}{mc} = \frac{1 \times 10^4\text{ cal}}{(1000\text{ g})[0.22\text{ cal/(g °C)}]} = 45.5\text{ °C} \qquad T_f = T_i + \Delta T = 23\text{ °C} + 45.5\text{ °C} = 68.5\text{ °C}$$

(c)

$$\Delta L = L\alpha\,\Delta T = (10\text{ cm})[2.6 \times 10^{-5}\text{ (°C)}^{-1}](45.5\text{ °C}) = 0.0118\text{ cm}$$

(d) The aluminum will lose heat to the ice–water mixture, which must be at 0 °C. The heat will go first into melting the ice; then, if there is enough heat to melt the ice completely, the heat will go into raising the temperature of the liquid water. Because the initial temperature of the aluminum is under the boiling point of water, the aluminum and water will reach their thermal equilibrium before any heat goes into converting the water to steam. First find the amount of heat needed to melt the ice.

$$Q = mh_f = (50\text{ g})(80\text{ cal/g}) = 4000\text{ cal}$$

Then find the amount by which the transfer of this much heat decreases the temperature of the aluminum.

$$\Delta T = \frac{Q}{mc} = \frac{4000\text{ cal}}{(1000\text{ g})[0.22\text{ cal/(g °C)}]} = 18.2\text{ °C}$$

The temperature of the aluminum is now

$$T_{Al} = 68.5\text{ °C} - 18.2\text{ °C} = 50.3\text{ °C}$$

There is still a temperature difference between the aluminum and now 100 g of liquid water, so the aluminum continues to lose heat to the water until the temperatures of the two substances are equal. The heat lost by the aluminum is the heat gained by the water, so

$$m_w c_w \Delta T_w = m_{Al} c_{Al} \Delta T_{Al}$$

If T_f represents the final equilibrium temperature,

$$m_w c_w (T_f - 0\text{ °C}) = m_{Al} c_{Al} (50.3\text{ °C} - T_f)$$

$$T_f = \frac{50.3\text{ °C}}{1 + \dfrac{m_w c_w}{m_{Al} c_{Al}}} = \frac{50.3\text{ °C}}{1 + \dfrac{(100\text{ g})[1\text{ cal/(g °C)}]}{(1000\text{ g})[0.22\text{ cal/(g °C)}]}} = 34.6\text{ °C}$$

(e)

$$v_w = \sqrt{\frac{E}{d}} = \sqrt{\frac{7.0 \times 10^{10}\text{ N/m}^2}{2.7 \times 10^3\text{ kg/m}^3}} = 5.09 \times 10^3\text{ m/s}$$

(f)

$$f_1 = \frac{v_w}{2L} = \frac{5.09 \times 10^3\text{ m/s}}{2(0.5\text{ m})} = 5.09 \times 10^3\text{ Hz}$$

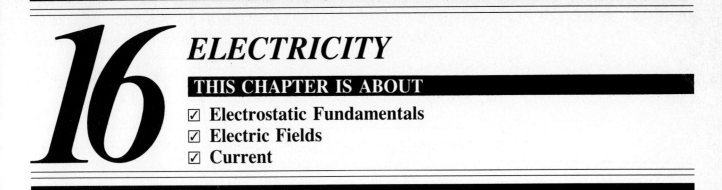

16 ELECTRICITY

THIS CHAPTER IS ABOUT

☑ Electrostatic Fundamentals
☑ Electric Fields
☑ Current

16-1. Electrostatic Fundamentals

• **Electrostatics** is the study of the effects of electrically charged particles at rest.

A. Fundamental subatomic particles

Atoms are composed of subatomic particles smaller than the smallest atom. Atoms, themselves, are electrically neutral, although some *subatomic* particles carry charge. We can use three of these particles to explain the atom's structure:

• A positively charged particle, called a **proton**;
• A negatively charged particle, called an **electron**;
• An electrically neutral particle, called a **neutron**.

You can think of the atom as having a very small core, the **nucleus**, which contains both protons and neutrons, with electrons forming a cloud around the nucleus. Do keep in mind that this is a very simplistic model of the atom, one that needs refining for more detailed study of atomic behavior.

B. Electrostatic forces

• Two charged particles exert a force on each other called the **electrostatic force.** Particles *attract* each other if their charges have *opposite* signs and *repel* each other if their charges have the *same* signs.

Electrostatic force has many similarities to gravitational force; both

• are mutual forces of interaction and obey Newton's third law (see Eq. 3-3);
• are inverse square law forces; that is, have magnitudes inversely proportional to the square of the distance between the particles exerting the forces;
• act through empty space;
• must be added vectorially.

There are also differences between the two kinds of force. The electrostatic force

• can be repulsive as well as attractive;
• is much stronger than the gravitational force.

C. Coulomb's Law

• The **coulomb** C is the SI unit of charge.

All electrons carry the same amount of *negative charge,* and all protons carry the same amount of *positive charge*. An electron carries a charge of -1.602×10^{-19} C, equal in magnitude to the charge carried by the proton. The charges on all other charged particles are integral multiples of the charge on the electron.

Coulomb's law states

• The magnitude of the electrostatic force between two charges at rest is directly proportional to the product of the magnitudes of the two charges and inversely proportional to the square of the distance between them:

COULOMB'S LAW

$$F = k\frac{Q_1 Q_2}{r^2} \quad \textbf{(16-1)}$$

where the constant $k = 9 \times 10^9$ N m^2/C^2. The direction of the force is along the line joining the particles.

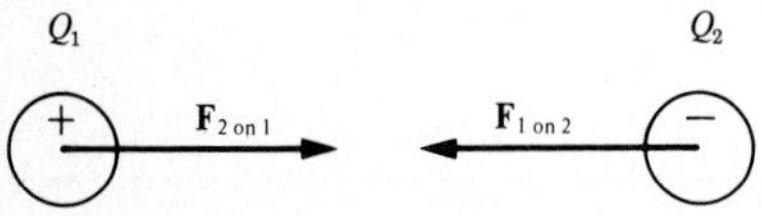

Figure 16-1. The attractive forces between a positve and a negative charge.

EXAMPLE 16-1: The two small spheres illustrated in Figure 16-1 are oppositely charged. The charged sphere on the right Q_2 has a negative charge equivalent to an excess of 5×10^6 electrons. The positively charged sphere Q_1 has a charge of 4×10^{-12} C. The distance between the centers of the spheres is 25 cm. Find the magnitude of the electrostatic force of attraction that one charged sphere exerts on the other.

Solution: First, express the negative charge of the excess electrons in coulombs:

$$Q_2 = (5 \times 10^6 \text{ electrons})(-1.60 \times 10^{-19} \text{ C/electron}) = -8 \times 10^{-13} \text{ C}$$

The electrostatic force measured outside a charged conducting sphere is the same as if all the charges on the sphere were placed at the sphere's center. So you can use Coulomb's law (Eq. 16-1) to calculate the magnitude of the force:

$$F = k\frac{Q_1 Q_2}{r^2} = \left(9 \times 10^9 \frac{\text{N m}^2}{\text{C}^2}\right)\frac{(4 \times 10^{-12} \text{ C})(-8 \times 10^{-13} \text{ C})}{(0.25 \text{ m})^2} = -4.61 \times 10^{-13} \text{ N}$$

16-2. Electric Fields

Just as we used the idea of a gravitational field to understand the gravitational force, we can use the idea of an electric field to understand the electrostatic force. [**recall**: An object modifies the space surrounding it by establishing a gravitational field which extends outward in all directions, falling in magnitude to zero at infinity. Any other mass located within this field experiences a force because of its location. (Sec. 8-2)]

The intensity or strength of an **electric field E** at a point in space is the force exerted on a positive test charge located at that point, divided by the magnitude of the charge:

ELECTRIC FIELD STRENGTH

$$\mathbf{E} = \frac{\mathbf{F}_{\text{elec}}(\text{on } Q)}{Q} \quad \textbf{(16-2)}$$

Note that the electric field strength is a vector quantity. The magnitude is $E = F/Q$. For multiple charges, the magnitude of the field at any location is the vector sum (see Sec. 1-4) of the fields produced. The direction of the field is the same as that of the force on the positive test charge.

EXAMPLE 16-2: Determine the magnitude of the electric field at the location of the charge Q_1 in Figure 16-1.

Solution: Use Eq. (16-2) to calculate the magnitude of the electric field. [Hint: You found the magnitude of the force in Example 16-1.]

$$E = \frac{F_{\text{elec}}(\text{on } Q_1)}{Q_1} = \frac{-4.61 \times 10^{-13} \text{ N}}{4 \times 10^{-12} \text{ C}} = -0.115 \text{ N/C}$$

The electric field at the location of charge Q_1 is produced by the negative charge Q_2.

We can use the definition of the electric field, Eq. (16-2), to find the magnitude of the field at a distance r from a small point charge Q:

$$E = \frac{F_{elec}(\text{on } Q_1)}{Q_1} = \frac{kQ_1Q/r^2}{Q_1} = \frac{kQ}{r^2}$$

ELECTRIC FIELD (POINT CHARGE)

$$E = \frac{kQ}{r^2} \qquad \textbf{(16-3)}$$

The electric field due to a charge Q can be calculated for any point in space, but the electrostatic force can be calculated only on another charge.

EXAMPLE 16-3: In Figure 16-2, $Q_1 = 4 \times 10^{-12}$ C at 0.30 m from point P, and $Q_2 = -8 \times 10^{-12}$ C at 0.20 m from P.
Calculate the magnitude of the total electric field on the positive unit test charge at point P. [**recall**: The field is the vector sum of the fields produced by Q_1 and Q_2.]

Solution: First, calculate the magnitude of E_1, which is the electric field produced by Q_1. Note that Q_1 is positive, so the direction of the field at P is away from Q_1:

$$E_1 = \frac{kQ_1}{r^2} = \frac{(9 \times 10^9 \text{ N m}^2/\text{C}^2)(4 \times 10^{-12} \text{ C})}{(0.30 \text{ m})^2} = 0.400 \text{ N/C}$$

Now, calculate the magnitude of E_2, the field produced by Q_2. Note that Q_2 is negative, so the direction of the field at P is toward Q_2:

$$E_2 = \frac{kQ_2}{r^2} = \frac{(9 \times 10^9 \text{ N m}^2/\text{C}^2)(8 \times 10^{-12} \text{ C})}{(0.20 \text{ m})^2} = 1.80 \text{ N/C}$$

Finally, use the law of cosines (Eq. 1-2) to find the magnitude of the total field at P:

$$E_{tot} = \sqrt{E_1^2 + E_2^2 - 2E_1E_2 \cos 100°} = \sqrt{(0.4)^2 + (1.8)^2 - 2(0.4)(1.8) \cos 100°}$$
$$= 1.91 \text{ N/C}$$

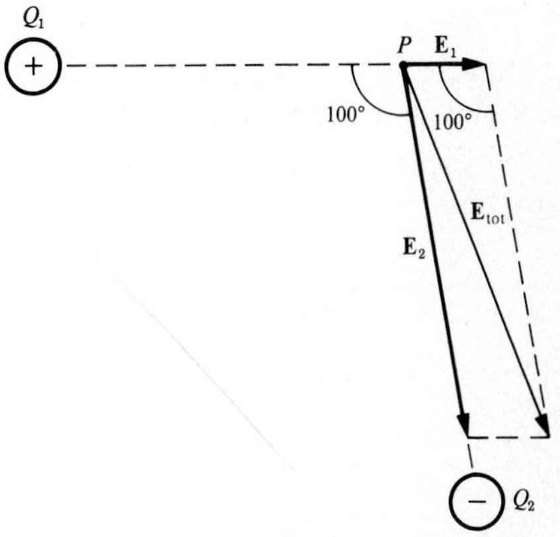

Figure 16-2. The total electric field produced by a positive charge and a negative charge.

EXAMPLE 16-4: A test charge of 5×10^{-11} C is placed at point P in Figure 16-2. Determine the magnitude of the electric force acting on it.

Solution: You found the magnitude of the field on a unit charge in Example 16-3, so use Eq. (16-2) to find the magnitude of the electric force on the new charge.

$$E = \frac{F(\text{on } Q)}{Q}$$

$$F(\text{on } Q) = EQ = (1.91 \text{ N/C})(5 \times 10^{-11} \text{ C}) = 9.55 \times 10^{-11} \text{ N}$$

16-3. Current

- **Electric current** is the rate at which charge flows past a point.

We measure electric current in **amperes** (A), one of the fundamental SI units. One ampere is the flow of one coulomb per second:

CURRENT OR CHARGE

$$\text{amperes} = \text{coulombs per second} \quad \text{or} \quad 1\text{A} = 1\frac{\text{C}}{\text{s}}$$
$$\text{coulombs} = \text{amperes} \times \text{seconds} \quad \text{or} \quad 1\text{C} = 1\text{A s} \qquad \textbf{(16-4)}$$

A. Conductivity and Conductors

- A material that offers little resistance to the flow of electric current is a good **conductor**, or has high **conductivity**.

Materials can be grouped by their conductive properties:

- In a **metallic conductor,** free electrons carry the electric current. Examples include copper, iron, aluminum, and mercury.
- In an **ionic conductor,** ions carry the electric current. An ion is a charged atom. Both molten ionic compounds and solutions of ionic compounds can be good conductors.
- Free electrons also carry the electric current in a **semiconductor**. Although semiconductors have low conductivity, their other properties make them enormously important in electronics. Examples include silicon and germanium.
- Very little current is carried in materials of extremely low conductivity called **insulators**. Examples include glass, paper, paraffin, and polyethylene.

An electric conductor contains charges that are free to move around within it; an insulator does not. If there is no net motion of charge within or on the surface of a conductor, the conductor is in **electrostatic equilibrium** and has these properties:

- The electric field is zero everywhere within the conductor.
- The electric charges on a conductor, because of their mutual repulsion, distribute themselves only on the surface of the conductor—there is no charge with a conductor.
- The electric field just outside of a charged conductor is perpendicular to the conductor's surface.
- Charge on an irregularly shaped conductor is not evenly distributed, but accumulates where the conductor's surface is most curved, for example, at sharp points.

If an electric field is present within a conductor, there is a steady flow of charges, producing a current. This is the situation when you connect a flashlight bulb to a battery by copper wires. The battery sets up an electric field that puts the electrons in the copper into motion. If electrons are not moving *within* a conductor, then the electric field is necessarily zero. This means that a conductor containing static charges on its surface has an electric field of zero everywhere within it.

B. Average and instantaneous current

- **Average current** I_{ave} in a conductor is the quantity of charge flowing through a cross section of the conductor divided by the time:

AVERAGE CURRENT

$$I_{\text{ave}} = \frac{\Delta Q}{\Delta t} \tag{16-5}$$

- If the current is not constant, the **instantaneous current** I is the limiting value (derivative) of $\Delta Q/\Delta t$ as Δt approaches zero:

INSTANTANEOUS CURRENT

$$I = \lim_{\Delta t \to 0} \frac{\Delta Q}{\Delta t} = \frac{dQ}{dt} \tag{16-6}$$

EXAMPLE 16-5: A current of 80 mA is flowing through the tungsten filament of a flashlight bulb.
How many electrons pass through the filament each minute?

Solution: The current is constant, so you can use Eq. (16-5) to find the charge passing through the bulb each minute:

$$I = \frac{\Delta Q}{\Delta t}$$

$$\Delta Q = I\Delta t = (80 \times 10^{-3}\ \text{C/s})(60\ \text{s/min}) = (4.8\ \text{C/min})\left(\frac{1\ \text{electron}}{1.60 \times 10^{-19}\ \text{C}}\right)$$

$$= 3.0 \times 10^{19}\ \text{electrons/min}$$

SUMMARY

Coulomb's law	$F = k\dfrac{Q_1 Q_2}{r^2}$	the force between two charges at rest is directly proportional to the product of the magnitudes of the two charges and inversely proportional to the square of the distance between them
Electric field strength	$\mathbf{E} = \dfrac{\mathbf{F}_{elec}(\text{on } Q)}{Q}$	the strength of an electric field at a point in space is the force exerted on a positive test charge located at that point, divided by the magnitude of the charge
Electric field (point charge)	$E = \dfrac{kQ}{r^2}$	the electric field at a distance r from a point charge Q is directly proportional to the charge on Q and inversely proportional to the square of the distance from Q
Current or charge	$1\text{A} = 1\dfrac{\text{C}}{\text{s}}$ $1\text{C} = 1\text{A s}$	one ampere is the flow of one coulomb per second
Average current	$I_{ave} = \dfrac{\Delta Q}{\Delta t}$	average current in a conductor is the quantity of charge flowing through a cross section of the conductor divided by the time required
Instantaneous current	$I = \lim\limits_{\Delta t \to 0} \dfrac{\Delta Q}{\Delta t}$	if the flow of electrons is not constant, the instantaneous current is the limiting value of the quantity of charge divided by the time required as the time approaches zero

RAISE YOUR GRADES

Can you define . . . ?

☑ an electron ☑ an electrostatic force ☑ an electric current

Can you . . . ?

☑ write the equation for the force between two charges separated by a distance r
☑ describe the direction of the force on one positive charge due to another positive charge
☑ describe the direction of the force on a negative charge due to a positive charge
☑ describe how the electrical force between two charges will change if their separation is doubled
☑ write the equation which defines the electric field at a point in space
☑ write the expression for the electric field at a distance r from a charge Q
☑ state the value of the electric field inside a conductor in which no current is present
☑ write the definition of current

SOLVED PROBLEMS

PROBLEM 16-1 A positive charge of 4×10^{-10} C exerts a force of 1.575×10^{-6} N on a negative charge of 7×10^{-10} C. Calculate the distance r from the positive charge to the negative charge.

Solution Solve Coulomb's law, Eq. (16-1), to find the distance between the charges:

$$F = k\frac{Q_1 Q_2}{r^2}$$

$$r = \sqrt{\frac{kQ_1 Q_2}{F}} = \sqrt{\frac{(9 \times 10^9 \text{ N m}^2/\text{C}^2)(4 \times 10^{-10}\text{ C})(7 \times 10^{-10}\text{ C})}{1.575 \times 10^{-6}\text{ N}}} = 0.040 \text{ m}$$

PROBLEM 16-2 Two equal positive charges of 5×10^{-11} C are separated by a distance of 70 cm. Find the magnitude of the electric field on a unit test charge at a point P, which lies on the straight line between the two charges, 30 cm from one charge and 40 cm from the other.

Solution First, use Eq. (16-3) to calculate the magnitude of the field produced by each charge:

$$E_1 = \frac{kQ_1}{r_1^2} = \frac{(9 \times 10^9 \text{ N m/C}^2)(5 \times 10^{-11}\text{ C})}{(0.30 \text{ m})^2} = 5.00 \text{ N}$$

$$E_2 = \frac{(9 \times 10^9 \text{ N m}^2/\text{C}^2)(5 \times 10^{-11}\text{ C})}{(0.40 \text{ m})^2} = 2.81 \text{ N}$$

Since these two fields are in opposite directions (each positive charge repels a positive test charge (see Sec. 16-2)), you can obtain the magnitude of the resultant field by subtracting E_2 from E_1:

$$E_{tot} = 5.00 \text{ N} - 2.81 \text{ N} = 2.19 \text{ N}$$

PROBLEM 16-3 When a test charge of 3.2×10^{-16} C is placed at a point in space, it is subject to an electrostatic force of 1.28×10^{-16} N. What is the magnitude of the electric field at this point?

Solution Eq. (16-2) provides the magnitude of the electric field directly:

$$E = \frac{F(\text{on } Q)}{Q} = \frac{1.28 \times 10^{-16}\text{ N}}{3.2 \times 10^{-16}\text{ C}} = 0.400 \text{ N/C}$$

PROBLEM 16-4 Two positive charges are located 20 cm and 15 cm from a unit test charge at point P, as illustrated in Fig. 16-3. Charge Q_1 is 6×10^{-12} C, and the total electric field produced by both charges is 1.806 N/C. Find the unknown charge Q_2.

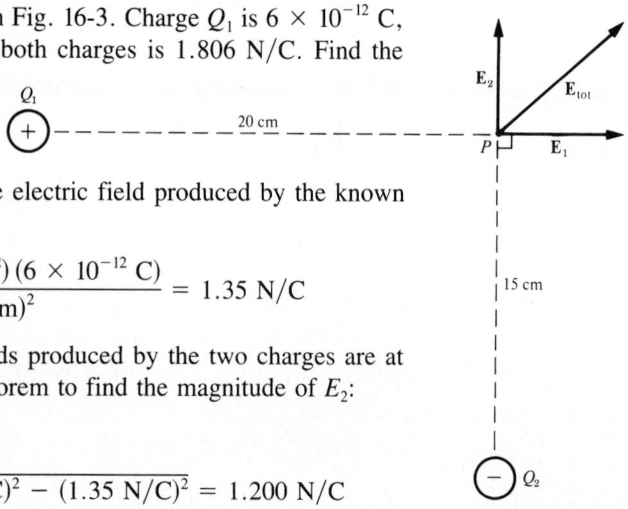

Figure 16-3

Solution Use Eq. (16-3) to calculate the electric field produced by the known charge:

$$E_1 = \frac{kQ_1}{r_1^2} = \frac{(9 \times 10^9 \text{ N m}^2/\text{C}^2)(6 \times 10^{-12}\text{ C})}{(0.20 \text{ m})^2} = 1.35 \text{ N/C}$$

You can see from Fig. 16-3 that the fields produced by the two charges are at right angles. So use the Pythagorean theorem to find the magnitude of E_2:

$$E_{tot}^2 = E_1^2 + E_2^2$$

$$E_2 = \sqrt{E_{tot}^2 - E_1^2} = \sqrt{(1.806 \text{ N/C})^2 - (1.35 \text{ N/C})^2} = 1.200 \text{ N/C}$$

Now, use Eq. (16-3) to solve for Q_2:

$$Q_2 = \frac{E_2 r^2}{k} = \frac{(1.200 \text{ N/C})(0.15 \text{ m})^2}{9 \times 10^9 \text{ N m}^2/\text{C}^2} = 3.00 \times 10^{-12}\text{ C}$$

PROBLEM 16-5 A copper rod 5-cm long exhibits a negative charge of 3.2×10^{-13} C. (**a**) How many excess electrons are present and (**b**) where are they located?

Solution

(**a**) The charge on an electron is 1.6×10^{-19} C. Let n represent the number of electrons on the rod, so that

$$n = \frac{3.2 \times 10^{-13}\ \text{C}}{1.6 \times 10^{-19}\ \text{C/electron}} = 2 \times 10^{6}\ \text{electrons}$$

(**b**) Since these electrons repel one another, the excess electrons all move to the outer surface of the conductor.

PROBLEM 16-6 When a wire is connected to a weak battery, 6.0 C pass through the wire in 30 s. Calculate the current in milliamperes.

Solution Use Eq. (16-5) to find the average current in the wire:

$$I_{\text{ave}} = \frac{\Delta Q}{\Delta t} = \frac{6.0\ \text{C}}{30\ \text{s}} = 0.20\ \text{A} = 200\ \text{mA}$$

PROBLEM 16-7 An unknown charge located 25 cm from a positive charge Q_1 of 6×10^{-10} C is attracted toward Q_1 by a force of magnitude 1.728×10^{-8} N. Calculate the magnitude of the unknown charge.

Solution Since the unknown charge is attracted toward the positive charge, the unknown charge is necessarily negative. Use Coulomb's law to find the magnitude of Q_2:

$$Q_2 = \frac{Fr^2}{kQ_1} = \frac{(1.728 \times 10^{-8}\ \text{N})(0.25\ \text{m})^2}{(9 \times 10^{9}\ \text{N m}^2/\text{C}^2)(6 \times 10^{-10}\ \text{C})} = 2.00 \times 10^{-10}\ \text{C}$$

PROBLEM 16-8 Two equal positive charges of 4×10^{-10} C are separated by 10 cm. A unit test charge is located at point P, 10 cm from each charge, as shown in Fig. 16-4. Calculate the magnitude of the total electric field at point P.

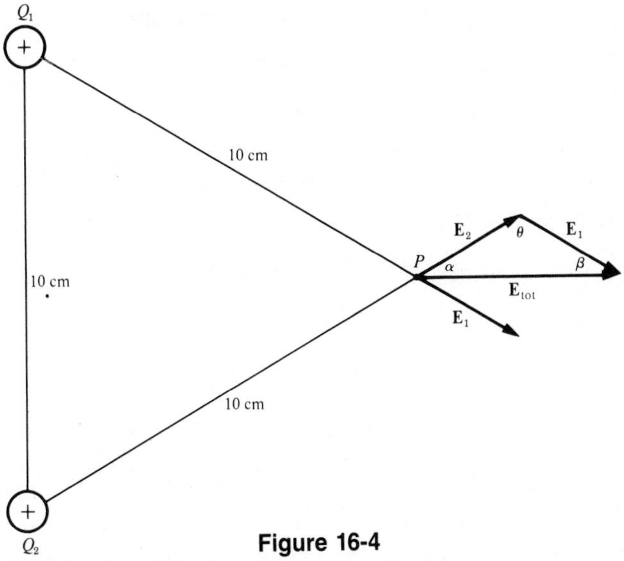

Figure 16-4

Solution You can see from Fig. 16-4 that the two point charges and the point P form an equilateral triangle, so the angle between the two vectors $\mathbf{E}_1$ and $\mathbf{E}_2$ at point P is 60°, and $\alpha = 30°$. Find the angle θ opposite E_{tot}:

$$\theta = 180° - \alpha - \beta = 180° - 30° - 30° = 120°$$

Next, calculate the magnitude of $\mathbf{E}_1$ and $\mathbf{E}_2$ (Eq. 16-3):

$$E_1 = E_2 = \frac{kQ_1}{r^2} = \frac{(9 \times 10^9 \text{ N m}^2/\text{C}^2)(4 \times 10^{-10} \text{ C})}{(0.10 \text{ m})^2} = 360 \text{ N/C}$$

Now, use the law of cosines to calculate the magnitude of the resultant field:

$$E_{tot} = \sqrt{E_1^2 + E_2^2 - 2E_1E_2 \cos\theta}$$

$$= \sqrt{(360 \text{ N/C})^2 + (360 \text{ N/C})^2 - 2(360 \text{ N/C})^2 \cos 120°} = 6.24 \times 10^2 \text{ N/C}$$

PROBLEM 16-9 Three charges, $Q_1 = -4 \times 10^{-12}$ C, $Q_2 = +6 \times 10^{-12}$ C, and $Q_3 = -2 \times 10^{-12}$ C are placed at three corners of a square 10 cm on a side, as shown in Figure 16-5. Find the magnitude and direction of the total electric field at point O.

Solution In Problem 16-8 you solved for the total electric field by using the law of cosines. The vector component method, however, is the best way to solve this three-particle problem. First, calculate the magnitude of each force.

$$E_1 = \frac{kQ_1}{r_1^2} = \frac{(9 \times 10^9 \text{ N m/C}^2)(4 \times 10^{-12} \text{ C})}{(0.1 \text{ m})^2} = 3.6 \text{ N/C}$$

$$E_2 = \frac{kQ_2}{r_2^2} = \frac{(9 \times 10^9 \text{ N m/C}^2)(6 \times 10^{-12} \text{ C})}{0.02 \text{ m}^2} = 2.7 \text{ N/C}$$

(Find r_2^2 by applying the Pythagorean theorem.)

$$E_3 = \frac{kQ_3}{r_3^2} = \frac{(9 \times 10^9 \text{ N m/C}^2)(2 \times 10^{-12} \text{ C})}{(0.1 \text{ m})^2} = 1.8 \text{ N/C}$$

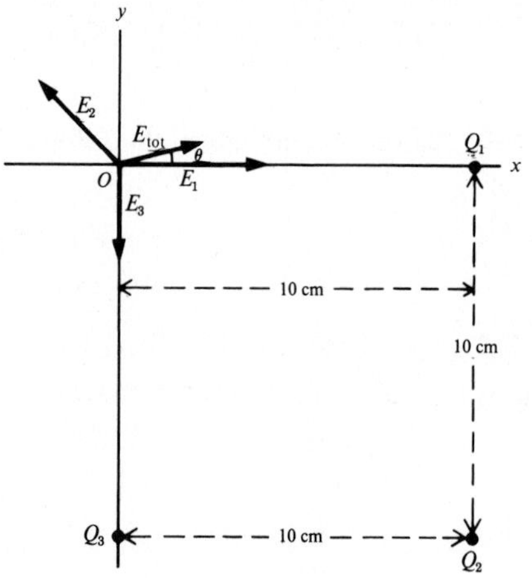

Figure 16-5

Next, resolve each vector into its x and y components. Remember that the sign of an electric field is the same as that of the force due to the field on a *positive* charge. Then add to find the components of E_{tot}.

vector	x component	y component
E_1	+3.6 N/C	0
E_2	−1.91 N/C	1.91 N/C
E_3	0	−1.80 N/C
E_{tot}	1.69 N/C	0.11 N/C

Now you can find the magnitude of E_{tot}:

$$E_{tot} = \sqrt{(1.69 \text{ N/C})^2 + (0.11 \text{ N/C})^2} = 1.69 \text{ N/C}$$

And the direction of E_{tot}:

$$\theta = \arctan(E_y/E_x) = \arctan(0.11/1.69) = 3.70°$$

PROBLEM 16-10 What is the current in milliamperes that corresponds to a flow of 1.20×10^{19} electrons per minute?

Solution Multiply by the charge of an electron in coulombs:

$$I = (1.20 \times 10^{19} \text{ electrons/min})(-1.6 \times 10^{-19} \text{ C/electron})$$

$$= (-1.92 \text{ C/min})\frac{(1 \text{ min})}{(60 \text{ s})} = -3.2 \times 10^{-2} \text{ C/s}$$

$$= -32 \times 10^{-3} \text{ A} = -32 \text{ mA}$$

Supplementary Exercises

PROBLEM 16-11 Two positive charges are separated by 12 cm. The force that Q_1 exerts on Q_2 is 4×10^{-8} N. What is this force if the distance between the two charges is reduced to 9 cm?

PROBLEM 16-12 Two uniformly charged spheres are separated by a distance of 20 cm from the center of one sphere to the center of the other. Each sphere contains a charge of 6×10^8 electrons. Calculate the force that one charged sphere exerts on the other.

PROBLEM 16-13 What is the magnitude of the electric field at a distance of 4 cm from a point charge of 8×10^{-10} C? By what factor is the electric field increased if the distance from the point charge is reduced from 4 cm to 2 cm?

PROBLEM 16-14 Two positive charges are separated by a distance of 30 cm. One charge is 3×10^{-10} C and the other is 5×10^{-10} C. What is the magnitude of the total electric field at a point halfway between the two charges?

PROBLEM 16-15 An electric field with a magnitude of 4×10^{-3} N/C due to Q_1 is directed toward the east. The electric field at the same point in space due to Q_2 has a magnitude of 6×10^{-3} N/C and is directed 20° north of east. What is the total electric field at this point?

PROBLEM 16-16 A current of 75 mA is flowing through the filament of a flashlight bulb. How many electrons flow through this filament in 3 minutes?

Answers to Supplementary Exercises

16-11: 7.11×10^{-8} N

16-12: 2.07×10^{-9} N

16-13: 4.50×10^3 N/C; 4

16-14: 80.0 N/C

16-15: 9.85×10^{-3} N/C

16-16: 8.44×10^{19} electrons

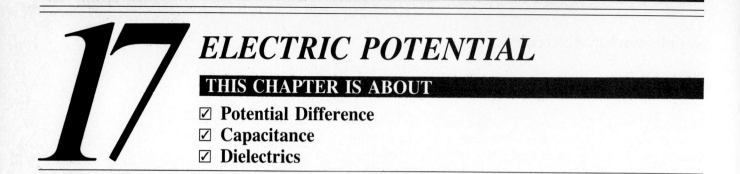

17 ELECTRIC POTENTIAL

THIS CHAPTER IS ABOUT

☑ Potential Difference
☑ Capacitance
☑ Dielectrics

17-1. Potential Difference

The concept of potential difference is easiest to see in an analogy. You know that a skier has more potential energy at the top of a ski jump than at the bottom. The difference in the amount of the skier's potential energy in the two positions is the *potential difference*. You can think of this difference in terms of the *work* done by the ski lift, which increases potential energy by moving the skier to the top of the jump. Analogously, *work* is required to transfer electrical charge within an electric field, so that

- The **potential difference** between two points A and B is the work required to move a positive charge Q from point B to point A, $W_{B\to A}$, divided by the magnitude of the charge:

POTENTIAL DIFFERENCE

$$V_{AB} = \frac{W_{B\to A}}{Q} \tag{17-1}$$

Potential difference is work per unit charge. We measure potential difference in the SI system by the **volt** V, defined as the potential difference between two points if one joule of work is required to move one coulomb of charge from one point to the other: 1 V = 1(J/C).

EXAMPLE 17-1: The work required to move a positive charge of 4.0×10^{-10} C from point B to point A is 2.0×10^{-7} J.
What is the potential difference between these two points?

Solution: Directly from Eq. (17-1),

$$V_{AB} = \frac{W_{B\to A}}{Q} = \frac{2.0 \times 10^{-7}\ \text{J}}{4.0 \times 10^{-10}\ \text{C}} = 500\ \text{V}$$

A. Equipotential surfaces

If points A and B are located on the surface of a conductor in electrostatic equilibrium, the work required to move a charge from point B to point A is *exactly* zero. This means that the potential difference between points A and B is zero, so these points are at the same potential.

- The actual surface of a conductor or an imaginary surface in space on which all points are at the same potential is called an **equipotential surface**.

Remember that if a conductor is in electrostatic equilibrium there is no electric field within it. As a result it takes no work to move a charge about inside a conductor, so the potential is the same everywhere in a conductor.

B. Potential of an isolated point charge

The potential at a distance r from an isolated point charge Q, or from the center of a uniformly charged sphere, is proportional to the charge and inversely proportional to the distance r:

POTENTIAL (ISOLATED POINT CHARGE)

$$V = \frac{kQ}{r} \qquad \textbf{(17-2)}$$

[**Recall**: The constant k equals $9 \times 10^9\ \text{N m}^2/\text{C}^2$ (Eq. (16-1), Coulomb's law)]

The potential at a distance r from a point charge is equal to the work required to bring a unit positive charge from an infinite distance up to a distance r from the point charge. Electric potential is a scalar quantity—it has no direction.

EXAMPLE 17-2: An isolated positive point charge Q has a magnitude of 6×10^{-9} C. How much work is required to move a charge of 1 C from infinity to a distance of 5 cm from the charge Q?

Solution: Directly from Eq. (17-2),

$$V = \frac{kQ}{r} = \frac{(9 \times 10^9\ \text{N m}^2/\text{C}^2)(6 \times 10^{-9}\ \text{C})}{0.05\ \text{m}} = 1080\ \text{V}$$

EXAMPLE 17-3: Point A is located 4 cm from an isolated positive charge Q of 8×10^{-10} C, and point B is 10 cm from Q, as shown in Figure 17-1. How many joules of work are required to move a positive charge Q' of 3×10^{-11} C from point B to point A?

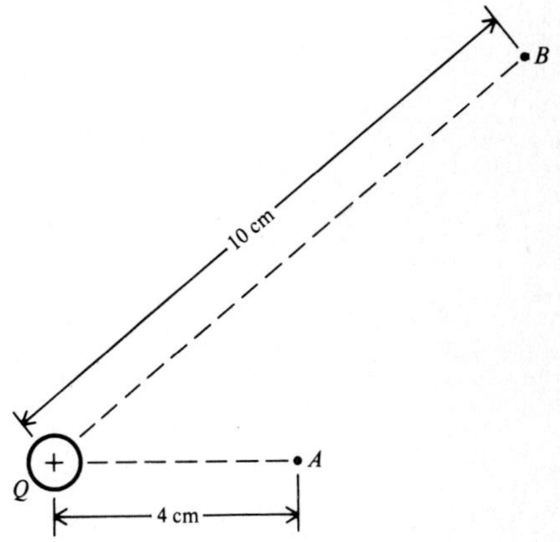

Figure 17-1

Solution: First, calculate the potential at points A and B directly from Eq. (17-2):

$$V_A = \frac{kQ}{r_A} = \frac{(9 \times 10^9\ \text{N m}^2/\text{C}^2)(8 \times 10^{-10}\ \text{C})}{0.04\ \text{m}} = 180\ \text{V}$$

$$V_B = \frac{kQ}{r_B} = \frac{(9 \times 10^9\ \text{N m}^2/\text{C}^2)(8 \times 10^{-10}\ \text{C})}{0.10\ \text{m}} = 72\ \text{V}$$

Now, calculate the potential difference between points A and B:

$$V_{AB} = V_A - V_B = (180 - 72)\ \text{V} = 108\ \text{V}$$

Finally, calculate the work required (Eq. 17-1):

$$W_{B \to A} = V_{AB}\,Q' = (108\ \text{V})(3 \times 10^{-11}\ \text{C}) = 3.24 \times 10^{-9}\ \text{J}$$

EXAMPLE 17-4: An isolated point charge has a value of 4×10^{-8} C. What is the shape of an equipotential surface in the vicinity of this point charge?

Solution: For a given charge, Equation (17-2) states that the potential at any point is proportional to the point's distance from the charge. Therefore, all points that are at the same distance from the charge have the same potential. The surface whose points are all at the same distance from another point is a sphere. Therefore, the equipotential surface is a sphere with its center at the point charge.

EXAMPLE 17-5: Two negative charges, $Q_1 = -5 \times 10^{-9}$ C and $Q_2 = -7 \times 10^{-9}$ C, are separated by 10 cm. What is the total potential at a point midway between Q_1 and Q_2?

Solution: The total potential is the sum of the individual potentials due to each charge. Since potential is a scalar quantity you don't have to worry about the direction of the forces due to each charge.

$$V_1 = \frac{kQ_1}{r_1} = \frac{(9 \times 10^9\ \text{C})(-5 \times 10^{-9}\ \text{C})}{0.05\ \text{m}} = -900\ \text{V}$$

and

$$V_2 = \frac{kQ_2}{r_2} = \frac{(9 \times 10^9\ \text{C})(-7 \times 10^{-9}\ \text{C})}{0.05\ \text{m}} = -1260\ \text{V}$$

so

$$V_{tot} = V_1 + V_2 = (-900 - 1260)\text{ V} = -2160\text{ V}$$

Notice that the potential is negative. This means that a positive charge moving from infinity to the final position between Q_1 and Q_2 could do work; that is, the force on it is attractive.

17-2. Capacitance

- A **capacitor** is a pair of conductors: one conductor with a negative charge and the other with an equal positive charge.

To describe the relation between the amount of charge on the conductors and the potential difference between them we use the term *capacitance*.

- The **capacitance** C of a pair of conductors is the charge on either conductor divided by the potential difference between the two conductors:

CAPACITANCE

$$C = \frac{Q}{V} \qquad \textbf{(17-3)}$$

The units of capacitance are coulombs/volt, or **farads** F. A capacitor has a capacitance of one farad if one coulomb of charge maintains a potential difference of one volt.

An effective way of storing charge is to use a pair of flat conductors placed close together and parallel to each other. Let A represent the area of either plate and d the distance between the two parallel plates. If the space between the plates is a vacuum, then the capacitance is

CAPACITANCE (PARALLEL PLATES; VACUUM)

$$C_{vac} = \frac{A}{4\pi kd} = \frac{\varepsilon_0 A}{d} \qquad \textbf{(17-4)}$$

The constant ε_0, called the **permittivity of free space,** is $1/4\pi k$, and has a value of $8.85 \times 10^{-12}\text{ C}^2/\text{N m}^2$.

EXAMPLE 17-6: For a given capacitor, the area of each parallel plate is 25 cm² and the separation is 0.5 mm.
Determine the capacitance.

Solution: Directly from Eq. (17-4),

$$C_{vac} = \frac{A}{4\pi kd} = \frac{(25\text{ cm}^2)(1\text{ m}^2/10^4\text{ cm}^2)}{4\pi(9 \times 10^9\text{ N m}^2/\text{C}^2)(0.5 \times 10^{-3}\text{ m})}$$

$$= 4.42 \times 10^{-11}\text{ F} = 4.42 \times 10^{-5}\ \mu\text{F}$$

One microfarad (μF) equals 1×10^{-6} farads.

EXAMPLE 17-7: The plates of the capacitor described in Example 17-5 have a potential difference of 800 V. How much charge is stored on each plate?

Solution: You can calculate the charge from Eq. (17-3):

$$C = \frac{Q}{V}$$

$$Q = CV = (4.42 \times 10^{-11}\text{ F})(800\text{ V}) = 3.54 \times 10^{-8}\text{ C}$$

17-3. Dielectrics

- A **dielectric** is a substance that does not conduct electricity; in other words, an *insulator*.

A dielectric placed between the two conductors of a parallel plate capacitor increases the capacitance.

- The **dielectric constant** K of an insulator is a measure of the increase in the capacitance of a pair of parallel plates when the vacuum between the plates is replaced by a dielectric medium:

DIELECTRIC CONSTANT

$$K = \frac{C_{\text{med}}}{C_{\text{vac}}} \qquad \textbf{(17-5)}$$

EXAMPLE 17-8: When the vacuum between the set of parallel plates in Example 17-6 is replaced by castor oil, the capacitance increases to 2.064×10^{-4} μF. What is the dielectric constant of castor oil?

Solution: Directly from Eq. (17-5),

$$K = \frac{C_{\text{med}}}{C_{\text{vac}}} = \frac{2.064 \times 10^{-4}\ \mu\text{F}}{4.42 \times 10^{-5}\ \mu\text{F}} = 4.67$$

Since parallel plate capacitors almost always have a dielectric between the two conductors, we can write the equation for capacitance as

CAPACITANCE (PARALLEL PLATES; DIELECTRIC)

$$C = \frac{KA}{4\pi kd} = \frac{K\varepsilon_0 A}{d} \qquad \textbf{(17-6)}$$

where d is the distance between the plates.

EXAMPLE 17-9: A pair of flat plates of 10 cm $\times$ 15 cm are separated by a thin film of wax ($K = 2.25$) 0.3 mm thick.
Calculate the capacitance.

Solution: Directly from Eq. (17-6),

$$C = \frac{K\varepsilon_0 A}{d} = \frac{(2.25)(8.85 \times 10^{-12}\ \text{C}^2/\text{N m}^2)(0.1\ \text{m} \times 0.15\ \text{m})}{0.3 \times 10^{-3}\ \text{m}}$$

$$= 9.96 \times 10^{-10}\ \text{F} = 9.96 \times 10^{-4}\ \mu\text{F}$$

EXAMPLE 17-10: A capacitor is connected to a nine-volt battery. Its capacitance is 4.0 μF.
How much charge is stored in this capacitor?

Solution: From Eq. (17-3),

$$Q = CV = (4.0 \times 10^{-6}\ \text{F})(9.0\ \text{V}) = 3.6 \times 10^{-5}\ \text{C}$$

EXAMPLE 17-11: Figure 17-2 shows a parallel plate capacitor (diagram symbol: ⊣⊢). The potential difference between the two conductors is 500 V, and the distance between the two plates is 1.0 cm.
How much work is required to move a charge of 1 C from the negatively charged conductor to the positively charged conductor?

Solution: Directly from Eq. (17-1),

$$W_{B\to A} = V_{AB}Q = (500\ \text{V})(1\ \text{C}) = 500\ \text{J}$$

EXAMPLE 17-12: The electric field between the two flat plates of the capacitor shown in Figure 17-2 is constant, and its magnitude is equal to the force on a unit positive charge placed between the plates.
What is the magnitude of the electric field between the plates of this capacitor?

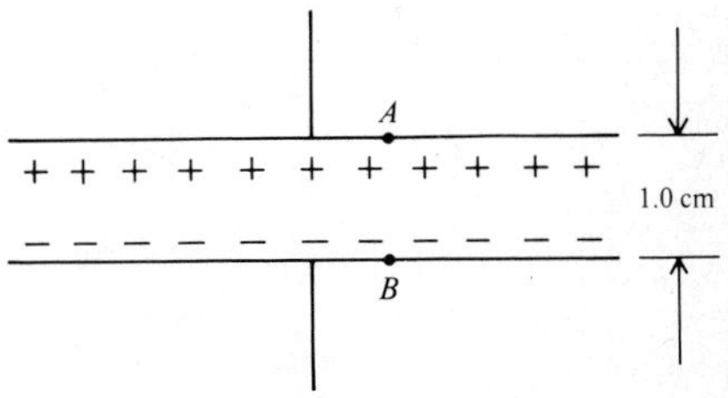

Figure 17-2

Solution: The force on a positive charge is downward because a positive charge is repelled by the positive charges on the upper plate and attracted by the negative charges on the lower plate. To move the charge vertically upward requires an external upward force F, and the work that this external force does is

$$W_{B\to A} = Fd$$

Then, solve for the magnitude of the electric force, which is equal to the force acting on the unit positive charge, and, by using Eq. (17-1), express $W_{B\to A}$ in terms of V_{AB} and Q.

$$F_{(\text{on } Q)} = \frac{W_{B\to A}}{d} = \frac{V_{AB}Q}{d}$$

The magnitude of the electric field is the force divided by the charge on which it acts (see Eq. 16-2). So,

$$E = \frac{F_{(\text{on } Q)}}{Q} = \frac{V_{AB}Q/d}{Q} = \frac{V_{AB}}{d}$$

ELECTRIC FIELD BETWEEN THE PARALLEL PLATES OF A CAPACITOR

$$E = \frac{V}{d} \qquad \textbf{(17-7)}$$

Now, you can calculate the magnitude of this particular electric field directly from Eq. (17-7):

$$E = \frac{500\ \text{V}}{0.01\ \text{m}} = 5 \times 10^4\ \text{V/m}$$

You can express the electric field either in units of N/C or V/m.

SUMMARY

Potential difference	$V_{AB} = \dfrac{W_{B\to A}}{Q}$	gives the potential difference between two points as the work required to move a positive charge from one point to the other, divided by the magnitude of the charge
Potential (isolated point charge)	$V = \dfrac{kQ}{r}$	gives the potential at a distance r from an isolated point charge, or from a uniformly charged sphere, as proportional to the charge and inversely proportional to the distance r
Capacitance	$C = \dfrac{Q}{V}$	gives the capacitance of a pair of conductors as the charge on either conductor divided by the potential difference between the two conductors
Capacitance (parallel plates; vacuum)	$C_{\text{vac}} = \dfrac{A}{4\pi kd} = \dfrac{\varepsilon_0 A}{d}$	gives the capacitance between a pair of parallel conductors separated by a vacuum
Dielectric constant	$K = \dfrac{C_{\text{med}}}{C_{\text{vac}}}$	gives the measure of the increase in the capacitance of a pair of parallel plates when the vacuum between the plates is replaced by a dielectric medium

Capacitance (parallel plates; dielectric)	$C = \dfrac{KA}{4\pi kd} = \dfrac{K\varepsilon_0 A}{d}$	gives the capacitance for a pair of parallel plates with a dielectric between the two conductors
Electric field in a parallel-plate capacitor	$E = \dfrac{V}{d}$	gives the magnitude of an electric field between the two parallel plates of a capacitor

RAISE YOUR GRADES

Can you define . . . ?

☑ a volt
☑ an equipotential surface
☑ a capacitor
☑ a dielectric

Can you . . . ?

☑ write the definition of potential difference between two points
☑ state the potential difference between two points on a conductor which has a static charge
☑ calculate the potential at a distance r from an isolated point charge
☑ calculate the amount of work required to move a test charge from point B to A if potentials V_B and V_A are both known
☑ calculate the potential due to two point charges
☑ describe an equipotential surface near an isolated point charge
☑ write the definition of capacitance for a pair of conductors with a potential difference V
☑ write the equation for the capacitance of a pair of flat plates separated by a distance d and containing a dielectric between the two plates
☑ calculate the amount of charge stored in a capacitor with a potential difference V
☑ write the definition of the dielectric constant K

SOLVED PROBLEMS

PROBLEM 17-1 A point P is located 5 cm from a positive charge Q_1 and 9 cm from a negative charge Q_2, as shown in Fig. 17-3. The magnitude of Q_1 is 6×10^{-10} C and the magnitude of Q_2 is -9×10^{-10} C. Calculate the total potential at point P.

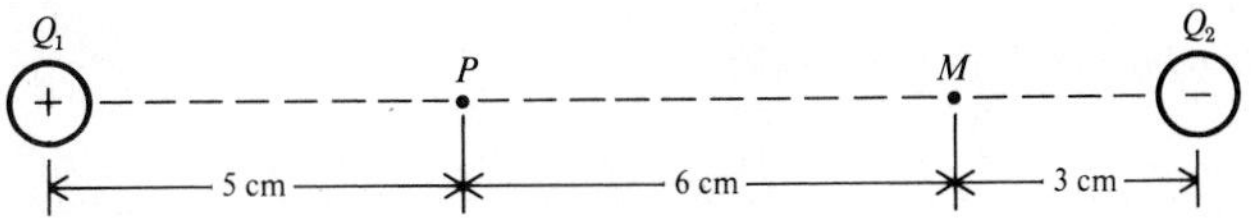

Figure 17-3

Solution Use Eq. (17-2) to calculate the potential due to each isolated point charge. For Q_1:

$$V_1 = \frac{kQ_1}{r_1} = \frac{(9 \times 10^9\ \text{N m}^2/\text{C}^2)(6 \times 10^{-10}\ \text{C})}{0.05\ \text{m}} = 108\ \text{V}$$

And for Q_2, which is a *negative* charge:

$$V_2 = \frac{kQ_2}{r_2} = \frac{(9 \times 10^9 \text{ N m}^2/\text{C}^2)(-9 \times 10^{-10} \text{ C})}{0.09 \text{ m}} = -90 \text{ V}$$

To find the *total* potential at P, simply add V_1 and V_2:

$$V_{tot} = V_1 + V_2 = (108 + (-90)) \text{ V} = 18.0 \text{ V}$$

PROBLEM 17-2 Using the data from Problem 17-1 and Fig. 17-3, determine how much work is required to move a charge of 3×10^{-8} C from point M a distance of 6 cm to point P.

Solution First, calculate the total potential at point M from Eq. (17-2). You can see from Fig. 17-3 that M is 11 cm from Q_1 and 3 cm from Q_2, so

$$V_1 = \frac{(9 \times 10^9 \text{ N m}^2/\text{C}^2)(6 \times 10^{-10} \text{ C})}{0.11 \text{ m}} = 49.1 \text{ V}$$

$$V_2 = \frac{(9 \times 10^9 \text{ N m}^2/\text{C}^2)(-9 \times 10^{-10} \text{ C})}{0.03 \text{ m}} = -270 \text{ V}$$

$$V_{tot} = (49.1 + (-270)) \text{ V} = -220.9 \text{ V}$$

Now, calculate the potential difference between points P and M:

$$V_{PM} = V_P - V_M = (18 - (-220.9)) \text{ V} = 238.9 \text{ V}$$

Finally, use Eq. (17-1) to calculate the work required to move a positive charge from point M to point P:

$$W_{M \to P} = V_{PM} Q = (238.9 \text{ V})(3 \times 10^{-8} \text{ C}) = 7.17 \times 10^{-6} \text{ J}$$

PROBLEM 17-3 Three charges ($Q_1 = 6 \times 10^{-6}$ C, $Q_2 = -20 \times 10^{-6}$ C, $Q_3 = -3 \times 10^{-6}$ C) are arranged in a right triangle so that the 14.1 cm distance from Q_1 to Q_3 is the hypotenuse and Q_2 is equidistant from Q_1 and Q_3. What is the potential at the point equidistant from all three charges?

Solution The total electric potential is the sum of the potentials due to each charge.

$$V_1 = \frac{kQ_1}{r_1} = \frac{(9 \times 10^9 \text{ N m}^2/\text{C}^2)(6 \times 10^{-6} \text{ C})}{0.10 \text{ m}} = 5.4 \times 10^5 \text{ V}$$

$$V_2 = \frac{kQ_2}{r_2} = \frac{(9 \times 10^9 \text{ N m}^2/\text{C}^2)(-20 \times 10^{-6} \text{ C})}{0.10 \text{ m}} = -18.0 \times 10^5 \text{ V}$$

$$V_3 = \frac{kQ_3}{r_3} = \frac{(9 \times 10^9 \text{ N m}^2/\text{C}^2)(-3 \times 10^{-6} \text{ C})}{0.10 \text{ m}} = -2.7 \times 10^5 \text{ V}$$

$$V_{tot} = V_1 + V_2 + V_3 = (5.4 + (-18.0) + (-2.7)) \times 10^5 \text{ V} = -15.3 \times 10^5 \text{ V}$$

PROBLEM 17-4 How much work is required to bring a charge of 5×10^{-9} C from an infinite distance up to a distance of 2 cm from an isolated point charge of 8×10^{-8} C?

Solution Use Eq. (17-2) to calculate the potential at a distance of 2 cm from the point charge:

$$V = \frac{kQ}{r} = \frac{(9 \times 10^9 \text{ N m}^2/\text{C}^2)(8 \times 10^{-8} \text{ C})}{0.02 \text{ m}} = 3.6 \times 10^4 \text{ V}$$

Now, use Eq. (17-1) to determine the work:

$$W = VQ = (3.6 \times 10^4 \text{ V})(5 \times 10^{-9} \text{ C}) = 1.80 \times 10^{-4} \text{ J}$$

PROBLEM 17-5 A copper sphere with a radius of 4 cm has a uniform charge of 6×10^{-8} C. Determine the potential of an imaginary equipotential spherical surface with a radius of 6 cm and a center that coincides with that of the copper sphere.

Solution Since the copper sphere has a uniform charge distribution, you can regard the charge as being concentrated at its center. Calculate the potential on the imaginary spherical surface from Eq. (17-2):

$$V = \frac{kQ}{r} = \frac{(9 \times 10^9 \text{ N m}^2/\text{C}^2)(6 \times 10^{-8} \text{ C})}{0.06 \text{ m}} = 9.00 \times 10^3 \text{ V}$$

PROBLEM 17-6 The plates of a parallel plate capacitor each have an area of 0.4 m^2. The space between the plates is filled with a slab of wax 0.2 mm thick. The dielectric constant of the wax is 2.25. How much charge can be stored in this capacitor if the potential difference between the two conductors is 120 V?

Solution First, use Eq. (17-6) to calculate the capacitance:

$$C = \frac{KA}{4\pi kd} = \frac{(2.25)(0.4\ \text{m}^2)}{4\pi(9 \times 10^9\ \text{N m}^2/\text{C}^2)(0.2 \times 10^{-3}\ \text{m})} = 3.979 \times 10^{-8}\ \text{F} = 3.979 \times 10^{-2}\ \mu\text{F}$$

Now, use Eq. (17-3) to calculate the charge that is stored:

$$C = \frac{Q}{V} \qquad Q = CV = (3.979 \times 10^{-8}\ \text{F})(120\ \text{V}) = 4.77 \times 10^{-6}\ \text{C}$$

PROBLEM 17-7 A parallel plate capacitor with dry air between the two plates has a capacitance of 3.0 μF. The dielectric constant of dry air is 1.0006. When the space between the two flat conductors is filled with methyl alcohol, the capacitance becomes 99.3 μF. Determine the dielectric constant of methyl alcohol.

Solution The dielectric constant of air is so close to that of a vacuum that you can use Eq. (17-5):

$$K = \frac{C_{\text{med}}}{C_{\text{vac}}} \cong \frac{C_{\text{med}}}{C_{\text{air}}} \cong \frac{99.3\ \mu\text{F}}{3.0\ \mu\text{F}} = 33.1$$

PROBLEM 17-8 By what factor will the capacitance of a parallel plate capacitor increase if the separation of the two conductors is changed from 1.0 mm to 0.5 mm?

Solution Write the formula for the capacitance for these two situations:

$$C_1 = \frac{KA}{4\pi k\ (1.0\ \text{mm})} \qquad C_2 = \frac{KA}{4\pi k\ (0.5\ \text{mm})}$$

Now, divide the second equation by the first:

$$\frac{C_2}{C_1} = \frac{1.0\ \text{mm}}{0.5\ \text{mm}} = 2$$

The capacitance doubles when the separation of the plates is reduced by one-half.

PROBLEM 17-9 A parallel plate capacitor with air between its two conductors has a capacitance of 6 μF. The potential difference between the two plates is 800 V and the separation is 0.5 cm. Calculate the magnitude of the electric field between these two parallel conductors.

Solution Use Eq. (17-7) directly to calculate the magnitude of the electric field:

$$E = \frac{V}{d} = \frac{800\ \text{V}}{5 \times 10^{-3}\ \text{m}} = 1.60 \times 10^5\ \text{V/m}$$

PROBLEM 17-10 A parallel plate capacitor has an area of 400 cm^2. The distance between the two plates is 0.5 mm. When the potential difference between the two conductors is 600 V, the charge stored is 1.92 μC. Calculate the dielectric constant of the insulator between the two conductors.

Solution First, use Eq. (17-3) to calculate the capacitance of this parallel plate capacitor:

$$C = \frac{Q}{V} = \frac{1.92 \times 10^{-6}\ \text{C}}{600\ \text{V}} = 3.2 \times 10^{-9}\ \text{F}$$

Then, solve Eq. (17-6) for the unknown dielectric constant:

$$K = \frac{4\pi Ckd}{A} = \frac{4\pi(3.2 \times 10^{-9}\ \text{F})(9 \times 10^9\ \text{N m}^2/\text{C}^2)(0.5 \times 10^{-3}\ \text{m})}{(400\ \text{cm}^2)(1\ \text{m}^2/10^4\ \text{cm}^2)} = 4.52$$

PROBLEM 17-11 A rather large parallel plate capacitor has a capacitance of 4.0 μF. The separation of the two plates is 1.0 cm and the electric field between the two parallel conductors is 1.2 × 10^5 V/m. Calculate the amount of charge that is stored.

Solution Use Eq. (17-7) to find the potential difference between the two plates:

$$E = \frac{V}{d}$$

$$V = Ed = (1.2 \times 10^5 \text{ V/m})(1.0 \times 10^{-2} \text{ m}) = 1.2 \times 10^3 \text{ V}$$

Now, use Eq. (17-3) to find the charge:

$$Q = CV = (4.0 \times 10^{-6} \text{ F})(1.2 \times 10^3 \text{ V}) = 4.80 \times 10^{-3} \text{ C}$$

Supplementary Exercises

PROBLEM 17-12 An isolated point charge of 3×10^{-10} C produces an electric field and an electric potential. What is the magnitude of the electric field at a distance of 2 cm from this point charge? What is the electric potential 2 cm from this charge?

PROBLEM 17-13 Point M is 8 cm from a positive point charge of 5×10^{-10} C. Point N is 12 cm from the same isolated point charge. Points M, N, and the charge lie in a line, with M between N and the charge. How much work is required to move a charge of 4×10^{-11} C from point N to point M?

PROBLEM 17-14 A copper sphere contains a static charge of 2×10^{-10} C. What is the potential difference between two points 4 cm apart on the surface of this sphere?

PROBLEM 17-15 What is the total potential at point P in Problem 16-8, 10 cm from two equal positive charges of 4×10^{-10} C (see Figure 16-4)?

PROBLEM 17-16 How much work is required to bring a charge of 4×10^{-9} C from infinity to a distance of 10 cm from an isolated point charge of 2×10^{-10} C?

PROBLEM 17-17 Point M is 12 cm from an isolated point charge of 8×10^{-9} C. Point N is only 9 cm from the point charge. Points M and N and the charge all lie on a straight line. What is the potential difference between points N and M?

PROBLEM 17-18 A 20-μF capacitor is connected to a 45-V battery. How much charge is stored in this capacitor?

PROBLEM 17-19 The 20-μF capacitor described in Problem 17-18 consists of two parallel plates separated by a distance of 0.2 mm. What is the magnitude of the electric field between the two flat conductors?

PROBLEM 17-20 A parallel plate capacitor is made of two sheets of aluminum separated by a sheet of wax 1 mm thick. The dielectric constant of the wax is 2.25. What area of each aluminum sheet will yield a capacitance of 9.95×10^{-10} F?

Answers to Supplementary Exercises

17-12: $E = 6.75 \times 10^3$ N/C; $V = 135$ V

17-13: 7.50×10^{-10} J

17-14: 0.0 V

17-15: 72.0 V

17-16: 7.20×10^{-8} J

17-17: 200 V

17-18: 9.00×10^{-4} C

17-19: 2.25×10^5 V/m

17-20: 500 cm^2

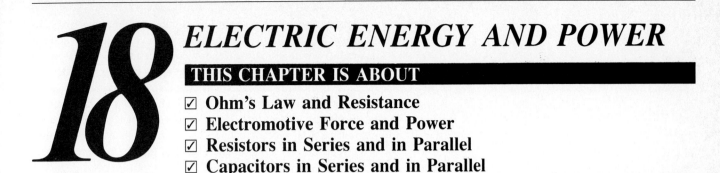

18 ELECTRIC ENERGY AND POWER

THIS CHAPTER IS ABOUT

☑ Ohm's Law and Resistance
☑ Electromotive Force and Power
☑ Resistors in Series and in Parallel
☑ Capacitors in Series and in Parallel

In Chapter 16, we introduced the idea of current, the flow of charge through a conductor. In the next several chapters, we'll extend this concept to **electrodynamics**, the study of moving charges, and **circuits**, networks of conductors and energy sources.

18-1. Ohm's Law and Resistance

A potential difference across a conductor produces an electric field that causes a current to flow. For most materials, the resulting current is proportional to the applied potential difference.

- **Ohm's law** states that the ratio of potential difference to current is constant for any given conductor—as long as the current is small enough that the conductor is not heated. The constant value is the conductor's **resistance**.

OHM'S LAW

$$R = \frac{V}{I} \tag{18-1}$$

- A **resistor** is a conductor that limits the flow of current, no matter in which direction the charges flow.

We use the symbol (⏦) to indicate a resistor in a drawing. The SI unit of resistance is the **ohm** Ω. A resistor has a resistance of one ohm if a potential difference of one volt maintains a constant current of one ampere: 1 Ω = 1 V/A.

EXAMPLE 18-1: A current of 50 mA flows through a piece of iron wire. The potential difference between the two ends of the iron wire is 2.0 V.
What is the resistance of the wire?

Solution: Directly from Eq. (18-1),

$$R = \frac{V}{I} = \frac{2.0\ \text{V}}{50 \times 10^{-3}\ \text{A}} = 40\ \Omega$$

EXAMPLE 18-2: The current for the wire of Example 18-1 is reduced to 20 mA. What is the potential difference between the two ends of the wire?

Solution: Use Eq. (18-1) to solve for the potential difference:

$$V = IR = (20 \times 10^{-3}\ \text{A})(40\ \Omega) = 0.8\ \text{V}$$

18.2. Electromotive Force and Power

A. Electromotive force

- A **seat** of **electromotive force** EMF is a device that *transforms* nonelectric energy into electric energy.

A continuous electric current will flow through a copper wire if the two ends of the wire are connected to a seat of electromotive force. Batteries, generators, and solar cells, which transform chemical, mechanical, and radiant energy, respectively, into electrical energy are examples of seats of EMF.

A seat of EMF does work on electrical charges. We measure the work done per unit charge by the electromotive force $\mathscr{E}$ in joules per coulomb, or volts. The flow of electric current through most materials generates heat, e.g., the heating elements on an electric range. These materials offer resistance to the flow of current, so that some of the electrical energy converts to heat energy.

Consider the conductor connected to a flashlight battery and a bulb, shown in Figure 18-1a. (The same circuit is shown in symbols in Figure 18-1b, where we use the symbol ⊣⊢ to represent the seat of EMF.) The current I that flows through the wire and the filament of the bulb depends on the EMF of the battery and the total resistance of the circuit:

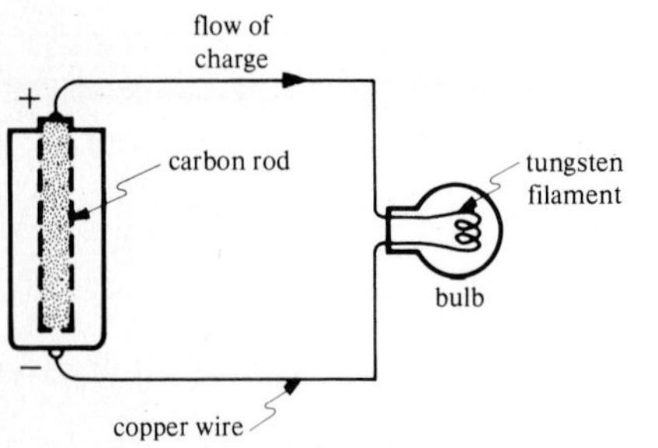

(a)

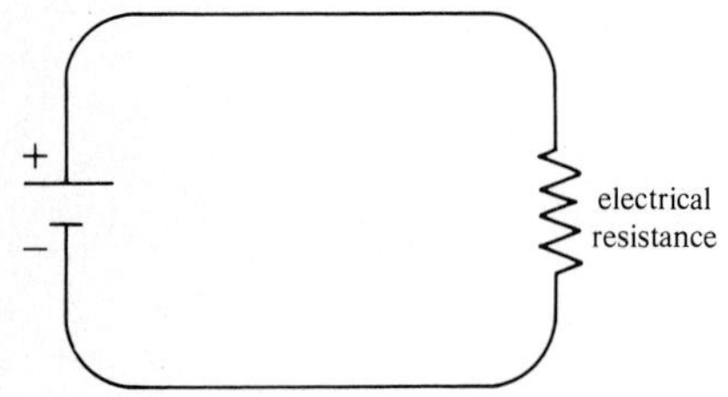

(b)

Figure 18-1

ELECTRIC CURRENT IN A CLOSED CIRCUIT

$$I = \frac{\mathscr{E}}{R_{\text{tot}}} \qquad \textbf{(18-2)}$$

The *direction* of the **conventional current** within the wire is the direction in which positive charge would have to flow in order to give the same electric effects as the flow of actual charge through the circuit, whether the actual moving charge is positive, negative, or both. This means that conventional current always flows from a point of higher potential to a point of lower potential.

You should always remember that when we say "circuit" we are describing a collection of conductors connected together in a special way. A circuit must have a conducting path that a charge can follow around a loop and back to its original position in the conductor, as in Figure 18-1. If there is no such path, as when we open a switch in a circuit, current no longer can flow and we say that the circuit is open.

In general, a seat of EMF has a small internal resistance. Figure 18-2 shows a battery with an internal resistance R_i connected to an external resistor. As a result of the internal resistance, the voltage across the terminals of the battery V_{AB} will be less than the EMF. The terminal voltage of a seat of EMF is

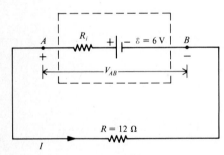

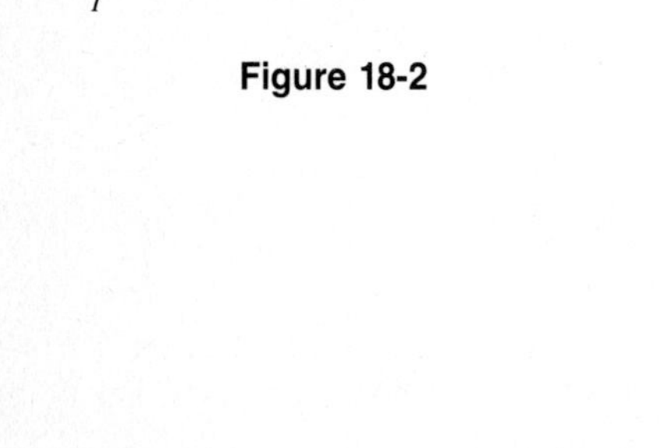

Figure 18-2

TERMINAL VOLTAGE OF A SEAT OF EMF

$$V_{AB} = \mathscr{E} - IR_i \qquad \textbf{(18-3)}$$

EXAMPLE 18-3: The voltage across the terminals of the battery diagrammed in Figure 18-2 is 5.76 V. Calculate the internal resistance.

Solution: First, use Ohm's law to calculate the current through the external resistor.

$$I = \frac{V}{R} = \frac{5.76\ \text{V}}{12\ \Omega} = 0.48\ \text{A}$$

Now, find the voltage drop across the internal resistance:

$$V_i = IR_i = \mathscr{E} - V = (6.0 - 5.76)\ \text{V} = 0.24\ \text{V}$$

And the internal resistance:

$$R_i = \frac{V_i}{I} = \frac{0.24\ \text{V}}{0.48\ \text{A}} = 0.5\ \Omega$$

B. Electric power

- A **resistor** is a conductor in which electric energy is *dissipated,* no matter in which direction the charges flow.

The total resistance of the circuit illustrated in Figure 18-1 is the resistance of the lamp filament plus the resistance of the wire and the internal resistance of the battery. The resistance of the copper wire and the battery is very small compared to the resistance of the tungsten filament. The flow of electrons through the filament causes it to become very hot, which means that the electrical energy associated with the current in the filament is transformed into heat energy, making the filament glow. The **electric power** dissipated in a resistor is the rate at which electric energy is transformed into heat:

POWER DISSIPATED IN A RESISTOR

$$P = IV \tag{18-4}$$

$$P = I^2R \tag{18-5}$$

$$P = \frac{V^2}{R} \tag{18-6}$$

EXAMPLE 18-4: The lamp filament in Figure 18-1 has a resistance of 12.5 Ω, and the electromotive force of the flashlight cell is $\mathscr{E} = 1.5$ V. The resistance of the copper wire and the battery is small enough to ignore.
What current flows through this circuit?

Solution: Directly from Eq. (18-2),

$$I = \frac{\mathscr{E}}{R_{tot}} = \frac{1.5\text{ V}}{12.5\ \Omega} = 0.12\text{ A} = 120\text{ mA}$$

EXAMPLE 18-5: Calculate the rate at which electric energy in the lamp filament of Example 18-1 is transformed into heat energy.

Solution: Directly from Eq. (18-5),

$$P = I^2R = (0.12\text{ A})^2(12.5\ \Omega) = 0.18\text{ W}$$

[**Recall**: The SI unit of power is the watt W, which is equal to one joule per second. (Sec. 6-5)]

18-3. Resistors in Series and in Parallel

A. Resistors connected in series

A closed circuit containing a battery and three resistors connected in series—i.e., the current flows through the resistors one after the other with no shortcuts—is shown in Figure 18-3. The *current* flowing through the battery and each resistor is exactly the same. We determine the total resistance of a circuit containing multiple **resistors in series** by adding the *potential difference* across each resistor and dividing by the constant current.

RESISTORS CONNECTED IN SERIES

$$R_{tot} = R_1 + R_2 + R_3 + \cdots \tag{18-7}$$

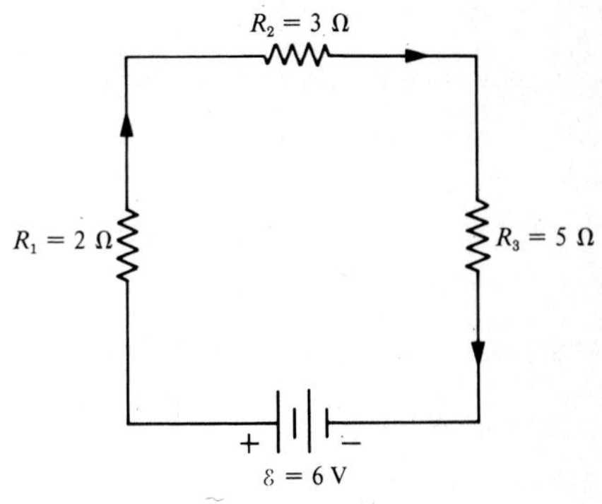

Figure 18-3

EXAMPLE 18-6: Calculate the current in the circuit illustrated in Figure 18-3.

Solution: Directly from Eq. (18-7),

$$R_{tot} = R_1 + R_2 + R_3 = (2 + 3 + 5)\ \Omega = 10\ \Omega$$

Now, calculate the current from Eq. (18-1):

$$I = \frac{\mathscr{E}}{R_{tot}} = \frac{6\text{ V}}{10\ \Omega} = 0.6\text{ A}$$

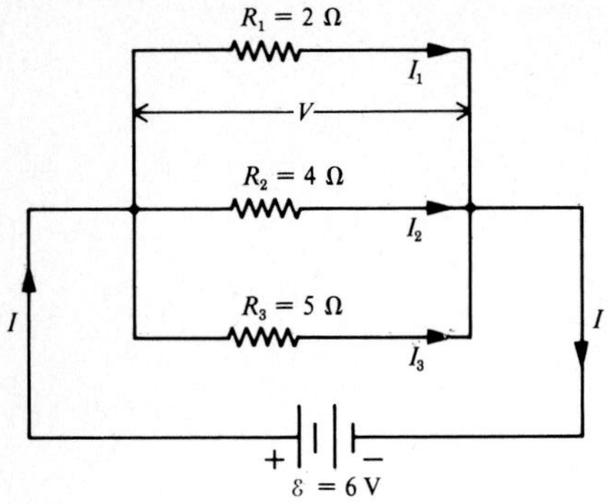

Figure 18-4

B. Resistors connected in parallel

A closed circuit containing a battery and three resistors connected in parallel—i.e., the incoming current divides, some going through each resistor—is shown in Figure 18-4. The *potential difference* across each resistor is exactly the same. The current flowing through the battery is the sum of the current in each of the resistors: $I = I_1 + I_2 + I_3$. We determine the total resistance of a circuit containing multiple **resistors in parallel** as follows:

RESISTORS CONNECTED IN PARALLEL

$$\frac{1}{R_{tot}} = \frac{1}{R_1} + \frac{1}{R_2} + \frac{1}{R_3} + \cdots \qquad \textbf{(18-8)}$$

EXAMPLE 18-7: Calculate the current that flows through the battery in the circuit illustrated in Figure 18-4.

Solution: Directly from Eq. (18-8),

$$\frac{1}{R_{tot}} = \frac{1}{R_1} + \frac{1}{R_2} + \frac{1}{R_3} = \frac{1}{2\ \Omega} + \frac{1}{4\ \Omega} + \frac{1}{5\ \Omega} = 0.95\ \Omega^{-1}$$

$$R_{tot} = 1.053\ \Omega$$

Now, you can calculate the current from Eq. (18-1):

$$I = \frac{6\text{V}}{1.053\ \Omega} = 5.70\ \text{A}$$

EXAMPLE 18-8: Calculate the current that flows through the 4-Ω resistor in Figure 18-4.

Solution: The three resistors are connected in parallel, so the potential difference V across each resistor is 6.0 V. From Eq. (18-1),

$$I_2 = \frac{6.0\ \text{V}}{4\ \Omega} = 1.5\ \text{A}$$

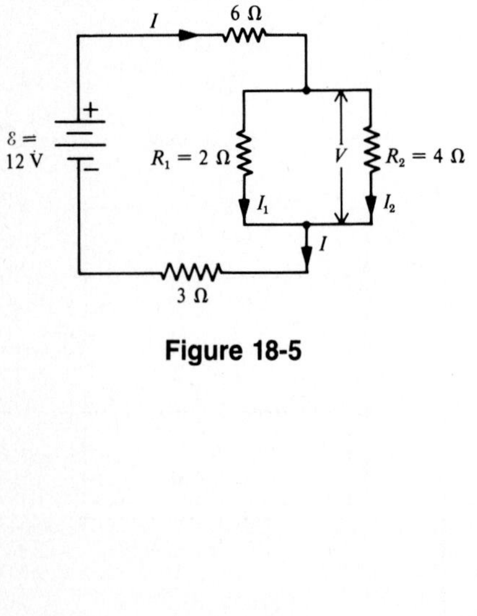

Figure 18-5

EXAMPLE 18-9: What current flows through the 4-Ω resistor in Figure 18-5?

Solution: This circuit has resistors both in parallel and in series. You must find the effective resistance of the two resistors in parallel *first, then* add the other resistances. You can not combine the 6-Ω resistor with R_1 first because the current flowing through them is not the same. The current in the 6-Ω resistor is split between R_1 and R_2. First, the effective resistance of the two resistors in parallel from Eq. (18-8):

$$\frac{1}{R_{eff}} = \frac{1}{2\ \Omega} + \frac{1}{4\ \Omega} = \frac{3}{4\ \Omega} \qquad R_{eff} = 1.333\ \Omega$$

Now, find the total resistance of the circuit containing resistors in series from Eq. (18-1):

$$R_{tot} = (6 + 3 + 1.333)\ \Omega = 10.333\ \Omega$$

Then, from Eq. (18-1), calculate the current in the circuit:

$$I = \frac{12\ \Omega}{10.333\ \Omega} = 1.161\ \text{A}$$

And, you can use Ohm's law (Eq. 18-5) to calculate the voltage across the two resistors connected in parallel:

$$R = \frac{V}{I} \qquad V = R_{eff}I = (1.333\ \Omega)(1.161\ \text{A}) = 1.548\ \text{V}$$

Finally, from Eq. (18-1), find the current in the 4-Ω resistors:

$$I_2 = \frac{1.548\ \text{V}}{4\ \Omega} = 0.387\ \text{A}$$

EXAMPLE 18-10: What current flows through the 2-Ω resistor in Figure 18-5?

Solution: Since the total current in the circuit is 1.161 A (from Example 18-9), you can subtract to find the current in the 2-Ω resistor:

$$I_1 = I - I_2 = (1.161 - 0.387)\ \text{A} = 0.774\ \text{A}$$

You can verify your result by using Eq. (18-1):

$$I_1 = \frac{V}{R_1} = \frac{1.548\ \text{V}}{2\ \Omega} = 0.774\ \text{A}$$

18-4. Capacitors in Series and in Parallel

A. Capacitors connected in series

A closed circuit containing a battery and three capacitors connected in series is shown in Figure 18-6a. No current is flowing in this circuit. The potential difference across the circuit is the sum of the potential differences across the capacitors. The effective capacitance in a circuit containing multiple **capacitors in series** is determined as follows:

CAPACITORS CONNECTED IN SERIES

$$\frac{1}{C_{\text{eff}}} = \frac{1}{C_1} + \frac{1}{C_2} + \frac{1}{C_3} + \cdots \qquad \textbf{(18-9)}$$

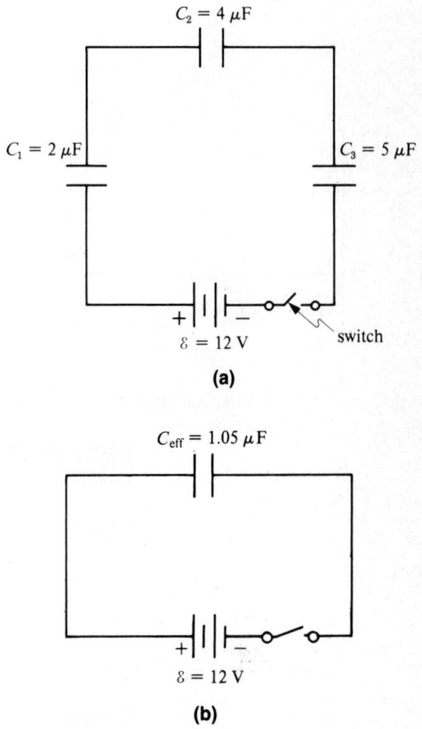

Figure 18-6

EXAMPLE 18-11: When the switch in Figure 18-6a is closed, the electromotive force of the battery causes charge to flow until the potential difference in the capacitors equals the electromotive force of the battery, divided by the total capacitance. Calculate the total charge on each capacitor when the switch is closed.

Solution: First, find the effective capacitance of the circuit directly from Eq. (18-9):

$$\frac{1}{C_{\text{eff}}} = \frac{1}{C_1} + \frac{1}{C_2} + \frac{1}{C_3} = \frac{1}{2\ \mu\text{F}} + \frac{1}{4\ \mu\text{F}} + \frac{1}{5\ \mu\text{F}} = 0.95\ \mu\text{F}^{-1}$$

$$C_{\text{eff}} = 1.053\ \mu\text{F}$$

When you substitute the effective capacitance for the individual capacitances, you can think of the circuit as being equivalent to the circuit shown in Figure 18-6b. Therefore, you can use Eq. (17-3) to calculate the charge:

$$C_{\text{eff}} = \frac{Q}{V} \qquad Q = C_{\text{eff}} V = (1.053\ \mu\text{F})(12\ \text{V}) = 12.6\ \mu\text{C}$$

This is equal to the charge on each individual capacitor.

B. Capacitors connected in parallel

A closed circuit containing a battery and three capacitors connected in parallel is shown in Figure 18-7. No current is flowing in this circuit. When the switch is closed, charge flows as in Example 18-10 and is stored in the capacitors. Each capacitor receives a different quantity of charge depending on the area and separation of the plates and on the dielectric material between the plates. The potential difference across each capacitor is the same. The effective capacitance in a circuit containing multiple **capacitors in parallel** is the sum of each individual *capacitance* (see Sec. 17-2):

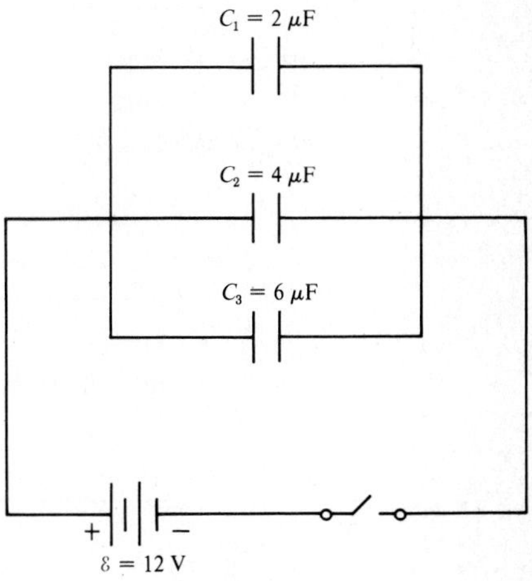

Figure 18-7

CAPACITORS CONNECTED IN PARALLEL

$$C_{\text{eff}} = C_1 + C_2 + C_3 + \cdots \qquad \textbf{(18-10)}$$

EXAMPLE 18-12: Calculate the total charge on the capacitors in Figure 18-7 when the switch is closed.

Solution: Find the effective capacitance of the circuit directly from Eq. (18-10):

$$C_{\text{eff}} = C_1 + C_2 + C_3 = (2 + 4 + 6)\ \mu\text{F} = 12\ \mu\text{F}$$

And Eq. (17-3):

$$Q = CV = (12\ \mu\text{F})(12\ \text{V}) = 144\ \mu\text{C}$$

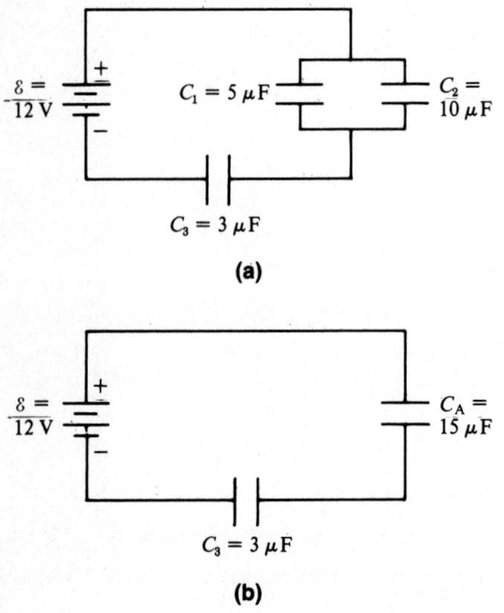

Figure 18-8

EXAMPLE 18-12: What is the effective capacitance of the circuit in Figure 18-8a?

Solution: This circuit has capacitors both in parallel and in series. Because the potential across C_1 and C_2 is the same, they can be added directly (Eq. 18-10). (You can not combine C_1 and C_3 first because the charges on them are not the same.)

$$C_A = C_1 + C_2 = (5 + 10)\ \mu\text{F} = 15\ \mu\text{F}$$

The circuit is thus equivalent to two capacitors in series (Fig. 18-8b), which have the same charges and can be added by Eq. (18-9).

$$\frac{1}{C_{\text{eff}}} = \frac{1}{C_A} + \frac{1}{C_3} = \frac{1}{15\ \mu\text{F}} + \frac{1}{3\ \mu\text{F}} = \frac{6}{15\ \mu\text{F}}$$

$$C_{\text{eff}} = \frac{15\ \mu\text{F}}{6} = 2.5\ \mu\text{F}$$

SUMMARY

Ohm's law	$R = \dfrac{V}{I}$	the ratio of potential difference to current is constant for any given resistor—if the current is so small that the resistor is not heated
Electric current in a closed circuit	$I = \dfrac{\mathscr{E}}{R_{\text{tot}}}$	the current that flows through a circuit depends on the electromotive force of the battery and the total resistance of the circuit
Terminal voltage of a seat of EMF	$V = \mathscr{E} - IR_i$	the terminal voltage of a seat of electromotive force is equal to the electromotive force minus the current times the internal resistance
Power dissipated in a resistor	$P = IV$ $P = I^2R$ $P = \dfrac{V^2}{R}$	electric power dissipated in a resistor is the rate at which electric energy is transformed into heat

Resistors connected in series	$R_{tot} = R_1 + R_2 + R_3 + \cdots$	the total resistance of a circuit containing multiple resistors in series is the sum of the resistances of each resistor
Resistors connected in parallel	$\frac{1}{R_{tot}} = \frac{1}{R_1} + \frac{1}{R_2} + \frac{1}{R_3} + \cdots$	the total resistance of a circuit containing multiple resistors in parallel is the reciprocal of the sum of the reciprocals of the resistance in each resistor
Capacitors connected in series	$\frac{1}{C_{eff}} = \frac{1}{C_1} + \frac{1}{C_2} + \frac{1}{C_3} + \cdots$	the effective capacitance in a circuit containing multiple capacitors connected in series is the reciprocal of the sum of the reciprocals of the capacitances of the individual capacitors
Capacitors connected in parallel	$C_{eff} = C_1 + C_2 + C_3 + \cdots$	the effective capacitance in a circuit containing multiple capacitors connected in parallel is the sum of each individual capacitance

RAISE YOUR GRADES

Can you define . . . ?

☑ electrodynamics
☑ circuits
☑ a seat of electromotive force
☑ a resistor

Can you . . . ?

☑ write the equation which relates the current in a circuit to the electromotive force and the total resistance
☑ write the equation which allows electric power to be calculated from the voltage across a resistor and the current through it
☑ calculate the rate at which electric energy in a resistor is transformed into heat energy
☑ state Ohm's Law
☑ calculate the total resistance of two or more resistors connected in series
☑ calculate the total resistance of two or more resistors connected in parallel
☑ find the current through a resistor connected in parallel with another resistor
☑ calculate the total EMF of a circuit that contains two seats of EMF in series
☑ write the equation for the effective capacitance of two or more capacitors connected in series
☑ write the equation for the effective capacitance of two or more capacitors connected in parallel

SOLVED PROBLEMS

PROBLEM 18-1 The circuit shown in Fig. 18-9 contains two batteries and three resistors. What current flows through the 6-Ω resistor?

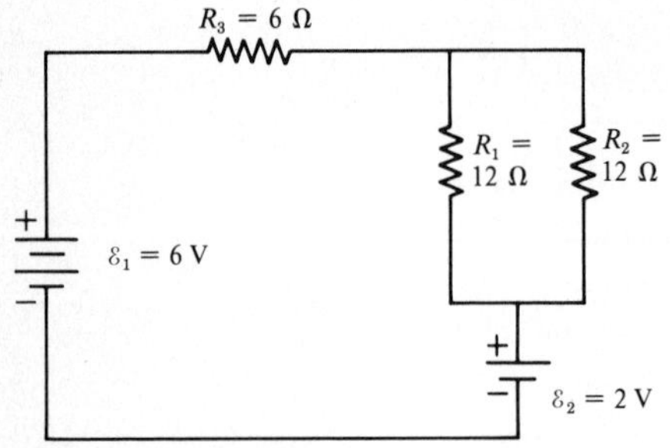

Figure 18-9

Solution Since the two batteries are connected in opposite directions, the total EMF of the circuit is the *difference* between the two EMFs. (EMFs in series are simply added.)

$$\mathscr{E}_{\text{tot}} = \mathscr{E}_1 - \mathscr{E}_2 = (6.0 - 2.0)\ \text{V} = 4.0\ \text{V}$$

Use Eq. (18-8) to calculate the effective resistance of the two resistors connected in parallel:

$$\frac{1}{R_{\text{eff}}} = \frac{1}{R_1} + \frac{1}{R_2} = \frac{1}{12\ \Omega} + \frac{1}{12\ \Omega} = \frac{2}{12\ \Omega} \qquad R_{\text{eff}} = 6\ \Omega$$

Since the 6-Ω resistor is connected in series with the two resistors in parallel, you can use Eq. (18-7) to find the total resistance of the circuit:

$$R_{\text{tot}} = R_{\text{eff}} + R_3 = (6 + 6)\ \Omega = 12\ \Omega$$

Now, use Eq. (18-2) to calculate the current:

$$I = \frac{\mathscr{E}_{\text{tot}}}{R_{\text{tot}}} = \frac{4.0\ \text{V}}{12\ \Omega} = 0.333\ \text{A}$$

PROBLEM 18-2 Fig. 18-10 shows a circuit which contains a capacitor and three resistors. Find the charge stored in the capacitor.

Solution No current flows through the capacitor. You must find the current in the circuit, however, to calculate the voltage across the capacitor. You'll have to calculate the total resistance of the two resistors connected in parallel. Use Eq. (18-8):

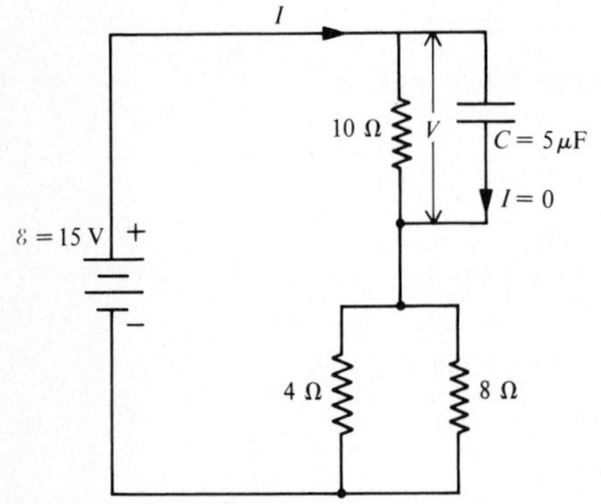

Figure 18-10

$$\frac{1}{R_{\text{eff}}} = \frac{1}{4\ \Omega} + \frac{1}{8\ \Omega} = \frac{3}{8\ \Omega} \qquad R_{\text{eff}} = 2.667\ \Omega$$

Then, find the total resistance in the circuit:

$$R_{\text{tot}} = (10 + 2.667)\ \Omega = 12.67\ \Omega$$

From Eq. (18-2),

$$I = \frac{15\ \text{V}}{12.67\ \Omega} = 1.184\ \text{A}$$

The potential difference between the two plates of the capacitor is equal to the voltage across the 10-Ω resistor. Use Eq. (18-1) to find this voltage:

$$R = \frac{V}{I} \qquad V = IR = (1.184\ \text{A})(10\ \Omega) = 11.84\ \text{V}$$

Now, use Eq. (17-3) to find the charge on the capacitor:

$$C = \frac{Q}{V} \qquad Q = CV = (5\ \mu\text{F})(11.84\ \text{V}) = 59.2\ \mu\text{C}$$

PROBLEM 18-3 At what rate is chemical energy in the seat of EMF in Fig. 18-10 being transformed into electrical energy?

Solution You can calculate the rate of transformation by using Eq. (18-4), but you'll need to replace V by the electromotive force $\mathscr{E}$ of the battery:

$$P = \mathscr{E}I = (15\ \text{V})(1.184\ \text{A}) = 17.8\ \text{W}$$

PROBLEM 18-4 Calculate the rate at which electrical energy is being transformed into heat energy in the two resistors in parallel in Fig. 18-10.

Solution You can find the rate at which energy is dissipated by using Eq. (18-5). Use the resistance of the two resistors in parallel that you found in Problem 18-2:

$$P = I^2R_{\text{tot}} = (1.184\ \text{A})^2(2.667\ \Omega) = 3.74\ \text{W}$$

PROBLEM 18-5 Electrical energy is transformed into heat energy in the 10-Ω resistor in Fig. 18-10. Show that the total rate at which electrical energy is transformed into heat energy is equal to the rate at which chemical energy in the battery is transformed into electrical energy.

Solution Use Eq. (18-5) to calculate the power within the 10-Ω resistor:

$$P = (1.184\ \text{A})^2(10\ \Omega) = 14.02\ \text{W}$$

The total rate at which electrical energy is transformed into heat energy is

$$P_{\text{tot}} = (3.74 + 14.02)\ \text{W} = 17.8\ \text{W}$$

This is equal to the rate at which chemical energy is being transformed into electrical energy within the battery (see Problem 18-3).

PROBLEM 18-6 Fig. 18-11 shows an electrical circuit that contains one capacitor and five resistors. Calculate the charge stored in the capacitor.

Solution There is no current flowing across the capacitor, so the voltage drop across the 5-Ω resistor is zero. The potential across the capacitor V is equal to that across the 10-Ω resistor plus the potential across the two 40-Ω resistors. Calculate the total resistance in the circuit:

$$\frac{1}{R_{\text{eff}}} = \frac{1}{40\ \Omega} + \frac{1}{40\ \Omega} = \frac{2}{40\ \Omega} \qquad R_{\text{eff}} = 20\ \Omega$$

$$R_{\text{tot}} = (10 + 10 + 20)\ \Omega = 40\ \Omega$$

Find the current:

$$I = \frac{20\ \text{V}}{40\ \Omega} = 0.5\ \text{A}$$

Now, calculate the voltage between points A and B:

$$V = IR = (0.5\ \text{A})(10 + 20)\ \Omega = 15\ \text{V}$$

Then, from Eq. (17-3), calculate the charge stored in the capacitor:

$$Q = CV = (5\ \mu\text{F})(15\ \text{V}) = 75.0\ \mu\text{C}$$

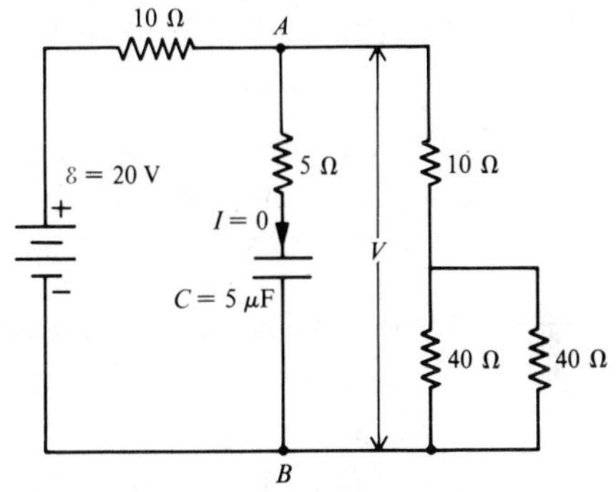

Figure 18-11

PROBLEM 18-7 A 12-V automobile battery has an internal resistance of 0.7 Ω. What is the terminal voltage when the battery produces a current of 6 A?

Solution Use Eq. (18-3):

$$V = \mathcal{E} - IR_i = 12.0\ \text{V} - (6\ \text{A})(0.7\ \Omega) = 7.8\ \text{V}$$

PROBLEM 18-8 Two capacitors connected in series with a battery have capacitances of 2 μF and 6 μF. The potential difference between the plates of the 6 μF capacitor is 12 volts. What is the EMF of the battery?

Solution Because the two capacitors are connected in series, they carry the same charge. First, calculate the charge on the 6-μF capacitor from Eq. (17-3):

$$Q = CV = (6\ \mu\text{F})(12\ \text{V}) = 72\ \mu\text{C}$$

Now, calculate the potential difference between the plates of the 2-μF capacitor:

$$V = \frac{Q}{C} = \frac{72\ \mu\text{C}}{2\ \mu\text{F}} = 36\ \text{V}$$

Since these two capacitors are connected in series, the total voltage across the two capacitors is the sum of the voltage across each one. This total voltage equals the EMF of the battery:

$$\mathscr{E} = V_1 + V_2 = (12 + 36)\text{ V} = 48\text{ V}$$

PROBLEM 18-9 Fig. 18-12 shows a circuit that contains a 12-V battery, two resistors, and two capacitors. What is the potential difference between the plates of the 6-μF capacitor?

Solution The current flows through the two resistors connected in series. Calculate the total resistance and the current:

$$R_{\text{tot}} = (15 + 5)\ \Omega = 20\ \Omega$$

$$I = \frac{12\text{ V}}{20\ \Omega} = 0.6\text{ A}$$

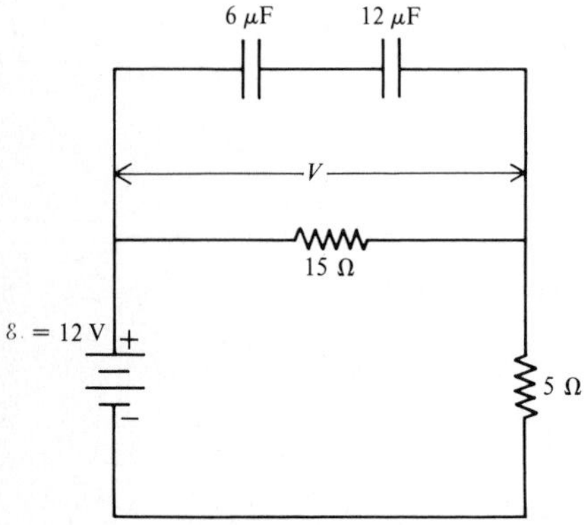

Figure 18-12

Now, find the voltage across the two capacitors:

$$V = (0.6\text{ A})(15\ \Omega) = 9.0\text{ V}$$

From Eq. (18-9), calculate the effective capacitance of the two capacitors connected in series:

$$\frac{1}{C_{\text{eff}}} = \frac{1}{C_1} + \frac{1}{C_2} = \frac{1}{6\ \mu\text{F}} + \frac{1}{12\ \mu\text{F}} = \frac{3}{12\ \mu\text{F}}$$

$$C_{\text{eff}} = 4\ \mu\text{F}$$

The charge on the 4-μF capacitor is the same as the charge on the two capacitors in series:

$$Q = C_{\text{eff}}V = (4\ \mu\text{F})(9\text{ V}) = 36\ \mu\text{C}$$

Finally, find the voltage across the 6-μF capacitor:

$$V_1 = \frac{Q_1}{C_1} = \frac{36\ \mu\text{C}}{6\ \mu\text{F}} = 6.00\text{ V}$$

Supplementary Exercises

PROBLEM 18-10 What current flows through the 2-Ω resistor in Fig. 18-4?

PROBLEM 18-11 What is the rate of transformation of chemical energy into electrical energy in the battery in Fig. 18-5? You found the current through the battery in Example 18-9.

PROBLEM 18-12 What is the potential difference across the plates of the 4-μF capacitor in Fig. 18-6? The charge stored in each of the three capacitors in series is 12.64 μC.

PROBLEM 18-13 What is the charge stored in the 6-μF capacitor in Fig. 18-7?

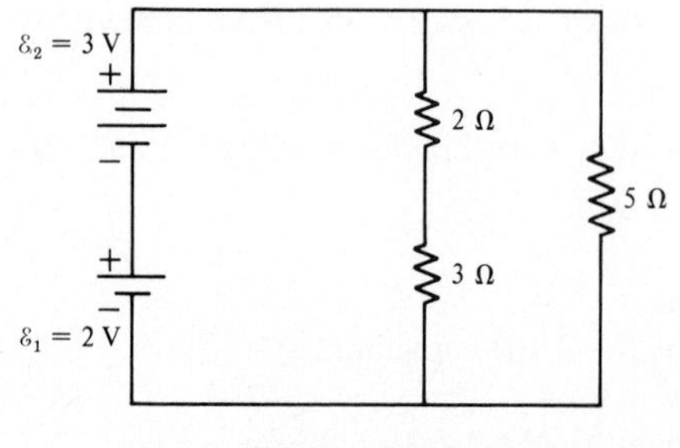

Figure 18-13

PROBLEM 18-14 What is the total EMF in the circuit in Figure 18-13?

PROBLEM 18-15 What is the total resistance of the circuit in Figure 18-13?

PROBLEM 18-16 What is the current in the 5-Ω resistor in Figure 18-13?

PROBLEM 18-17 What is the electric power dissipated in the 5-Ω resistor in Figure 18-13?

PROBLEM 18-18 Calculate the rate at which electric energy is transformed into heat energy in the 3-Ω resistor in Fig. 18-8.

PROBLEM 18-19 The voltage across the 10-Ω resistor in Fig. 18-10 is 11.84 V. What current flows through the 4-Ω resistor?

PROBLEM 18-20 Find the current that flows through the 8-Ω resistor in Fig. 18-10.

PROBLEM 18-21 A battery having an EMF of 12.0 V is connected to an 80-Ω resistor. The voltage measured across the terminals of the battery is 11.707 V. What is the internal resistance of this battery?

Answers to Supplementary Exercises

18-10: 3.00 A

18-11: 13.9 W

18-12: 3.16 V

18-13: 72.0 μC

18-14: 5.00 V

18-15: 2.50 Ω

18-16: 1.00 A

18-17: 5.00 W

18-18: 3.00 W

18-19: 0.790 A

18-20: 0.395 A

18-21: 2.00 Ω

19 DIRECT CURRENT CIRCUITS

THIS CHAPTER IS ABOUT

- ☑ Kirchhoff's Laws
- ☑ Resistivity
- ☑ The Ammeter and Voltmeter
- ☑ The Wheatstone Bridge
- ☑ The Potentiometer

19-1. Kirchhoff's Laws

- In a **direct current (DC) circuit,** the current flows in a constant direction with a magnitude that does not vary with time.

For DC circuits, the basic conservation laws of charge and energy are often expressed as **Kirchhoff's laws:**

KIRCHHOFF'S FIRST LAW (BRANCH THEOREM) The algebraic sum of the currents into and out of any junction point in a circuit is zero. (Currents entering are considered positive; currents leaving are considered negative).

$$I_1 + I_2 + I_3 + \cdots = 0 \qquad \textbf{(19-1)}$$

KIRCHHOFF'S SECOND LAW (LOOP THEOREM) The algebraic sum of the changes in potential (voltage increases and decreases) around any closed path is zero. (The flow of current through a resistor is always from a higher to a lower potential.)

$$\mathscr{E}_1 + \mathscr{E}_2 + \cdots + V_1 + V_2 + \cdots = 0 \qquad \textbf{(19-2)}$$

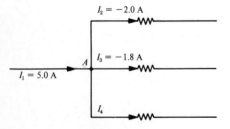

Figure 19-1

EXAMPLE 19-1: Figure 19-1 shows a junction point in a circuit. The current I_1 entering junction point A is 5.0 A. Two of the three currents leaving the junction are shown.
Find the unknown current I_4.

Solution: Kirchhoff's first law (Eq. 19-1), written for junction point A, is

$$I_1 + I_2 + I_3 + I_4 = 0$$

I_2 and I_3 are negative because they are leaving the junction.

$$(5.0 - 2.0 - 1.8)\text{ A} + I_4 = 0$$

$$I_4 = (-5.0 + 2.0 + 1.8)\text{ A} = -1.2\text{ A}$$

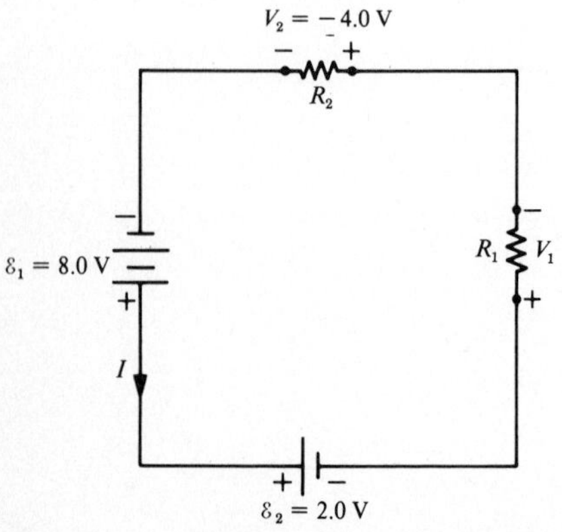

Figure 19-2

EXAMPLE 19-2: Figure 19-2 shows a closed circuit that contains two seats of EMF and two resistors.
Find the **voltage drop** (change in potential) across resistor R_1.

Solution: Use Kirchhoff's second law (Eq. 19-2). Notice that the two seats of EMF are connected opposite each other, so the value of one is negative and the value of the other is positive. Choose the one with the larger value as positive so the net EMF for the entire circuit will be positive. The moving charges lose energy as they go through the resistors, so V_1 and V_2 are negative.

$$\mathscr{E}_1 + \mathscr{E}_2 + V_1 + V_2 = 0$$

$$(8.0 - 2.0 - 4.0)\text{ V} + V_1 = 0$$

$$V_1 = (-8.0 + 2.0 + 4.0)\text{ V} = -2.0\text{ V}$$

19-2. Resistivity

The resistance of a conductor depends on the material it is made of, its shape, and its size. If we want to describe the material only, we use the concept of *resistivity*.

- **Resistivity** ρ is a measure of how easily electric current flows through a substance.

The resistance of a resistor is proportional to its resistivity and its length, and inversely proportional to the cross-sectional area:

RELATIONSHIP BETWEEN RESISTIVITY AND RESISTANCE

$$R = \rho\frac{L}{A} \qquad \textbf{(19-3)}$$

Table 19-1 lists the resistivity, in units of Ω m (ohm-meters), of several conductors and insulators. Note that the resistivity of good conductors like aluminum is low and that of insulators like wood is high. We also sometimes speak of a material's **conductivity** σ, defined as $\sigma = 1/\rho$. Thus aluminum has a high conductivity; wood, a low one.

TABLE 19-1: The Resistivity of Various Substances

Substance	ρ (Ω m)
aluminum	2.60×10^{-8}
carbon	4×10^{-5}
Celluloid™	4×10^{12}
copper	1.59×10^{-8}
fused quartz	5×10^{17}
gold	2.27×10^{-8}
iron	1.1×10^{-7}
Nichrome™	1.0×10^{-6}
silicon	3×10^{4}
silver	1.47×10^{-8}
tungsten	5.0×10^{-8}
wood (maple)	3×10^{8}

EXAMPLE 19-3: A copper wire has a diameter of 0.3 mm. How long is a piece of this wire if it has a resistance of 4.0 Ω? [**recall**: The wire is cylindrical, so $A = 1/4\pi d^2$. (Example 9-2)]

Solution: Use Eq. (19-3) to find the unknown length:

$$R = \frac{\rho L}{A} \qquad L = \frac{RA}{\rho} = \frac{(4.0\ \Omega)(\pi/4)(0.3 \times 10^{-3}\text{ m})^2}{1.59 \times 10^{-8}\ \Omega\text{ m}} = 17.8\text{ m}$$

EXAMPLE 19-4: Two pieces of iron wire have the same length. Wire R_1 has a diameter of 0.5 mm and wire R_2 has a diameter of 0.2 mm. Calculate the ratio of the resistance of R_1 to R_2.

Solution: Use Eq. (19-3) to find the resistance of each wire:

$$R_1 = \frac{\rho L}{(\pi/4)\,d_1^2} \qquad R_2 = \frac{\rho L}{(\pi/4)\,d_2^2}$$

Now, divide the first equation by the second:

$$\frac{R_1}{R_2} = \left(\frac{d_2}{d_1}\right)^2 = \left(\frac{0.2\text{ mm}}{0.5\text{ mm}}\right)^2 = 0.160$$

EXAMPLE 19-5: The tungsten filament of a flashlight bulb has a length of 1.0 cm and a resistance of 20 Ω. Find the diameter of this wire.

Solution: Solve Eq. (19-3) for the cross-sectional area A, and then solve for the diameter:

$$A = \frac{\rho L}{R} \qquad \frac{\pi d^2}{4} = \frac{\rho L}{R}$$

$$d = \sqrt{\frac{4\rho L}{\pi R}} = \sqrt{\frac{4(5.0 \times 10^{-8}\ \Omega\text{ m})(1.0 \times 10^{-2}\text{ m})}{\pi(20\ \Omega)}} = 5.64 \times 10^{-6}\text{ m}$$

$$= 5.64 \times 10^{-3}\text{ mm}$$

19-3. The Ammeter and Voltmeter

- An **ammeter** measures the current in a circuit.

A high-quality ammeter has a very low resistance and is connected *in series* so that the current flows through the ammeter. There is only a very small potential difference across an ammeter. Incorporated in an ammeter are a sensitive **galvanometer**, a current-indicating device, and a low-resistance **shunt** connected in parallel. The shunt provides a bypass that carries most of the current, making the ammeter less sensitive, allowing larger currents to be measured.

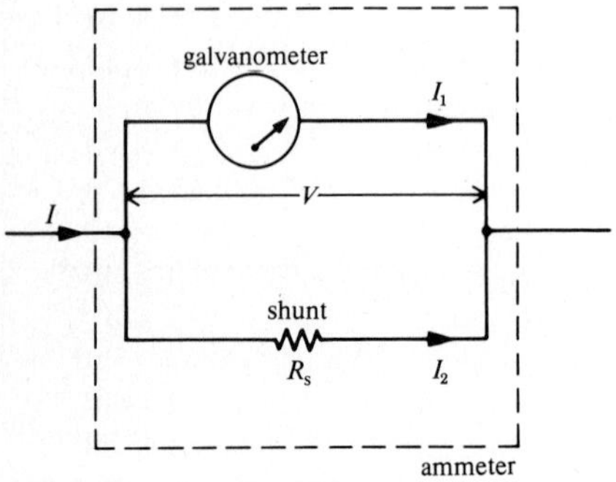

Figure 19-3

EXAMPLE 19-6: The ammeter shown in Figure 19-3 incorporates a galvanometer with an internal resistance R_G of 800 Ω. A full-scale deflection of the dial occurs when 50 μA pass through it.
Calculate the resistance of the shunt connected in parallel so that an input current of 5.0 A will cause a full-scale deflection of the dial.

Solution: First, calculate the voltage drop across the galvanometer from Eq. (18-1):

$$R = \frac{V}{I}$$

$$V = I_1 R_G = (50 \times 10^{-6}\ \text{A})(800\ \Omega) = 0.04\ \text{V}$$

Next, find the current passing through the shunt (Eq. 19-1):

$$I_2 = I - I_1 = (5.0 - 50 \times 10^{-6})\ \text{A} \cong 5.00\ \text{A}$$

Now, use Eq. (18-1) to find the resistance of the shunt:

$$R_S = \frac{V}{I_2} = \frac{0.04\ \text{V}}{5.00\ \text{A}} = 0.008\ \Omega$$

- A **voltmeter** measures the potential difference between two points in a circuit.

A high-quality voltmeter has a very high resistance that does not alter the voltage being measured very much. Very little current flows through a good voltmeter. The voltmeter is connected *in parallel* across a potential difference (resistor or seat of EMF). A sensitive galvanometer and a large resistance connected in series are incorporated in a voltmeter.

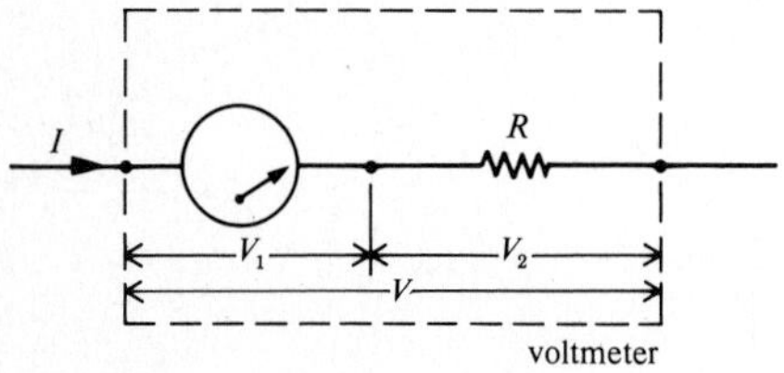

Figure 19-4

EXAMPLE 19-7: The voltmeter shown in Figure 19-4 incorporates a galvanometer with an internal resistance R_G of 800 Ω and a full-scale dial deflection at 50 μA.
Calculate the value of the series resistor so that a full-scale dial deflection occurs when $V = 50$ V.

Solution: Simply divide the voltage by the current and subtract the resistance of the galvanometer:

$$R = \frac{V}{I} - R_G = \frac{50\ \text{V}}{50 \times 10^{-6}\ \text{A}} - 800\ \Omega = 9.992 \times 10^5\ \Omega \cong 1.0 \times 10^6\ \Omega$$

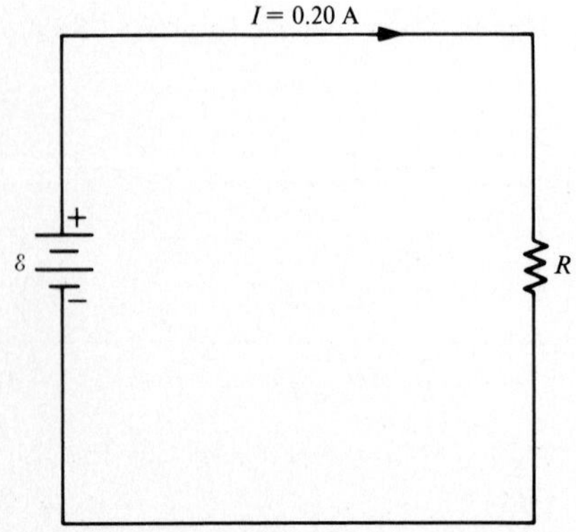

Figure 19-5

EXAMPLE 19-8: Figure 19-5 shows a circuit with a 2.0-V battery and a 10-Ω resistor. The current in this circuit is 0.2 A. Assume that the ammeter described in Example 19-6 is placed in series with the 10-Ω resistor.
Determine the resultant current.

Solution: Use Eq. (18-8) to calculate the total resistance of the ammeter:

$$\frac{1}{R_A} = \frac{1}{R_G} + \frac{1}{R} = \frac{1}{800\ \Omega} + \frac{1}{0.008\ \Omega}$$

$$R_A = 8.000 \times 10^{-3}\ \Omega$$

Now, use Eq. (18-2) to calculate the current:

$$I = \frac{\mathcal{E}}{R_{tot}} = \frac{2.0\ \text{V}}{(10 + 8.0 \times 10^{-3})\ \Omega} = 0.1998\ \text{A} \cong 0.200\ \text{A}$$

You can see that the ammeter does not significantly decrease the current.

EXAMPLE 19-9: Figure 19-6 shows a circuit with a 25-V battery and a 10-Ω resistor. Assume that the voltmeter described in Example 19-7 is connected across the 10-Ω resistor.

Determine the potential difference across the 10-Ω resistor.

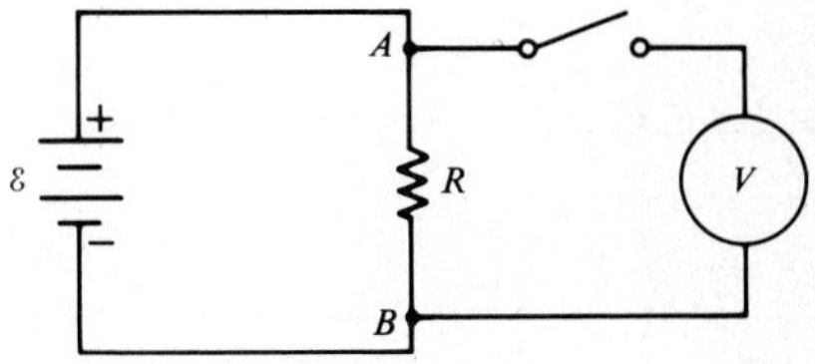

Figure 19-6

Solution: First, use Eq. (18-8) to calculate the total resistance between points A and B in the circuit:

$$\frac{1}{R_{tot}} = \frac{1}{10\ \Omega} + \frac{1}{9.992 \times 10^5\ \Omega} = 10.00\ \Omega$$

Next, use Eq. (18-2) to calculate the current through the voltmeter connected in parallel:

$$I = \frac{25\ \text{V}}{9.992 \times 10^5\ \Omega} = 2.5 \times 10^{-5}\ \text{A}$$

The current flowing through the 10-Ω resistor is

$$I = (2.5 - 2.5 \times 10^{-5})\text{A} \cong 2.50\ \text{A}$$

Finally, calculate the potential difference across the 10-Ω resistor:

$$V = IR = (2.50\ \text{A})(10\ \Omega) = 25.0\ \text{V}$$

It is clear that the voltmeter does not significantly reduce the voltage being measured.

19-4. The Wheatstone Bridge

- The **Wheatstone bridge** circuit, illustrated in Figure 19-7, is a very accurate device for measuring an unknown resistance.

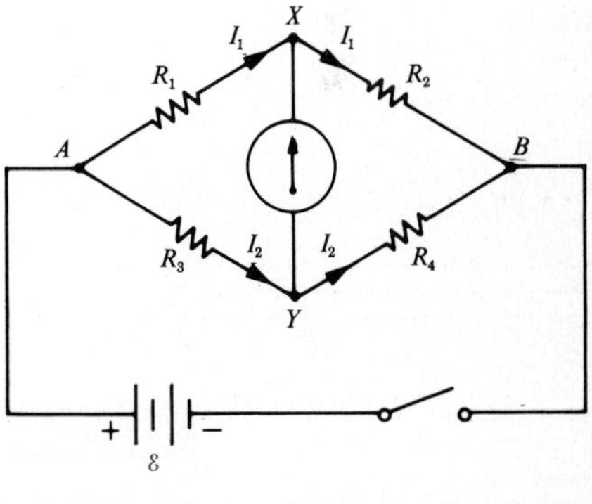

Figure 19-7. The Wheatstone bridge.

The bridge is said to be *balanced* when the current from X to Y through the galvanometer is zero. Under these conditions, there is no potential difference between X and Y and the potential difference across R_1 is the same as that across R_3:

$$I_1 R_1 = I_2 R_3$$

Also, the potential difference across R_2 is equal to that across R_4:

$$I_1 R_2 = I_2 R_4$$

When you divide the first equation by the second, the currents cancel and the result is

BALANCED WHEATSTONE BRIDGE

$$\frac{R_1}{R_2} = \frac{R_3}{R_4} \qquad \textbf{(19-4)}$$

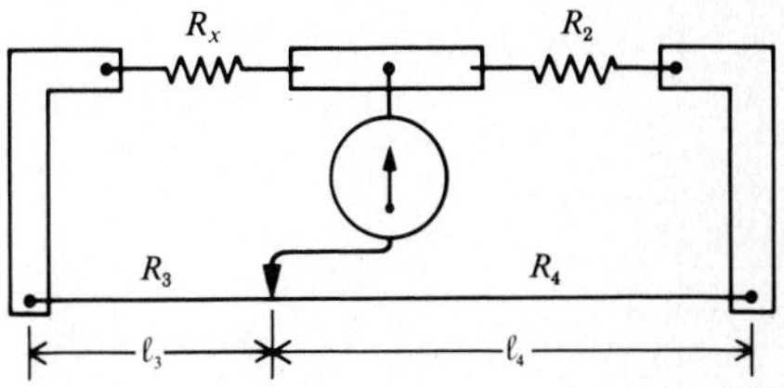

Figure 19-8. A slide-wire Wheatstone bridge.

If you know any three of the resistances, you can calculate the fourth one. A common form of the Wheatstone bridge, illustrated in Figure 19-8, contains a stretched wire with a sliding contact. Because the wire has a constant diameter, the ratio of the resistances R_3/R_4 is equal to the ratio of the two lengths ℓ_3/ℓ_4. If R_x is the unknown resistance, you can find it from the following equation:

BALANCED WHEATSTONE BRIDGE (SLIDE-WIRE)

$$\frac{R_x}{R_2} = \frac{\ell_3}{\ell_4} \tag{19-5}$$

EXAMPLE 19-10: The Wheatstone bridge shown in Figure 19-8 has a slide-wire 100 cm long. The known resistance R_2 has a value of 875 Ω, and the current through the galvanometer is zero when the sliding contact is 34.5 cm from the left end of the wire.
Calculate the unknown resistance R_x.

Solution: Use Eq. (19-5) to find the unknown resistor:

$$R_x = R_2\frac{\ell_3}{\ell_4} = (875\ \Omega)\frac{34.5\text{ cm}}{(100 - 34.5)\text{ cm}} = 461\ \Omega$$

19-5. The Potentiometer

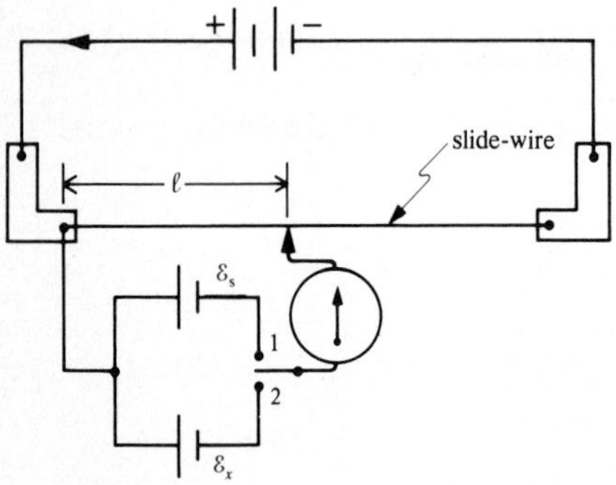

Figure 19-9. Potentiometer circuit.

- The **potentiometer** is the most accurate device for measuring the EMF of a cell.

Unlike a voltmeter, the potentiometer draws no current from the cell under study. A typical potentiometer circuit, shown in Figure 19-9, contains a standard cell, with a known EMF of $\mathscr{E}_s$, and a cell with an unknown EMF of $\mathscr{E}_x$. When the switch is connected to position 1 and the potentiometer is balanced by means of the standard cell, the length of the slide-wire is ℓ_1. We place the unknown EMF in the circuit by moving the switch from position 1 to position 2. When the potentiometer is balanced against the unknown EMF, the length of the slide wire is ℓ_2. We calculate the unknown EMF as follows:

POTENTIOMETER EMF COMPARISON

$$\mathscr{E}_x = \mathscr{E}_s\left(\frac{\ell_2}{\ell_1}\right) \tag{19-6}$$

EXAMPLE 19-11: The potentiometer illustrated in Figure 19-9 is balanced first with the standard cell ($\mathscr{E}_s = 1.0183$ V) at a length of 31.5 cm. When the unknown cell is used, the circuit is balanced when the length of the wire is 45.5 cm.
Calculate the unknown EMF.

Solution: Use Eq. (19-6) to find the unknown EMF:

$$\mathscr{E}_x = \mathscr{E}_s\left(\frac{\ell_2}{\ell_1}\right) = (1.0183\text{ V})\frac{45.5\text{ cm}}{31.5\text{ cm}} = 1.47\text{ V}$$

SUMMARY

Kirchhoff's first law (branch theorem)	$I_1 + I_2 + I_3 + \cdots = 0$	the algebraic sum of the currents into and out of any junction point in a circuit is zero
Kirchhoff's second law (loop theorem)	$\mathscr{E}_1 + \mathscr{E}_2 + \cdots + V_1 + V_2 + \cdots = 0$	the algebraic sum of the changes in potential around any closed path is zero
Relationship between resistivity and resistance	$R = \frac{\rho L}{A}$	gives the resistance of a resistor as proportional to its resistivity and its length, and inversely proportional to the cross-sectional area

Balanced Wheatstone bridge	$\frac{R_1}{R_2} = \frac{R_3}{R_4}$	when a Wheatstone bridge is balanced, the ratio of the first resistance to the second is equal to the ratio of the third resistance to the fourth
Balanced Wheatstone bridge (slide-wire)	$\frac{R_x}{R_2} = \frac{\ell_3}{\ell_4}$	when a slide-wire Wheatstone bridge is balanced, the ratio of the resistances is equal to the ratio of the lengths
Potentiometer EMF comparison	$\mathscr{E}_x = \mathscr{E}_s\left(\frac{\ell_2}{\ell_1}\right)$	when a potentiometer is balanced, the unknown EMF is equal to the known EMF times the ratio of the lengths

RAISE YOUR GRADES

Can you define . . . ?

- ☑ resistivity
- ☑ an ammeter
- ☑ a voltmeter
- ☑ a potentiometer

Can you . . . ?

- ☑ state Kirchhoff's laws
- ☑ write the equation for the resistance of a wire, given its length and cross-sectional area
- ☑ compare the resistances of two iron wires of the same length if one wire has half the diameter of the other
- ☑ draw a diagram showing how a resistor is connected to a galvanometer to make an ammeter and a voltmeter
- ☑ write the equation for a balanced Wheatstone bridge
- ☑ calculate an unknown EMF, given potentiometer data

SOLVED PROBLEMS

PROBLEM 19-1 Fig. 19-10 shows a circuit that contains two resistors in parallel and a third resistor in series with the battery. The current entering junction point P is 2.0 A. The current I_1 leaving the junction is -1.2 A. Calculate (**a**) the current I_2 and (**b**) the unknown resistor R_2.

Solution

(**a**) Find the unknown current by using Kirchhoff's first law, Eq. (19-1):

$$I + I_1 + I_2 = 0$$

$$I_2 = -I - I_1$$

$$= -2.0\ \text{A} - (-1.2\ \text{A})$$

$$= -0.8\ \text{A}$$

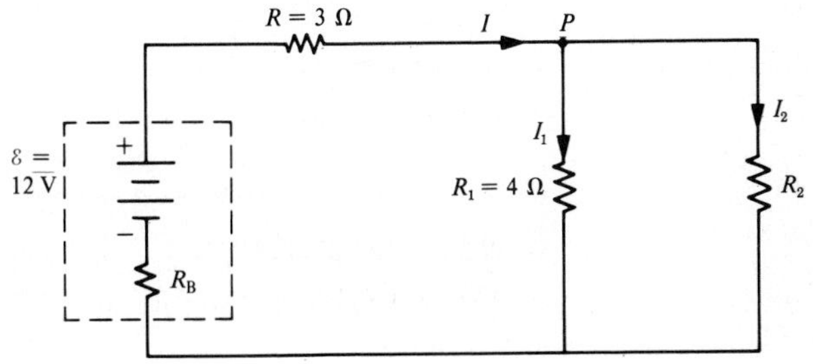

Figure 19-10

(**b**) From Eq. (18-2), calculate the voltage across the two resistors in parallel:

$$V = I_1 R_1 = (1.2\ \text{A})(4\ \Omega) = 4.8\ \text{V}$$

Then, use Ohm's law (Eq. 18-1) to find the unknown resistor:

$$R_2 = \frac{V}{I_2} = \frac{4.8\ \text{V}}{0.8\ \text{A}} = 6.00\ \Omega$$

PROBLEM 19-2 The battery in Fig. 19-10 has an unknown internal resistance. Find (**a**) the voltage across the terminals of the battery and (**b**) the internal resistance.

Solution (**a**) Use Kirchhoff's second law (Eq. 19-2) to find the voltage V_B across the battery. Because the resistors cause a drop in voltage, V and V_1 are negative.

$$V_B + -(V + V_1) = V_B + -(IR + I_1 R_1) = 0$$

$$V_B = -(-IR - I_1 R_1) = -[-(2.0\ \text{A})(3\ \Omega) - (1.2\ \text{A})(4\ \Omega)] = 10.8\ \text{V}$$

(**b**) Use Eq. (18-3) to find the voltage drop across the battery's internal resistance:

$$V_B = \mathscr{E} - IR \qquad IR_B = \mathscr{E} - V_B = 12.0\ \text{V} - 10.8\ \text{V} = 1.2\ \text{V}$$

Since you know the current, you can calculate the unknown resistance:

$$IR_B = 1.2\ \text{V} \qquad R_B = \frac{1.2\ \text{V}}{2.0\ \text{A}} = 0.600\ \Omega$$

PROBLEM 19-3 Find the currents through each of the resistors in Figure 19-11.

Solution Because there is an EMF in each of the loops, no two resistors have the same current through them or the same voltage across them, so you cannot combine the resistors in series or parallel combinations. Use Kirchhoff's second law (Eq. 19-2) for each loop. Notice that both currents go through R_2, so $V_2 = (I_1 + I_2)R_2$. Use negative values for the voltages across the resistors, as you did in Example 19-2. In the left loop

$$\mathscr{E}_1 - V_1 - V_2 = \mathscr{E}_1 - I_1 R_1 - (I_1 + I_2)R_2 = 0$$

$$\mathscr{E}_1 = I_1 R_1 + (I_1 + I_2)R_2$$

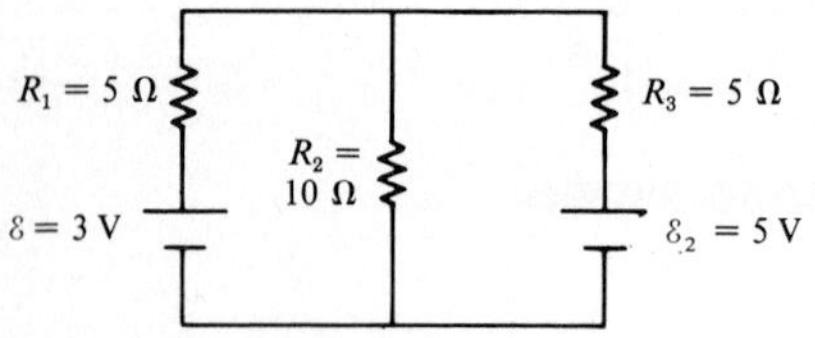

Figure 19-11

In the right loop

$$\mathscr{E}_2 - V_3 - V_2 = \mathscr{E}_2 - I_2 R_3 - (I_1 + I_2)R_2 = 0$$

$$\mathscr{E}_2 = I_2 R_3 + (I_1 + I_2)R_2$$

Now you have two equations relating two unknowns, so you can solve for the two currents. Solve the equation for the left loop for I_1, then substitute for I_1 in the equation for the right loop.

$$I_1 = \frac{\mathscr{E}_1 - I_2 R_2}{R_1 + R_2} = \frac{3\ \text{V} - I_2(10\ \Omega)}{(5 + 10)\ \Omega} = 0.2\ \text{V}/\Omega - \frac{2}{3} I_2$$

$$\mathscr{E}_2 = I_2 R_3 + (0.2\ \text{V}/\Omega - \tfrac{2}{3} I_2 + I_2)R_2$$

$$I_2 = \frac{\mathscr{E}_2 - (0.2\ \text{V}/\Omega)R_2}{R_3 + \frac{1}{3}R_2} = \frac{5\ \text{V} - 0.2\ \text{V}/\Omega)(10\ \Omega)}{5\ \Omega + (\frac{1}{3})(10\ \Omega)} = 0.360\ \text{A} \quad \text{in} \quad R_3$$

$$I_1 = \frac{\mathscr{E}_1 - I_2 R_2}{R_1 + R_2} = \frac{3\ \text{V} - (0.36\ \text{A})(10\ \Omega)}{(5 + 10)\ \Omega} = -0.040\ \text{A} \quad \text{in} \quad R_1$$

The negative result indicates that the current I_1 in the left loop *opposes* the EMF in that loop; that is, I_1 is counterclockwise in Figure 19-11.

$$I_1 + I_2 = -0.040\ \text{A} + 0.360\ \text{A} = 0.320\ \text{A} \quad \text{in} \quad R_2$$

PROBLEM 19-4 The iron slide-wire of a Wheatstone bridge is 100 cm long and has a resistance of 20.0 Ω. What is the diameter of this wire?

Solution Use Eq. (19-3) to find the cross-sectional area of the slide-wire:

$$A = \frac{\rho L}{R}$$

Now, solve for the unknown diameter:

$$\left(\frac{\pi}{4}\right)d^2 = \frac{\rho L}{R}$$

$$d = \sqrt{\frac{4\rho L}{\pi R}} = \sqrt{\frac{4(1.1 \times 10^{-7}\ \Omega\text{m})(1.0\ \text{m})}{\pi(20.0\ \Omega)}}$$

$$= 8.37 \times 10^{-5}\ \text{m} = 8.37 \times 10^{-2}\ \text{mm}$$

PROBLEM 19-5 A strip of Celluloid is 1.0 cm wide, 0.1 mm thick, and 6.0 cm long. What is the resistance of this strip of Celluloid measured from one end to the other over a length of 6.0 cm?

Solution From Eq. (19-3),

$$R = \frac{\rho L}{A} = \frac{(4 \times 10^{12}\ \Omega\ \text{m})(0.06\ \text{m})}{(0.01\ \text{m})(1.0 \times 10^{-4}\ \text{m})} = 2.4 \times 10^{17}\ \Omega$$

PROBLEM 19-6 An ammeter with a full-scale reading of 2.0 A is constructed with a galvanometer whose full-scale reading is 200 μA. The internal resistance of the galvanometer is 1500 Ω. What is the value of the shunt resistor?

Solution Calculate the voltage drop across the galvanometer from Eq. (18-1):

$$V = IR_G = (0.2 \times 10^{-3}\ \text{A})(1.5 \times 10^3\ \Omega) = 0.3\ \text{V}$$

Now, from Eq. (19-1), calculate the current through the resistor connected in parallel with the galvanometer:

$$I_2 = I - I_1 = 2.0\ \text{A} - 0.2 \times 10^{-3}\ \text{A} = 1.9998\ \text{A} \cong 2.00\ \text{A}$$

The voltage across the shunt is 0.3 V, so

$$I_2 R = 0.3\ \text{V}$$

$$R = \frac{0.3\ \text{V}}{2.00\ \text{A}} = 0.150\ \Omega$$

PROBLEM 19-7 The galvanometer described in Problem 19-5 is used to construct a voltmeter whose full-scale reading is 6.0 V. Calculate the series resistor.

Solution Calculate the potential difference across the series resistor:

$$V_2 = 6.0\ \text{V} - 0.3\ \text{V} = 5.7\ \text{V}$$

Since you know from Problem 19-5 that the full-scale current is 200 μA,

$$IR_s = 5.7\ \text{V}$$

$$R_s = \frac{5.7\ \text{V}}{0.20 \times 10^{-3}\ \text{A}} = 2.85 \times 10^4\ \Omega$$

PROBLEM 19-8 The voltmeter described in Problem 19-6 is used to measure the potential difference across the terminals of a 12-V battery whose internal resistance is $R_B = 1.2\ \Omega$. Calculate the measured voltage.

Solution You'll have to find the total resistance in the circuit (Eq. 18-7):

$$R_{tot} = R_s + R_G + R_B = 2.85 \times 10^4\ \Omega + 1.50 \times 10^3\ \Omega + 1.2\ \Omega = 3.00 \times 10^4\ \Omega$$

Now, calculate the current (Eq. 18-2):

$$I = \frac{12.0\ \text{V}}{3.00 \times 10^4\ \Omega} = 4.0 \times 10^{-4}\ \text{A}$$

Then, use Eq. (18-6) to calculate the terminal voltage:

$$V = \mathscr{E} - IR = 12.0\ \text{V} - (4.0 \times 10^{-4}\ \text{A})(1.2\ \Omega) = 11.99952\ \text{V}$$

The current through the voltmeter is so small that the measured voltage is very close to the EMF of the battery.

PROBLEM 19-9 The slide-wire Wheatstone bridge illustrated in Fig. 19-8 is used to measure an unknown resistance. The known resistor R_2 has a value of 22 kΩ. The current through the galvanometer is zero when the contact point is 28.5 cm from the left end of the 100-cm slide-wire. Calculate the unknown resistance.

Solution Use Eq. (19-5):

$$R_x = R_2\frac{\ell_3}{\ell_4} = 22 \times 10^3\ \Omega\frac{28.5\ \text{cm}}{(100 - 28.5\ \text{cm})} = 8.77\ \text{k}\Omega$$

PROBLEM 19-10 The two resistors in Fig. 19-8, R_x and R_2, have the values 15 kΩ and 19 kΩ, respectively. Where should you place the contact point to reduce the current through the galvanometer to zero?

Solution Let x represent the distance from the contact point to the left end of the slide-wire, and use Eq. (19-5) to find the unknown distance:

$$\frac{R_x}{R_2} = \frac{\ell_3}{\ell_4}$$

$$\frac{15\ \text{k}\Omega}{19\ \text{k}\Omega} = \frac{x}{100\ \text{cm} - x}$$

$$15 \times 10^5\ \Omega\ \text{cm} - x(15 \times 10^3\ \Omega) = x(19 \times 10^3\ \Omega)$$

$$x(34 \times 10^3\ \Omega) = 15 \times 10^5\ \Omega\ \text{cm}$$

$$x = 44.1\ \text{cm}$$

PROBLEM 19-11 The potentiometer illustrated in Fig. 19-9 is used with a known EMF of 1.20 V. When the known EMF is balanced, the length of the wire is 23.5 cm from the left end. The unknown EMF is balanced when the length is 29.4 cm. Calculate the unknown EMF.

Solution Use Eq. (19-6):

$$\mathscr{E}_x = \mathscr{E}_s\left(\frac{\ell_2}{\ell_1}\right) = (1.20\ \text{V})\left[\frac{29.4\ \text{cm}}{23.5\ \text{cm}}\right] = 1.50\ \text{V}$$

Supplementary Exercises

PROBLEM 19-12 The two currents entering the junction point in Fig. 19-12 are $I_1 = 1.5$ A and $I_2 = 2.5$ A. One of the currents leaving the junction is $I_3 = -1.8$ A. Find the unknown current I_4.

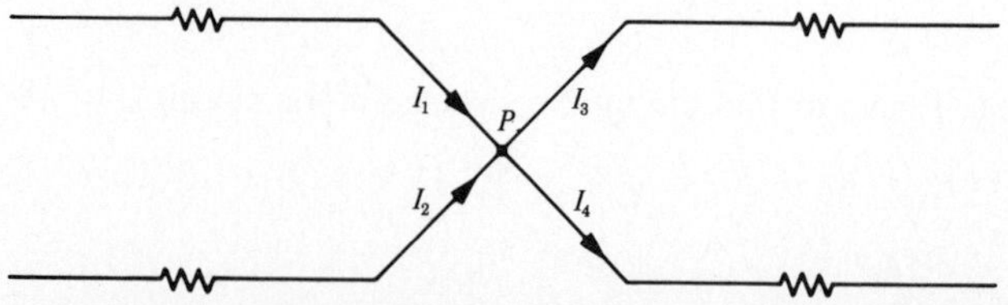

Figure 19-12

PROBLEM 19-13 Two resistors connected in series are wired to a battery whose EMF is 6.0 V. The internal resistance of the battery is 0.5 Ω and the values of the two external resistors are 2.5 Ω and 3.0 Ω. **(a)** Calculate the voltage across the 3.0-Ω resistor. **(b)** Calculate the potential difference across the terminals of the battery.

PROBLEM 19-14 Calculate the resistance of a carbon rod 5 mm in diameter and 15 cm long.

PROBLEM 19-15 Two aluminum wires have lengths of $L_1 = 50$ cm and $L_2 = 80$ cm. The shorter wire has a diameter of 0.2 mm and the longer wire has a diameter of 0.15 mm. Calculate the ratio of the resistances R_2/R_1.

PROBLEM 19-16 A galvanometer has a full-scale deflection of 0.5 mA. Its internal resistance is 2000 Ω. Calculate the value of the shunt resistor that will convert this galvanometer into an ammeter with a full-scale reading of 1.0 A.

PROBLEM 19-17 The galvanometer described in Problem 19-15 is converted into a voltmeter with a full-scale reading of 10 V. Find the value of the required series resistor.

PROBLEM 19-18 A voltmeter has a total resistance of 2000 Ω. When this voltmeter is connected to the terminals of a battery with an EMF of 24 V and an internal resistance of 4 Ω, what will the terminal voltage be?

PROBLEM 19-19 The three known resistances of a Wheatstone bridge are $R_2 = 850\ \Omega$, $R_3 = 1200\ \Omega$, and $R_4 = 950\ \Omega$. Calculate the unknown resistor R_1.

PROBLEM 19-20 A slide-wire Wheatstone bridge is balanced when the contact point is 27.0 cm from the left end. The known resistor is $R_2 = 1255\ \Omega$ and the slide-wire is 100-cm long. Calculate the unknown resistor.

PROBLEM 19-21 The slide-wire potentiometer illustrated in Fig. 19-9 has a standard cell whose EMF is 1.25 V. The standard cell is balanced when ℓ_1 is 25 cm. The unknown EMF is balanced when ℓ_2 is 33 cm. Calculate the unknown EMF.

Answers to Supplementary Exercises

19-12: −2.20 A

19-13: (a) 3.00 V (b) 5.50 V

19-14: 0.306 Ω

19-15: 2.84

19-16: 1.0005 Ω

19-17: 18 kΩ

19-18: 23.95 V

19-19: 1.07×10^3

19-20: 464 Ω

19-21: 1.65 V

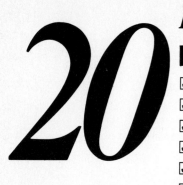

20 ELECTROMAGNETISM

THIS CHAPTER IS ABOUT

☑ The Magnetic Field
☑ Magnetic Force on a Current-Carrying Conductor
☑ Forces and Torques on a Current Loop
☑ Sources of Magnetic Field
☑ The Ampere Revisited
☑ Magnetic Flux and Induced EMF

20-1. The Magnetic Field

In Chapters 8 and 16, we introduced the concept of a **field** to describe gravitational and electrostatic (Coulomb) forces. In Section 8-2, we defined the gravitational field strength on a test mass m as

$$\mathbf{g} = \frac{\mathbf{F}_{(\text{grav})}(\text{on } m)}{m}$$

where the units of **g** are force per unit mass (N/kg). In Section 16-2, we introduced a similar equation for the electric field strength on a positive test charge Q,

$$\mathbf{E} = \frac{\mathbf{F}_{(\text{elec})}(\text{on } Q)}{Q}$$

where the units of **E** are force per unit charge (N/C). In neither case does the force depend on the motion of the test object. Now, let's turn our attention to a type of force that *does* depend on the motion of the test object.

Consider a positive test charge Q in space. If we detect a force on Q when it is at rest, we know that an electrostatic field is present. If we detect an additional force on Q when it is *moving,* we say that Q experiences a **magnetic force $\mathbf{F}_{\text{mag}}$** and that a **magnetic field B** is present. Just as with gravitational and electrostatic fields, we express the value of the magnetic field at any point as a vector. The SI unit for the magnitude of **B** is the tesla T where

$$\text{T} = \frac{\text{N}}{\text{C(m/s)}} = \frac{\text{N s}}{\text{C m}}$$

If we examine a magnetic field by observing its effects on a moving charge Q, we find that

(1) The direction of the force is perpendicular to the direction of Q's velocity **v**.
(2) The magnitude of the force $\mathbf{F}_{\text{mag}}$ is proportional to the charge on Q.
(3) For any given direction of **v**, the force is proportional to the magnitude of **v**.
(4) For any given magnitude of **v**, the force varies with the direction of **v**, being zero when **v** is parallel to **B** and maximum when **v** is perpendicular to **B**.

We can summarize all these observations in one equation that expresses $\mathbf{F}_{\text{mag}}$ as a vector cross product (see Sec. 1-5B).

FORCE ON A MOVING CHARGE IN A MAGNETIC FIELD

$$\mathbf{F}_{\text{mag}} = Q\mathbf{v} \times \mathbf{B} \tag{20-1}$$

The cross product specifies that the force is at right angles to both the magnetic field and the velocity of the charge experiencing the magnetic force. We use the right-hand-screw rule (see Example 1-11) to find the direction of $\mathbf{F}_{\text{mag}}$ on a moving *positive* test charge: The direction of $\mathbf{F}_{mag}$ on a moving positive charge is the direction a right-hand screw would advance if rotated from **v** toward **B**. The magnitude of the force is

MAGNITUDE OF A MAGNETIC FORCE

$$F_{mag} = QvB \sin\theta \qquad (20\text{-}2)$$

where θ is the angle between **v** and **B**. If a test charge is moving at right angles to the magnetic field, then $\theta = 90°$ and $\sin\theta = 1$, and we have the maximum magnetic force on a moving charge:

MAXIMUM MAGNETIC FORCE (v ⊥ B)

$$F_{max} = QvB \qquad (20\text{-}3)$$

If a test charge is moving parallel to the magnetic field, then $\theta = 0$ and $\sin\theta = 0$, so the charge experiences zero force.

EXAMPLE 20-1: A positive charge of 2×10^{-8} C moves in a direction 30° above the x axis at a speed of 300 m/s through a 5-T magnetic field directed in the positive x direction (see Figure 20-1). Find the magnitude and direction of the magnetic force on the moving charge.

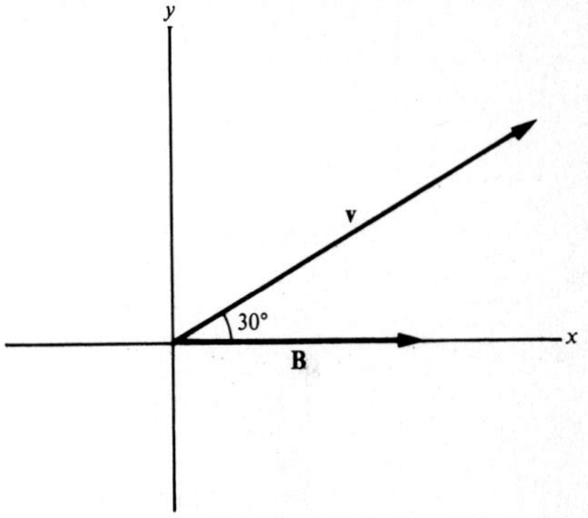

Figure 20-1

Solution: Equation (20-2) gives the magnitude directly:

$$F_{mag} = QvB \sin\theta = (2 \times 10^{-8}\ \text{C})(300\ \text{m/s})(5\ \text{N s/C m}) \sin 30°$$
$$= 1.5 \times 10^{-5}\ \text{N}$$

Imagine a right-hand screw rotated in the direction that will move **v** toward **B**. This screw will move downward, perpendicular to the plane of the drawing. So, the direction of $\mathbf{F}_{mag}$ is into the paper.

EXAMPLE 20-2: When a positive charge of 4×10^{-8} C moves through a magnetic field at a speed of 500 m/s in different directions, the maximum force that acts on the charge is 1.6×10^{-4} N.
Determine the magnitude of this magnetic field.

Solution: Directly from Eq. (20-3):

$$B = \frac{F_{max}}{Qv} = \frac{1.6 \times 10^{-4}\ \text{N}}{(4 \times 10^{-8}\ \text{C})(500\ \text{m/s})} = 8.00\ \text{T}$$

20-2. Magnetic Force on a Current-Carrying Conductor

We can readily extend our discussion of magnetic force on a single moving charge Q to N charges confined to a conductor such as a segment of wire. Recall that current I is charge per time (see Sec. 16-3B), and that the time required for any charge to move a distance $\Delta\ell$ in a wire is $\Delta\ell/v$. For N charges, we can write

$$I = \frac{NQ}{\Delta\ell/v} \quad \text{and} \quad I\Delta\ell = NQv$$

If we express Eq. (20-1) for multiple charges as $\mathbf{F}_{mag} = NQ\mathbf{v} \times \mathbf{B}$ and substitute $I\,\Delta\ell$ for $NQ\mathbf{v}$ we get

MAGNETIC FORCE ON A CURRENT-CARRYING CONDUCTOR

$$\mathbf{F}_{mag} = I\Delta\ell \times \mathbf{B} \qquad (20\text{-}4)$$

where $\Delta\ell$ has the direction of the positive current through the conductor; i.e., the conventional current. The directions of the magnetic field **B** and the positive **current segment** $I\,\Delta\ell$ are usually not perpendicular.

Just as we did with a single charge in space, we can evaluate the cross product in the vector equation to produce the magnitude of $\mathbf{F}_{mag}$,

$$F_{mag} = I\,\Delta\ell B\,\sin\theta \qquad \textbf{(20-4a)}$$

and the maximum force on a wire carrying a current,

$$F_{max} = I\,\Delta\ell B \qquad \textbf{(20-5)}$$

note: We can also define the tesla in terms of current:

$$\mathrm{T} = \frac{\mathrm{N}}{\mathrm{C(m/s)}} = \frac{\mathrm{N}}{\mathrm{(C/s)m}} = \frac{\mathrm{N}}{\mathrm{A\ m}}$$

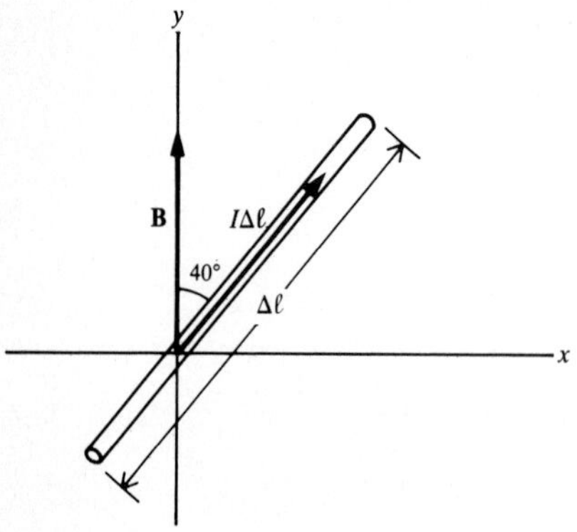

Figure 20-2

EXAMPLE 20-3: A 12-cm segment of copper wire that is part of a circuit carries a current of 5 A. The wire segment lies along a line 50° above the x axis and in a 5-T magnetic field that is directed in the positive y direction (see Figure 20-2). Find the magnitude and direction of the magnetic force on the conductor.

Solution: You can use Eq. (20-4a) to find the magnitude of the magnetic force:

$$F_{mag} = I\,\Delta\ell B\,\sin\theta = (5\ \mathrm{A})(0.12\ \mathrm{m})(3\ \mathrm{N/A\ m})\,\sin 40° = 1.16\ \mathrm{N}$$

Now, imagine a right-hand screw rotated in a direction that will twist $I\,\Delta\ell$ toward **B**. Since the screw has to be rotated counterclockwise, it will move upward out of the drawing, which indicates that the magnetic force is vertically upward.

EXAMPLE 20-4: A 25-cm current segment carries a current of 4 A. The maximum magnetic force on this current segment is 7.0 N.
Calculate the magnitude of the magnetic field.

Solution: Use Eq. (20-5) to solve for B:

$$B = \frac{F_{max}}{I\,\Delta\ell} = \frac{7.0\ \mathrm{N}}{(4\ \mathrm{A})(0.25\ \mathrm{m})} = 7.0\ \mathrm{T}$$

20-3. Forces and Torques on a Current Loop

Circular or rectangular **current loops** give us a practical indication of the direction of a magnetic field. Consider the rectangular loop shown in Figure 20-3a. It carries current in the counterclockwise direction and is in a magnetic field **B** that points to the left. By using Eq. (20-4a), $F_{mag} = I\,\Delta\ell B\,\sin\theta$, we can calculate the force on each segment. The forces along the top and bottom of the loop are zero because **B** is parallel to $\Delta\ell$; that is, $\sin\theta = 0$. The force $F_1 = IbB\,\sin 90° = IbB$ on the left vertical segment points into the page, as indicated by ⊗. (Use the right-hand rule, with $\Delta\ell$ down and **B** to the left.) Similarly, the force F_2 on the right side of the loop points out of the paper, as indicated by ⊙, and has the same magnitude. These forces each cause a torque (= moment arm × force, Eq. 4-15), which has the magnitude $\tau = Fa/2$ (the moment arm is $a/2$, the distance from the axis of rotation to the vertical segment) and tries to lift the right edge of the loop off the page and to force the left edge into the page. These torques are more easily understood from Figure 20-3b, which is an edge view looking up from the bottom on the same coil. The torques due to F_1 and F_2 have the same magnitude and direction, so the total torque is

$$\tau = \tau_1 + \tau_2 = F_1\frac{a}{2} + F_2\frac{a}{2} = 2IbB\frac{a}{2} = IBA$$

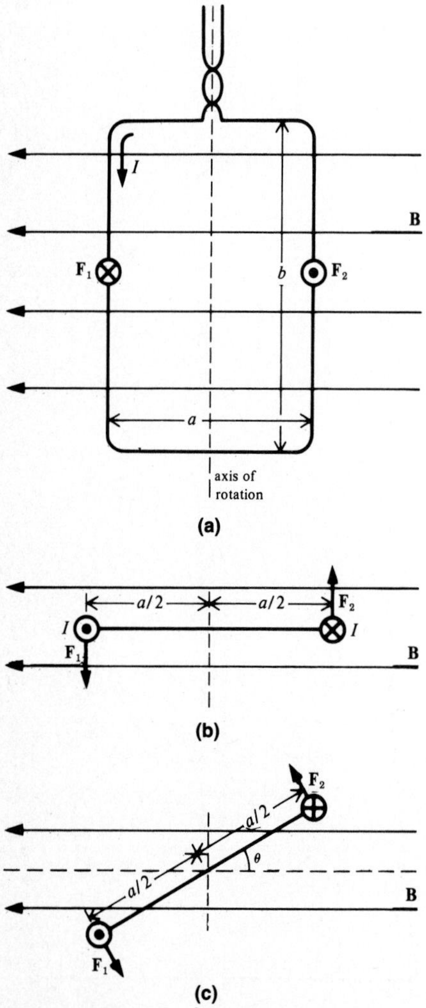

Figure 20-3. A rectangular loop within a magnetic field: (a) side view; (b) top view.

where $A = ab$ is the area of the loop. The torque causes the loop to rotate counterclockwise, as shown in Figure 20-3b. Notice that because the forces are of equal magnitude but opposite direction the net force is zero. This is true for a current loop with any shape. In general, the loop could be oriented at an angle to the magnetic field, as shown in Figure 20-3c. The moment arm, and thus the torque, then depends on the angle θ:

$$\tau = IAB \cos \theta$$

note: You'll encounter several conventions for indicating the direction of a vector into or out of the plane of a drawing. We'll use a dot (· or ⊙) to represent the head of a vector arrow coming out of the plane of the paper, and a cross (× or ⊗) to represent the tail of a vector arrow going into the plane of the paper.

EXAMPLE 20-5: For the rectangular loop shown in Figure 20-3, $a = 6$ cm and $b = 10$ cm. The current is 2 A and the magnitude of **B** is 5 T. Calculate the torque acting on the loop when the plane of the loop makes an angle of 30° with respect to **B**.

Solution: The vectors $I\,\Delta\ell$ and **B** are at right angles, so the vertical sides of the loop experience the maximum magnetic force. Use Eq. (20-5) to find this force:

$$F_1 = I_1\,\Delta\ell B = (2\text{ A})(0.10\text{ m})(5\text{ T}) = 1.0\text{ N}$$

Now, use Eq. (4-15) to calculate the torque produced by F_1:

$$\tau_1 = (\text{moment arm})F_1 = \left(\frac{a}{2}\cos 30°\right)F_1$$

$$= (0.03\text{ m})(1.0\text{ N})\cos 30° = 0.026\text{ N m}$$

The force $\mathbf{F}_2$ acting on the opposite side of the loop has the same magnitude as $\mathbf{F}_1$ and the same moment arm, so the total torque is $2\tau_1 = 0.052$ N m.

EXAMPLE 20-6: What orientation of the loop shown in Figure 20-3 would produce the maximum possible torque?

Solution: You can see from the figure that the torque is at a maximum when the plane of the loop is parallel to the magnetic field **B**. In this orientation, the moment arm has its maximum value of $a/2$, ($\theta = 0$; $\cos\theta = 1$), so the total torque is

$$\tau_{max} = 2(a/2)(F_1) = 2(0.03\text{ m})(1.0\text{ N}) = 0.0600\text{ N m}$$

Our formula for the torque caused by a current flowing through a rectangular coil in a magnetic field can be generalized for any coil of area A and any number of turns N.

TORQUE (COIL OF *N* TURNS)

$$\tau = NIBA \cos \theta \qquad \textbf{(20-6)}$$

The angle θ is the angle between the magnetic field **B** and the plane of the coil.

20-4. Sources of Magnetic Field

A. Long straight conductors

Current-carrying conductors produce magnetic fields. For a long straight wire, the magnitude of **B** at distance a from the wire is

MAGNETIC FIELD (LONG STRAIGHT CONDUCTOR)

$$B = k'\frac{2I}{a} = \frac{\mu_0 I}{2\pi a} \qquad \textbf{(20-7)}$$

where $k' = 1 \times 10^{-7}\text{ N/A}^2 = \mu_0/4\pi$. The constant μ_0 is called the **permeability of free space** and has a value of $4\pi \times 10^{-7}$ T m/A. The magnetic field produced encircles the wire, as shown in Figure 20-4. You can determine the direction of **B** with a simple right-hand rule: Point your right thumb in the

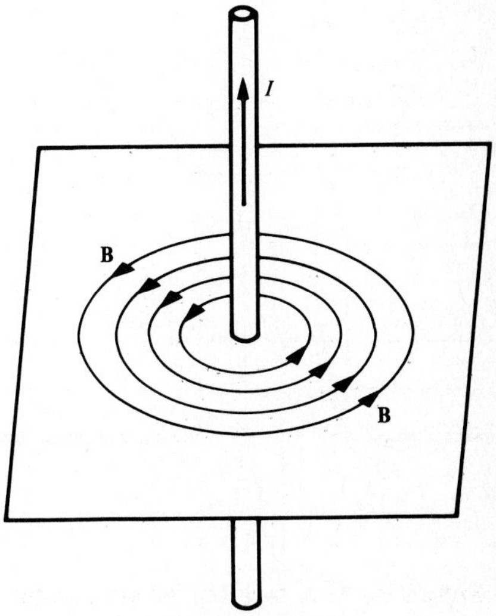

Figure 20-4

direction of the current I and your fingers will curl in the direction of the magnetic field **B**.

EXAMPLE 20-7: A long straight wire carries a current of 12 A.
At what distance from this wire will the magnitude of **B** have the value of 1.2×10^{-4} T?

Solution: Use Eq. (20-7) and solve for the unknown distance:

$$a = \frac{2Ik'}{B} = \frac{24\ \text{A}(10^{-7}\ \text{N/A}^2)}{1.2 \times 10^{-4}\ \text{T}} = 0.0200\ \text{m}$$

B. Circular loops

The current in a **circular loop** also produces a magnetic field. The magnitude of **B** at the center of a loop of radius r is

MAGNETIC FIELD (CENTER OF A CIRCULAR LOOP)

$$B = k'\frac{2\pi I}{r} \qquad \textbf{(20-8)}$$

Similarly, a current in a circular coil produces a magnetic field. The magnitude of the magnetic field at the center of a coil of radius r is

$$B = k'\frac{N2\pi I}{r} \qquad \textbf{(20-8a)}$$

where N is the number of turns in the coil. The direction of the magnetic field **B** can be found by using the right-hand rule. Curl your fingers in the direction of the current flow in the coil, and your thumb will point in the direction of **B**.

EXAMPLE 20-8: A circular loop of one turn has a radius of 1.5 cm.
What current is required to produce a magnetic field at the center of the loop with a magnitude of 1.257×10^{-4} T?

Solution: Use Eq. (20-8) and solve for the unknown current:

$$I = \frac{Br}{2\pi k'} = \frac{(1.257 \times 10^{-4}\ \text{T})(0.015\ \text{m})}{2\pi(10^{-7}\ \text{N/A}^2)} = 3.00\ \text{A}$$

C. Permanent magnets

Permanent magnets—including the earth—also produce magnetic fields. The magnetic field is directed away from the magnet's *north pole* and towards its *south pole*.

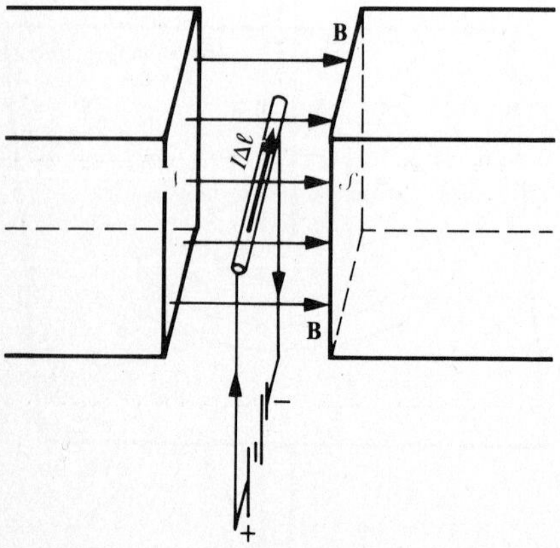

Figure 20-5

EXAMPLE 20-9: Figure 20-5 shows a 20-cm current segment located between the north and south poles of two permanent magnets, at right angles to **B**. When a current of 7 A flows through this conductor, it experiences a downward force of 1.12 N.
Calculate the magnitude of the magnetic field between the two magnets.

Solution: Since the conductor is perpendicular to the magnetic field, use Eq. (20-4) to find the magnetic force:

$$B = \frac{F_{\text{mag}}}{I\,\Delta\ell} = \frac{1.12\ \text{N}}{(7\ \text{A})(0.2\ \text{m})} = 0.800\ \text{T}$$

20-5. The Ampere Revisited

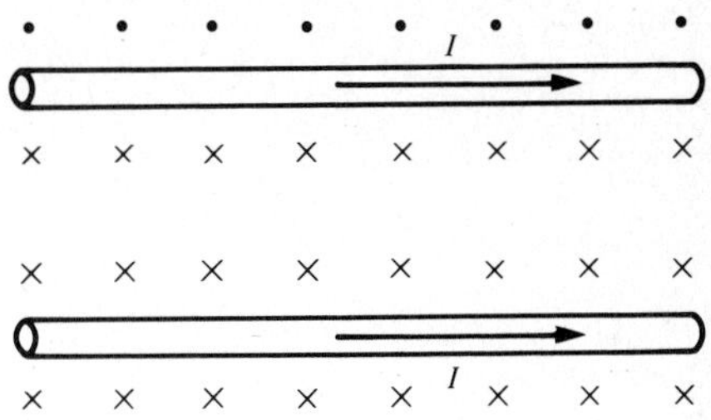

Figure 20-6

In Section 16-3, we defined the ampere in terms of derived SI units as coulombs per second. We defined the coulomb in terms of the charge on the electron. Given the properties of a magnetic field, we can now define the ampere (and thus the coulomb) in mechanical terms. The definition is based on the physical situation shown in Figure 20-6, where two long parallel wires carry current in the same direction and the symbols • and × indicate the magnetic field due to the current in the upper wire.

AMPERE Two parallel conductors one meter apart experience a mutual force of attraction of exactly 2×10^{-7} N/m if each conductor carries a current of one ampere in the same direction.

We can express this definition in SI base units of meters, kilograms, and seconds; and it allows us to link our electrical standards of current, charge, and potential difference to mechanical standards.

EXAMPLE 20-10: The parallel wires shown in Figure 20-6 are each 150 cm long and are 1 cm apart. Each wire carries a current of 10 A in the same direction. Calculate the magnetic force acting on each wire.

Solution: First, find the magnitude of **B** at a distance of 1 cm from the upper wire by using Eq. (20-7):

$$B = k'\frac{2I}{a} = (10^{-7}\ \text{N/A}^2)\frac{2(10\ \text{A})}{(0.01\ \text{m})} = 2.0 \times 10^{-4}\ \text{T}$$

The lower wire is perpendicular to the magnetic field produced by the upper wire, so you can calculate the force from Eq. (20-5):

$$F_{\text{max}} = I\,\Delta\ell B = (10\ \text{A})(1.5\ \text{m})(2.0 \times 10^{-4}\ \text{T}) = 3.0 \times 10^{-3}\ \text{N}$$

Each wire is attracted to the other with a force of 3.0×10^{-3} N.

20-6. Magnetic Flux and Induced EMF

If a ring made of a conductor lies at rest next to a permanent magnet no current will flow in the ring. The same is true if the magnet is replaced by a current-carrying circuit, which creates its own magnetic field. However, if either the ring or the magnet is moved, there will be a current in the ring *while* the motion persists. We say that a changing magnetic field *induces* an EMF, which causes the current in the ring. (Remember that the magnetic field **B** is a vector quantity, so a change in direction with respect to the ring represents a change in **B**.) To express the relation between the magnetic field and the induced EMF we use the concept of *magnetic flux* to represent the amount of magnetic field within an area.

- The **magnetic flux** Φ through a flat conducting loop is the product of the magnitude of the magnetic field inside the loop, the area inside the loop, and the cosine of the angle ϕ between **B** and a line *perpendicular* to the plane of the loop (see Figure 20-7).

MAGNETIC FLUX $$\Phi = BA\cos\phi \qquad \textbf{(20-9)}$$

Now we can express an induced EMF as the rate of change of magnetic flux:

FARADAY'S LAW (INDUCED EMF) $$\mathscr{E} = \lim_{\Delta t \to 0} -\frac{\Delta\Phi}{\Delta t} = -\frac{d\Phi}{dt} \qquad \textbf{(20-10)}$$

Any motion of the loop or **B** with respect to each other that changes ϕ— and therefore Φ— will induce a current in the loop. Notice that a change in the magnitude of **B** or of the area A of the loop will also change the magnetic flux and induce an EMF.

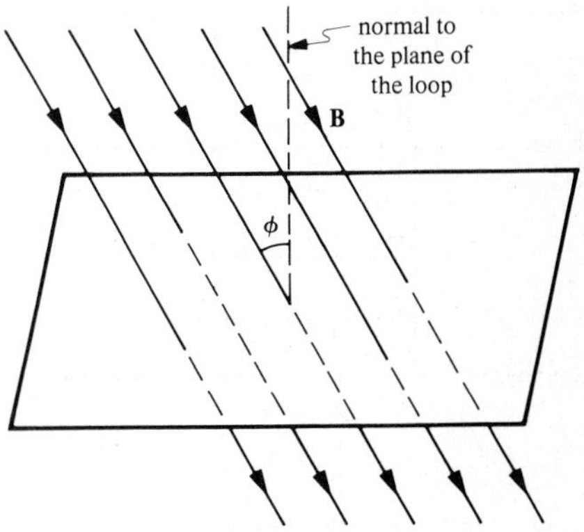

Figure 20-7. The magnetic flux through a rectangular loop.

- The **weber** Wb is the derived SI unit for magnetic flux, where

$$1\ \text{T} = 1\frac{\text{Wb}}{\text{m}^2} \quad \text{and} \quad 1\ \text{V} = 1\frac{\text{Wb}}{\text{s}}$$

EXAMPLE 20-11: A 0.02-T magnetic field makes an angle ϕ of 30° with respect to the normal of a 0.4-m^2 rectangular loop. When the orientation of the conducting loop changes so that ϕ equals 0°, the flux increases.
If the loop is rotated through 30° within 0.2 s, what average EMF is induced in the loop?

Solution: Calculate the two values of the flux from Eq. (20-9):

$$\Phi_1 = BA \cos \phi_1 = (0.02\ \text{T})(0.4\ \text{m}^2) \cos 30° = 6.928 \times 10^{-3}\ \text{Wb}$$

$$\Phi_2 = BA \cos \phi_2 = (0.02\ \text{T})(0.4\ \text{m}^2) \cos 0° = 8.000 \times 10^{-3}\ \text{Wb}$$

Now, divide the change in flux by the time interval to find the average EMF induced:

$$\mathscr{E}_{\text{ave}} = -\frac{\Delta\Phi}{\Delta t} = -\frac{(8.000 - 6.928) \times 10^{-3}\ \text{Wb}}{0.2\ \text{s}} = -5.36 \times 10^{-3}\ \text{V}$$

$$= -5.36\ \text{mV}$$

A. Lenz's law

An EMF induced within a conducting loop causes a current to flow within the loop. **Lenz's law** gives the direction of the induced EMF and the resulting current:

- An induced EMF produces a current that tends to oppose the change of flux that caused it.

The minus sign in Eq. (20-10) is the result of Lenz's law.

EXAMPLE 20-12: The rectangular loop illustrated in Figure 20-7 rotates through 30° so that the flux through the loop increases.
What is the direction of the induced current as viewed from above?

Solution: You find the direction of the induced current from Lenz's law. Since the magnetic flux through the loop increased, the direction of the induced current will tend to oppose this increase in flux. A counterclockwise current will produce an upward flux through the loop.

B. Induced EMF in a conductor moving through a magnetic field

If any conductor, whether or not it is part of a circuit, moves through a magnetic field, there is a force (Eq. 20-1) on each of the current carriers, usually the electrons, in the conductor. This force results in an induced EMF in the conductor. We can find this induced EMF from

INDUCED EMF (MOVING CONDUCTOR)

$$\mathscr{E} = B\ell v \sin \theta \qquad \textbf{(20-11)}$$

where θ is the angle between **B** and **v**, and ℓ is the component of the conductor's length that is perpendicular to both **B** and **v**.

EXAMPLE 20-13: A piece of copper wire 10 cm long moves through a magnetic field at a speed of 25 m/s (see Figure 20-8). The wire is perpendicular to both **B** and **v** and so lies at right angles to the plane of the page. The magnitude of **B** is 0.5 T and the angle θ between **B** and **v** is 20°.
Calculate the induced EMF within this moving conductor.

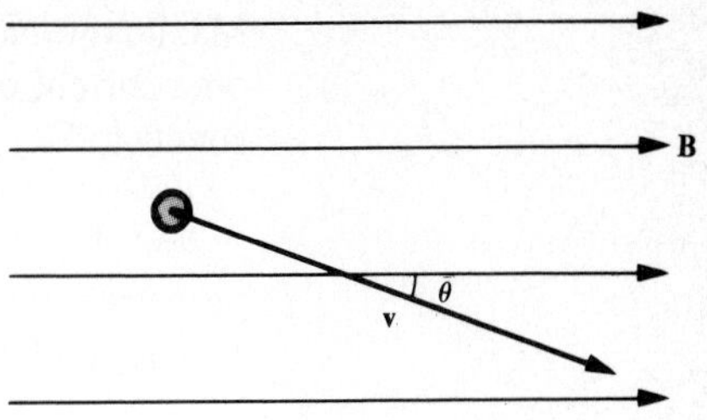

Figure 20-8

Solution: Directly from Eq. (20-11),

$$\mathscr{E} = B\ell v \sin\theta = (0.5 \text{ T})(0.1 \text{ m})(25 \text{ m/s}) \sin 20° = 0.428 \text{ V}$$

EXAMPLE 20-14: If we change the direction of motion in Figure 20-8,
(**a**) what direction of the conductor's motion will produce a maximum EMF and
(**b**) what direction will produce an EMF of zero?

Solution:

(**a**) Use Eq. (20-11). The EMF will have its maximum value of $B\ell v$ when $\sin\theta = 1$ and $\theta = 90°$; that is, when the conductor is moving perpendicular to the magnetic field:

MAXIMUM INDUCED EMF (MOVING CONDUCTOR)

$$\mathscr{E}_{max} = B\ell v \qquad \textbf{(20-12)}$$

(**b**) When the conductor is moving parallel to **B**, $\theta = 0$ and the induced EMF is zero.

SUMMARY

Magnetic force on a moving charge	$\mathbf{F}_{mag} = Q\mathbf{v} \times \mathbf{B}$	the magnetic force on a test charge is the charge times the cross product of the charge's velocity and the magnetic field
Magnitude of a magnetic force	$F_{mag} = QvB \sin\theta$	the magnitude of the magnetic force on a test charge is the product of the charge, the charge's speed, the strength of the magnetic field, and the sine of the angle between them
Maximum magnetic force ($\mathbf{v} \perp \mathbf{B}$)	$F_{max} = QvB$	the maximum magnetic force on a moving charge exists when a test charge is moving at right angles to the magnetic field
Magnetic force on a current-carrying conductor	$\mathbf{F}_{mag} = I\,\Delta\boldsymbol{\ell} \times \mathbf{B}$	the magnetic force on multiple charges confined to a conductor is the cross product of the current segment and the magnetic field
Magnitude of a magnetic force on a current-carrying conductor	$F_{mag} = I\,\Delta\ell B \sin\theta$	the magnitude of the magnetic force is the product of the current segment, the strength of the magnetic field, and the sine of the angle between them

Maximum magnetic force on a current-carrying conductor	$F_{max} = I\,\Delta\ell B$	the maximum force on a current-carrying wire is the product of the current segment times the strength of the magnetic field
Torque (coil of *N* turns)	$\tau = NIBA\ \cos\theta$	the torque acting on a coil is the product of the number of turns, the current, the strength of the magnetic field, the area inside the coil, and the cosine of the angle between the magnetic field and the plane of the coil
Magnetic field (long straight conductor)	$B = k'\dfrac{2I}{a}$	the magnitude of a magnetic field at a given distance from a long straight conductor is proportional to the current and inversely proportional to the distance from the conductor
Magnetic field (center of a circular loop)	$B = k'\dfrac{2\pi I}{r}$	the magnitude of a magnetic field at the center of a loop is proportional to the current and inversely proportional to the radius of the loop
Magnetic flux	$\Phi = BA\ \cos\phi$	magnetic flux is the product of the magnitude of the magnetic field, the cosine of angle the field makes with the normal to the loop, and the area of the loop
Faraday's law (induced EMF)	$\mathscr{E} = \lim\limits_{\Delta t \to 0} -\dfrac{\Delta\Phi}{\Delta t}$	induced EMF equals the rate of change of magnetic flux
Induced EMF (moving conductor)	$\mathscr{E} = B\ell v\ \sin\theta$	an induced EMF within a conductor depends on the length of the conductor, its speed, and the angle between the magnetic field and the velocity
Maximum induced EMF (moving conductor)	$\mathscr{E}_{max} = B\ell v$	an induced EMF has its maximum value when the conductor is moving at right angles to the magnetic field

RAISE YOUR GRADES

Can you . . . ?

- ☑ write the equation for the magnetic force on a charge moving through a magnetic field
- ☑ state the direction of a moving charge that will give rise to a maximum magnetic force
- ☑ describe how a charge can move through a magnetic field and experience zero magnetic force
- ☑ write the equation for the magnetic force on a conductor carrying a current in the presence of a magnetic field
- ☑ describe how a conductor carrying a current within a magnetic field can experience zero magnetic force
- ☑ write the equation for the torque on a loop of N turns placed in a magnetic field
- ☑ describe the orientation of a current-carrying loop within a magnetic field when the torque on the loop is zero
- ☑ write the expression for the magnitude of a magnetic field at a distance a from a long straight wire carrying a current I
- ☑ write the equation for the magnitude of the magnetic field at the center of a circular loop carrying a current I
- ☑ calculate the force of repulsion between two parallel conductors carrying currents in opposite directions
- ☑ write the equation for the magnetic flux through a loop of area A
- ☑ write the equation for the induced EMF within a conducting loop caused by a changing flux
- ☑ write Lenz's law for the direction of an induced current
- ☑ describe the direction of the induced current in a horizontal current loop when the downward flux through the loop is decreased
- ☑ write the equation for the induced EMF in a conductor moving through a magnetic field
- ☑ describe the direction of motion of a conductor through a magnetic field so that the induced EMF is zero

SOLVED PROBLEMS

PROBLEM 20-1 A positive charge of 40 μC moves at a speed of 600 m/s through a uniform magnetic field. The charge is moving toward the west and experiences no magnetic force. What is the direction of **B**?

Solution The charge must be moving either parallel or antiparallel (parallel but in the opposite direction) to the magnetic field to experience a force of zero, so the direction of **B** is either due west or due east.

PROBLEM 20-2 When the positive charge described in Problem 20-1 moves along the north-south axis, it experiences a magnetic force of 0.288 N. What is the magnitude of the magnetic field?

Solution Since the velocity is perpendicular to **B**, you can find the maximum value of the magnetic force from Eq. (20-3). Solve for the magnitude of **B**:

$$F_{\text{max}} = QvB$$

$$B = \frac{F_{\text{mag}}}{Qv} = \frac{0.288\ \text{N}}{(4.0 \times 10^{-5}\ \text{C})(600\ \text{m/s})} = 12.0\ \text{T}$$

PROBLEM 20-3 A uniform magnetic field of 12 T is directed toward the east. A positive charge of 40 μC moves toward the northeast at a speed of 600 m/s, experiencing a magnetic force of 0.144 N. Find the angle θ between the vectors **v** and **B**.

Solution Use Eq. (20-2) and solve for the unknown angle:

$$F_{mag} = QvB \sin\theta$$

$$\sin\theta = \frac{F_{mag}}{QvB} = \frac{0.144 \text{ N}}{(4 \times 10^{-5} \text{ C})(600 \text{ m/s})(12 \text{ T})} = 0.5$$

$$\theta = \arcsin 0.5 = 30°$$

So, the charge is moving 30° north of east.

PROBLEM 20-4 A proton has a positive charge of 1.6×10^{-19} C and a mass of 1.67×10^{-27} kg. A uniform vertical magnetic field, directed downward, with a magnitude of 4 T, causes the proton to move counterclockwise in a horizontal circular path with a radius of 6 cm. Calculate the speed of the proton.

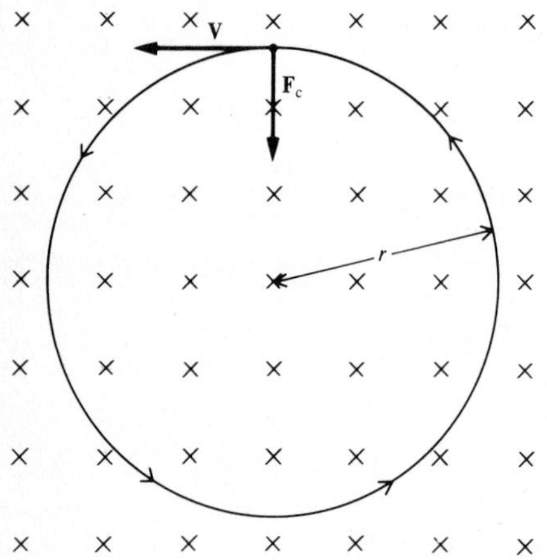

Figure 20-9

Solution The charge is moving perpendicular to the magnetic field, so the force on the proton is given by Eq. (20-3), $F_{max} = QvB$. The path of the proton through the magnetic field is shown in Fig. 20-9. Apply the right-hand rule and you'll see that the direction of the magnetic force is directed toward the center of the circular path. In other words, $\mathbf{F}_{mag}$ is the centripetal force. [**Recall:** The acceleration of an object toward the center of its circular path is the centripetal acceleration. (Eq. 4-12)]

$$F_c = m\frac{v^2}{r}$$

Since F_{max} is equal to F_c,

$$QvB = m\frac{v^2}{r}$$

Divide the equation by v, and solve for the unknown speed:

$$v = \frac{QBr}{m} = \frac{(1.6 \times 10^{-19} \text{ C})(4 \text{ T})(0.06 \text{ m})}{(1.67 \times 10^{-27} \text{ kg})} = 2.30 \times 10^7 \text{ m/s}$$

PROBLEM 20-5 A conductor carrying a current of 12 A is perpendicular to a magnetic field whose magnitude is 8 T. The length of the conductor is 5 cm. Calculate the magnitude of the magnetic force on this current segment.

Solution Because the current segment $I\,\Delta\ell$ is perpendicular to **B**, you can use Eq. (20-5):

$$F_{max} = I\,\Delta\ell B = (12 \text{ A})(0.05 \text{ m})(8 \text{ T}) = 4.80 \text{ N}$$

PROBLEM 20-6 What will the magnetic force become if the conductor described in Problem 20-5 is at an angle of 40° with respect to the magnetic field?

Solution Directly from Eq. (20-4a):

$$F_{mag} = I\,\Delta\ell B \sin\theta = (12 \text{ A})(0.05 \text{ m})(8 \text{ T}) \sin 40° = 3.09 \text{ N}$$

PROBLEM 20-7 A small rectangular coil is used to measure the magnitude of a magnetic field between the north and south poles of two bar magnets. The area of the coil is 2 cm^2 and the current is 0.8 A. The maximum torque on this 15-turn coil is 1.2×10^{-2} N m. Calculate the magnitude of the magnetic field.

Solution Use Eq. (20-6) and solve for the magnitude of **B**. The maximum torque is present when θ is 0°:

$$\tau = NIBA \cos\theta$$

$$B = \frac{\tau}{NIA} = \frac{1.2 \times 10^{-2} \text{ N m}}{(15)(0.8 \text{ A})(2 \text{ cm}^2)}\left(\frac{10^4 \text{ cm}^2}{1 \text{ m}^2}\right) = 5.00 \text{ T}$$

PROBLEM 20-8 The magnitude of a magnetic field 2 cm from a long straight wire is found to be 6×10^{-5} T. What current is flowing through the wire?

Solution Use Eq. (20-7) and solve for the unknown current:

$$B = k'\frac{2I}{a} \qquad I = \frac{Ba}{2k'} = \frac{(6 \times 10^{-5}\ \text{T})(0.02\ \text{m})}{2(10^{-7}\ \text{N/A}^2)} = 6.00\ \text{A}$$

PROBLEM 20-9 A circular loop carries a current of 4 A. The magnetic field at the center of this loop has a magnitude of 1.257×10^{-4} T. Calculate the radius of the circular loop.

Solution Use Eq. (20-8) and solve for the unknown radius:

$$B = k'\frac{2\pi I}{r} \qquad R = \frac{k'2\pi I}{B} = \frac{(10^{-7}\ \text{N/A}^2)2\pi(4\ \text{A})}{1.257 \times 10^{-4}\ \text{T}} = 0.0200\ \text{m}$$

PROBLEM 20-10 Two parallel wires have equal currents in the same direction. The wires are 2.0 m long and only 5 mm apart. The magnetic force on each wire is 1.152×10^{-2} N. Calculate the current in the two wires.

Solution Use Eq. (20-7) to calculate the magnetic field produced by one of the wires:

$$B = k'\frac{2I}{a} = \frac{(10^{-7}\ \text{N/A}^2)2I}{5 \times 10^{-3}\ \text{m}}$$

Since the current is perpendicular to **B**, use Eq. (20-5) to find the magnetic force:

$$F = I\,\Delta\ell B = \frac{I(2.0\ \text{m})(10^{-7}\ \text{N/A}^2)2I}{5 \times 10^{-3}\ \text{m}} = (8 \times 10^{-5}\ \text{N/A}^2)I^2 = 1.152 \times 10^{-2}\ \text{N}$$

Now, solve for the unknown current:

$$I^2 = \frac{1.152 \times 10^{-2}\ \text{N}}{8 \times 10^{-5}\ \text{N/A}^2} = 144\ \text{A}^2 \qquad I = 12.0\ \text{A}$$

PROBLEM 20-11 A conducting loop lies on a horizontal surface within a uniform magnetic field that is directed vertically upward, as shown in Fig. 20-10. The magnitude of the field is 4.0 T and the area of the loop is 0.3 m². When the left end of the conducting loop is pulled, the area is reduced to 0.1 m². This change in area takes place within 0.2 s. Calculate the magnitude of the average EMF that is induced within the loop.

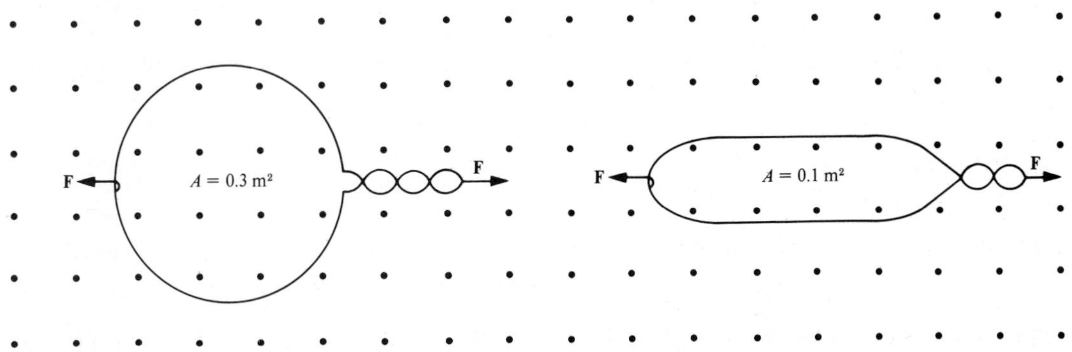

Figure 20-10

Solution Use Eq. (20-9) to calculate the flux through the loop before and after the shape of the loop is changed:

$$\Phi_1 = BA_1 \cos\theta = (4.0\ \text{T})(0.3\ \text{m}^2)\cos 0° = 1.2\ \text{Wb}$$

$$\Phi_2 = BA_2 \cos\theta = (4.0\ \text{T})(0.1\ \text{m}^2)\cos 0° = 0.4\ \text{Wb}$$

Now, use Eq. (20-10) to calculate the average EMF induced within the loop:

$$\mathscr{E}_{ave} = \frac{\Delta\Phi}{\Delta t} = \frac{(1.2 - 0.4)\ \text{Wb}}{0.2\ \text{s}} = 4.00\ \text{V}$$

PROBLEM 20-12 What is the direction of the induced current in the conducting loop described in Problem 20-11?

Solution Since the flux through the loop decreases, the induced current must produce a flux in the same direction to oppose the reduction in flux. A counterclockwise current will produce an upward flux within the loop.

Supplementary Exercises

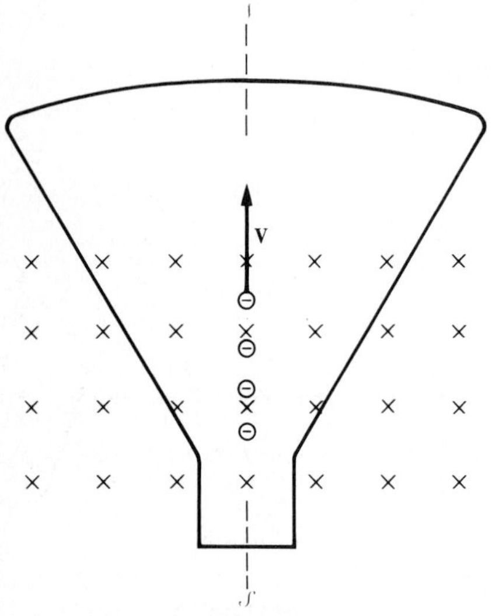

Figure 20-11

PROBLEM 20-13 A stream of electrons moves due north within a television tube. A magnetic field, directed vertically downward, is present, as shown in Fig. 20-11. What is the direction of the magnetic force on the moving electrons? [Hint: The force of a magnetic field on a negative charge carrier is in the direction opposite that on a positive charge carrier.]

PROBLEM 20-14 The moving electrons described in Problem 20-13 experience a force of 7.68×10^{-16} N when the magnitude of the magnetic field is 6.0 T. The charge of the electrons is 1.6×10^{-19}. How fast are the electrons moving?

PROBLEM 20-15 A magnetic field whose magnitude is 3.0 T is directed toward the south. A conductor 30 cm long is placed within this magnetic field. The conductor is horizontal and makes an angle of 25° with respect to the north-south axis. The magnitude of the magnetic force on the conductor is 4.564 N. What current is flowing through the conductor?

PROBLEM 20-16 What will the force on the conductor described in Problem 20-15 be if the conductor is aligned with the east-west axis?

PROBLEM 20-17 A rectangular loop carrying a current of 5 A is placed within a uniform magnetic field with a magnitude of 1 T, as illustrated in Fig. 20-3. The dimensions of the loop are: $a = 4$ cm and $b = 7$ cm. Calculate the total torque acting on the loop when the plane of the loop makes an angle of 30° with respect to B.

PROBLEM 20-18 The magnitude of the magnetic field 1 cm from a long straight conductor is $B = 2.6 \times 10^{-4}$ T. What is the current in this conductor?

PROBLEM 20-19 A circular loop containing 4 turns has a current of 6 A. The magnitude of the magnetic field at the center of this loop is 1.508×10^{-3} T. What is the radius of the loop?

PROBLEM 20-20 The magnetic flux within a conducting loop increases from 8.0 Wb to 12.0 Wb within 0.05 s. Calculate the induced EMF within the conducting loop.

Answers to Supplementary Exercises

20-13: $\mathbf{F}_{mag}$ is toward the east

20-14: 800 m/s

20-15: 12.0 A

20-16: 10.8 N

20-17: 1.21×10^{-2} N m

20-18: 13.0 A

20-19: 1.00 cm

20-20: 80.0 V

21 ALTERNATING CURRENT CIRCUITS

THIS CHAPTER IS ABOUT

☑ Alternating Current
☑ Purely Resistive, Capacitive, or Inductive AC Circuits
☑ Series *RLC* Circuits
☑ Transformers

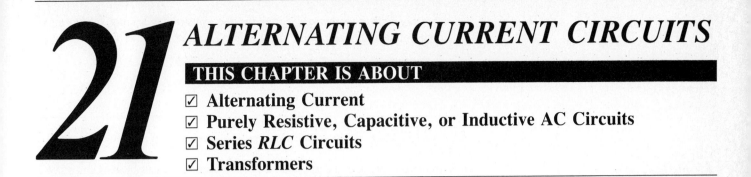

21-1. Alternating Current

Unlike a battery, which produces a direct current (DC), an electromechanical generator (symbol —⊖—) produces an *alternating* current whose output voltage varies as a sine function.

- **Alternating current** (AC) is an electric current that reverses direction at regular intervals and has a magnitude that varies sinusoidally.
- The **frequency** f of alternating current is the number of times per second that the current changes from the + direction to the − direction and back again.

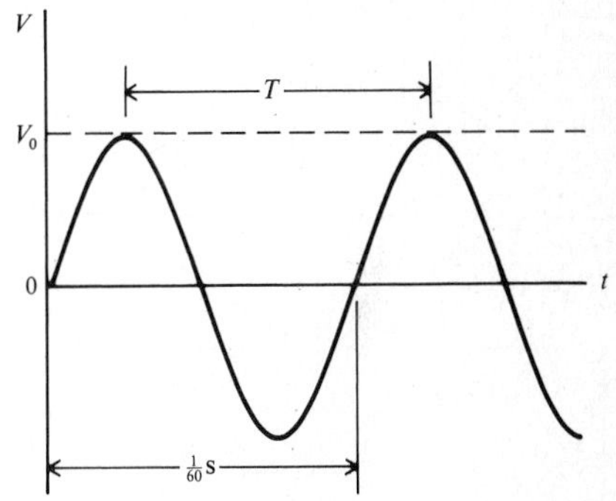

Figure 21-1

Figure 21-1 shows a graph over time of the voltage supplied from a standard household outlet in the United States. The frequency is 60 cycles per second (60 Hz), so the period T is 1/60 s.

The variation in the voltage of an AC generator is a sine function that depends on the frequency:

AC VOLTAGE $$V = V_0 \sin 2\pi ft \tag{21-1}$$

where V_0 is the *peak amplitude* of the voltage, f is the frequency, and t is the time.

21-2. Purely Resistive, Capacitive, or Inductive AC Circuits

A. Purely resistive AC circuits

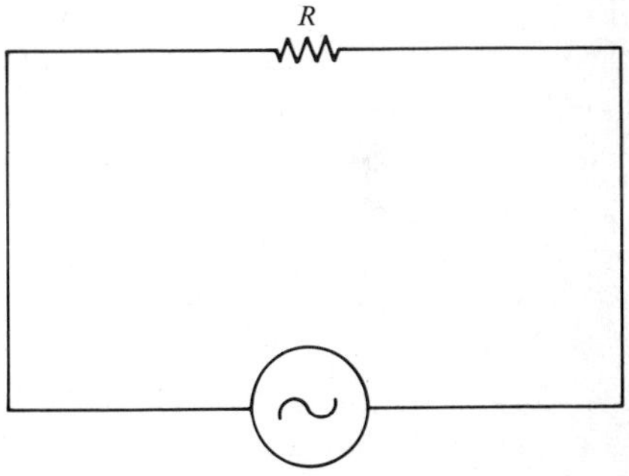

Figure 21-2

When an AC generator is placed in a circuit containing a single resistor, as shown in Figure 21-2, the AC current in the circuit is *in phase* with the applied voltage. This means that I is positive when V is positive, I is negative when V is negative, and the maximum and minimum values of I and V occur at the same time (see Figure 21-3).

All of the power used in AC circuits is in the resistive elements (see Sec. 19-2). The current oscillates from zero to maximum current and then back to zero every half cycle.

- **Power transfer** is the rate at which electrical energy is transformed to heat energy within the resistive elements:

The power transfer changes as the current varies. But we can calculate the *average* rate at which energy is transformed over one or many cycles.

AVERAGE POWER TRANSFER IN A RESISTOR $$P_{ave} = \frac{1}{2}I_0^2R \tag{21-2}$$

$$P_{ave} = I_{eff}^2R \tag{21-3}$$

The average current, called the **effective** (or root-mean-square) **current** is given by

EFFECTIVE CURRENT (RESISTIVE) $$I_{eff} = \frac{I_0}{\sqrt{2}} \tag{21-4}$$

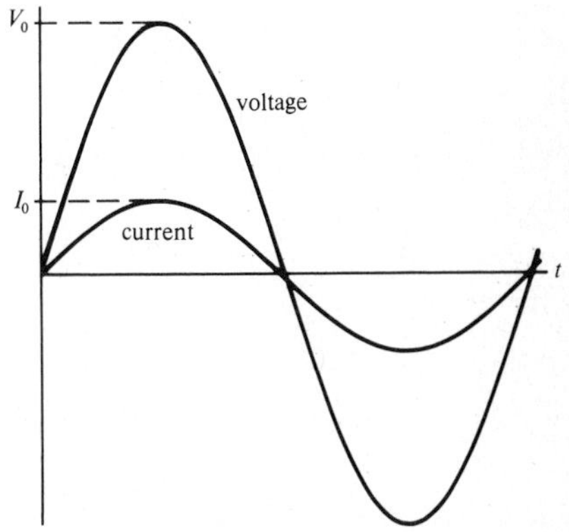

Figure 21-3

where I_0 is the peak current. Similarly, the voltage varies from zero to maximum voltage and back again every half cycle. The **effective voltage** is

EFFECTIVE VOLTAGE (RESISTIVE)

$$V_{\text{eff}} = \frac{V_0}{\sqrt{2}} \tag{21-5}$$

EXAMPLE 21-1: The maximum voltage of the AC circuit shown in Figure 21-2 is 5.0 V.
Determine the average rate at which a 20-Ω resistor produces heat. (Ohm's law is also valid for AC circuits.)

Solution: First, calculate the maximum current for the given voltage and resistance from Ohm's law (Eq. 18-1):

$$I_0 = \frac{V_0}{R} = \frac{5.0\ \text{V}}{20\ \Omega} = 0.25\ \text{A}$$

Now, you can find the average power transfer (heat produced) within the resistor from Eq. (21-2):

$$P_{\text{ave}} = \frac{1}{2}I_0^2R = \frac{1}{2}(0.25\ \text{A})^2(20\ \Omega) = 0.625\ \text{W}$$

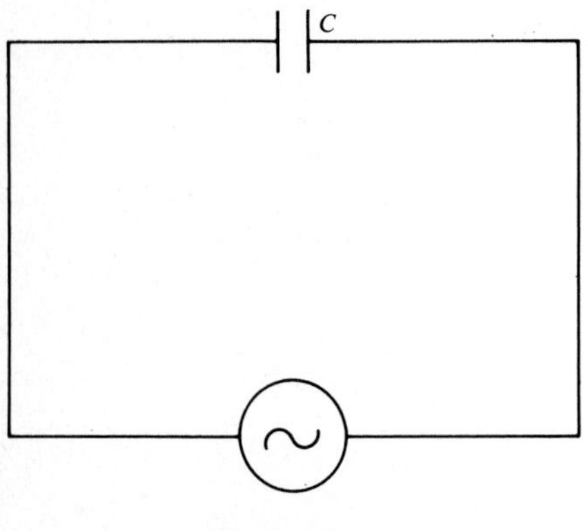

Figure 21-4

B. Purely capacitive AC circuits

When a DC circuit with a capacitor is turned on, the current flows only until charge accumulates on the capacitor plates. The capacitor then blocks the flow of current. In an AC circuit, a capacitor impedes the free flow of current but doesn't stop it entirely. In an AC circuit with a capacitor, the charges flow onto the capacitor plate when the current is in one direction, then flow off again when the current changes direction. When an AC circuit contains a single capacitor (see Sec. 17-2), as shown in Figure 21-4, the current sine wave leads the applied voltage sine wave by $\frac{1}{4}$ cycle; that is, I_0 precedes V_0 by 90°. This means that the current reaches its peak value *before* the applied voltage reaches its peak value. The effective current depends on the capacitance C and the frequency of the generator:

EFFECTIVE CURRENT (CAPACITIVE)

$$I_{\text{eff}} = V_{\text{eff}}(2\pi fC) \tag{21-6}$$

The capacitor impedes the flow of the alternating current, especially if the frequency is low and the capacitance is small.

- The **reactance** X_C of a capacitor is its ability to impede the flow of an AC current.

If we think of Eq. (21-6) as a form of Ohm's law, $I = V/R$, the equations will look the same if we define reactance as

CAPACITIVE REACTANCE

$$X_C = \frac{1}{2\pi fC} \tag{21-7}$$

You can calculate the effective current through a capacitor from the reactance of the capacitor, which is measured in ohms, and the effective voltage across the two plates of the capacitor:

EFFECTIVE CURRENT (CAPACITIVE)

$$I_{\text{eff}} = \frac{V_{\text{eff}}}{X_C} \tag{21-8}$$

The maximum current is related to the maximum voltage as follows:

MAXIMUM CURRENT (CAPACITIVE)

$$I_0 = \frac{V_0}{X_C} \tag{21-9}$$

EXAMPLE 21-2: The capacitance of the circuit shown in Figure 21-4 is 6 μF, the effective AC voltage is 12 V, and the voltage source has a frequency of 60 Hz. Calculate (**a**) the reactance of the capacitor and (**b**) the effective current in the circuit.

Solution:
(**a**) Directly from Eq. (21-7):

$$X_C = \frac{1}{2\pi f C} = \frac{1}{2\pi(60\ \text{Hz})(6 \times 10^{-6}\ \text{F})} = 442.1\ \Omega$$

(**b**) Directly from Eq. (21-8):

$$I_{\text{eff}} = \frac{V_{\text{eff}}}{X_C} = \frac{12\ \text{V}}{442.1\ \Omega} = 2.71 \times 10^{-2}\ \text{A}$$

EXAMPLE 21-3: For the circuit shown in Figure 21-4, $V_{\text{eff}} = 12$ V and $C = 6\ \mu$F.
Determine the maximum current.

Solution: Use Eq. (21-5) and solve for the maximum voltage of the AC generator:

$$V_{\text{eff}} = \frac{V_0}{\sqrt{2}} \qquad V_0 = \sqrt{2}\,V_{\text{eff}} = \sqrt{2}(12\ \text{V}) = 16.97\ \text{V}$$

You found X_C in Example 21-2, so you can calculate the maximum current from Eq. (21-9):

$$I_0 = \frac{V_0}{X_C} = \frac{16.97\ \text{V}}{442.1\ \Omega} = 3.84 \times 10^{-2}\ \text{A}$$

C. Purely inductive AC circuits

When the last switch in a circuit is closed the current does not jump instantly to its final value. Instead, when the current flows it sets up a magnetic field and thus a magnetic flux through the circuit. Because the current is increasing, the flux is changing. Lenz's law (Chapter 20) tells us that this changing magnetic flux causes a current in the direction opposite the initial current. The net effect is to reduce the rate at which the current increases. This limitation of current buildup is called **inductance** and can be greatly enhanced by including a closely wound coil of wire in the circuit.

- A **pure inductor** (symbol ‿‿‿‿) is a coil whose resistance is negligible.

We measure inductance in **henries** H, where an inductance of 1 H induces an EMF of 1 V in a coil when the current changes at a rate of 1 A/s. The coil shown in Figure 21-5 consists of several closely spaced turns of copper wire. When an AC circuit contains a single pure inductor with inductance L, the coil impedes the flow of current and the current lags behind the applied voltage sine wave by $\frac{1}{4}$ cycle; that is, V_0 precedes I_0 by 90°. This means that the current reaches its peak value *after* the applied voltage reaches its peak value.

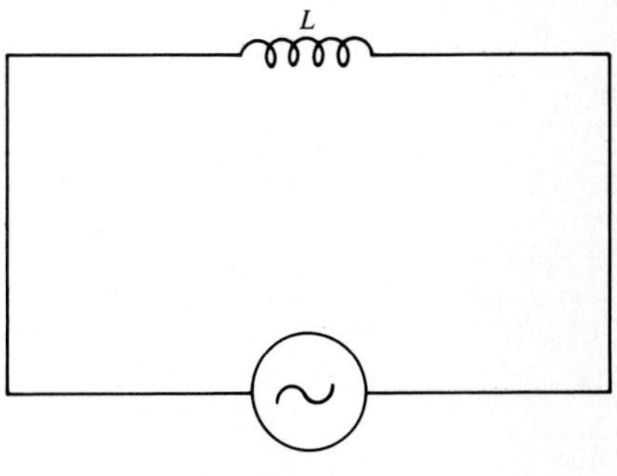

Figure 21-5

The reactance of the inductor is given by

INDUCTIVE REACTANCE

$$X_L = 2\pi f L \tag{21-10}$$

We can use an equation that again looks like Ohm's law to give the effective current as

EFFECTIVE CURRENT (INDUCTIVE)

$$I_{\text{eff}} = \frac{V_{\text{eff}}}{X_L} \tag{21-11}$$

EXAMPLE 21-4: The inductance of the circuit shown in Figure 21-5 is 0.2 H, the frequency of the generator is 50 Hz, and the maximum voltage of the generator is 169.7 V.
Determine the effective current in the circuit.

Solution: To find the effective current, you must know the inductive reactance and the effective voltage. So, first use Eq. (21-10) to calculate the inductive reactance:

$$X_L = 2\pi fL = 2\pi(50\ \text{Hz})(0.2\ \text{H}) = 62.83\ \Omega$$

Then, from Eq. (21-5):

$$V_{\text{eff}} = \frac{V_0}{\sqrt{2}} = \frac{169.7\ \text{V}}{\sqrt{2}} = 120\ \text{V}$$

Now, you can calculate the effective current from Eq. (21-11):

$$I_{\text{eff}} = \frac{V_{\text{eff}}}{X_L} = \frac{120\ \text{V}}{62.83\ \Omega} = 1.91\ \text{A}$$

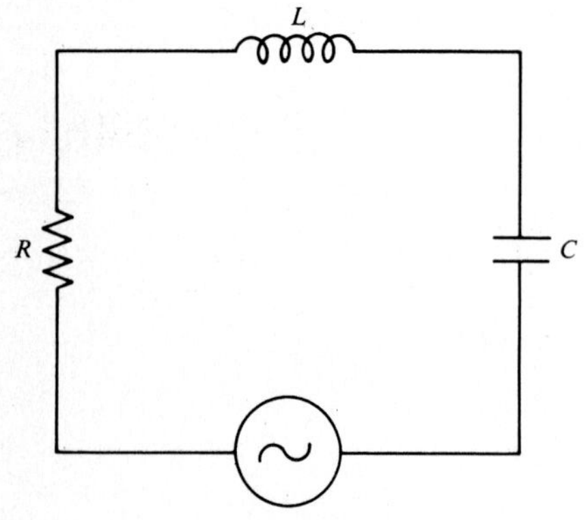

Figure 21-6

21-3. Series *RLC* Circuits

- A **series *RLC*** circuit contains resistive, inductive, and capacitive components connected in series (see Figure 21-6).

[**recall**: In a circuit with components connected in series, the current runs through the components one after the other. (Sec. 18-2A)] All three components impede the flow of the alternating current in the circuit.

The effective current in a series *RLC* circuit is

EFFECTIVE CURRENT (SERIES *RLC*)

$$I_{\text{eff}} = \frac{V_{\text{eff}}}{Z} \tag{21-12}$$

The **impedance** Z of the circuit is given by

IMPEDANCE (SERIES *RLC*)

$$Z = \sqrt{R^2 + (X_L - X_C)^2} \tag{21-13}$$

EXAMPLE 21-5: For the series *RLC* circuit shown in Figure 21-6, $R = 75\ \Omega$, $L = 0.4$ H, $C = 12\ \mu$F, $f = 60$ Hz, and $V_0 = 50$ V.
What is the effective current in the circuit?

Solution: To find the effective current, you must know the effective voltage and the impedance of the circuit. To find the impedance, you must know the resistance, the inductive reactance, and the capacitive reactance. Begin by calculating the inductive reactance from Eq. (21-10):

$$X_L = 2\pi fL = 2\pi(60\ \text{Hz})(0.4\ \text{H}) = 150.8\ \Omega$$

Next, find the capacitive reactance from Eq. (21-7):

$$X_C = \frac{1}{2\pi(60\ \text{Hz})(12 \times 10^{-6}\ \text{F})} = 221.05\ \Omega$$

Now, you can use Eq. (21-13) to find the impedance of the circuit:

$$Z = \sqrt{R^2 + (X_L - X_C)^2} = 102.8\ \Omega$$

Then, calculate the effective voltage from Eq. (21-5):

$$V_{\text{eff}} = \frac{50\ \text{V}}{\sqrt{2}} = 35.36\ \text{V}$$

Finally, use Eq. (21-12) to find the effective current in the circuit:

$$I_{\text{eff}} = \frac{V_{\text{eff}}}{Z} = \frac{35.36\text{ V}}{102.8\ \Omega} = 0.344\text{ A}$$

The relationship between impedance, resistance, and reactance of an *RLC* circuit is illustrated in Figure 21-7. The impedance Z is the hypotenuse of a right triangle whose other two sides are the resistance R and the combined reactance $(X_L - X_C)$ of the inductor and the capacitor.

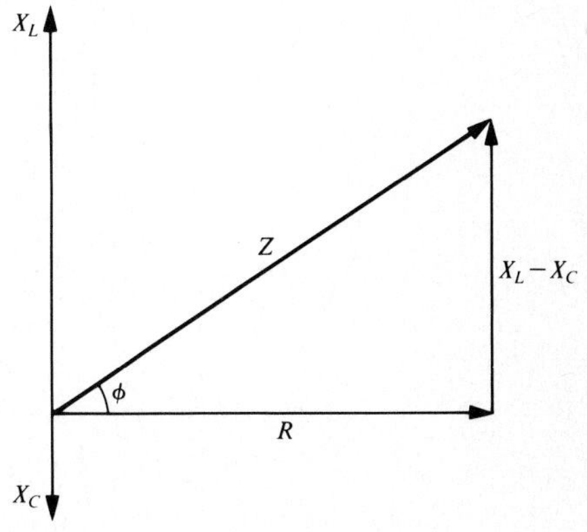

Figure 21-7

- The angle ϕ is the **phase difference** between the current and the voltage of an *RLC* circuit:

PHASE DIFFERENCE

$$\phi_R = \arctan\left(\frac{X_L - X_C}{R}\right) \qquad \textbf{(21-14)}$$

note: If $(X_L - X_C)$ is positive, the voltage *leads* the current.

EXAMPLE 21-6: Determine the phase difference for the *RLC* circuit described in Example 21-5.

Solution: You can find the angle ϕ directly from Eq. (21-14):

$$\phi = \arctan\left(\frac{X_L - X_C}{R}\right) = \arctan\left(\frac{150.8\ \Omega - 221\ \Omega}{75\ \Omega}\right)$$
$$= \arctan(-0.936) = -43.1^\circ.$$

The negative sign indicates that the voltage *lags behind* the current by 43.1°

EXAMPLE 21-7: For the *RLC* circuit described in Example 21-5, determine the inductance of a coil that, when added in series with the 0.4-H inductor, will put the voltage and current in phase. (Hint: The reactances of inductors in series are added.)

Solution: You can see from Figure 21-7 that the phase difference will be zero if $X_L - X_C$ is zero. Let X_L' represent the reactance of the inductor to be added in series. Then,

$$X_L' + X_L = X_C$$
$$X_L' = X_C - X_L = (221 - 150.8)\ \Omega = 70.2\ \Omega$$

And, you can use Eq. (21-10) to solve for the inductance of the coil to be added:

$$L' = \frac{X_L'}{2\pi f} = \frac{70.2\ \Omega}{2\pi(60\text{ Hz})} = 0.186\text{ H}$$

EXAMPLE 21-8: An AC circuit contains a 40-Ω resistor and a pure inductor connected in series with a generator. The effective voltage of the generator is 50 V and its frequency is 60 Hz. The voltage across the resistor and inductor leads the current by 30°.
Calculate the inductance.

Solution: Since the circuit contains no capacitance, Eq. (21-14) reduces to $\tan\phi = X_L/R$, and you can solve for the reactance of the inductor:

$$X_L = R\tan 30^\circ = (40\ \Omega)(0.5774) = 23.09\ \Omega$$

Use Eq. (21-11) to solve for the inductance:

$$L = \frac{X_L}{2\pi f} = \frac{23.09\ \Omega}{2\pi(60\text{ Hz})} = 0.0613\text{ H}$$

A. Series resonance

The impedance of an *RLC* circuit depends on the frequency, and so, for a given applied voltage, the current changes when the frequency changes.

- A series *RLC* circuit is **in resonance** when the current in the circuit has its maximum possible value.

The current is equal to the voltage divided by the impedance (V/Z), so the maximum current corresponds to the minimum impedance. Since the impedance has its minimum value when $X_L = X_C$ ($\phi_R = 0$), the condition of resonance is $2\pi fL = 1/2\pi fC$.

- The **resonant frequency** of a circuit is the frequency at which the current is in phase with the voltage:

RESONANT FREQUENCY (SERIES *RLC* CIRCUIT)

$$f_{\text{res}} = \frac{1}{2\pi\sqrt{LC}} \quad \textbf{(21-15)}$$

EXAMPLE 21-9: Calculate the resonant frequency for the *RLC* circuit described in Example 21-5.

Solution: Directly from Eq. (21-15):

$$f_{\text{res}} = \frac{1}{2\pi\sqrt{LC}} = \frac{1}{2\pi\sqrt{(0.4\text{ H})(12 \times 10^{-6}\text{ F})}} = 72.6\text{ Hz}$$

EXAMPLE 21-10: Calculate the maximum current that flows through the *RLC* circuit described in Example 21-5.

Solution: The current is at a maximum when the frequency of the generator has the value found in Example 21-9. At this frequency, the impedance of the circuit is equal to R because $X_L - X_C = 0$. Use Eq. (21-12) to calculate the current at resonance:

$$I_{\text{eff}} = \frac{(50\text{ V})/\sqrt{2}}{75\ \Omega} = 0.471\text{ A}$$

B. Parallel resonance

Figure 21-8 shows a radio antenna coupled to a parallel tuned circuit. The parallel tuned circuit must have a resonant frequency equal to the radio station's broadcast frequency so that the radio receiver can amplify the broadcast frequency. The following equation gives the resonant frequency:

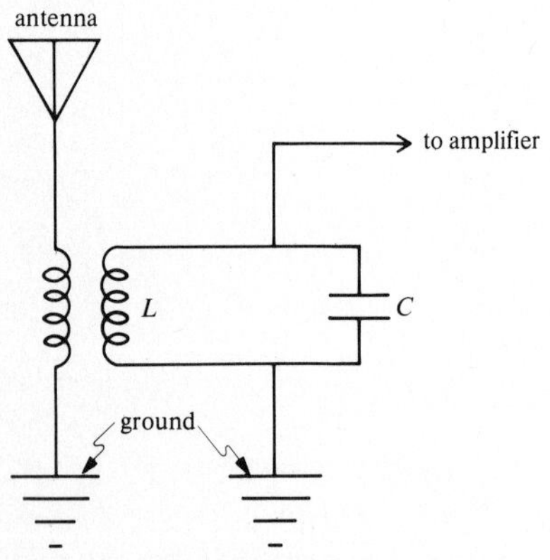

Figure 21-8

RESONANT FREQUENCY (PARALLEL *RLC* CIRCUIT)

$$f_{\text{res}} = \frac{1}{2\pi\sqrt{LC}} \quad \textbf{(21-16)}$$

EXAMPLE 21-11: The broadcast frequency of a radio station is 1500 kHz. The inductance of the parallel resonant circuit is 5 mH.
What value of the capacitance will allow this broadcast signal to be received?

Solution: Use Eq. (21-16) and solve for the unknown capacitance:

$$f = \frac{1}{2\pi\sqrt{LC}}$$

$$C = \frac{1}{4\pi^2 f^2 L} = \frac{1}{4\pi^2(1.5 \times 10^6\text{ Hz})^2(5 \times 10^{-3}\text{ H})} = 2.25 \times 10^{-12}\text{ F}$$

The transmitter of a radio signal also contains a parallel tuned circuit.

EXAMPLE 21-12: The inductance in the parallel tuned circuit of a radio transmitter is 8 mH. The capacitance is adjusted to 4×10^{-12} F.
What is the resonant frequency of this parallel tuned circuit?

Solution: Directly from Eq. (21-16):

$$f = \frac{1}{2\pi\sqrt{(8 \times 10^{-3}\ \text{H})(4 \times 10^{-12}\ \text{F})}} = 890\ \text{kHz}$$

21-4. Transformers

- A **transformer** is a device capable of increasing or reducing the AC voltage of a generator without appreciable loss of power; that is, a transformer can change the effective value of the EMF in a circuit.

Most transformers consist of an iron core with two coils wrapped around the core (see Figure 21-9). The *primary coil* is the coil connected to the generator. The effective voltage produced by the *secondary coil* depends on the number of turns in the secondary coil N_s and the number of turns in the primary coil N_p:

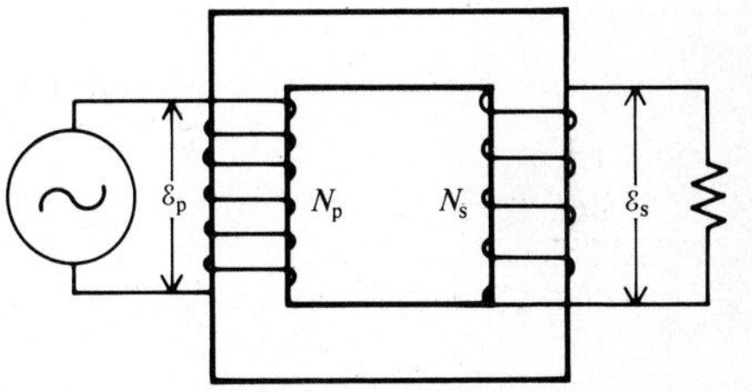

Figure 21-9

EFFECTIVE VOLTAGE IN THE SECONDARY COIL OF A TRANSFORMER

$$\mathscr{E}_s = \mathscr{E}_p \frac{N_s}{N_p} \qquad \textbf{(21-17)}$$

note: The value of voltage or current for an AC circuit always refers to the *effective* value, unless otherwise stated.

EXAMPLE 21-13: A transformer used for a doorbell has a secondary voltage of 12 V. The primary coil, which is connected to a 120-V outlet, has 600 turns.
How many turns does the secondary coil have?

Solution: Use Eq. (21-17) and solve for the number of turns in the secondary coil:

$$\mathscr{E}_s = \mathscr{E}_p \frac{N_s}{N_p}$$

$$N_s = N_p \frac{\mathscr{E}_s}{\mathscr{E}_p} = 600 \frac{12\ \text{V}}{120\ \text{V}} = 60$$

EXAMPLE 21-14: The primary coil of a transformer is connected to a 110-V generator with a frequency of 60 Hz. The effective current through the primary coil is 0.2 A.
Calculate the inductance of the primary coil of the transformer.

Solution: First, calculate the reactance of the primary coil from Eq. (21-11):

$$X_L = \frac{V_{\text{eff}}}{I_{\text{eff}}} = \frac{110\ \text{V}}{0.2\ \text{A}} = 550\ \Omega$$

Then, calculate the inductance from Eq. (21-10):

$$L = \frac{X_L}{2\pi f} = \frac{550\ \Omega}{2\pi(60\ \text{Hz})} = 1.46\ \text{H}$$

SUMMARY

AC voltage	$V = V_0 \sin 2\pi ft$	the variation in the voltage of an AC generator is a sine function that depends on the frequency
Average power transfer in a resistor	$P_{\text{ave}} = \frac{1}{2} I_0^2 R$ $P_{\text{ave}} = I_{\text{eff}}^2 R$	power transfer is the rate at which electrical energy is transformed to heat energy within the resistive elements
Effective current (resistive)	$I_{\text{eff}} = \frac{I_0}{\sqrt{2}}$	effective current in a purely resistive AC circuit is the root mean square of the maximum current
Effective voltage (resistive)	$V_{\text{eff}} = \frac{V_0}{\sqrt{2}}$	effective voltage is the root mean square of the maximum voltage
Effective current (capacitive)	$I_{\text{eff}} = V_{\text{eff}}(2\pi fC)$	effective current in a purely capacitive AC circuit depends on the capacitance and the frequency of the generator
Capacitive reactance	$X_C = \frac{1}{2\pi fC}$	reactance of a capacitor is the rate at which it impedes the flow of an AC current
Effective current (capacitive)	$I_{\text{eff}} = \frac{V_{\text{eff}}}{X_C}$	the effective current of a purely capacitive AC circuit is the ratio of the effective voltage across the two plates of the capacitor and the capacitive reactance
Maximum current (capacitive)	$I_0 = \frac{V_0}{X_C}$	the maximum current of a purely capacitive AC circuit is the ratio of the maximum voltage and the capacitive reactance
Inductive reactance	$X_L = 2\pi fL$	reactance of an inductor is the rate at which it impedes the flow of an AC current
Effective current (inductive)	$I_{\text{eff}} = \frac{V_{\text{eff}}}{X_L}$	the effective current of a purely inductive AC circuit is the ratio of the effective voltage and the inductive reactance
Effective current (series *RLC*)	$I_{\text{eff}} = \frac{V_{\text{eff}}}{Z}$	the effective current in a series *RLC* circuit is the ratio of the effective voltage and the impedance of the circuit

Impedance (series *RLC*)	$Z = \sqrt{R^2 + (X_L - X_C)^2}$	the impedance of a series *RLC* circuit depends on the resistance and reactances
Phase difference	$\phi = \text{arc tan}\left(\frac{X_L - X_C}{R}\right)$	phase difference is the angle between the voltage and the current of an *RLC* circuit
Resonant frequency (series *RLC* circuit)	$f_{\text{res}} = \frac{1}{2\pi\sqrt{LC}}$	frequency at which the current is in phase with the voltage in a series *RLC* circuit
Resonant frequency (parallel *RLC* circuit)	$f_{\text{res}} = \frac{1}{2\pi\sqrt{LC}}$	frequency at which the current is in phase with the voltage in a parallel *RLC* circuit
Effective voltage in the secondary coil of a transformer	$\mathscr{E}_s = \mathscr{E}_p \frac{N_s}{N_p}$	the effective voltage produced by the secondary coil is the ratio of the number of turns in the secondary coil and the number of turns in the primary coil

RAISE YOUR GRADES

Can you define . . . ?

- ☑ power transfer
- ☑ reactance
- ☑ resonant frequency
- ☑ transformer

Can you . . . ?

- ☑ write the relationship between the frequency and the period of an AC generator
- ☑ write the equation for the voltage of an AC generator as a function of time
- ☑ calculate the effective value of an alternating current from its maximum value
- ☑ write the relationship between the effective voltage of an AC generator and the maximum voltage
- ☑ write the expression for the rate at which electrical energy is transformed into heat energy within a pure resistor
- ☑ calculate the reactance of a capacitor if you know the frequency and capacitance
- ☑ write the equation for the effective AC current through a capacitor whose reactance is known
- ☑ write the equation for the reactance of an inductor
- ☑ state the relationship between the effective current through an inductor and the effective voltage
- ☑ write the relationship between the effective current through an *RLC* circuit and the effective voltage of the AC generator
- ☑ calculate the impedance of an *RLC* circuit if you know the voltage of the generator and the current
- ☑ write the expression for the impedance Z of a circuit containing resistance, inductance, and capacitance
- ☑ state the phase angle between current and voltage for a circuit containing a pure capacitance
- ☑ write the relationship between the voltages of the secondary coil of a transformer and the primary coil
- ☑ write the relationship between X_L and X_C when a series *RLC* circuit is in resonance

SOLVED PROBLEMS

PROBLEM 21-1 An AC generator with a frequency of 80 Hz and a maximum voltage of 50 V is connected in series with a resistor of 40 Ω. Calculate the effective current in this circuit.

Solution Calculate the effective voltage directly from Eq. (21-5):

$$V_{\text{eff}} = \frac{V_0}{\sqrt{2}} = \frac{50\ \text{V}}{\sqrt{2}} = 35.36\ \text{V}$$

Now, use Ohm's law (Eq. 18-1) to calculate the effective current:

$$I_{\text{eff}} = \frac{V_{\text{eff}}}{R} = \frac{35.36\ \text{V}}{40\ \Omega} = 0.884\ \text{A}$$

PROBLEM 21-2 Calculate the rate at which electrical energy is transformed into heat energy in the circuit described in Problem 21-1.

Solution Directly from Eq. (21-3):

$$P = I_{\text{eff}}^2 R = (0.884\ \text{A})^2(40\ \Omega) = 31.3\ \text{W}$$

PROBLEM 21-3 A circuit consists of a capacitor connected in series with a generator whose frequency is 60 Hz. The effective voltage of the generator is 85 V. The effective current in this circuit is 0.641 A. Find the value of the capacitance.

Solution Use Eq. (21-8) and solve for the reactance of the capacitor:

$$I_{\text{eff}} = \frac{V_{\text{eff}}}{X_C}$$

$$X_C = \frac{V_{\text{eff}}}{I_{\text{eff}}} = \frac{85\ \text{V}}{0.641\ \text{A}} = 132.6\ \Omega$$

Now, use Eq. (21-7) to calculate the capacitance:

$$X_C = \frac{1}{2\pi f C}$$

$$C = \frac{1}{2\pi f X_C} = \frac{1}{2\pi(60\ \text{Hz})(132.6\ \Omega)} = 20.0\ \mu\text{F}$$

PROBLEM 21-4 A circuit consists of a pure inductor connected in series with an AC generator whose frequency is 50 Hz. The effective voltage of the generator is 120 V and the effective current through the inductor is 6.366 A. Calculate the inductance.

Solution Use Eq. (21-11) to find the reactance of the coil:

$$I_{\text{eff}} = \frac{V_{\text{eff}}}{X_L}$$

$$X_L = \frac{V_{\text{eff}}}{I_{\text{eff}}} = \frac{120\ \text{V}}{6.366\ \text{A}} = 18.85\ \Omega$$

Now, use Eq. (21-10) to calculate the inductance:

$$X_L = 2\pi f L$$

$$L = \frac{X_L}{2\pi f} = \frac{18.85\ \Omega}{2\pi(50\ \text{Hz})} = 0.0600\ \text{H}$$

PROBLEM 21-5 An AC generator is connected in series with a 30-Ω resistor and a 12-μF capacitor. The frequency of the generator is 80 Hz and its maximum voltage is 150 V. Calculate the effective current in this circuit.

Solution Calculate the reactance of the capacitor from Eq. (21-7):

$$X_C = \frac{1}{2\pi(80\text{ Hz})(12 \times 10^{-6}\text{ F})} = 165.8\ \Omega$$

Now, calculate the impedance of the circuit from Eq. (21-13):

$$Z = \sqrt{R^2 + (X_L - X_C)^2} = \sqrt{(30\ \Omega)^2 + (0 - 165.8\ \Omega)^2} = 168.5\ \Omega$$

Then, find the effective current from Eq. (21-12). (Hint: Notice that you're given the maximum voltage, so use Eq. (21-5), $V_{\text{eff}} = V_0/\sqrt{2}$, to obtain the effective voltage.)

$$I_{\text{eff}} = \frac{V_{\text{eff}}}{Z} = \frac{(150\text{ V})/\sqrt{2}}{168.5\ \Omega} = 0.630\text{ A}$$

PROBLEM 21-6 What is the phase difference between the voltage and the current in the circuit described in Problem 21-5?

Solution Use Eq. (21-14) to find the phase angle:

$$\theta = \arctan\left(\frac{X_L - X_C}{R}\right) = \arctan\left(\frac{0 - 165.8\ \Omega}{30\ \Omega}\right) = \arctan(-5.527) = -79.7°$$

So, you know that the voltage *lags* the current by 79.7°.

PROBLEM 21-7 The *RLC* circuit illustrated in Fig. 21-10 contains a 30-Ω resistor, a 0.1-H inductor and two capacitors in parallel. The two capacitors have capacitances of 2 μF and 4 μF. The frequency of the AC generator is 100 Hz and its maximum voltage is 40 V. Calculate the effective current in this circuit.

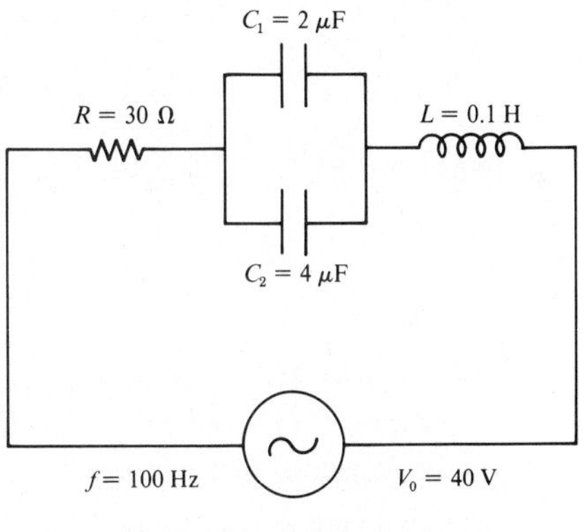

Figure 21-10

Solution You'll need the effective voltage and the impedance to calculate the effective current. So, use Eq. (21-5) to calculate the effective voltage of the generator:

$$V_{\text{eff}} = \frac{40\text{ V}}{\sqrt{2}} = 28.28\text{ V}$$

To find the impedance, first calculate the reactance of the capacitors and the inductor. Since the capacitors are connected in parallel, the total capacitance is

$$C_{\text{tot}} = C_1 + C_2 = 2\ \mu\text{F} + 4\ \mu\text{F} = 6\ \mu\text{F}$$

Find the reactance of the total capacitance from Eq. (21-7):

$$X_C = \frac{1}{2\pi fC} = \frac{1}{2\pi(100\text{ Hz})(6 \times 10^{-6}\text{ F})} = 265.26\ \Omega$$

And, the reactance of the inductor from Eq. (21-10):

$$X_L = 2\pi fL = 2\pi(100\text{ Hz})(0.1\text{ H}) = 62.83\ \Omega$$

Then, calculate the impedance of the circuit from Eq. (21-13):

$$Z = \sqrt{R^2 + (X_L - X_C)^2} = \sqrt{(30\ \Omega)^2 + (62.83\ \Omega - 265.26\ \Omega)^2} = 204.64\ \Omega$$

Now, you can calculate the effective current from Eq. (21-12):

$$I_{\text{eff}} = \frac{V_{\text{eff}}}{Z} = \frac{28.28\text{ V}}{204.64\ \Omega} = 0.138\text{ A}$$

PROBLEM 21-8 The current in the *RLC* circuit shown in Fig. 21-10 will become greater if the frequency of the generator is changed to the resonant frequency. Calculate the effective current in this circuit at resonance.

Solution Since X_L is equal to X_C at the resonant frequency (see Sec. 21-3A), the impedance becomes equal to R. Use Eq. (21-12) to calculate the effective current:

$$I_{\text{eff}} = \frac{28.28\ \text{V}}{30\ \Omega} = 0.943\ \text{A}$$

PROBLEM 21-9 A transformer with 900 turns in its primary coil is designed to reduce 120 V to 40 V. How many turns are required in the secondary coil?

Solution Use Eq. (21-17) to find the number of turns in the secondary coil:

$$\mathscr{E}_{\text{s}} = \mathscr{E}_{\text{p}}\frac{N_{\text{s}}}{N_{\text{p}}}$$

$$N_{\text{s}} = N_{\text{p}}\frac{\mathscr{E}_{\text{s}}}{\mathscr{E}_{\text{p}}} = 900\left(\frac{40\ \text{V}}{120\ \text{V}}\right) = 300$$

PROBLEM 21-10 A 50-Ω resistor is placed in series with an inductor. The AC generator has a frequency of 200 Hz and an effective voltage of 30 V. The voltage in this circuit leads the current by 45°. Calculate the inductance of the inductor.

Solution Since the tangent of 45° is equal to one, the reactance of the inductor is exactly equal to the resistance. From Eq. (21-14):

$$\tan\phi = \frac{X_L - X_C}{R}$$

$$\tan 45° = \frac{X_L - 0}{R}$$

$$X_L = R\tan 45° = (50\ \Omega)(1.0) = 50\ \Omega$$

Now, use Eq. (21-10) to find the inductance:

$$L = \frac{X_L}{2\pi f} = \frac{50\ \Omega}{2\pi(200\ \text{Hz})} = 0.0398\ \text{H}$$

PROBLEM 21-11 If a capacitor of the proper value is added in series with the resistor and inductor described in Problem 21-10, the circuit can have a resonant frequency of 200 Hz. Find the value of the capacitor that is needed.

Solution Since resonance occurs when $X_C = X_L$, X_C must equal 50 Ω. Use Eq. (21-7) to solve for the capacitance:

$$C = \frac{1}{2\pi(200\ \text{Hz})(50\ \Omega)} = 15.9\ \mu\text{F}$$

PROBLEM 21-12 An *RLC* circuit containing resistive, inductive, and capacitive elements in series is connected to an AC generator whose effective voltage is 75 V. The effective current in this circuit is 5 A. Calculate the impedance of the circuit.

Solution Use Eq. (21-12) and solve for the impedance:

$$Z = \frac{V_{\text{eff}}}{I_{\text{eff}}} = \frac{75\ \text{V}}{5\ \text{A}} = 15\ \Omega$$

Supplementary Exercises

PROBLEM 21-13 The maximum voltage of an AC generator is 25 V. The frequency of this generator is 100 Hz. At $t = 0$, the voltage is zero. How long will it take for the voltage to reach 25 V?

PROBLEM 21-14 An AC generator with a frequency of 50 Hz and a maximum voltage of 120 V is connected to a 25-Ω resistor. How much electrical energy is transformed into heat in one minute? [Hint: Remember that energy is power × time.]

PROBLEM 21-15 An AC generator has a maximum voltage of 120 V. What is the effective voltage of this generator?

PROBLEM 21-16 An AC generator with a maximum voltage of 80 V is connected in series with a pure resistor. The maximum current in this circuit is 4.0 A. What is the resistance?

PROBLEM 21-17 An AC generator is connected in series with a pure inductor whose inductance is 20 mH. The frequency of the generator is 60 Hz and its maximum voltage is 240 V. What is the effective current in this circuit?

PROBLEM 21-18 An AC generator is connected in series with a pure inductor. There is no resistance in the circuit. By what angle does the voltage across the inductor lead the current?

PROBLEM 21-19 An AC generator whose frequency is 80 Hz is connected in series with a pure capacitor. The maximum voltage of the generator is 120 V and the capacitance is 12 μF. What is the effective current through this capacitor?

PROBLEM 21-20 A 50-Hz generator is connected in series with a 30-Ω resistor, a 0.2-H inductor and a 15-μF capacitor. The effective voltage of the generator is 90 V. What is the effective current through this circuit?

PROBLEM 21-21 Calculate the resonant frequency of the *RLC* circuit described in Problem 21-20.

PROBLEM 21-22 What is the effective current in the *RLC* circuit described in Problem 21-20 when the frequency of the generator is changed to the resonant frequency?

PROBLEM 21-23 A transformer has its primary coil connected to an AC generator whose effective voltage is 120 V. The primary coil contains 1200 turns. If the secondary coil contains 300 turns, what will the effective voltage be across the secondary coil?

PROBLEM 21-24 The effective current in the primary coil of the transformer described in Problem 21-23 is 2.0 A. What is the inductance of the primary coil if the frequency of the generator is 60 Hz?

Answers to Supplementary Exercises

21-13: 0.00250 s

21-14: 1.73×10^4 J

21-15: 84.9 V

21-16: 20.0 Ω

21-17: 22.5 A

21-18: 90°

21-19: 0.512 A

21-20: 0.591 A

21-21: 91.9 Hz

21-22: 3.00 A

21-23: 30.0 V

21-24: 0.159 H

22 ELECTRICAL APPLICATIONS

THIS CHAPTER IS ABOUT

☑ **Motors, Generators, and Back EMF**
☑ **Energy Transmission**
☑ **Temperature Measurement**
☑ **Energy Storage**
☑ **Current Amplification**

22-1. Motors, Generators, and Back EMF

- An electric **motor** is a device that changes electrical energy into rotational mechanical energy.

A motor is built by wrapping many turns of wire (a **coil**) around a rotor or **armature**. This armature is mounted on a shaft or axle and is in a magnetic field formed by a permanent magnet or electromagnet. When a current flows through the coil, the magnetic force causes the shaft to rotate (Sec. 20-3).

- A **generator** is a device that changes mechanical energy into electrical energy.

A generator can be built just like a motor: a coil of wire wrapped around an armature attached to an axle and in a magnetic field. In a generator, the axle is turned mechanically and the rotation of the conducting coil through the magnetic field induces an EMF. Thus a motor and a generator have a reciprocal nature:

- If the axle is turned mechanically, the device *generates* electricity.
- If current is supplied to the coil, the axle turns and the device runs as a *motor*.

When a motor is turned on, the magnetic force on the coil causes the armature to accelerate. But the armature does not accelerate indefinitely even though there is always a force on it. As the rotation speeds up, the magnetic flux through the coil changes, inducing an EMF (Eq. 20-10). According to Lenz's law (Sec. 20-6A), this induced EMF (called **back EMF** or **counter EMF**) opposes the EMF causing the current in the coil. The back EMF increases as the speed of the motor increases, until a balance is reached and the speed on the motor remains constant. Thus the back EMF controls the speed of the motor.

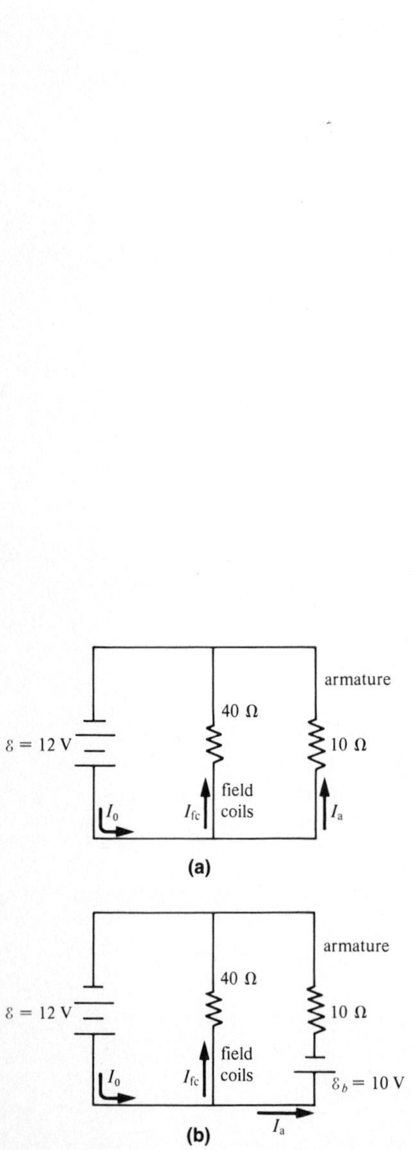

Figure 22-1. Circuit diagram of a motor with field coils: (a) not running; (b) full speed.

EXAMPLE 22-1: The motor illustrated in Figure 22-1a has an armature of resistance 10 Ω. The field coils (resistance 40 Ω) supply the magnetic field required to turn the armature. Both the armature and the field coils are connected to a 12-V battery. Determine the current through the motor at start-up.

Solution: The initial current I_0 with the armature at rest is the sum of the currents through the field coils I_{fc} and the armature I_a. Use Eq. (18-2) to find the electric current in this closed circuit:

$$I_{fc} = \frac{\mathscr{E}}{R} = \frac{12\ \text{V}}{40\ \Omega} = 0.3\ \text{A} \qquad I_a = \frac{12\ \text{V}}{10\ \Omega} = 1.2\ \text{A}$$

Then,

$$I_0 = I_{fc} + I_a = (0.3 + 1.2)\ \text{A} = 1.5\ \text{A}$$

EXAMPLE 22-2: For the motor described in Example 22-1, the full-speed back EMF is 10 V (see Figure 22-1b).

Determine the current through the motor at full speed.

Solution: The current through the field coils is independent of the speed of the armature's rotation, so $I_{fc} = 0.3$ A. Because the back EMF opposes the EMF of the battery (which is what causes the armature to rotate), the voltage across the armature is

$$V = \mathscr{E} - \mathscr{E}_b = (12 - 10)\ \text{V} = 2\ \text{V}$$

Now, use Ohm's law (Eq. 18-1) to calculate the current through the rotating armature:

$$R = \frac{V}{I} \qquad I_a = \frac{V}{R} = \frac{2\ \text{V}}{10\ \Omega} = 0.2\ \text{A}$$

So, the total current is

$$I = I_{fc} + I_a = (0.3 + 0.2)\ \text{A} = 0.5\ \text{A}$$

22-2. Energy Transmission

Any power line through which electric energy is transmitted has some resistance. Current flowing through a line of given resistance R transforms electric energy into heat (a loss of usable energy) at the rate of $P = I^2R$ (Eq. 21-3). This results in a lower voltage at the end of the power line and costs the utility company money to generate the energy that is wasted. Electric power is equal to the product of the current and voltage, $P = IV$ (Eq. 18-4), so one can transmit the same amount of power by using a higher voltage and smaller current. The reduced current results in less energy loss in the lines. Electric companies use **transformers** (Sec. 21-4) to achieve higher voltages.

EXAMPLE 22-3: A two-wire transmission line, with a total resistance of 0.3 Ω, supplies 50 A at 200 V to a transformer. Calculate (**a**) the power supplied to the transformer and (**b**) the power loss due to the resistance of the transmission lines.

Solution:
(**a**) Directly from Eq. (18-4):

$$P_1 = IV = (50\ \text{A})(200\ \text{V}) = 10\,000\ \text{W}$$

(**b**) You can figure the rate at which energy is lost directly from Eq. (21-3):

$$P_2 = I^2R = (50\ \text{A})^2(0.3\ \Omega) = 750\ \text{W}$$

EXAMPLE 22-4: What voltage must be supplied at the beginning of the transmission line in Example 22-3?

Solution: There must always be a voltage difference across a resistor, here the transmission line, for a current flow. The power dissipated by the resistor is $P_2 = IV_2$, so the voltage drop across the transmission line is

$$V_2 = P_2/I = 750\ \text{W}/50\ \text{A} = 15\ \text{V}$$

The voltage at the beginning of the line is the voltage at the end of the line plus the voltage drop across the line.

$$V_b = V + V_2 = (22 + 15)\ \text{V} = 215\ \text{V}$$

EXAMPLE 22-5: What percentage of the power supplied to the transformer in Example 22-3 is lost to heat?

Solution: First, calculate the total power supplied by the power source:

$$P_{tot} = P_1 + P_2 = 10\,000\ \text{W} + 750\ \text{W} = 10\,750\ \text{W}$$

Now, calculate the percentage of the total power lost to heat:

$$\frac{P_2}{P_{tot}} \times 100\% = \frac{750\ \text{W}}{10\,750\ \text{W}} \times 100\% = 6.98\%$$

EXAMPLE 22-6: The voltage delivered by the transmission line of Examples 22-3 and 22-4 is increased from 200 V to 1000 V, but the power supplied to the transformer remains the same. What percentage of power is lost to heat?

Solution: From Problem 22-3(a), you know that the power supplied to the transformer is 10 000 W, so you can calculate the current through the transmission line from Eq. (18-3):

$$I = \frac{P_1}{V} = \frac{10\,000\ \text{W}}{1000\ \text{V}} = 10\ \text{A}$$

Now, from Eq. (18-5), calculate the rate at which energy is lost:

$$P_2 = I^2R = (10\ \text{A})^2(0.3\ \Omega) = 30\ \text{W}$$

The total power delivered is

$$P_{tot} = P_1 + P_2 = (10\,000 + 30)\ \text{W} = 10\,030\ \text{W}$$

The percentage of power lost to heat when the power line voltage is increased to 1000 V is

$$\frac{P_2}{P_{tot}} \times 100\% = \frac{30\ \text{W}}{10\,030\ \text{W}} \times 100\% = 0.3\%$$

Notice that the higher voltage reduces the current and the heat generated.

22-3. Temperature Measurement

The resistance of a conductor varies with temperature, so we can determine the temperature of a wire by measuring its resistance at two temperatures:

RELATIONSHIP OF RESISTANCE TO TEMPERATURE

$$R_T = R_0(1 + \alpha T) \qquad \textbf{(22-1)}$$

where T is Celsius temperature, R_T is the resistance of the wire at T °C, R_0 is the resistance at 0 °C, and α is the **temperature coefficient of resistance.** Table 22-1 lists the temperature coefficient of resistance for several types of wire.

TABLE 22-1: The Temperature Coefficient of Resistance

Substance	$(°C)^{-1}$
aluminum	4.0×10^{-3}
copper	3.9×10^{-3}
iron	5.2×10^{-3}
Nichrome	2.0×10^{-4}
platinum	3.92×10^{-3}
tungsten	4.6×10^{-3}

EXAMPLE 22-7: A piece of platinum wire is used to measure the melting point of an alloy. The resistance of the wire is 150.0 Ω at 0 °C, and 285.0 Ω when the alloy starts to melt. What is the alloy's melting point?

Solution: Solve Eq. (22-1) for the unknown temperature:

$$T = \frac{R_T - R_0}{\alpha R_0} = \frac{(285.0 - 150.0)\ \Omega}{(3.92 \times 10^{-3}\ (°C)^{-1})(150.0\ \Omega)} = 230\ °C$$

EXAMPLE 22-8: A pure metal wire has a resistance of 250 Ω at 0 °C, and a resistance of 410 Ω at 160 °C. What might this metal be?

Solution: Solve Eq. (22-1) for α:

$$\alpha = \frac{R_T - R_0}{TR_0} = \frac{(410 - 250)\ \Omega}{(160\ °C)(250\ \Omega)} = 4.00 \times 10^{-3}\ (°C)^{-1}$$

From Table 22-1, you can see that aluminum has an equivalent temperature coefficient of resistance.

22-4. Energy Storage

A. Capacitors

A capacitor and resistor connected in series with a battery are shown in Figure 22-2. When the switch is closed, electrons move from the left-hand plate of the capacitor through the battery to the right-hand plate of the capacitor until the capacitor is fully charged. The energy stored in the capacitor is

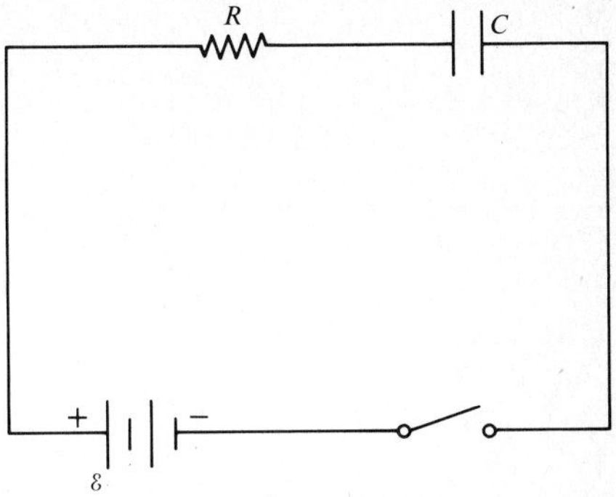

Figure 22-2

ENERGY IN A CAPACITOR

$$E_C = \frac{1}{2}QV \tag{22-2}$$

$$E_C = \frac{1}{2}CV^2 \tag{22-3}$$

EXAMPLE 22-9: Calculate the energy stored in a 5-μF capacitor connected in series to a 10-V battery.

Solution: Directly from Eq. (22-3):

$$E_C = \frac{1}{2}CV^2 = \frac{1}{2}(5 \times 10^{-6}\text{ F})(10\text{ V})^2 = 2.5 \times 10^{-4}\text{ J}$$

Capacitors are not charged instantaneously; that is, the current in the circuit is not steady-state, but rather, it varies with time. Figure 22-3 shows a time-varying current i through a battery and series resistor as a function of time. (Note: We'll use lower-case letters for time-variable quantities.) As the charge builds up on the capacitor plates, the current decreases according to the following equations:

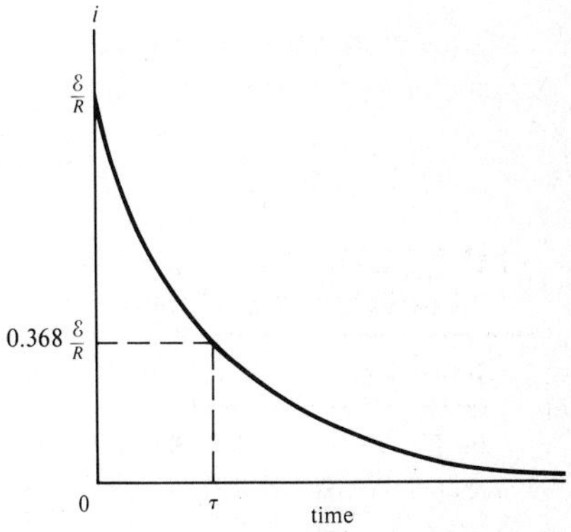

Figure 22-3. Current through a circuit containing a capacitor and resistor.

CHARGING A CAPACITOR (THROUGH A SERIES RESISTOR)

$$i = \frac{\mathscr{E}}{R}e^{-t/RC} \tag{22-4}$$

where $e \cong 2.7183$. When the switch is closed at $t = 0$, the initial current is $i_0 = \mathscr{E}/R$. The **capacitive time constant** τ is the time required to reduce the current to $1/e \cong 0.368$ of its initial value.

TIME CONSTANT (CAPACITOR AND RESISTOR IN SERIES)

$$\tau = RC \tag{22-5}$$

The larger the value of τ, the slower the decrease in current.

EXAMPLE 22-10: Calculate the current through a circuit containing a 40-Ω resistor, a 5-μF capacitor, and a 10-V battery connected in series at time $t = 2\tau$.

Solution: Use Eq. (22-4) and substitute τ for RC and 2τ for t:

$$i = \frac{\mathscr{E}}{R}e^{-t/RC} = \frac{\mathscr{E}}{R}e^{-t/\tau} = \frac{\mathscr{E}}{R}e^{-2\tau/\tau}$$

$$= \frac{10\text{ V}}{40\ \Omega}e^{-2} = (0.25)(0.1353)\text{ A} = 3.38 \times 10^{-2}\text{ A}$$

B. Inductors

An inductor and a resistor connected in series with a battery are shown in Figure 22-4. When the switch is closed, the initial current through the circuit is zero. In time, the current reaches its maximum value:

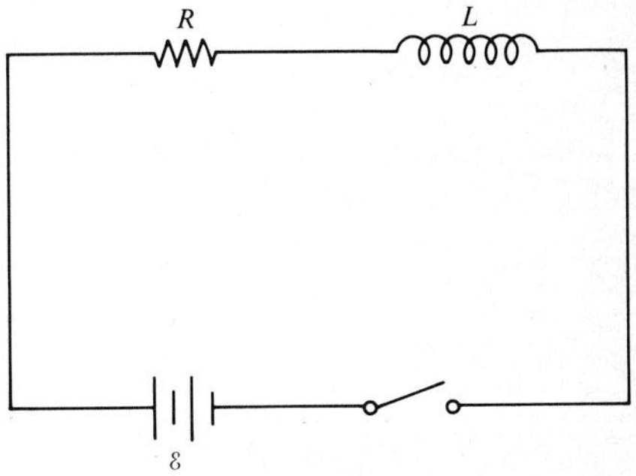

Figure 22-4

MAXIMUM CURRENT THROUGH AN INDUCTOR

$$I_{max} = \frac{\mathscr{E}}{R} \tag{22-6}$$

After the current reaches its maximum value, the energy stored in the inductor is

ENERGY IN AN INDUCTOR

$$E_L = \frac{1}{2}LI^2 \quad \textbf{(22-7)}$$

EXAMPLE 22-11: Given that $R = 30\ \Omega$, $L = 0.2$ H, and $\mathscr{E} = 12$ V, calculate the maximum energy stored in the inductor illustrated in Figure 22-4.

Solution: Directly from Eq. (22-7):

$$E_L = \frac{1}{2}LI^2 = \frac{1}{2}(0.2\ \text{H})\left(\frac{12\ \text{V}}{30\ \Omega}\right)^2 = 0.016\ \text{J}$$

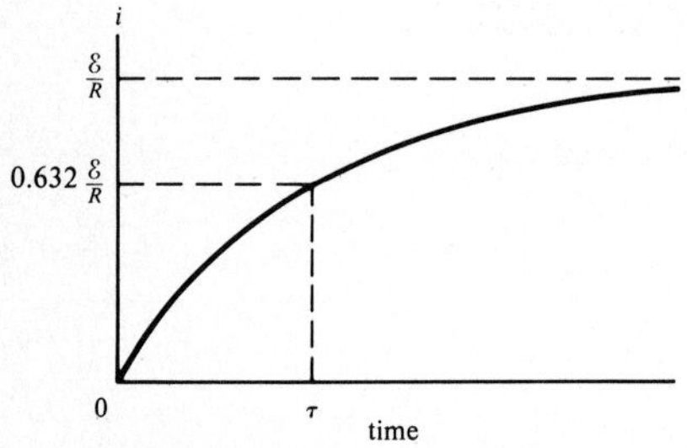

Figure 22-5. Current through an inductor with a resistor in series.

Maximum current doesn't flow through an inductor instantaneously. Figure 22-5 shows how the current through an inductor increases with time, as described by the following equation:

CURRENT THROUGH AN INDUCTOR (WITH A SERIES RESISTOR)

$$i = \frac{\mathscr{E}}{R}(1 - e^{-(R/L)t}) \quad \textbf{(22-8)}$$

When the switch is closed at $t = 0$, the initial current i_0 is zero. The **inductive time constant** τ_L is the time required to increase the current through the inductor to $(1 - 1/e) \cong 0.632$ of its maximum value:

TIME CONSTANT (INDUCTOR AND RESISTOR IN SERIES)

$$\tau_L = \frac{L}{R} \quad \textbf{(22-9)}$$

The larger the value of τ_L, the slower the increase in current.

EXAMPLE 22-12: Find the current through a 0.2-H inductor connected in series to a 30-Ω resistor and a 12-V battery at time $t = 2\tau$.

Solution: Use Eq. (22-8) with $\tau_L = R/L$ and $t = 2\tau$:

$$i = \frac{\mathscr{E}}{R}(1 - e^{-(R/L)t}) = \frac{\mathscr{E}}{R}(1 - e^{-2\tau/\tau})$$

$$= \frac{12\ \text{V}}{30}(1 - e^{-2}) = 0.346\ \text{A}$$

22-5. Current Amplification

- **Current amplification** is a large gain in current output resulting from a small change in current input.
- A **transistor** acts as a current amplifier.

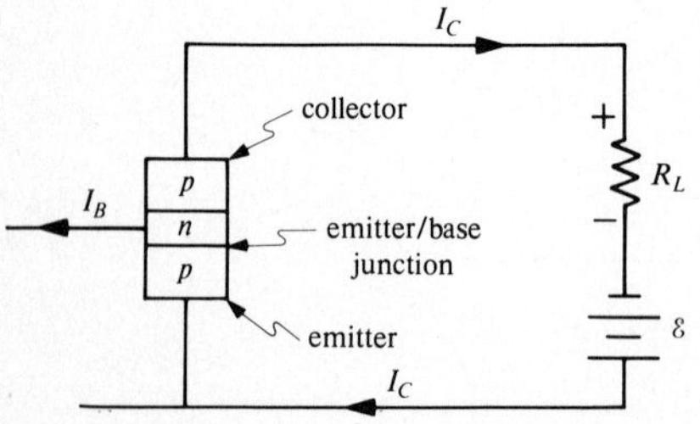

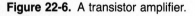

Figure 22-6. A transistor amplifier.

Figure 22-6 shows a transistor amplifier. When the base current I_B changes slightly, the collector current I_C through the load resistor R_L changes significantly. This current amplification or current gain β of a transistor is given by

CURRENT AMPLIFICATION (TRANSISTOR)

$$\beta = \frac{\Delta I_C}{\Delta I_B} \quad \textbf{(22-10)}$$

EXAMPLE 22-13: The transistor in Figure 22-6 has a current amplification of 120 and a collector current of 2 mA when the base current is 80 μA. Calculate the collector current through the load resistor when the base current is increased to 90 μA.

Solution: Use Eq. (22-10) and solve for the change in the collector current:

$$\Delta I_C = \beta\,\Delta I_B = (120)(90 - 80)\ \mu\text{A} = 1.2 \times 10^{-3}\ \text{A}$$
$$= 1.2\ \text{mA}$$

Now, you can calculate the resulting collector current I'_C. [Hint: The change in the collector current is simply its later value minus its earlier value.]

$$I'_C - I_C = \Delta I_C$$
$$I'_C = I_C + \Delta I_C = (2 + 1.2)\ \text{mA} = 3.2\ \text{mA}$$

SUMMARY

Relationship of resistance to temperature	$R_T = R_0(1 + \alpha T)$	the resistance of a wire at any temperature is proportional to its resistance at 0 °C and its temperature coefficient of resistance
Energy in a capacitor	$E_C = \frac{1}{2}QV$ or $E_C = \frac{1}{2}CV^2$	gives energy stored in a capacitor
Charging a capacitor (through a series resistor)	$i = \frac{\mathscr{E}}{R}e^{-t/RC}$	gives the current decrease of a time-varying current through a battery and series resistor as a function of time
Time constant (capacitor and resistor in series)	$\tau = RC$	gives the time required to reduce the current through a capacitor to $1/e$ of its initial value
Maximum current through an inductor	$I_{\text{max}} = \frac{\mathscr{E}}{R}$	gives the maximum value of a current through an inductor as the ratio of the induced EMF and the resistance
Energy in an inductor	$E_L = \frac{1}{2}LI^2$	gives the energy stored in an inductor
Current through an inductor (with a series resistor)	$i = \frac{\mathscr{E}}{R}(1 - e^{-(R/L)t})$	gives the current increase through an inductor as a function of time
Time constant (inductor and resistor in series)	$\tau_L = \frac{L}{R}$	gives the time required to increase the current through an inductor to 63% of its maximum value
Current amplification (transistor)	$\beta = \frac{\Delta I_C}{\Delta I_B}$	gives the current gain of a transistor

RAISE YOUR GRADES

Can you define . . . ?

- ☑ a motor
- ☑ a generator
- ☑ a back EMF
- ☑ a transistor
- ☑ current amplification

Can you . . . ?

- ☑ calculate the starting current through an electric motor if you know the resistance of the armature and the field coils
- ☑ calculate the loss of electrical energy in a transmission line
- ☑ state why transmission lines use very high voltages
- ☑ write the relationship between the resistance of a wire at a temperature T °C and at 0 °C
- ☑ calculate the energy stored in a capacitor if you know its voltage and charge
- ☑ write the equation for the current in a circuit that contains a capacitor and resistor in series with a seat of EMF
- ☑ calculate the time constant of a capacitor and resistor in series
- ☑ calculate the maximum current through an inductor
- ☑ calculate the energy stored in an inductor
- ☑ calculate the current through an inductor connected in series with a resistor
- ☑ calculate the time constant of a circuit containing an inductor and resistor in series
- ☑ write the expression for the current amplification of a transistor

SOLVED PROBLEMS

PROBLEM 22-1 A DC motor has field coils with a resistance of 30 Ω and an armature with a resistance of 6 Ω. What is the initial current through this motor when it is connected to a 12-V battery?

Solution First, use Eq. (18-2) to calculate the current through the field coils and the current through the armature:

$$I_{\text{fc}} = \frac{\mathscr{E}}{R_{\text{fc}}} = \frac{12\text{ V}}{30\ \Omega} = 0.4\text{ A}$$

$$I_{\text{a}} = \frac{\mathscr{E}}{R_{\text{a}}} = \frac{12\text{ V}}{6\ \Omega} = 2.0\text{ A}$$

Now, you can calculate the total current:

$$I_{\text{tot}} = I_{\text{fc}} + I_{\text{a}} = 0.4\text{ A} + 2.0\text{ A} = 2.4\text{ A}$$

PROBLEM 22-2 When the motor described in Problem 22-1 reaches full speed, the total current through the motor is reduced to 0.9 A. What is the back EMF generated by the rotating armature?

Solution Calculate the current through the armature. [**recall:** I_{fc} is independent of the speed of the armature's rotation (see Example 22-2).]

$$I_{\text{a}} = I_{\text{tot}} - I_{\text{fc}} = 0.9\text{ A} - 0.4\text{ A} = 0.5\text{ A}$$

Now, calculate the voltage across the rotating armature from Ohm's law (Eq. 18-1):

$$R = \frac{V}{I}$$

$$\Delta V = I_a R_a = (0.5\ \text{A})(6\ \Omega) = 3\ \text{V}$$

This voltage is equal to the battery EMF minus the back EMF:

$$\Delta V = \mathscr{E} - \mathscr{E}_b$$

So, you can solve for the back EMF:

$$\mathscr{E}_b = \mathscr{E} - \Delta V = 12\ \text{V} - 3\ \text{V} = 9\ \text{V}$$

PROBLEM 22-3 A power plant producing 800 V is connected to transmission lines that have a total resistance of 0.2 Ω. The current through the transmission lines is 120 A, and the lines deliver 95 kW of power to a factory. Calculate the fraction of energy that is wasted because of heating of the transmission lines.

Solution Calculate the loss of power due to the resistance of the transmission lines directly from Eq. (21-3):

$$P_2 = I^2 R = (120\ \text{A})^2 (0.2\ \Omega) = 2880\ \text{W}$$

Then, calculate the total power produced by the generating station:

$$P_{\text{tot}} = 95\ \text{kW} + 2.88\ \text{kW} = 97.88\ \text{kW}$$

Now, you can calculate the fraction of the electric power that is wasted:

$$\frac{P_2}{P_{\text{tot}}} \times 100\% = \frac{2.88\ \text{kW}}{97.88\ \text{kW}} \times 100\% = 2.94\%$$

PROBLEM 22-4 If the voltage of the power plant and transmission lines in Problem 22-3 is doubled, reducing the current to 60 A, what fraction of the electric energy is lost because of heating of the transmission lines?

Solution Calculate the loss of power within the transmission lines:

$$P_2 = (60\ \text{A})^2 (0.2\ \Omega) = 720\ \text{W}$$

The power supplied to the factory remains the same, so the total power produced is

$$P_{\text{tot}} = 95\ \text{kW} + 0.72\ \text{kW} = 95.72\ \text{kW}$$

Now, calculate the fraction of the total power that is wasted:

$$\frac{0.72\ \text{kW}}{95.72\ \text{kW}} \times 100\% = 0.752\%$$

PROBLEM 22-5 A piece of iron wire at 0 °C has a resistance of 80 Ω. What will the resistance be when the iron wire is heated to 350 °C?

Solution Use Eq. (22-1) and Table 22-1 to find the resistance at 350 °C:

$$R_T = R_0(1 + \alpha T) = (80\ \Omega)[1 + (5.2 \times 10^{-3}\ {}^\circ\text{C}^{-1})(350\ {}^\circ\text{C})] = (80\ \Omega)(2.82) = 226\ \Omega$$

PROBLEM 22-6 A piece of tungsten wire has a resistance of 110 Ω at 0 °C. When this wire is placed in a furnace, its resistance becomes 322.5 Ω. Calculate the temperature of the furnace.

Solution Solve Eq. (22-1) for the unknown temperature:

$$T = \frac{R_T - R_0}{\alpha R_0} = \frac{(332.5 - 110)\ \Omega}{4.6 \times 10^{-3}\ ({}^\circ\text{C})^{-1}(110\ \Omega)} = 440\ {}^\circ\text{C}$$

PROBLEM 22-7 A 20-μF capacitor is connected in series with a 2500-Ω resistor and a 15-V battery. How much electrical energy is stored in the capacitor when it is fully charged?

Solution When the capacitor is fully charged, the potential difference between the two plates will be 15 V (see Sec. 17-1). So, directly from Eq. (22-3):

$$E_C = \frac{1}{2}CV^2 = \frac{1}{2}(20 \times 10^{-6}\ \text{F})(15\ \text{V})^2 = 2.25 \times 10^{-3}\ \text{J}$$

PROBLEM 22-8 When the capacitor described in Problem 22-7 is connected to the battery, what is the initial current?

Solution Use Eq. (22-4) and set $t = 0$:

$$i_0 = \frac{\mathscr{E}}{R}e^0 = \frac{\mathscr{E}}{R} = \frac{15\ \text{V}}{2500\ \Omega} = 6.00\ \text{mA}$$

PROBLEM 22-9 As the capacitor described in Problem 22-7 becomes charged, the current decreases. At what time after the switch is closed will the current be equal to 5% of its initial value?

Solution Use Eq. (22-4) to find the time when $i = 0.05\ i_0$:

$$i = \frac{\mathscr{E}}{R}e^{-t/RC}$$

$$0.05\ i_0 = i_0 e^{-t/RC}$$

Now, cancel i_0 and take the natural logarithm of both sides of the equation:

$$\ln(0.05) = -t/RC$$

$$t = -\ln(0.05)RC = -(-2.996)(2500\ \Omega)(20 \times 10^{-6}\ \text{F}) = 0.150\ \text{s}$$

PROBLEM 22-10 When an inductor has a current of 6 A, the energy stored in the inductor is 0.216 J. Calculate the inductance.

Solution Solve Eq. (22-7) for the inductance:

$$E_L = \frac{1}{2}LI^2$$

$$L = \frac{2\ E_L}{I^2} = \frac{2(0.216\ \text{J})}{(6\ \text{A})^2} = 12\ \text{mH}$$

PROBLEM 22-11 A closely spaced coil with an inductance of 12 mH is connected in series with a 50-Ω resistor and a 20-V battery. How long will it take for the current in this circuit to increase to 63% of its final value?

Solution The time required for the current to build up to 63% of its final value is the time constant τ. Directly from Eq. (22-9):

$$\tau = \frac{L}{R} = \frac{12 \times 10^{-3}\ \text{H}}{50\ \Omega} = 2.4 \times 10^{-4}\ \text{s}$$

PROBLEM 22-12 How long will it take for the current through the inductor described in Problem 22-11 to increase to 99% of its final value?

Solution The final value of the current through an inductor is $i_f = \mathscr{E}/R$, Eq. (22-6). First, find the time required for i to become $0.99\ i_f$ from Eq. (22-8):

$$0.99\ i_f = i_f(1 - e^{-t(R/L)})$$

Then, divide both sides of the equation by i_f:

$$0.99 = 1 - e^{-t(R/L)}$$

Subtract -1

$$-0.01 = -e^{-t(R/L)}$$

and take the natural logarithm of both sides of the equation:

$$\ln(0.01) = -t\left(\frac{R}{L}\right)$$

Now, solve for t:

$$t = -\ln(0.01)\left(\frac{L}{R}\right) = \frac{-(-4.605)(12 \times 10^{-3}\ \text{H})}{50\ \Omega} = 1.11 \times 10^{-3}\ \text{s}$$

PROBLEM 22-13 Assume the current gain of the transistor illustrated in Fig. 22-6 is 150. When the base current is 80 μA, the current through the load resistor is 2 mA. How much will the voltage across the 1000-Ω load resistor change if the base current is increased to 120 μA?

Solution Use Eq. (22-10) to calculate the increase in the collector current:

$$\beta = \frac{\Delta I_C}{\Delta I_B}$$

$$\Delta I_C = \beta\, \Delta I_B = (150)(40\ \mu\text{A}) = 6\ \text{mA}$$

Now, find the collector current corresponding to a base current of 120 μA:

$$I'_C = I_C + \Delta I_C = (2 + 6)\ \text{mA} = 8\ \text{mA}$$

Then, calculate the voltage across the 1000-Ω load resistor corresponding to the collector current of 2 mA and 8 mA:

$$\Delta V = IR = (2\ \text{mA})(1000\ \Omega) = 2\ \text{V}$$

$$\Delta V' = (8\ \text{mA})(1000\ \Omega) = 8\ \text{V}$$

$$\Delta V' - \Delta V = (8 - 2)\ \text{V} = 6\ \text{V}$$

So, the voltage across the load resistor increases by 6 V.

Supplementary Exercises

PROBLEM 22-14 When the electric motor whose circuit is shown in Fig. 22-1 is connected to an 8-V battery, the back EMF at full speed is only 6 V. Calculate the total current through this motor.

PROBLEM 22-15 A power station produces an output voltage of 500 V. The current through the transmission line is 60 A. The total resistance of this transmission line is 0.4 Ω. Calculate the power loss due to heating of the transmission lines.

PROBLEM 22-16 A power plant produces 27 kW of electricity at 500 V and delivers it to a small factory through the transmission line described in Problem 22-15. Calculate the percentage of the total power wasted because of heating in the transmission lines.

PROBLEM 22-17 A piece of iron wire has a resistance of 60 Ω at 0 °C. When this wire is placed in a furnace, its resistance becomes 200.4 Ω. Calculate the temperature of the furnace.

PROBLEM 22-18 A piece of wire has a resistance of 45.0 Ω at 0 °C. When this wire is heated to 210 °C, its resistance is 82.04 Ω. What is the wire's temperature coefficient of resistance?

PROBLEM 22-19 The potential difference between the two plates of a capacitor is 12 V. The charge on the capacitor is 240 μC. Calculate the energy stored in the capacitor.

PROBLEM 22-20 Calculate the capacitance of the capacitor described in Problem 22-19.

PROBLEM 22-21 A 20-μF capacitor is connected in series with a 50-Ω resistor, a 20-V battery, and a switch. What is the initial value of the current when the switch is closed?

PROBLEM 22-22 How long will it take for current in the circuit described in Problem 22-21 to be reduced to $1/e$ of its initial value?

PROBLEM 22-23 How long will it take for the current in the circuit described in Problem 22-21 to be reduced to 0.1% of its initial value?

PROBLEM 22-24 A pure inductor is connected in series with a 40-Ω resistor and a 20-V battery. The inductance is 0.6 H. What is the maximum current through this inductor?

PROBLEM 22-25 What is the maximum energy stored in the inductor described in Problem 22-24?

PROBLEM 22-26 The initial current through the inductor described in Problem 22-24 is zero. How long does it take for the current through the inductor to reach 63% of its maximum value?

PROBLEM 22-27 How long will it take for the current through the inductor described in Problem 22-24 to reach 99.9% of its final value?

PROBLEM 22-28 The current amplification of a transistor amplifier is 180. What change in the base current will cause the collector current to increase by 25 mA?

Answers to Supplementary Exercises

22-14: $I_{tot} = 0.4$ A

22-15: 1440 W

22-16: 4.32%

22-17: 450 °C

22-18: 3.92×10^{-3} (°C)$^{-1}$

22-19: 1.44×10^{-3} J

22-20: 20 μF

22-21: 0.4 A

22-22: 1.0×10^{-3} s

22-23: 6.91×10^{-3} s

22-24: 0.5 A

22-25: 0.075 J

22-26: 0.015 s

22-27: 0.104 s

22-28: 1.39×10^{-4} A

MIDSEMESTER EXAM (Chapters 16–22)

1. Two point charges are 4 cm apart. The first point charge is $+8 \times 10^{-2}$ C and the second point charge is -6×10^{-3} C. Calculate the force of attraction that each charge exerts on the other. **[Ch. 16]**

2. A charge of 4 μC experiences an electric force of 2×10^{-3} N. Calculate the magnitude of the electric field. **[Ch. 16]**

3. A current of 120 mA is flowing through a small lamp bulb. How many electrons pass through this lamp in 30 s? **[Ch. 16]**

4. The work required to move a positive charge of 12 μC from point B to point A is 6.0×10^{-5} J. Calculate the potential difference between points B and A. **[Ch. 17]**

5. A point charge of 12 μC is 4 cm from point P. What is the potential at P due to this charge? **[Ch. 17]**

6. The potential difference between two parallel conductors is 250 V. One of the conductors has a positive charge of 50 μC and the other conductor has a negative charge of 50 μC. Calculate the capacitance of this pair of conductors. **[Ch. 17]**

7. A circuit contains a battery whose EMF is 12 V. The internal resistance of this battery is 0.5 Ω. The circuit connected to the battery contains a resistor of 5.5 Ω. What current flows through this circuit? **[Ch. 18]**

8. Calculate the rate at which electric energy is transformed into heat in the 5.5-Ω resistor in Problem 7. **[Ch. 18]**

9. A current of 1.2 A is flowing through the filament of a lamp. The voltage across the filament is 8.0 V. Calculate the rate at which electric energy is transformed into heat in this lamp filament. **[Ch. 18]**

10. A 6-Ω resistor and a 10-Ω resistor are connected in parallel. What is the resistance of these two resistors? **[Ch. 18]**

11. Three conductors are connected at a junction point B. A current of 250 mA is moving toward the junction in one of the conductors. A second conductor has a current of 450 mA moving toward the junction. What current is present in the third conductor? **[Ch. 19]**

12. The resistivity of aluminum is 2.60×10^{-8} Ω m. Calculate the resistance of an aluminum wire 80 cm long and 0.2 mm in diameter. **[Ch. 19]**

13. A galvanometer has an internal resistance of 1200 Ω. Its full-scale current is 100 μA. Calculate the resistance of a resistor connected in parallel with this galvanometer so that the full-scale current will be 2.0 A. **[Ch. 19]**

14. The galvanometer described in Problem 13 is to be converted to a voltmeter with a full-scale deflection of 5.0 V. Calculate the value of the series resistor that is needed. **[Ch. 19]**

15. A point charge of 5 μC is moving through a magnetic field of 6.0 T at a speed of 20 m/s. The charge is moving perpendicular to the magnetic field. Calculate the magnetic force on this moving charge. **[Ch. 20]**

16. What will the magnetic force become if the angle between the velocity vector and the magnetic field described in Problem 15 is 25°? **[Ch. 20]**

17. A copper wire 40 cm long is carrying a current of 6.0 A and is perpendicular to a magnetic field of 5.0 T. Calculate the magnetic force acting on this conductor. **[Ch. 20]**

18. What is the magnitude of the magnetic field at a distance of 2 cm from a long straight conductor carrying a current of 8 A? **[Ch. 20]**

19. An alternating current generator producing a maximum voltage of 40 V is connected to a 10-Ω resistor. What is the *maximum* current through this resistor? **[Ch. 18]**

20. At what rate is electric energy transformed into heat in the 10-Ω resistor in Problem 19? **[Ch. 18]**

21. What is the effective or root-mean-square current in the circuit described in Problem 19? **[Ch. 21]**

22. An AC generator whose frequency is 60 Hz and whose effective voltage is 50 V is connected to a capacitor whose value is 10 μF. Calculate the alternating current that flows through this capacitor. **[Ch. 21]**

23. An AC generator whose effective voltage is 100 V is connected to a pure inductor. The frequency of the generator is 60 Hz and the inductance of the coil is 8 mH. Calculate the effective current through this inductor. **[Ch. 21]**

24. A transmission line consists of two parallel wires, each of which has a resistance of 0.5 Ω. The current through this transmission line is 4.0 A. Calculate the rate at which electric energy is transformed into heat. **[Ch. 18]**

25. The resistance of an aluminum wire at 0 °C is 5 Ω. The temperature coefficient of resistance of aluminum is 0.0040 $(°C)^{-1}$. Calculate the resistance of this wire at 150 °C. **[Ch. 22]**

26. A capacitor contains a charge of 40 μC. The potential difference between the two plates is 600 V. Calculate the energy stored within this capacitor. **[Ch. 22]**

27. **(a)** Find the currents through R_1, R_2, and point A in Figure E-4. **[Chs. 18, 19]**
(b) What is the power dissipated in each resistor? **[Ch. 18]**
(c) R_1 and R_2 are replaced with capacitors C_1 and C_2 with capacitances of 10 μF and 20 μF, respectively. What is the charge on each capacitor? **[Ch. 18]**
(d) C_1 and C_2 are now replaced by an equivalent single capacitor. What must the capacitance of this new capacitor be? **[Ch. 18]**
(e) How much charge does the new capacitor hold? **[Ch. 17]**
(f) The battery is now replaced by a 10-V AC generator whose frequency is 60 Hz, and a 10-Ω resistor is inserted in the circuit. What is the effective current in the circuit? **[Ch. 21]**

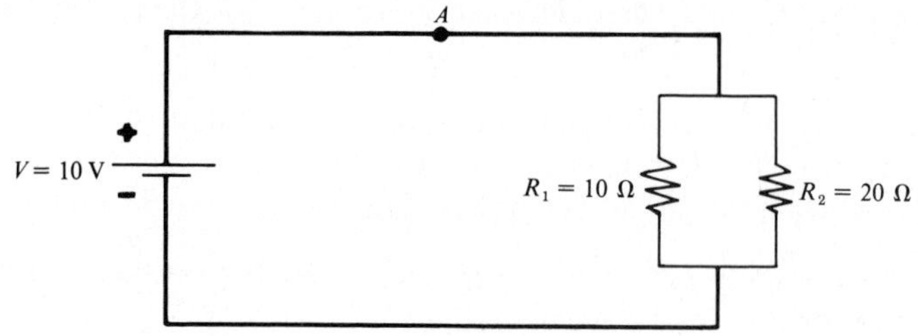

Figure E-4

Solutions to Midsemester Exam

1. Use Coulomb's law.

$$F = k\frac{Q_1 Q_2}{r^2} = \left(9 \times 10^9 \frac{\text{N m}^2}{\text{C}^2}\right)\frac{(8 \times 10^{-2}\text{ C})(-6 \times 10^{-3}\text{ C})}{(4 \times 10^{-2}\text{ m})^2} = -2.70 \times 10^9\text{ N}$$

2.
$$E = \frac{F_{\text{elec}}}{Q} = \frac{2 \times 10^{-3}\text{ N}}{4 \times 10^{-6}\text{ C}} = 500\text{ N/C}$$

3. The average current is $I_{\text{ave}} = \Delta Q/\Delta t$, so

$$\Delta Q = I_{\text{ave}}\,\Delta t = (120 \times 10^{-3}\text{ A})(30\text{ s}) = 3.6\text{ C}$$

Now divide this charge by the charge on one electron to find the number of electrons.

$$3.6\text{ C}\left(\frac{1\text{ electron}}{1.6 \times 10^{-19}\text{ C}}\right) = 2.25 \times 10^{19}\text{ electrons}$$

4. The potential difference between two points is the work required to move a positive charge from one point to the other, divided by the charge.

$$V_{\text{AB}} = \frac{W_{\text{B}\to\text{A}}}{Q} = \frac{6 \times 10^{-5}\text{ J}}{12 \times 10^{-6}\text{ C}} = 5.00\text{ V}$$

5.
$$V = \frac{kQ}{r} = \frac{(9 \times 10^9\text{ N m}^2/\text{C}^2)(12 \times 10^{-6}\text{ C})}{4 \times 10^{-2}\text{ m}} = 2.70 \times 10^6\text{ V}$$

6.
$$C = \frac{Q}{V} = \frac{50 \times 10^{-6}\text{ C}}{250\text{ V}} = 2.00 \times 10^{-7}\text{ F} = 0.200\ \mu\text{F}$$

7. Resistors connected in series can be added directly, so

$$R_{\text{tot}} = R_i + R = 0.5\ \Omega + 5.5\ \Omega = 6.0\ \Omega$$

Now divide the EMF by the total resistance to find the current.

$$I = \frac{\mathcal{E}}{R_{\text{tot}}} = \frac{12\text{ V}}{6\ \Omega} = 2.00\text{ A}$$

8.
$$P = I^2R = (2\text{ A})^2(5.5\ \Omega) = 22.0\text{ J/s}$$

9.
$$P = IV = (1.2\text{ A})(8\text{ V}) = 9.60\text{ J/s}$$

10. Resistors in parallel must be added by means of their reciprocals.

$$\frac{1}{R_{\text{tot}}} = \frac{1}{R_1} + \frac{1}{R_2} = \frac{1}{6\ \Omega} + \frac{1}{10\ \Omega} = 0.267\ \Omega^{-1} \qquad R_{\text{tot}} = \frac{1}{0.267\ \Omega^{-1}} = 3.75\ \Omega$$

11. Use Kirchhoff's first law.

$$I_1 + I_2 + I_3 = 0 \qquad I_3 = -I_1 - I_2 = (-250 - 450)\text{ mA} = -700\text{ mA}$$

The negative sign indicates that the direction of the current is away from the junction.

12. The cross-sectional area of the wire is

$$A = \pi r^2 = \pi(0.2\text{ mm}/2)^2 = 3.14 \times 10^{-2}\text{ mm}^2 = 3.14 \times 10^{-8}\text{ m}^2$$

so the resistance of the wire is

$$R = \frac{\rho L}{A} = \frac{(2.6 \times 10^{-8}\ \Omega\text{ m})(8 \times 10^{-1}\text{ m})}{3.14 \times 10^{-8}\text{ m}^2} = 0.662\ \Omega$$

13. First, find the voltage drop across the galvanometer by using Ohm's law.

$$V = I_G R_G = (100 \times 10^{-6}\text{ A})(1200\ \Omega) = 0.12\text{ V}$$

Second, use Kirchhoff's first law to find the current through the shunt.

$$I_G + I_S = I_{\text{tot}} \qquad I_S = I_{\text{tot}} - I_G = 2.0\text{ A} - 100 \times 10^{-6}\text{ A} \approx 2.00\text{ A}$$

Now you can find the resistance of the shunt by using Ohm's law again.

$$R_S = \frac{V}{I_S} = \frac{0.12 \text{ V}}{2.00 \text{ A}} = 6.00 \times 10^{-2} \ \Omega$$

14. Since the resistors are connected in series, the resistance of the galvanometer can be simply subtracted from the total resistance. Use Ohm's law to find the total resistance of the voltmeter.

$$R = R_{tot} - R_G = \frac{V}{I} - R_G = \frac{5.0 \text{ V}}{100 \times 10^{-6} \text{ A}} - 1200 \ \Omega = 4.88 \times 10^4 \ \Omega$$

15. $$F_{mag} = QvB \sin\theta = (5 \times 10^{-6} \text{ C})(20 \text{ m/s})(6.0 \text{ T})(\sin 90°) = 6.00 \times 10^{-4} \text{ N}$$

16. $$F_{mag} = QvB \sin\theta = (5 \times 10^{-6} \text{ C})(20 \text{ m/s})(6.0 \text{ T})(\sin 25°) = 2.54 \times 10^{-4} \text{ N}$$

17. $$F_{mag} = I \, \Delta\ell B \sin\theta = (6 \text{ A})(4 \times 10^{-1} \text{ m})(5.0 \text{ T})(\sin 90°) = 12.0 \text{ N}$$

18. $$B = k'\frac{2I}{a} = (1 \times 10^{-7} \text{ N/A}^2)\frac{2(8 \text{ A})}{2 \times 10^{-2} \text{ m}} = 8.00 \times 10^{-5} \text{ T}$$

19. Use Ohm's law, $V = IR$

$$I = \frac{V}{R} = \frac{40 \text{ V}}{10 \ \Omega} = 4.00 \text{ A}$$

20. $$P_{ave} = \tfrac{1}{2}I_0^2R = \tfrac{1}{2}(4.00 \text{ A})^2(10 \ \Omega) = 80.0 \text{ W}$$

21. $$I_{eff} = \frac{I_0}{\sqrt{2}} = \frac{4.00 \text{ A}}{\sqrt{2}} = 2.83 \text{ A}$$

22. $$I_{eff} = V_{eff}(2\pi fC) = (50 \text{ V})(2\pi)(60 \text{ Hz})(10 \times 10^{-6} \text{ F}) = 0.188 \text{ A}$$

23. First find the inductive reactance, then divide the voltage by the reactance to find the effective current.

$$X_L = 2\pi fL = 2\pi(60 \text{ Hz})(8 \times 10^{-3} \text{ H}) = 3.016 \ \Omega$$

$$I_{eff} = \frac{V_{eff}}{X_L} = \frac{100 \text{ V}}{3.016 \ \Omega} = 33.2 \text{ A}$$

24. $$P = I^2R_{tot} = (4 \text{ A})^2(0.5 \ \Omega + 0.5 \ \Omega) = 16.0 \text{ W}$$

25. $$R_T = R_0(1 + \alpha T) = 5 \ \Omega[1 + (0.0040 \ (°\text{C})^{-1})(150 \ °\text{C})] = 8.00 \ \Omega$$

26. $$E_C = \tfrac{1}{2}QV = \tfrac{1}{2}(40 \times 10^{-6} \text{ C})(600 \text{ V}) = 0.0120 \text{ J}$$

27. **(a)** Since R_1 and R_2 are connected in parallel, both have 10 V across them.

$$I_1 = \frac{V}{R_1} = \frac{10 \text{ V}}{10 \ \Omega} = 1.00 \text{ A} \qquad I_2 = \frac{V}{R_2} = \frac{10 \text{ V}}{20 \ \Omega} = 0.500 \text{ A}$$

To find the current through point A, use Kirchhoff's first law.

$$I_A = I_1 + I_2 = (1.00 + 0.50) \text{ A} = 1.50 \text{ A}$$

(b) $$P_1 = I_1^2R_1 = (1 \text{ A})^2(10 \ \Omega) = 10.0 \text{ W} \qquad P_2 = I_2^2R_1 = (0.5 \text{ A})^2(20 \ \Omega) = 5.00 \text{ W}$$

(c) Since C_1 and C_2 are connected in parallel, both have 10 V across them.

$$Q_1 = C_1V = (10 \times 10^{-6} \text{ F})(10 \text{ V}) = 1.00 \times 10^{-4} \text{ C} \quad Q_2 = C_2V = (20 \times 10^{-6} \text{ F})(10 \text{ V}) = 2.00 \times 10^{-4} \text{ C}$$

(d) The capacitances of capacitors in parallel may be added directly.

$$C = C_1 + C_2 = (10 + 20) \ \mu\text{F} = 30 \ \mu\text{F}$$

(e) $$Q = CV = (30 \times 10^{-6} \text{ F})(10 \text{ V}) = 3.00 \times 10^{-4} \text{ C}$$

(f) First find the capacitive reactance X_C, then combine the capacitive reactance with the resistance R to find the circuit's total impedance Z. Finally, divide the voltage by the impedance to find the effective current.

$$X_C = \frac{1}{2\pi fC} = \frac{1}{2\pi(60 \text{ Hz})(30 \times 10^{-6} \text{ F})} = 88.4 \ \Omega$$

$$Z = \sqrt{R^2 + (X_L - X_C)^2} = \sqrt{(10 \ \Omega)^2 + (0 - 88.4 \ \Omega)^2} = 89.0 \ \Omega$$

(X_L, the inductive reactance, is zero because there is no inductor in the circuit.)

$$I_{eff} = \frac{V_{eff}}{Z} = \frac{10 \text{ V}}{89.0 \ \Omega} = 0.112 \text{ A}$$

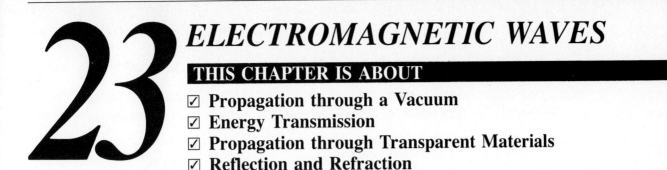

23 ELECTROMAGNETIC WAVES

THIS CHAPTER IS ABOUT

- ☑ Propagation through a Vacuum
- ☑ Energy Transmission
- ☑ Propagation through Transparent Materials
- ☑ Reflection and Refraction

note: Our discussion of electromagnetic waves necessarily follows from the laws of electricity and magnetism. So, before you begin this chapter, take a moment to review the discussion of the electric field **E** (Section 16-2) and the magnetic field **B** (Section 20-1). Pay close attention to the similarity of the arguments.

23-1. Propagation through a Vacuum

A changing electric field induces a magnetic field, and a changing magnetic field induces an electric field. This symmetrical relationship provides the interaction that propagates an electromagnetic wave.

- An **electromagnetic** (E-M) **wave** can be represented graphically as transverse **E**- and **B**-fields oscillating in phase (see Sec. 21-2A), as shown in Figure 23-1. [**recall:** A transverse wave is generated when the particles in a medium oscillate at right angles to the direction of propagation of the wave (see Sec. 11-1).]

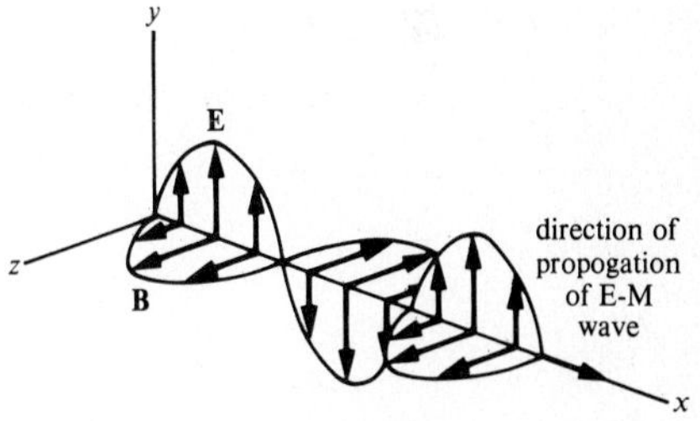

Figure 23-1. Propagation of an electromagnetic wave.

In an electromagnetic wave, the electric and magnetic fields are perpendicular to each other and both are perpendicular to the direction of propagation. In Figure 23-1, the direction of propagation is along the x axis and extends without bound. The length of the vector arrows under each curve represents the *magnitude* of the **E**- and **B**-fields at any given point along the wave.

The *speed* of propagation of an electromagnetic wave through a vacuum is related to the electrical constant k of Coulomb's law (Eq. 16-2) and the magnetic constant k', which appears in the equations for the magnetic field (Eqs. 20-7 and 20-8):

SPEED OF PROPAGATION (E-M WAVE: VACUUM)

$$c = \sqrt{\frac{k}{k'}} \qquad \textbf{(23-1)}$$

Radio waves, light, and X rays are familiar examples of electromagnetic waves.

EXAMPLE 23-1: Calculate the speed of propagation of electromagnetic waves through a vacuum.

Solution: You know that $k \approx 9 \times 10^9\ \text{N m}^2/\text{C}^2$ and that $k' \approx 10^{-7}\ \text{N/A}^2$. Substitute these values directly into Eq. (23-1):

$$c = \sqrt{\frac{k}{k'}} = \sqrt{\frac{9 \times 10^9\ \text{N m}^2/\text{C}^2}{10^{-7}\ \text{N/A}^2}} = \sqrt{9 \times 10^{16}\ \text{m}^2\ \text{A}^2/\text{C}^2}$$

$$= 3.00 \times 10^8\ \text{m/s}$$

We can also express the speed of propagation of electromagnetic waves through a vacuum in terms of the **permittivity constant** ε_0 and the **permeability constant** μ_0, which are related to k and k' as follows:

PERMITTIVITY CONSTANT

$$\varepsilon_0 = \frac{1}{4\pi k} \approx 8.854 \times 10^{-12}\ \text{C}^2/\text{N m}^2 \qquad \textbf{(23-2)}$$

PERMEABILITY CONSTANT

$$\mu_0 = 4\pi k' \approx 1.257 \times 10^{-6}\ \text{N/A}^2 \qquad \textbf{(23-3)}$$

If we solve Eqs. (23-2 and 23-3) for k and k' respectively, and substitute into Eq. (23-1), we can express the speed of propagation of an electromagnetic wave as

SPEED OF PROPAGATION (E-M WAVE; VACUUM)

$$c = \frac{1}{\sqrt{\varepsilon_0 \mu_0}} \qquad \textbf{(23-4)}$$

The magnitude of the **B**-field is related to the magnitude of the **E**-field as follows:

$$B^2 = \varepsilon_0 \mu_0 E^2 = \frac{E^2}{c^2} \qquad \textbf{(23-5)}$$

23-2. Energy Transmission

Electromagnetic waves transmit energy from their source to other points in space. The power transmitted per second by electromagnetic waves through an area of 1 m^2, perpendicular to the direction of propagation, is given by

POWER PER UNIT AREA

$$\frac{P}{A} = \frac{c}{2}\left(\frac{B^2}{\mu_0} + \varepsilon_0 E^2\right) \qquad \textbf{(23-6)}$$

Since $\sqrt{\varepsilon_0 \mu_0} = 1/c$ and $B = E/c$, we can also express Eq. (23-6) as

POWER PER UNIT AREA

$$\frac{P}{A} = c\,\varepsilon_0 E^2 \qquad \textbf{(23-7)}$$

EXAMPLE 23-2: When the sun is directly overhead on a bright clear day, the solar power received at the earth's surface is approximately 1000 W/m^2. Calculate the effective magnitude of the electric field under these conditions.

Solution: You can find the effective (or root-mean-square) value of **E** from Eq. (23-7):

$$\frac{P}{A} = c\,\varepsilon_0 E^2$$

$$E = \sqrt{\frac{P/A}{c\,\varepsilon_0}} = \sqrt{\frac{1000\ \text{W/m}^2}{(3 \times 10^8\ \text{m/s})(8.854 \times 10^{-12}\ \text{C}^2/\text{N m}^2)}} = 614\ \text{N/C}$$

EXAMPLE 23-3: Use the result of Example 23-2 to find the effective value of the magnetic field produced by the sun's radiation.

Solution: Use Eq. (23-5) and solve for **B**:

$$B^2 = \varepsilon_0 \mu_0 E^2$$

$$B = \sqrt{\varepsilon_0 \mu_0}\,E = \frac{1}{c}E = \frac{614\ \text{N/C}}{3.00 \times 10^8\ \text{m/s}} = 2.05 \times 10^{-6}\ \text{T}$$

23-3. Propagation through Transparent Materials

Electromagnetic waves have different wavelengths, but they all travel at the speed of light c. The wavelength, frequency, and speed of propagation of light in a vacuum are related as follows:

$$c = f\lambda \qquad \textbf{(23-8)}$$

Visible light, which is just a small portion of the electromagnetic spectrum, has wavelengths within the range of 3.8×10^{-7} to 7.6×10^{-7} m.

EXAMPLE 23-4: The minimum wavelength that can be detected by the human eye is 3.80×10^{-7} m.
At what frequency is this wavelength propagated through empty space?

Solution: Solve Eq. (23-8) for the frequency:

$$c = f\lambda \qquad f = \frac{c}{\lambda} = \frac{3.00 \times 10^{8}\ \text{m/s}}{3.80 \times 10^{-7}\ \text{m}} = 7.89 \times 10^{14}\ \text{Hz}$$

When light waves enter a transparent material such as glass, plastic, or water, both the speed of propagation and the wavelength decrease, but the *frequency* stays the same.

- The **index of refraction** is a constant that measures the optical density of a material.

The speed of light c through a vacuum divided by the speed of light v in a transparent substance provides the index of refraction for that substance:

INDEX OF REFRACTION (TRANSPARENT SUBSTANCE)

$$n = \frac{c}{v} \qquad \textbf{(23-9)}$$

The *higher* the value of n, the more *dense* the material. Table 23-1 lists the index of refraction for several transparent substances.

TABLE 23-1: Index of Refraction

Substance	n
air at 0°	1.0003
benzene	1.50
crown glass	1.52
diamond	2.42
flint glass	1.66
water	1.333

EXAMPLE 23-5: Light waves are propagated through clear water at a speed of 2.25×10^{8} m/s. What is the index of refraction of water?

Solution: Directly from Eq. (23-9):

$$n = \frac{c}{v} = \frac{3.00 \times 10^{8}\ \text{m/s}}{2.25 \times 10^{8}\ \text{m/s}} = 1.333$$

EXAMPLE 23-6: The wavelength of red light in vacuum is 6.50×10^{-7} m. What is the wavelength of this electromagnetic wave when it enters clear water ($v = 2.25 \times 10^{8}$ m/s)?

Solution: First, calculate the frequency of red light in a vacuum from Eq. (23-8):

$$c = f\lambda \qquad f = \frac{c}{\lambda} = \frac{3.00 \times 10^{8}\ \text{m/s}}{6.50 \times 10^{-7}\ \text{m}} = 4.62 \times 10^{14}\ \text{Hz}$$

Since the frequency does not change when the light enters the water, you can also calculate the wavelength from Eq. (23-8):

$$\lambda = \frac{v}{f} = \frac{2.25 \times 10^{8}\ \text{m/s}}{4.62 \times 10^{14}\ \text{s}^{-1}} = 4.87 \times 10^{-7}\ \text{m}$$

23-4. Reflection and Refraction

When a light ray strikes the surface of a substance, several things may happen:

- The ray can be *reflected* off the surface of the substance.
- The ray can be *transmitted* through the substance if the substance is transparent. The transmitted ray changes direction as it enters the substance. This bending of a ray is called **refraction**.
- The ray can be *absorbed* by the substance. Most substances absorb some of the radiation incident on them, increasing their internal energy.

In general, reflection and absorption occur whenever light strikes a substance. If the substance is transparent, refraction may occur also.

There are two kinds of reflection of light rays.

- **Diffuse reflection** occurs when light strikes a rough surface and is reflected in all directions.

For example, if you shine a flashlight on this book in a darkened room, you will be able to see the book from any part of the room. Diffuse reflection scatters rays leaving the book in all directions.

- **Regular** or **specular reflection** occurs when light strikes a mirror or other smooth surface and is reflected in only one particular direction.

For example, if you shine a flashlight on a mirror in a darkened room, you can see the mirror only if your eyes are in the direction of the reflected rays. We always measure the direction of a light ray with respect to a line *normal* (perpendicular) to the reflecting or refracting surface.

- The ray striking the surface is the **ray of incidence.**
- The ray reflected from the surface is the **ray of reflection.** The **angle of reflection** is the angle that the ray of reflection makes with the normal. The incident ray, the reflected ray, and the normal all lie in the same plane, as shown in Figure 23-2. For regular reflection, the angle of incidence always equals the angle of reflection.

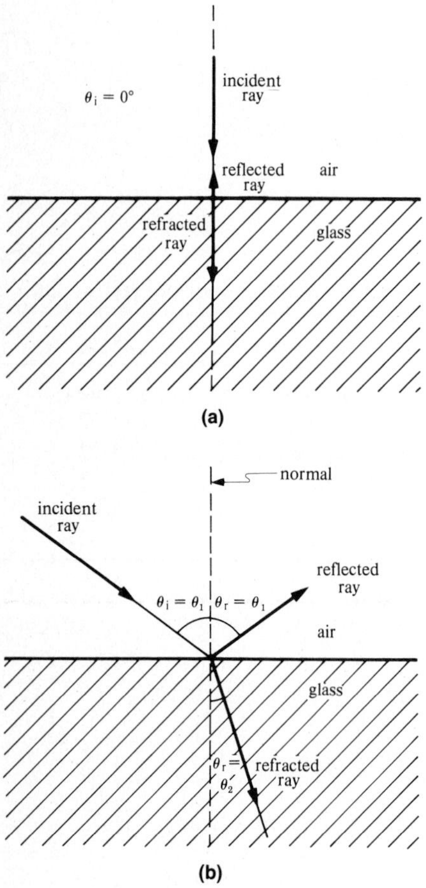

Figure 23-2

LAW OF REFLECTION $\qquad \theta_i = \theta_r \qquad$ **(23-10)**

- The ray that is transmitted when a light ray strikes a transparent substance is the **ray of refraction.** The **angle of refraction** (θ_2 in Fig. 23-2b) is the angle that the ray of refraction makes with the normal.

The incident ray, the refracted ray, and the normal all lie in the same plane. The angle of refraction depends on the indices of refraction of the two substances in which the rays travel, as given by **Snell's law.**

SNELL'S LAW $\qquad n_1 \sin \theta_1 = n_2 \sin \theta_2 \qquad$ **(23-11)**

note: Light bends *toward* the normal when it enters an optically denser medium (as shown in Fig. 23-2b) and *away from* the normal when it enters an optically less dense medium (not shown). There is no bending only in the special case where $\theta_i = 0°$ (Fig. 23-2a).

EXAMPLE 23-7: A ray of light traveling in air strikes the surface of crown glass. The angle θ_1 measured with respect to the normal is 30°.
What is the angle of the ray of refraction?

Solution: Find the index of refraction for each of the two media from Table 23-1 and then use Snell's law:

$$n_1 \sin \theta_1 = n_2 \sin \theta_2$$

$$\sin \theta_2 = \left(\frac{1.0003}{1.52}\right) \sin 30° = 0.329$$

$$\theta_2 = \arcsin(0.329) = 19.2°$$

EXAMPLE 23-8: A ray of light is incident upon the surface of a transparent material. The angle of incidence in air is 40° and the angle of refraction is 22.79°. What might this material be?

Solution: From Snell's law:

$$n_2 = n_1 \frac{\sin \theta_1}{\sin \theta_2} = (1.0003) \frac{\sin 40°}{\sin 22.79°} = 1.66$$

You can see from Table 23-1 that the material could be flint glass.

EXAMPLE 23-9: An electromagnetic wave with a wavelength of 5.3×10^{-7} m travels through a piece of flint glass. This ray of light is incident on the lower surface of the glass at an angle of 20° with respect to the normal. The ray enters a second transparent medium with an index of refraction of 1.25.
Calculate the angle of refraction within the second medium.

Solution: You can find the angle of refraction θ_2 from Eq. (23-10):

$$\sin \theta_2 = (n_1/n_2) \sin \theta_1 = \left(\frac{1.66}{1.25}\right) \sin 20° = 0.4542$$

$$\theta_2 = \text{arc sin}(0.4542) = 27.0°$$

Suppose we increase the angle of incidence in Example 23-9. If $\theta_1 = 40°$, $\theta_2 = 58.6°$; if $\theta_1 = 45°$, $\theta_2 = 70.0°$; if $\theta_1 = 48°$, $\theta_2 = 80.7°$. If we try $\theta_1 = 50°$ though, we find ourselves trying to calculate arc sin(1.02). But there is no angle θ for which arc sin $\theta > 1$! Refraction is physically impossible at this angle; the angle of refraction can't be larger than 90°. When light goes from a medium with a large index of refraction to one with a smaller index, there is a certain angle, called the **critical angle** θ_c, above which refraction can't occur. If light strikes the surface at an angle greater than θ_c, it is reflected back into the first medium, and we say that there is **total internal reflection**. From Snell's law and the fact that sin 90° = 1, we can find the critical angle for any combination of transparent substances.

CRITICAL ANGLE

$$\sin \theta_c = \frac{n_2}{n_2} \quad (\text{for } n_2 < n_1) \qquad \textbf{(23-12)}$$

note: If a light ray passes from a less dense medium to a more dense medium ($n_1 < n_2$), no critical angle exists. For $n_1 < n_2$, all angles of incidence give angles of refraction.

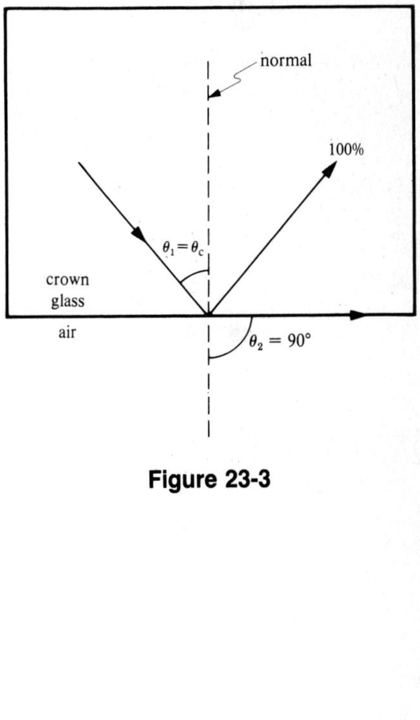

Figure 23-3

EXAMPLE 23-10: Figure 23-3 shows a ray of light traveling in crown glass. What is the critical angle?

Solution: Apply Eq. (23-12).

$$\theta_c = \text{arc sin}(n_2/n_1) = \text{arc sin}(1.003/1.52) = 41.2°$$

The speeds of propagation of light within two transparent media are $v_1 = f\lambda_1$ and $v_2 = f\lambda_2$. Because the frequency does not change when light passes from one medium to another, the ratio of the wavelengths in the two media is the same as the ratio of the speeds of propagation:

RATIO OF THE WAVELENGHTS IN TWO DIFFERENT MEDIA

$$\frac{\lambda_2}{\lambda_1} = \frac{v_2}{v_1} \qquad \textbf{(23-13)}$$

Eq. (23-9) gives the speed of propagation in each medium: $v_2 = c/n_2$ and $v_1 = c/n_1$. Dividing the first equation by the second gives

RATIO OF THE SPEEDS OF LIGHT IN TWO DIFFERENT MEDIA

$$\frac{v_2}{v_1} = \frac{n_1}{n_2} \qquad \textbf{(23-14)}$$

When you combine Eqs. (23-11) and (23-12), the result is

RELATIONSHIP BETWEEN WAVELENGTHS AND INDICES OF REFRACTION

$$\frac{\lambda_2}{\lambda_1} = \frac{n_1}{n_2} \qquad \textbf{(23-15)}$$

EXAMPLE 23-11: Calculate the wavelength of the ray of light within the second medium described in Example 23-9.

Solution: Use Eq. (23-13):

$$\lambda_2 = \lambda_1 \frac{n_1}{n_2} = (5.3 \times 10^{-7}\ \text{m})\frac{1.66}{1.25} = 7.04 \times 10^{-7}\ \text{m}$$

SUMMARY

Speed of propagation (E-M wave; vacuum)	$c = \sqrt{\dfrac{k}{k'}}$	gives the speed of propagation of an E-M wave through a vacuum as a ratio of the electrical constant k and the magnetic constant k'
Permittivity constant	$\varepsilon_0 = \dfrac{1}{4\pi k} \approx 8.854 \times 10^{-12}\ \text{C}^2/\text{N m}^2$	
Permeability constant	$\mu_0 = 4\pi k' \approx 1.257 \times 10^{-6}\ \text{N/A}^2$	
Speed of propagation (E-M wave; vacuum)	$c = \dfrac{1}{\sqrt{\varepsilon_0\mu_0}}$	gives the speed of propagation of an E-M wave through a vacuum in terms of the permittivity and permeability constants
	$B^2 = \varepsilon_0\mu_0 E^2$	gives the relationship of the magnitude of the magnetic field to the magnitude of the electric field in an E-M wave
Power per unit area	$\dfrac{P}{A} = \dfrac{c}{2}\left(\dfrac{B^2}{\mu_0} + \varepsilon_0 E^2\right)$	gives the power transmitted by an E-M wave as a function of its accompanying electric and magnetic fields
Power per unit area	$\dfrac{P}{A} = c\,\varepsilon_0 E^2$	gives an alternative expression for the power transmistted by an E-M wave
Definition of frequency and wavelength	$c = f\lambda$	gives the relationship of the wavelength, frequency, and speed of propagation of light in a vacuum
Index of refraction (transparent substance)	$n = \dfrac{c}{v}$	gives a substance's index of refraction as the ratio of the speed of light through a vacuum and the speed of light through the substance
Law of reflection	$\theta_i = \theta_r$	in regular reflection, the angle of incidence equals the angle of reflection

Snell's law	$n_1 \sin \theta_1 = n_2 \sin \theta_2$	the angle of refraction of a light ray as it passes from one transparent medium to another is determined by the indices of refraction of the two media
Critical angle	$\sin \theta_c = \frac{n_2}{n_1}$	for light striking the surface of a substance with an index of refraction lower than that of the substance in which the light is traveling, gives the angle of incidence above which there is total internal reflection
Ratio of the wavelengths in two different media	$\frac{\lambda_2}{\lambda_1} = \frac{v_2}{v_1}$	the ratio of the speeds of propagation of light within two transparent media is equal to the ratio of the wavelengths in the two media
Ratio of the speeds of light in two different media	$\frac{v_2}{v_1} = \frac{n_1}{n_2}$	the ratio of the speeds of light in two different media is equivalent to the ratio of the indices of refraction
Relationship between wavelengths and indices of refraction	$\frac{\lambda_2}{\lambda_1} = \frac{n_1}{n_2}$	the ratio of the wavelengths is equivalent to the inverse of the ratio of the indices of refraction

RAISE YOUR GRADES

Can you define . . . ?

- ☑ index of refraction
- ☑ refraction
- ☑ critical angle

Can you . . . ?

- ☑ calculate the speed of propagation of an electromagnetic wave in a vacuum from the constants k and k'
- ☑ write the equation for the permittivity constant ε_0
- ☑ write the equation for the permeability constant μ_0
- ☑ calculate the speed of propagation of light waves in vacuum, using the constants ε_0 and μ_0
- ☑ write the relationship between the magnitudes of the magnetic field and the electric field of an electromagnetic wave
- ☑ write the equation for the energy transmitted per unit area per second by an electromagnetic wave
- ☑ write the relationship between frequency, wavelength, and speed of propagation of an electromagnetic wave
- ☑ calculate the index of refraction of a transparent material if you know the speed of propagation of light within the material
- ☑ write Snell's law for the bending of light as it passes from one medium to another
- ☑ write the law of reflection

SOLVED PROBLEMS

PROBLEM 23-1 The average distance from the sun to the earth is 1.49×10^{11} m. How long does it take for electromagnetic waves to travel from the sun to the earth?

Solution Since the speed of propagation of light in empty space is constant, you can use the equation c = distance/time = s/t and solve for the time. [Hint: The speed of propagation of electromagnetic waves through a vacuum is 3.00×10^8 m/s (see Example 23-1).]

$$t = \frac{s}{c} = \frac{1.49 \times 10^{11}\ \text{m}}{3.00 \times 10^8\ \text{m/s}} = 496.7\ \text{s} = (496.7\ \text{s})\left(\frac{1\ \text{min}}{60\ \text{s}}\right) = 8.28\ \text{min}$$

PROBLEM 23-2 Calculate the speed of propagation of light through a vacuum from the permittivity constant and the permeability constant.

Solution Directly from Eq. (23-4):

$$c = \frac{1}{\sqrt{\varepsilon_0 \mu_0}} = \frac{1}{\sqrt{(8.854 \times 10^{-12}\ \text{C}^2/\text{N m}^2)(1.257 \times 10^{-6}\ \text{N/A}^2)}} = 3.00 \times 10^8\ \text{m/s}$$

PROBLEM 23-3 The solar power that strikes the upper atmosphere is approximately 1350 W/m^2. Calculate the electric field strength of the solar radiation falling on the earth's atmosphere.

Solution Use Eq. (23-7) and solve for the magnitude of the electric field:

$$\frac{P}{A} = c\varepsilon_0 E^2$$

$$E = \sqrt{\frac{P/A}{c\varepsilon_0}} = \sqrt{\frac{1350\ \text{W/m}^2}{(3.0 \times 10^8\ \text{m/s})(8.854 \times 10^{-12}\ \text{C}^2/\text{N m}^2)}} = 713\ \text{N/C}$$

PROBLEM 23-4 What is the magnitude of the magnetic field associated with solar radiation striking the earth's atmosphere? Use the data from Problem 23-3.

Solution The magnitude of the magnetic field is given by Eq. (23-5):

$$B^2 = \varepsilon_0 \mu_0 E^2$$

$$B = \sqrt{\varepsilon_0 \mu_0} E = \frac{E}{c} = \frac{713\ \text{N/C}}{3.0 \times 10^8\ \text{m/s}} = 2.38 \times 10^{-6}\ \text{T}$$

PROBLEM 23-5 The speed of propagation of electromagnetic waves through the atmosphere is very close to the speed of propagation through a vacuum. Calculate the wavelength of a radio wave whose frequency is 1.04 MHz.

Solution Use Eq. (23-8) and solve for the unknown wavelength:

$$c = f\lambda \qquad \lambda = \frac{c}{f} = \frac{3.00 \times 10^8\ \text{m/s}}{1.04 \times 10^6\ \text{s}^{-1}} = 288\ \text{m}$$

PROBLEM 23-6 When light enters a diamond it slows down considerably. Calculate the speed of light within a diamond.

Solution Since the index of refraction of a transparent substance is the ratio of the speed of light through a vacuum and the speed of light in a transparent substance, you can find the speed of propagation from Table 23-1 and Eq. (23-9).

$$n = \frac{c}{v} \qquad v = \frac{c}{n} = \frac{3.00 \times 10^8\ \text{m/s}}{2.42} = 1.24 \times 10^8\ \text{m/s}$$

PROBLEM 23-7 A light ray in air is incident on the surface of a diamond at an angle of 50°. Calculate the angle of refraction of this light ray within the diamond.

Solution Use Snell's law (Eq. 23-10) to find the angle of refraction:

$$n_1 \sin \theta_1 = n_2 \sin \theta_2$$

$$\sin \theta_2 = (n_1/n_2) \sin \theta_1 = \frac{1.0003}{2.42} \sin 50^\circ = 0.3166$$

$$\theta_2 = \text{arc} \sin(0.3166) = 18.5^\circ$$

PROBLEM 23-8 The light entering the diamond in Problem 23-7 had a wavelength in air of 6.50×10^{-7} m. What is the wavelength of this light within the diamond?

Solution From Eq. (23-13) the ratio of the two wavelengths is given by

$$\frac{\lambda_2}{\lambda_1} = \frac{n_1}{n_2}$$

Solve for the unknown wavelength;

$$\lambda_2 = \lambda_1 \frac{n_1}{n_2} = (6.50 \times 10^{-7} \text{ m}) \frac{1.0003}{2.42} = 2.69 \times 10^{-7} \text{ m}$$

PROBLEM 23-9 A ray of light in air enters a transparent plastic at an angle of 40°. The angle of refraction within the plastic is 26.32°. Find the index of refraction of this plastic material.

Solution From Snell's law:

$$n_2 = n_1 \frac{\sin \theta_1}{\sin \theta_2} = (1.0003) \frac{\sin 40^\circ}{\sin 26.32^\circ} = 1.45$$

PROBLEM 23-10 The ray of light described in Problem 23-9 has a wavelength in air of 6.0×10^{-7} m. Calculate the wavelength of this light in the transparent plastic.

Solution From Eq. (23-13):

$$\lambda_2 = \lambda_1 \left(\frac{n_1}{n_2}\right) = (6.0 \times 10^{-7} \text{ m}) \left(\frac{1.0003}{1.45}\right) = 4.14 \times 10^{-7} \text{ m}$$

PROBLEM 23-11 A light ray strikes a flint glass bottle containing benzene. If the angle of incidence is 30°, what is the angle of refraction in the benzene?

Solution The ray is refracted twice, once when it passes from air to glass and again when it passes from glass to benzene, so solve the problem in two steps. First use Snell's law to find the angle of refraction θ_2 from air to flint glass.

$$n_1 \sin \theta_1 = n_2 \sin \theta_2$$

$$\sin \theta_2 = n_1 \sin \theta_1 / n_2 = 1.0003 \sin 30^\circ / 1.66 = 0.301$$

$$\theta_2 = \text{arc} \sin(0.301) = 17.5^\circ$$

This θ_2 is the angle of incidence on the glass–benzene interface. Now solve for the angle of refraction θ_3 in the benzene.

$$n_2 \sin \theta_2 = n_3 \sin \theta_3$$

$$\theta_3 = \text{arc} \sin(n_2 \sin \theta_2 / n_3) = \text{arc} \sin[(1.66)(0.301)/1.50] = 19.5^\circ$$

PROBLEM 23-12 Calculate the critical angle for total internal reflection for flint glass. Let the external medium be air.

Solution

$$\sin \theta_c = (n_2/n_1) \sin \theta_2 = \frac{1.0003}{1.66} \sin 90^\circ = 0.6026$$

$$\theta_c = \text{arc} \sin(0.6026) = 37.1^\circ$$

Supplementary Exercises

PROBLEM 23-13 On a cloudy day, the solar energy incident upon a surface of 1.0 m^2 perpendicular to the light rays from the sun is only 750 W/m^2. Calculate the effective magnitude of the electric field under these conditions.

PROBLEM 23-14 When the effective value of the electric field of an electromagnetic wave is equal to 1.20×10^3 N/C, what is the effective value of the magnetic field?

PROBLEM 23-15 The longest wavelength that can be detected by the human eye is approximately 7.6×10^{-7} m. What is the frequency when this wavelength is propagated through a vacuum?

PROBLEM 23-16 What is the speed of propagation of light when it enters a transparent medium whose index of refraction is 1.55?

PROBLEM 23-17 A ray of light whose wavelength in air is 5.4×10^{-7} m enters a block of crown glass. What is the wavelength of this light within the glass block?

PROBLEM 23-18 A ray of light traveling in air falls on the surface of a glass block. The angle of incidence is 50° and the angle of refraction within the glass is 28.61°. Calculate the index of refraction of this glass.

PROBLEM 23-19 What is the speed of propagation of light within a transparent material whose index of refraction is 1.600?

PROBLEM 23-20 A ray of light traveling in air has a wavelength of 6.4×10^{-7} m. What does this wavelength become when the light enters a medium whose index of refraction is 1.66?

Answers to Supplementary Exercises

23-13: 531 N/C

23-14: 4.00×10^{-6} T

23-15: 3.95×10^{14} Hz

23-16: 1.94×10^8 m/s

23-17: 3.55×10^{-7} m

23-18: 1.60

23-19: 1.88×10^8 m/s

23-20: 3.86×10^{-7} m

24 GEOMETRICAL OPTICS

THIS CHAPTER IS ABOUT

☑ Lenses
☑ Mirrors

24-1. Lenses

According to Snell's law of refraction (Eq. 23-10), a light ray passing through a lens is bent *twice:* toward the perpendicular when it enters the lens and away from the perpendicular when it leaves the lens.

- In a **thin lens,** where the thickness of the lens is negligible in comparison to the distance from the lens to other objects, we can accurately approximate the refraction of a ray by considering that the ray bends only once, at the plane passing through the center of the lens, as illustrated in Figure 24-1. We measure the distance from an object to a thin lens as the distance from the object to this plane.

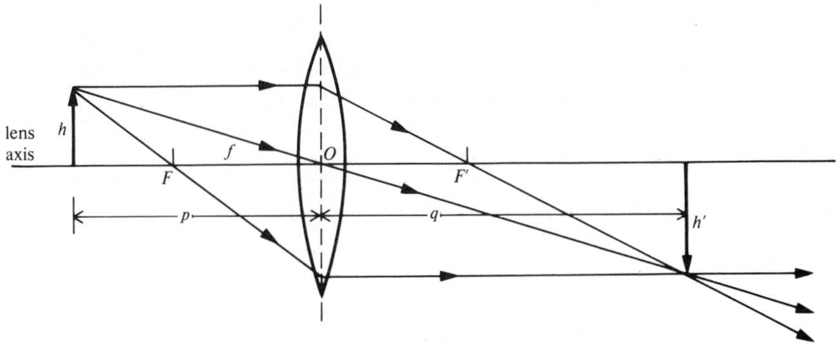

Figure 24-1

We'll use the following terms and notation in our discussion of thin lenses.

- The **object distance** p is the perpendicular distance from an object to the lens.
- The **image distance** q is the perpendicular distance from the image of the object to the lens.
- The **focal length** f is the image distance of an object at an infinite distance from the lens.
- The **radius of curvature** R is the radius of the circle whose circumference most closely matches the curvature of the lens surface.

We'll also use the following terms to describe lens shape and function.

- The **lens axis** is an imaginary line through the center of a lens and perpendicular to the plane of the lens.
- A **converging** or **positive lens** is thicker at its center than at its outer edge. Parallel rays *converge* after they pass through a positive lens. A converging lens has a *positive* focal length.
- The **focal point** F of a positive lens is the point where incoming rays parallel to the lens axis converge. The distance between the focal point and the lens is the focal length.

- The actual convergence of light rays forms a **real image.** A real image has a *positive* image distance q.
- A **diverging** or **negative lens** is thinner at its center than at its outer edge. Parallel rays *diverge* after they pass through a negative lens. A diverging lens has a *negative* focal length.
- The **focal point** F' of a diverging lens is the point from which diverging rays *appear* to originate.
- The divergence of light rays forms a **virtual image**—the rays do not actually reach the image location. A virtual image has a *negative* image distance q.

We'll examine each of these in the following sections.

A. Positive thin lenses

Figure 24-1 shows an object of height h placed in front of a thin positive lens so that its base is flush with the lens axis. The lens produces an inverted real image of height h'. When we place a piece of photographic film at the location of the inverted image, the image is recorded on the film.

The relationship between the object distance p and the image distance q depends on the focal length f of the lens:

THIN-LENS EQUATION

$$\frac{1}{p} + \frac{1}{q} = \frac{1}{f} \tag{24-1}$$

note: If an object is farther from a positive thin lens than the lens' focal point, q is positive and the lens forms a *real image* of the object. But if the object lies between the lens and the focal point, q is negative and the resulting image in *virtual*.

EXAMPLE 24-1: The positive thin lens illustrated in Figure 24-1 has a focal length of 3.0 cm. The object is located 5.0 cm from point O.
Calculate the image distance.

Solution: Use Eq. (24-1) and solve for the image distance:

$$\frac{1}{p} + \frac{1}{q} = \frac{1}{f}$$

$$\frac{1}{q} = \frac{1}{f} - \frac{1}{p} = \frac{p - f}{fp}$$

$$q = \frac{fp}{p - f} = \frac{(3\text{ cm})(5\text{ cm})}{5\text{ cm} - 3\text{ cm}} = 7.50\text{ cm}$$

Since q is positive, the image is real.

EXAMPLE 24-2: An object is placed 6.0 cm in front of a positive thin lens, which produces an inverted image 12.0 cm from the center of the lens.
Determine the focal length of this lens.

Solution: From Eq. (24-1),

$$\frac{1}{f} = \frac{1}{p} + \frac{1}{q} = \frac{p + q}{pq}$$

$$f = \frac{pq}{p + q} = \frac{(6\text{ cm})(12\text{ cm})}{6\text{ cm} + 12\text{ cm}} = 4.00\text{ cm}$$

EXAMPLE 24-3: The positive thin lens in Figure 24-2 has a focal length of 6.0 cm. An object is placed 3.0 cm to the left of the lens.
Find the location of the image.

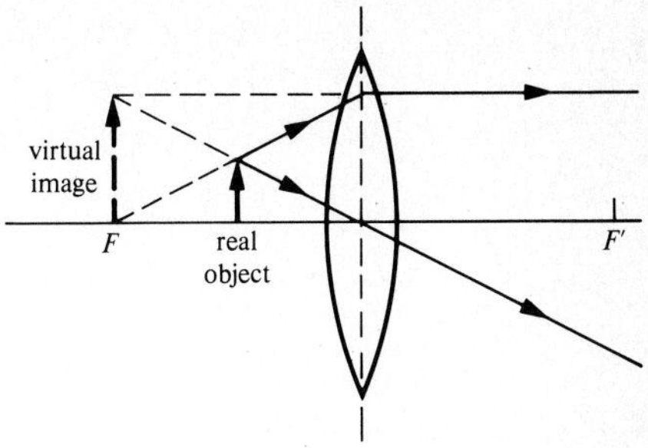

Figure 24-2

Solution: From Eq. (24-1),

$$\frac{1}{q} = \frac{1}{f} - \frac{1}{p} = \frac{p - f}{fp}$$

$$q = \frac{fp}{p - f} = \frac{(6\text{ cm})(3\text{ cm})}{3\text{ cm} - 6\text{ cm}} = -6.00\text{ cm}$$

The value of q is *negative,* which means that a *virtual* image is formed 6.0 cm to the *left* of the lens.

B. Negative thin lenses

In Figure 24-3, an object of height h is placed in front of a negative thin lens. Because the rays of light *diverge* after they pass through the lens, no real image is formed. If an observer to the right of the lens looks through the lens at the object, he will see an *erect* virtual image of height h'. This image cannot be recorded on a piece of film.

EXAMPLE 24-4: The negative thin lens shown in Figure 24-3 has a focal length of -4.0 cm. The object is 8.0 cm to the left of the lens.
Calculate the image distance.

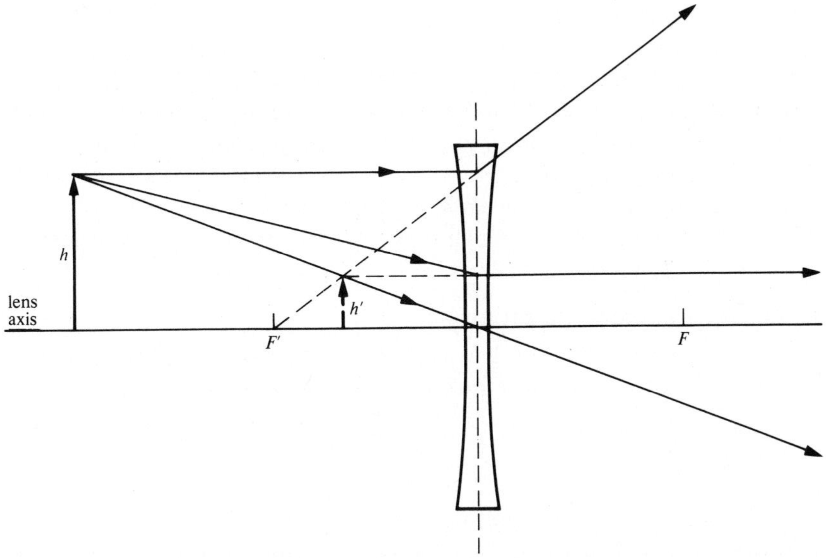

Figure 24-3

Solution: From Eq. (24-1):

$$\frac{1}{q} = \frac{1}{f} - \frac{1}{p} = \frac{p - f}{fp}$$

$$q = \frac{fp}{p - f} = \frac{(-4\text{ cm})(8\text{ cm})}{8\text{ cm} - (-4\text{ cm})} = -2.67\text{ cm}$$

The value of q is negative, so the virtual image is 2.67 cm to the left of the lens.

EXAMPLE 24-5: An object is placed 6.0 cm to the left of a lens. The lens produces a virtual image 2.0 cm to the left of the lens.
Calculate the focal length of this lens.

Solution: From Eq. (24-1):

$$\frac{1}{f} = \frac{1}{p} + \frac{1}{q} = \frac{p + q}{pq}$$

$$f = \frac{pq}{p + q} = \frac{(6 \text{ cm})(-2 \text{ cm})}{6 \text{ cm} + (-2 \text{ cm})} = -3.00 \text{ cm}$$

Since its focal length is negative, this lens is a diverging lens.

C. Ray tracing

- **Ray tracing** is a graphical technique for locating the image produced by a lens.

Look again at Figures 24-1 and 24-3. Let's trace three rays of light leaving the top of our object of height h and passing through a converging lens. A ray parallel to the lens axis is bent as its passes through the lens so that it intersects the focal point to the *right* of the lens and continues in a straight line. A second ray passing through point O, the center of the lens, is not bent, but continues in a straight line and eventually intersects the first ray. A third ray that passes through the focal point to the *left* of the lens is bent so that it emerges parallel to the lens axis—and intersects the other two rays at the point where they cross. The intersection of the three rays defines the location of the real image. You can see that the intersection of any two of these rays is sufficient to locate the image.

Now let's see how a negative lens affects these three rays. For a negative lens, a ray parallel to the lens axis diverges at an angle such that the ray appears to have passed through the focal point to the *left* of the lens. Another ray directed toward the focal point to the *right* of the lens is bent so that it is parallel to the lens axis. The ray that passes through the center of the lens is not bent. The three rays appear to have intersected at one point—the location of the virtual image—although only one, the ray through the center of the lens, actually passed through it.

D. The lens-maker's equation

The focal length of a thin lens depends on the radii of curvature of the two surfaces and the index of refraction (Eq. 23-9) of the lens material:

LENS-MAKER'S EQUATION

$$\frac{1}{f} = (n - 1)\left(\frac{1}{R_1} + \frac{1}{R_2}\right) \qquad \textbf{(24-2)}$$

where the radii of curvature of the first and second surfaces are R_1 and R_2, respectively. R is positive if the surface is convex, negative if the surface is concave, and infinite if the surface is planar ($1/R = 0$). Figure 24-4 shows examples of several lenses.

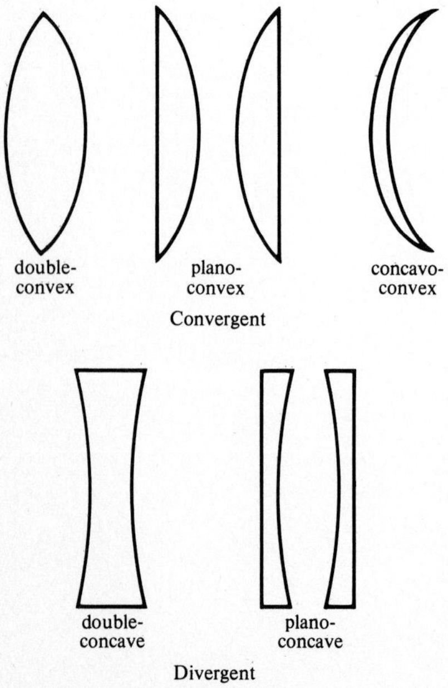

Figure 24-4

EXAMPLE 24-6: The convex surface of a plano-convex lens has a radius of curvature of 6.0 cm. The index of refraction of the glass is 1.50.
Calculate the focal length of the lens.

Solution: Directly from Eq. (24-2),

$$\frac{1}{f} = (n - 1)\left(\frac{1}{R_1} + \frac{1}{R_2}\right) = (1.50 - 1.00)\left(\frac{1}{6.0 \text{ cm}} + \frac{1}{\infty}\right)$$

$$f = 12.0 \text{ cm}$$

EXAMPLE 24-7: A double-convex lens ($n = 1.50$) has two surfaces with the same radius of curvature R and a focal length of 4.0 cm.
Find R.

Solution: From Eq. (24-2),

$$\frac{1}{f} = (n - 1)\left(\frac{1}{R} + \frac{1}{R}\right) = (n - 1)\frac{2}{R}$$

$$R = 2f(n - 1) = 2(4 \text{ cm})(1.50 - 1.00) = 4.00 \text{ cm}$$

EXAMPLE 24-8: A plano-concave lens ($n = 1.50$) has a focal length of -12 cm.
Find the radius of curvature of the concave surface.

Solution: From Eq. (24-2),

$$\frac{1}{f} = (n - 1)\left(\frac{1}{R_1} + \frac{1}{R_2}\right) = (n - 1)\left(0 + \frac{1}{R_2}\right)$$

$$R_2 = f(n - 1) = (-12.0 \text{ cm})(1.50 - 1.00) = -6.00 \text{ cm}$$

E. Magnification

- The **magnification** of a lens is the negative of the image size h' divided by the object size h, which is equal to the image distance divided by the object distance:

MAGNIFICATION $$m = -\frac{h'}{h} = -\frac{q}{p} \qquad \textbf{(24-3)}$$

The negative sign ensures that an inverted image will give a negative magnification. A negative magnification indicates an inverted image; a positive magnification indicates an erect image.

EXAMPLE 24-9: **(a)** Determine the magnification of a double-convex thin lens for a 7.5-cm image distance and a 5.0-cm object distance. **(b)** Is the image erect or inverted?

Solution:
(a) From Eq. (24-3):

$$m = -\frac{q}{p} = -\frac{7.5 \text{ cm}}{5.0 \text{ cm}} = -1.50$$

(b) The negative sign means the image is inverted.

EXAMPLE 24-10: **(a)** Calculate the magnification of a double-convex thin lens for a -6.0-cm image distance and a 3.0-cm object distance. **(b)** Is the image real or virtual; erect or inverted?

Solution:
(a)

$$m = -\left(\frac{-6.0 \text{ cm}}{3.0 \text{ cm}}\right) = 2.00$$

(b) The image distance is negative, so the image is virtual; the magnification is positive, so the image is erect.

EXAMPLE 24-11: (**a**) Determine the magnification of a double-concave thin lens for a −2.67-cm image distance and an 8.00-cm object distance. (**b**) Is the image real or virtual; erect or inverted; larger or smaller than the object?

Solution:
(**a**)

$$m = -\left(\frac{-2.67\ \text{cm}}{8.00\ \text{cm}}\right) = 0.334$$

(**b**) This negative lens produces an erect virtual image that is only one third as large as the object.

F. Systems of two lenses

Many optical instruments include combinations of two or more lenses. In systems of multiple lenses, the image formed by one lens becomes the object for the next lens. Figure 24-5 shows a second positive lens placed to the right of the real, inverted image formed by the first lens. The second lens focuses the rays leaving the image just as if they were rays leaving a real object. But suppose the second lens is inserted *between* the first lens and image. The original image disappears, and the second lens may or may not form a new image, but the image that the first lens *would have formed* still serves as the object for the second lens. Because the first image is not actually formed but still functions as the object for the second lens, we call it a **virtual object.** The object distance p of a virtual object is *negative*.

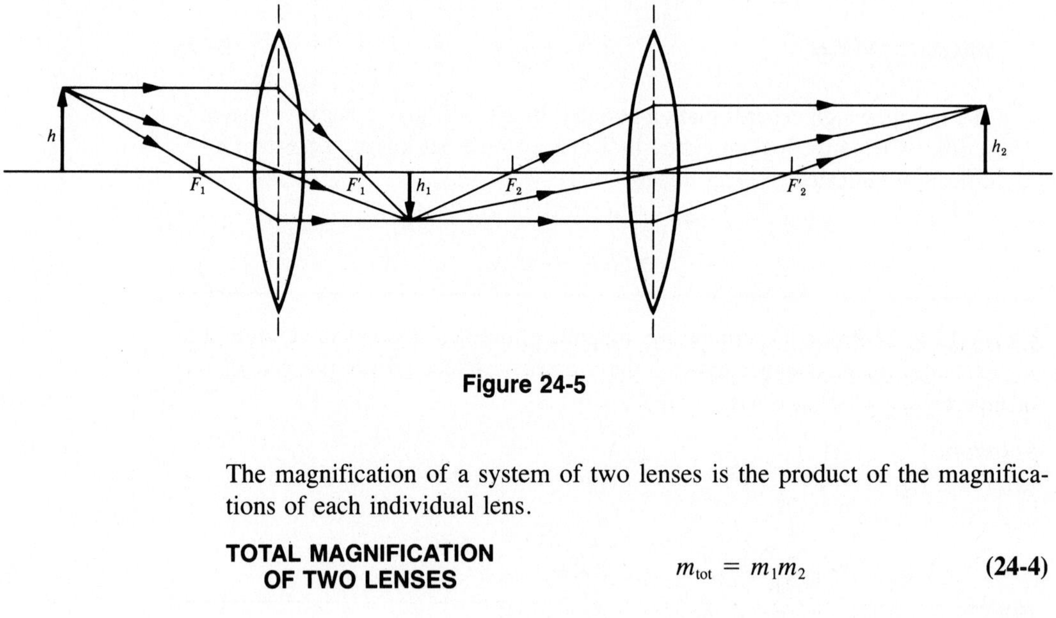

Figure 24-5

The magnification of a system of two lenses is the product of the magnifications of each individual lens.

TOTAL MAGNIFICATION OF TWO LENSES

$$m_{tot} = m_1 m_2 \tag{24-4}$$

EXAMPLE 24-12: Figure 24-6a shows a real, inverted image produced 7.25 cm from a positive lens. When a negative lens is placed 3 cm in front of this image, the original image disappears and a new real image is created 9 cm to the right of the negative lens (Figure 24-6b). Find the focal length of the negative lens.

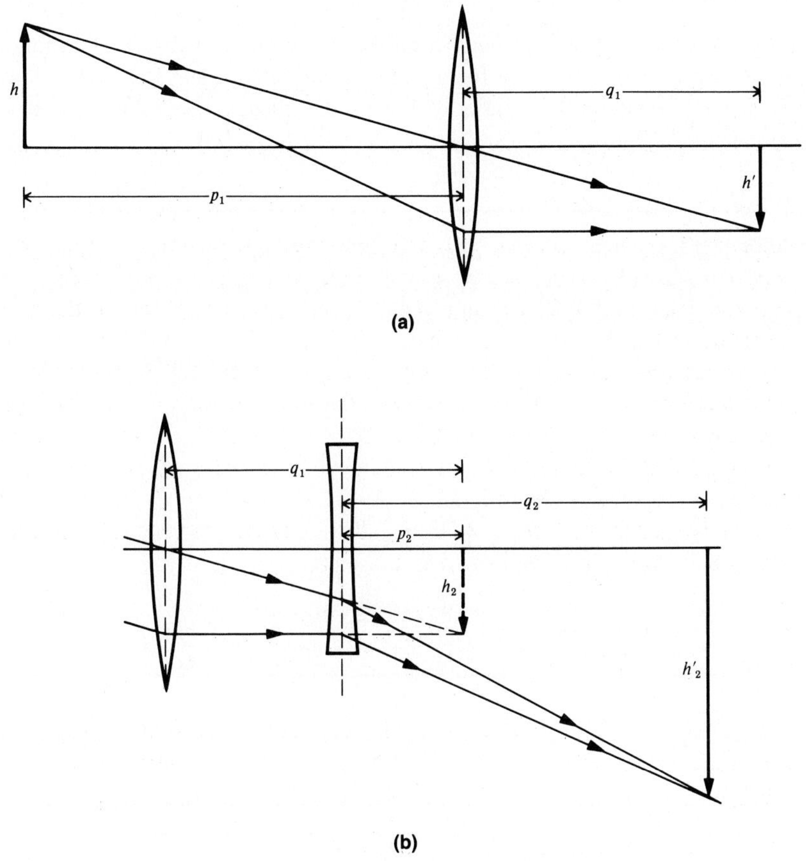

Figure 24-6

Solution: The original image functions as a virtual object for the second lens, and, because the second lens is placed 3 cm in front of the original image, the object distance p_2 for the second lens is -3 cm.

$$\frac{1}{f_2} = \frac{1}{p_2} + \frac{1}{q_2} = \frac{p_2 + q_2}{p_2 q_2}$$

$$f_2 = \frac{p_2 q_2}{p_2 + q_2} = \frac{(-3\text{ cm})(9\text{ cm})}{-3\text{ cm} + 9\text{ cm}} = -4.5\text{ cm}$$

EXAMPLE 24-13: If the distance between the real object and the first lens in Example 24-12 is 10.9 cm, what is the total magnification for the two lenses?

Solution:

$$m_{\text{tot}} = m_1 m_2 = \left(\frac{-q_1}{p_1}\right)\left(\frac{-q_2}{p_2}\right) = \left(\frac{-7.25\text{ cm}}{10.9\text{ cm}}\right)\left(\frac{-9\text{ cm}}{-3\text{ cm}}\right) = -2.00$$

The negative sign specifies that the image is inverted. This is an example of a virtual object producing a real image.

24-2. Mirrors

Spherical mirrors produce images, which may be either real or virtual, by reflection. They too obey the thin-lens equation (24-1). The focal length of a mirror of radius of curvature R is

FOCAL LENGTH OF A MIRROR

$$f = \frac{R}{2} \tag{24-5}$$

Since light rays do not pass through a mirror, the numeric value of the radius of curvature of a mirror is positive for a concave surface and negative for a convex one. Concave mirrors produce either real or virtual images; convex mirrors usually produce virtual images.

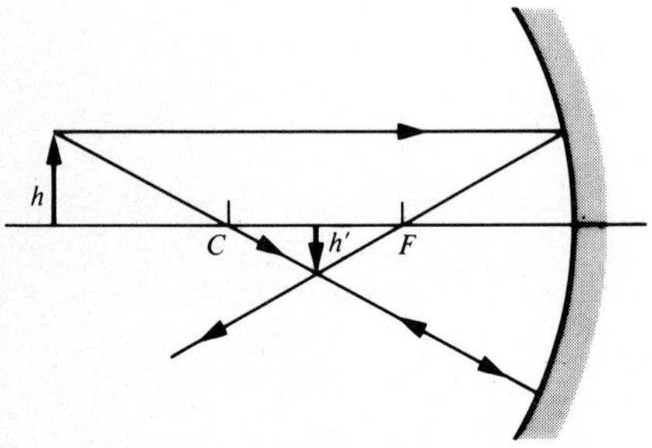

Figure 24-7

EXAMPLE 24-14: The concave mirror shown in Figure 24-7 has a radius of curvature of 8.0 cm. A real object h is placed 12.0 cm in front of the mirror.
Find the location of the image h' this mirror produces.

Solution: From Eq. (24-5), $f = R/2$, so $1/f = 2/R$. Substitute this value into Eq. (24-1):

$$\frac{1}{p} + \frac{1}{q} = \frac{2}{R}$$

$$\frac{1}{q} = \frac{2}{R} - \frac{1}{p} = \frac{2p - R}{Rp}$$

$$q = \frac{Rp}{2p - R} = \frac{(8 \text{ cm})(12 \text{ cm})}{2(12 \text{ cm}) - 8 \text{ cm}} = 6.00 \text{ cm}$$

For mirrors, a positive image distance means the image is real and located in front of the mirror.

EXAMPLE 24-15: For the convex mirror shown in Figure 24-8, $R = -10.0$ cm. Object h is 8.0 cm in front of the mirror.
Find the location of the image h' produced by this mirror.

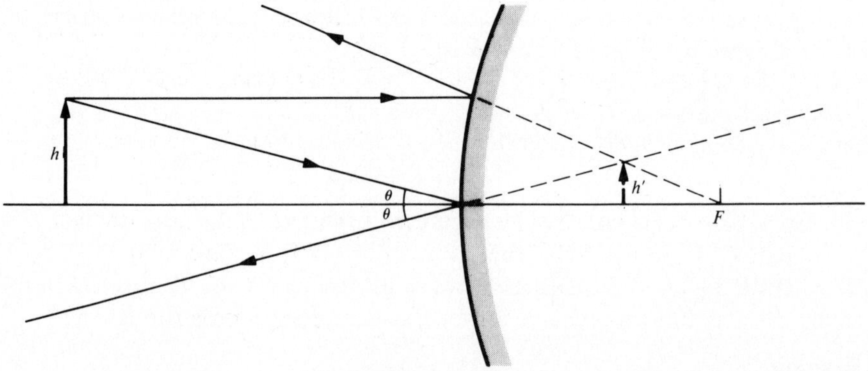

Figure 24-8

Solution: Use the relationship you derived in Example 24-14, and remember that the radius of curvature of a convex mirror is negative.

$$\frac{1}{q} = \frac{2}{R} - \frac{1}{p} = \frac{2p - R}{Rp}$$

$$q = \frac{Rp}{2p - R} = \frac{(-10\text{ cm})(8\text{ cm})}{2(8\text{ cm}) - (-10\text{ cm})} = -3.08\text{ cm}$$

The negative sign means the image is located behind the mirror. Since there are no light rays behind the mirror, the image is virtual.

EXAMPLE 24-16: What is the magnification of the concave mirror in Example 24-14?

Solution:

$$m = -\frac{h'}{h} = -\frac{6.0\text{ cm}}{12.0\text{ cm}} = -0.5$$

The image is inverted.

Figure 24-9 shows an object h in front of a plane mirror. Two rays have been drawn from the top of the object. Ray 1 is incident on the mirror at an angle of 0° with respect to the normal. This ray is reflected from the mirror at the same angle (Eq. 23-10). Ray 2 strikes the mirror at an angle θ and is reflected from the mirror at the same angle. The reflected rays do not intersect, so there is no real image—but there is a virtual image h' behind the mirror.

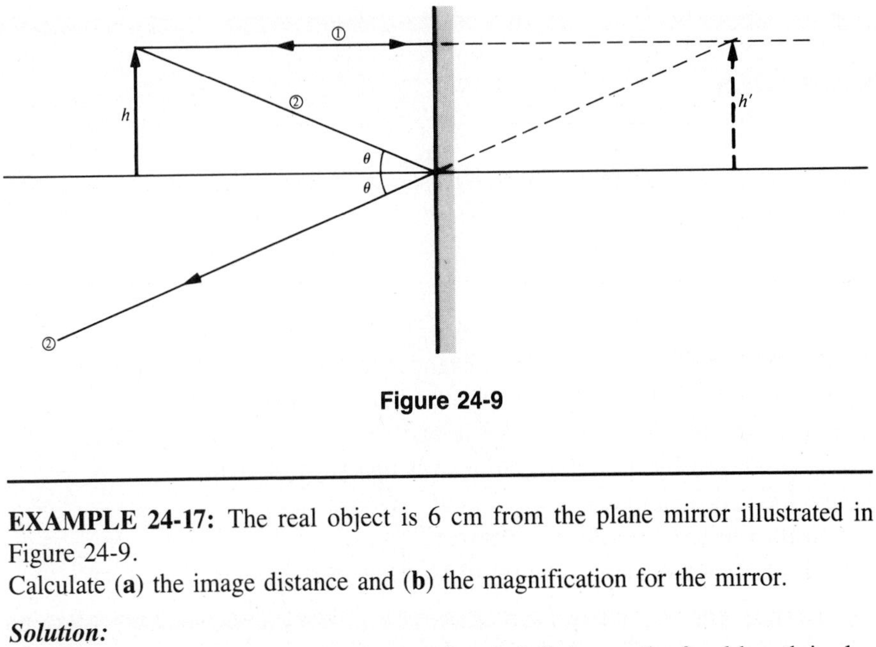

Figure 24-9

EXAMPLE 24-17: The real object is 6 cm from the plane mirror illustrated in Figure 24-9.
Calculate (**a**) the image distance and (**b**) the magnification for the mirror.

Solution:

(**a**) The radius of curvature of a plane surface is infinite, so the focal length is also infinite ($f = \infty$; $1/f = 0$). You can use Eq. (24-1) to solve for q:

$$\frac{1}{q} = \frac{1}{f} - \frac{1}{p} = 0 - \frac{1}{6.0\text{ cm}} \qquad q = -6.0\text{ cm}$$

The negative sign means that the virtual image is behind the mirror.

(**b**) Directly from Eq. (24-3),

$$m = -\left(\frac{-6.0\text{ cm}}{6.0\text{ cm}}\right) = 1.0$$

so $h' = h$ for plane mirrors. A positive magnification means the image is erect. When you look at yourself in a plane mirror, you see a virtual image exactly the same size as you are.

SUMMARY

Thin-lens equation	$\frac{1}{p} + \frac{1}{q} = \frac{1}{f}$	the relationship between the object distance and the image distance depends on the focal length of the lens
Lens-maker's equation	$\frac{1}{f} = (n - 1)\left(\frac{1}{R_1} + \frac{1}{R_2}\right)$	the focal length of a thin lens depends on the radii of curvature of the two surfaces and the index of refraction of the lens material
Magnification	$m = -\frac{h'}{h} = -\frac{q}{p}$	magnification of a lens is the ratio of image size and object size
Total magnification of two lenses	$m_{tot} = m_1 m_2$	total magnification of two lenses if the product of the magnification of each lens
Focal length of a mirror	$f = \frac{R}{2}$	focal length of a mirror is one half of the radius of curvature

RAISE YOUR GRADES

Can you define . . . ?

- ☑ object distance
- ☑ image distance
- ☑ focal length
- ☑ radius of curvature
- ☑ a converging lens
- ☑ a negative lens
- ☑ a real image
- ☑ a virtual image

Can you . . . ?

- ☑ write the equation relating object and image distances to the focal length of a lens
- ☑ calculate the focal length of a lens if you know the lens' radii of curvature
- ☑ write the equation for the magnification of an image
- ☑ describe an image produced by the intersection of actual rays of light
- ☑ describe an image located behind a mirror
- ☑ write the equation for the focal length of a mirror
- ☑ write the equation for the total magnification of two lenses

SOLVED PROBLEMS

PROBLEM 24-1 An object is placed to the left of a positive lens whose focal length is 8.0 cm. The image is formed 24 cm to the right of the lens. Find the distance from the object to the lens.

Solution Use Eq. (24-1) to find the object distance:

$$\frac{1}{p} + \frac{1}{q} = \frac{1}{f}$$

$$\frac{1}{p} = \frac{1}{f} - \frac{1}{q} = \frac{q - f}{fq}$$

$$p = \frac{fq}{q - f} = \frac{(8 \text{ cm})(24 \text{ cm})}{24 \text{ cm} - 8 \text{ cm}} = 12.0 \text{ cm}$$

Since p is positive, the image is real.

PROBLEM 24-2 What is the magnification of the image described in Problem 24-1?

Solution From Eq. (24-3),

$$m = -\frac{q}{p} = -\frac{24 \text{ cm}}{12 \text{ cm}} = -2$$

Since m is negative, the image is inverted.

PROBLEM 24-3 An object is located in front of a positive lens. An inverted, real image is behind the lens. The magnification of this image is -4, and the distance from the object to the image is 30 cm. Find the object distance and the image distance.

Solution Use Eq. (24-3) to calculate p and q. (Hint: $p + q = 30$ cm.)

$$m = -\frac{q}{p}$$

$$m = -\frac{(30 \text{ cm} - p)}{p} = -\frac{30 \text{ cm}}{p} + 1$$

$$p = \frac{-30 \text{ cm}}{m - 1} = \frac{-30 \text{ cm}}{-4 - 1} = \frac{-30 \text{ cm}}{-5} = 6.00 \text{ cm}$$

$$q = 30 \text{ cm} - p = 30 \text{ cm} - 6 \text{ cm} = 24.0 \text{ cm}$$

PROBLEM 24-4 What is the focal length of the lens described in Problem 24-3?

Solution From Eq. (24-1),

$$\frac{1}{f} = \frac{1}{p} + \frac{1}{q} = \frac{q + p}{pq}$$

$$f = \frac{pq}{q + p} = \frac{(6 \text{ cm})(24 \text{ cm})}{24 \text{ cm} + 6 \text{ cm}} = 4.80 \text{ cm}$$

PROBLEM 24-5 An object is 10 cm in front of a negative lens whose focal length is -5 cm. Find the image distance.

Solution Use Eq. (24-1):

$$\frac{1}{q} = \frac{1}{f} - \frac{1}{p} = \frac{p - f}{fp} \qquad q = \frac{fp}{p - f} = \frac{(-5 \text{ cm})(10 \text{ cm})}{10 \text{ cm} - (-5 \text{ cm})} = -3.33 \text{ cm}$$

The image is virtual.

PROBLEM 24-6 What is the magnification of the image in Problem 24-5?

Solution From Eq. (24-3),

$$m = \left(\frac{-3.33 \text{ cm}}{10 \text{ cm}}\right) = 0.333$$

This image is erect.

PROBLEM 24-7 A plano-convex lens made of flint glass has an index of refraction of 1.66 and a focal length of 6.0 cm. Find the radius of curvature of the convex surface.

Solution Use Eq. (24-2) to find the unknown radius of curvature. Remember that the radius of a planar surface is infinite, so $1/R_2 = 0$.

$$\frac{1}{f} = (n-1)\left(\frac{1}{R_1} + \frac{1}{R_2}\right) = (n-1)\left(\frac{1}{R_2} + 0\right) = \frac{n-1}{R_1}$$

$$R_1 = f(n-1) = 6.0 \text{ cm}(1.66 - 1.00) = 3.96 \text{ cm}$$

PROBLEM 24-8 A negative lens is made of flint glass with an index of refraction of 1.66. Its focal length is -4.5 cm and its two concave surfaces have the same radius of curvature. Calculate R.

Solution Use Eq. (24-2):

$$\frac{1}{f} = (n-1)\left(\frac{1}{R} + \frac{1}{R}\right) = (n-1)\left(\frac{2}{R}\right)$$

$$R = 2f(n-1) = 2(-4.5 \text{ cm})(1.66 - 1.00) = -5.94 \text{ cm}$$

The negative sign means that each surface must be concave.

PROBLEM 24-9 An object is placed in front of a concave mirror whose radius of curvature is 10 cm. The inverted image produced by the mirror has exactly the same size as the object. Find the location of the object and the image.

Solution Use Eq. (24-3) to find the relationship between object and image distances.

$$m = -\frac{q}{p} \qquad -1 = -\frac{q}{p} \qquad p = q$$

Now, use Eq. (24-4) to find the focal length:

$$f = \frac{1}{2}R = \frac{1}{2}(10 \text{ cm}) = 5 \text{ cm}$$

Finally, use Eq. (24-1):

$$\frac{1}{p} + \frac{1}{q} = \frac{1}{f}$$

$$\frac{1}{p} + \frac{1}{p} = \frac{1}{f} = \frac{2}{p}$$

$$p = q = 2f = 2(5 \text{ cm}) = 10.0 \text{ cm}$$

If the object is placed at the center of curvature of the concave mirror, the image will also be located there. Of course, the image will be inverted.

PROBLEM 24-10 A convex mirror produces a virtual image 4 cm behind the mirror. The object is located 12 cm in front of the mirror. Find the radius of curvature of this mirror.

Solution First find f from Eq. (24-1):

$$\frac{1}{f} = \frac{1}{p} + \frac{1}{q} = \frac{q+p}{pq} \qquad f = \frac{pq}{q+p} = \frac{(12 \text{ cm})(-4 \text{ cm})}{-4 \text{ cm} + 12 \text{ cm}} = -6 \text{ cm}$$

Now, use Eq. (24-4) to calculate R:

$$R = 2f = 2(-6 \text{ cm}) = -12.0 \text{ cm}$$

PROBLEM 24-11 Two positive lenses have focal lengths of 8 cm and 6 cm. A real object is 16 cm to the left of the first lens as shown in Fig. 24-10. A distance of 4 cm separates the two lenses. Find the location of the image produced by the second lens.

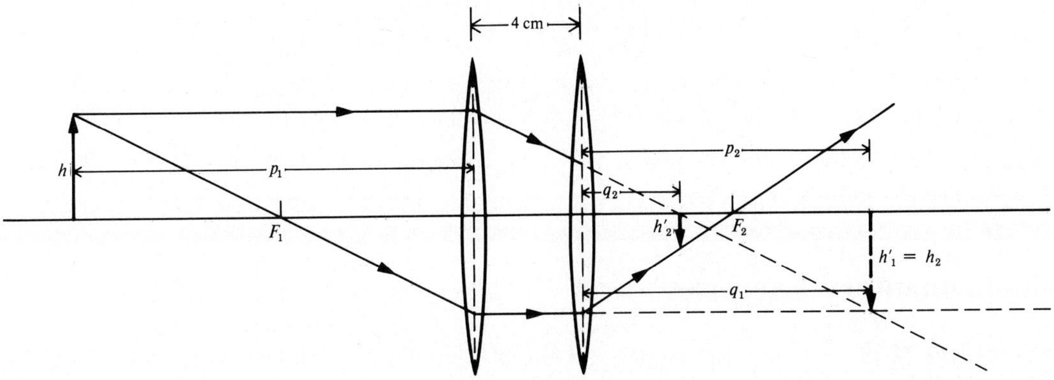

Figure 24-10

Solution You can solve this problem by using Eq. (24-1) twice, since the *real* image of the first lens is the *virtual object* of the second lens ($h'_1 = h_2$). Calculate the distance q_1 of the image produced by the first lens:

$$\frac{1}{q_1} = \frac{1}{f_1} - \frac{1}{p_1} = \frac{p_1 - f_1}{f_1 p_1}$$

$$q_1 = \frac{f_1 p_1}{p_1 - f_1} = \frac{(8 \text{ cm})(16 \text{ cm})}{16 \text{ cm} - 8 \text{ cm}} = 16 \text{ cm}$$

Because q_1 is located to the right of the second lens, the object distance p_2 (p_1 − 4 cm) is negative, so $p_2 = -12$ cm. Now, find the image produced by the second lens:

$$\frac{1}{q_2} = \frac{1}{f_2} - \frac{1}{p_2} = \frac{p_2 - f_2}{f_2 p_2}$$

$$q_2 = \frac{f_2 p_2}{p_2 - f_2} = \frac{(6 \text{ cm})(-12 \text{ cm})}{-12 \text{ cm} - 6 \text{ cm}} = 4.00 \text{ cm}$$

PROBLEM 24-12 Calculate the overall magnification of the two positive lenses in Figure 24-10.

Solution Solve this problem by combining the equation relating object and image distances to magnification (Eq. 24-3) with the equation for total magnification of two lenses (Eq. 24-4).

$$m_{\text{tot}} = m_1 m_2 = \left(-\frac{q_1}{p_1}\right)\left(-\frac{q_2}{p_2}\right) = \left(-\frac{16 \text{ cm}}{16 \text{ cm}}\right)\left(-\frac{4 \text{ cm}}{-12 \text{ cm}}\right) = -0.333 \text{ cm}$$

The negative result means that the image is inverted.

PROBLEM 24-13 A real object is placed 4 cm in front of a positive lens, which produces an inverted image three times as large as the object ($m = -3$). Find **(a)** the image distance and **(b)** the focal length of the lens.

Solution

(a)

$$m = -\frac{q}{p} \qquad q = -pm = -(4 \text{ cm})(-3) = 12.0 \text{ cm}$$

(b)

$$\frac{1}{f} = \frac{1}{p} + \frac{1}{q} = \frac{p + q}{pq} \qquad f = \frac{pq}{p + q} = \frac{(4 \text{ cm})(12 \text{ cm})}{4 \text{ cm} + 12 \text{ cm}} = 3.00 \text{ cm}$$

PROBLEM 24-14 The positive lens in Problem 24-13 is made of flint glass with an index of 1.66. It is a double convex lens with equal radii. Find the radius of curvature of each surface that will give a focal length of 3 cm.

Solution Use Eq. (24-2):

$$\frac{1}{f} = (n - 1)\left(\frac{1}{R} + \frac{1}{R}\right) = (n - 1)\left(\frac{2}{R}\right)$$

$$R = 2f(n - 1) = 2(3 \text{ cm})(1.66 - 1) = 3.96 \text{ cm}$$

Supplementary Exercises

PROBLEM 24-15 A real object is placed 8 cm in front of a positive lens with a focal length of 12 cm. Find the image distance.

PROBLEM 24-16 What is the magnification of the image in Problem 24-15?

PROBLEM 24-17 A real object is 6 cm in front of a negative lens whose focal length is −10 cm. Find the image distance.

PROBLEM 24-18 What is the magnification of the image in Problem 24-17?

PROBLEM 24-19 An object is placed in front of a positive lens with a focal length of 10 cm. The inverted image is four times as large as the object. Find the object distance and the image distance.

PROBLEM 24-20 A plano-convex lens has a focal length of 10 cm. The index of refraction is 1.50. Find the radius of curvature of the convex surface.

PROBLEM 24-21 A plano-concave lens has a focal length of −10 cm. The index of refraction is 1.66. Find the radius of curvature of the concave surface.

PROBLEM 24-22 A virtual object is 6 cm to the right of a negative lens whose focal length is −12 cm. Find the image distance.

PROBLEM 24-23 Find the magnification of the image in Problem 24-22.

PROBLEM 24-24 A virtual object is 5 cm behind a concave mirror whose radius of curvature is 20 cm. Find the image distance.

PROBLEM 24-25 What is the magnification of the image in Problem 24-24?

Answers to Supplementary Exercises

24-15: $q = -24.0$ cm

24-16: $m = 3.00$

24-17: $q = -3.75$ cm

24-18: $m = 0.625$

24-19: $p = 12.5$ cm; $q = 50.0$ cm

24-20: $R = 5.00$ cm

24-21: $R = -6.60$ cm

24-22: $q = 12.0$ cm

24-23: $m = 2.00$

24-24: $q = 3.33$ cm

24-25: $m = 0.667$

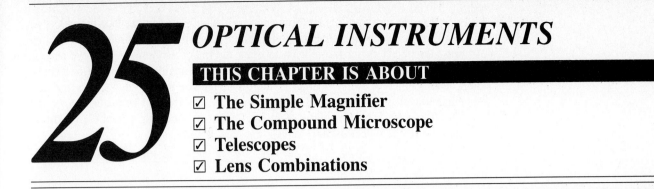

25 OPTICAL INSTRUMENTS

THIS CHAPTER IS ABOUT

- ☑ The Simple Magnifier
- ☑ The Compound Microscope
- ☑ Telescopes
- ☑ Lens Combinations

25-1. The Simple Magnifier

- A **simple magnifier** (e.g., a reading glass) is a positive lens, usually either double-convex or plano-convex, that has the ability to enlarge the angular size of an object.
- The **angular size** of an object is the object's *apparent* size—the *proportion* of the observer's field of view that the object covers.

For a small object, this is very nearly its height divided by its distance from the lens. An object to be viewed through a simple magnifier is placed *between* the lens and the lens' focal point, so that the image formed is erect and *virtual*. This is the situation described in Example 24-3 and illustrated in Figure 24-2.

- The angular magnification M of a magnifier is the ratio of the angular size of the image to the angular size of the object:

ANGULAR MAGNIFICATION

$$M = \frac{\text{angular size of image}}{\text{angular size of object}} = \frac{\theta}{\theta_0} \tag{25-1}$$

We'll assume the observer's eye is close to the magnifier, so that the distance from the eye to the image is practically the same as the distance from the lens to the image. To measure the strength of a magnifier, we compare the angular size of the image viewed through the magnifier with the object's maximum angular size that we could have seen without the magnifier. By moving an object closer to your eye, of course, you increase its angular size, but there is a limit to how close the human eye can focus.

- The **near point,** the shortest distance at which the eye can form sharp images on its retina, averages about 25 cm, and we'll use that figure as our value for the near point.

The greatest practical angular size of an object, therefore, is its angular size when it is 25 cm from the eye (Figure 25-1).

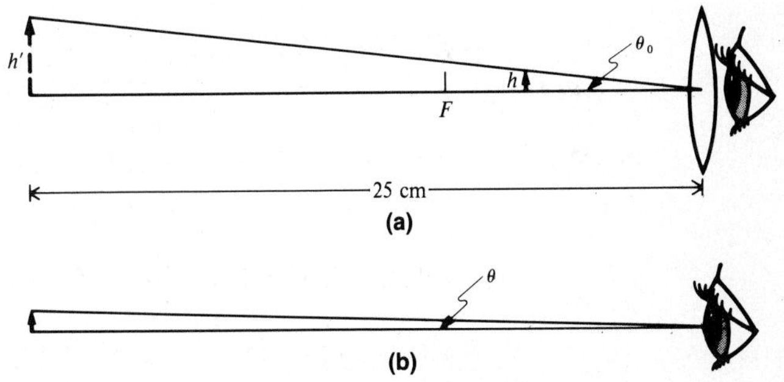

Figure 25-1. (a) Maximum magnification of a simple magnifier. (b) Maximum angular size of unmagnified object.

STRENGTH OF A MAGNIFIER

$$M = \frac{\text{angular size of image seen through magnifier}}{\text{angular size of object at near point}} \tag{25-1a}$$

Because angular size is determined by the ratio of image size to image distance, we can express Eq. (25-1a) as

$$M = \frac{h'/q}{h/25\ \text{cm}} \tag{25-1b}$$

where h is the height of the object, h' is the height of the image, and q is the image distance.

How do you find the maximum magnification? You could place the object at the focal point, to make the image infinitely large (but also infinitely distant), or bring the object closer to the lens, to bring the image as close to your eye as you can focus. It may seem strange, but the minimum angular magnification occurs when the object is at the focal point and the image size is infinite. We can find this minimum magnification by (1) rearranging Eq. (25-1b), (2) substituting q/p for h'/h (the definition of magnification, Eq. 24-3), (3) canceling q, and (4) substituting f for p (the object distance equals the focal length).

$$M = \frac{h'/q}{h/25\text{ cm}} = \frac{h'}{h}\frac{25\text{ cm}}{q} = \frac{q}{p}\frac{25\text{ cm}}{q} = \frac{25\text{ cm}}{p} = \frac{25\text{ cm}}{f}$$

MINIMUM MAGNIFICATION (MAGNIFIER)

$$M = \frac{25\text{ cm}}{f} \qquad \textbf{(25-2)}$$

When the image is at the near point, 25 cm, the angular magnification has its maximum value. We can find this maximum magnification in the same way we found the minimum magnification if we express p in terms of f and remember that a simple magnifier forms a virtual image, so the value of q is negative.

$$\frac{1}{p} = \frac{1}{f} - \frac{1}{q} = \frac{q-f}{fq} \qquad p = \frac{fq}{q-f} = \frac{f(-25\text{ cm})}{-25\text{ cm} - f}$$

$$M = \frac{h'/q}{h/25\text{ cm}} = \frac{h'}{h}\frac{25\text{ cm}}{q} = \frac{q}{p}\frac{25\text{ cm}}{q} = \frac{25\text{ cm}}{p} = \frac{(25\text{ cm})(-25\text{ cm} - f)}{f(-25\text{ cm})}$$

$$= \frac{-25\text{ cm} - f}{-f} = \frac{25\text{ cm} + f}{f} = \frac{25\text{ cm}}{f} + 1$$

MAXIMUM MAGNIFICATION (MAGNIFIER)

$$M = \frac{25\text{ cm}}{f} + 1 \qquad \textbf{(25-3)}$$

EXAMPLE 25-1: A magnifier has two convex surfaces, each with a radius of curvature of 10 cm. The index of refraction of the glass is 1.50.
What is the maximum magnification of this magnifier?

Solution: Use the lens-maker's equation (24-2) to calculate f:

$$\frac{1}{f} = (n-1)\left(\frac{1}{R} + \frac{1}{R}\right) = (n-1)\left(\frac{2}{R}\right)$$

$$f = \frac{R}{2(n-1)} = \frac{10\text{ cm}}{2(1.5-1)} = 10\text{ cm}$$

Then, from Eq. (25-3):

$$M = \frac{25\text{ cm}}{f} + 1 = \frac{25\text{ cm}}{10\text{ cm}} + 1 = 3.50$$

EXAMPLE 25-2: What is the magnification of a magnifier whose focal length is 10 cm when an object is placed 10 cm in front of the lens?

Solution: Directly from Eq. (25-2),

$$M = \frac{25\text{ cm}}{f} = \frac{25\text{ cm}}{10\text{ cm}} = 2.50$$

25-2. The Compound Microscope

A compound microscope consists of two positive lenses. Light entering the microscope passes first through the **objective** lens, which has a short focal length and forms a real image. The second lens, the **eyepiece** or **ocular**, is a simple magni-

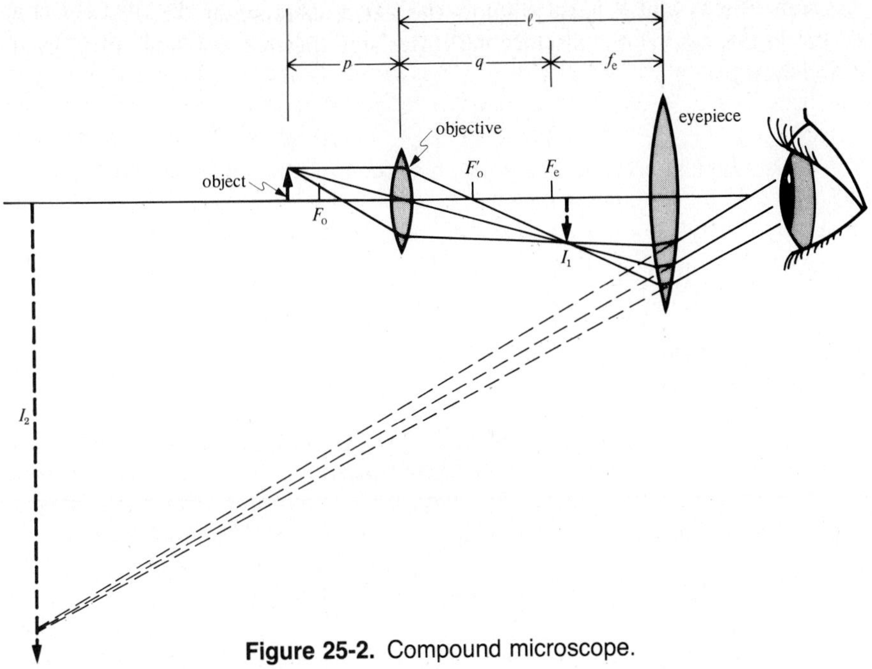

Figure 25-2. Compound microscope.

fier. The real image formed by the objective lies between the ocular and the ocular's focal point, so that the ocular forms a magnified virtual image (see Figure 25-2). The total magnification of the compound microscope is the product of the magnification of each lens. If the eyepiece is adjusted so that the final image I_2 is at infinity, the first image is at the focal point of the eyepiece and the total magnification of the microscope is

MAGNIFICATION (MICROSCOPE)

$$M = \left(\frac{-q}{p}\right)\left(\frac{25\ \text{cm}}{f_e}\right) \tag{25-4}$$

EXAMPLE 25-3: The objective lens of a microscope has a focal length f_o of 1.8 cm. The eyepiece has a focal length f_e of 2.0 cm. An object is placed 2.0 cm in front of the objective lens and the final image is at infinity.
What is the magnification of the microscope?

Solution: First, find the image distance of the objective lens from Eq. (24-1):

$$\frac{1}{q} = \frac{1}{f} - \frac{1}{p} = \frac{p - f}{fp}$$

$$q = \frac{fp}{p - f} = \frac{(1.8\ \text{cm})(2\ \text{cm})}{2\ \text{cm} - 1.8\ \text{cm}} = 18\ \text{cm}$$

Then, directly from Eq. (25-4):

$$M = \left(\frac{-q}{p}\right)\left(\frac{25\ \text{cm}}{f_e}\right) = \left(\frac{-18\ \text{cm}}{2\ \text{cm}}\right)\left(\frac{25\ \text{cm}}{2\ \text{cm}}\right) = -112.5$$

The negative value of M means that the final image is inverted.

EXAMPLE 25-4: A compound microscope has an objective lens with a focal length of 2.0 cm and an eyepiece with a focal length of 2.5 cm. The lenses are separated by a distance of 14.5 cm. The final image is at infinity.
Calculate the magnification of the microscope.

Solution: First, you'll need to know the image distance of the objective lens which, in this case, is the distance the lenses are separated minus the focal length of the eyepiece:

$$q = \ell - f_e = (14.5 - 2.5)\text{ cm} = 12.0\text{ cm}$$

You'll also have to calculate the object distance p of the objective lens:

$$\frac{1}{p} = \frac{1}{f} - \frac{1}{q} = \frac{q - f}{fq}$$

$$p = \frac{fq}{q - f} = \frac{(2\text{ cm})(12\text{ cm})}{12\text{ cm} - 2\text{ cm}} = 2.4\text{ cm}$$

Then, from Eq. (25-4),

$$M = \left(\frac{-12\text{ cm}}{2.4\text{ cm}}\right)\left(\frac{25\text{ cm}}{2.5\text{ cm}}\right) = -50.0$$

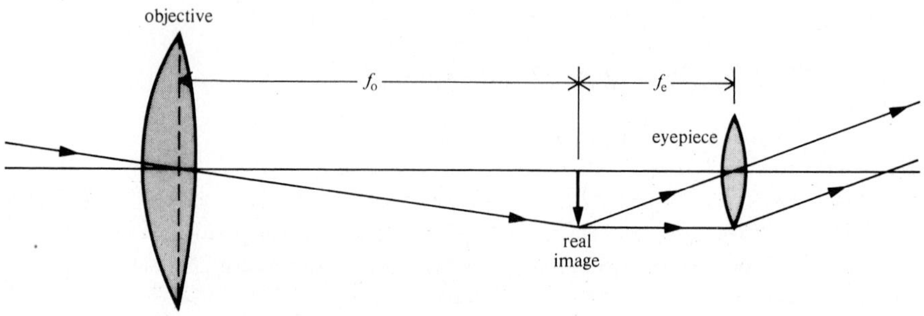

Figure 25-3. Astronomical telescope.

25-3. Telescopes

A. The astronomical telescope

Figure 25-3 illustrates an astronomical telescope used to view very distant objects. The astronomical telescope is similar to a microscope except that in the telescope the real image is smaller than the object. Since the object is very far away, the incident rays are practically parallel, and the *real image* is at the focal point of the objective lens. When the eyepiece is adjusted for minimum eyestrain, the eyepiece's focal point coincides with the real image, so the final, virtual image produced by the eyepiece is at infinity. The total magnification of an astronomical telescope is the ratio of the focal lengths of the two lenses:

MAGNIFICATION (ASTRONOMICAL TELESCOPE)

$$M = -\frac{f_o}{f_e} \qquad \textbf{(25-5)}$$

EXAMPLE 25-5: An astronomical telescope has an objective lens with a focal length of 30 cm and an eyepiece with a focal length of 1.5 cm. What is the magnification of this telescope?

Solution: Directly from Eq. (25-5):

$$M = -\frac{f_o}{f_e} = -\left(\frac{30\text{ cm}}{1.5\text{ cm}}\right) = -20.0$$

The negative sign means the image is inverted.

EXAMPLE 25-6: An astronomical telescope has an objective lens and an ocular separated by a distance of 42 cm. The ocular has a focal length of 2.0 cm. Find the angular magnification of the telescope.

Solution: First, find the focal length of the objective lens:

$$f_o = 42\text{ cm} - f_e = (42 - 2)\text{ cm} = 40\text{ cm}$$

Then, use Eq. (25-5):

$$M = -\left(\frac{40\text{ cm}}{2\text{ cm}}\right) = -20.0$$

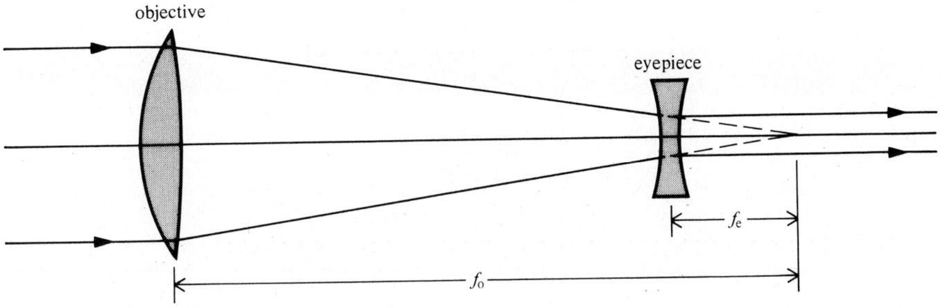

Figure 25-4. Galilean telescope.

B. The Galilean telescope

A Galilean telescope (Figure 25-4) consists of a positive objective lens and a negative eyepiece that forms a virtual image. An important advantage of a Galilean telescope is that the image is erect. The equation for the angular magnification is

MAGNIFICATION (GALILEAN TELESCOPE)

$$M = -\frac{f_o}{f_e} \tag{25-6}$$

EXAMPLE 25-7: A pair of opera glasses has objective lenses with a focal length of 12 cm and oculars with focal lengths of −3 cm.
What is the angular magnification of the opera glasses?

Solution: Directly from Eq. (25-6),

$$M = -\left(\frac{12\text{ cm}}{-3\text{ cm}}\right) = 4$$

25-4. Lens Combinations

In many optical instruments, two thin lenses are placed in contact, as shown in Figure 25-5. Their effective focal length is

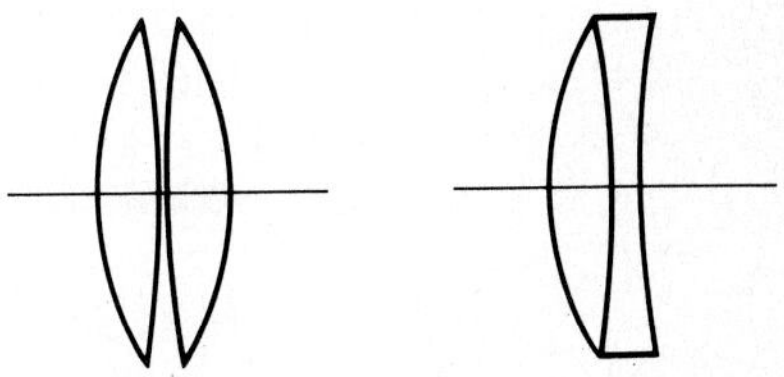

Figure 25-5. Lens combinations.

EFFECTIVE FOCAL LENGTH OF LENSES IN CONTACT

$$\frac{1}{f_{eff}} = \frac{1}{f_1} + \frac{1}{f_2} \tag{25-7}$$

EXAMPLE 25-8: A thin lens with a focal length of 20 cm is placed in contact with a second lens whose focal length is 25 cm.
What is the effective focal length of these two lenses?

Solution: Directly from Eq. (25-7),

$$\frac{1}{f_{eff}} = \frac{1}{f_1} + \frac{1}{f_2} = \frac{f_1 + f_2}{f_1 f_2}$$

$$f_{eff} = \frac{f_1 f_2}{f_1 + f_2} = \frac{(20\text{ cm})(25\text{ cm})}{20\text{ cm} + 25\text{ cm}} = 11.1\text{ cm}$$

EXAMPLE 25-9: A positive lens with a focal length of 18 cm is placed in contact with a negative lens. The effective focal length of this lens combination is −36 cm. Calculate the focal length of the negative lens.

Solution: Use Eq. (25-7) and solve for the focal length of the second lens:

$$\frac{1}{f_2} = \frac{1}{f_{\text{eff}}} - \frac{1}{f_1} = \frac{f_1 - f_{\text{eff}}}{f_{\text{eff}} f_1}$$

$$f_2 = \frac{f_{\text{eff}} f_1}{f_1 - f_{\text{eff}}} = \frac{(-36\text{ cm})(18\text{ cm})}{18\text{ cm} - (-36\text{ cm})} = -12.0\text{ cm}$$

SUMMARY

Angular magnification	$M = \dfrac{\text{angular size of image}}{\text{angular size of object}}$	angular magnification is the ratio of the angular size of the image to the angular size of the object
Minimum magnification (magnifier)	$M = \dfrac{25\text{ cm}}{f}$	minimum angular magnification occurs when an object is placed in front of the lens at the focal point, so that the image is at infinity
Maximum magnification (magnifier)	$M = \dfrac{25\text{ cm}}{f} + 1$	angular magnification has its maximum value when the image is at the nearpoint
Magnification (microscope)	$M = \left(\dfrac{-q}{p}\right)\left(\dfrac{25\text{ cm}}{f_e}\right)$	total magnification of a compound microscope is the product of the magnification of each lens when the eyepiece is adjusted so that the final image is at infinity
Magnification (astronomical telescope and Galilean telescope)	$M = -\dfrac{f_o}{f_e}$	total magnification of a telescope is the ratio of the focal lengths of the two lenses
Effective focal length of lenses in contact	$\dfrac{1}{f_{\text{eff}}} = \dfrac{1}{f_1} + \dfrac{1}{f_2}$	the reciprocals of focal lengths of lenses in contact are additive

RAISE YOUR GRADES

Can you define . . . ?

☑ angular size ☑ near point ☑ objective ☑ eyepiece

Can you . . . ?

☑ write the equation for the minimum angular magnification of a magnifier
☑ write the equation for the maximum angular magnification of a magnifier
☑ write the equation for the magnification of a compound microscope
☑ calculate the magnification of an astronomical telescope if you know the focal lengths of the objective lens and the eyepiece
☑ write the equation for the magnification of a Galilean telescope
☑ calculate the effective focal length of two lenses placed in contact

SOLVED PROBLEMS

PROBLEM 25-1 A simple magnifier consists of a plane surface and a convex surface whose radius of curvature is 5 cm. What is the maximum magnification of this magnifier? ($n = 1.50$) [**recall:** R is infinite if the surface is planar. (Sec. 24-10)]

Solution Calculate the focal length from Eq. (24-2):

$$\frac{1}{f} = (n - 1)\left(\frac{1}{R_1} + \frac{1}{R_2}\right) = (n - 1)\left(\frac{1}{R_1} + 0\right) = \frac{n - 1}{R_1}$$

$$f = \frac{R}{n - 1} = \frac{5\text{ cm}}{1.50 - 1} = 10\text{ cm}$$

Then, use Eq. (25-3) to find the maximum magnification:

$$M = \frac{25\text{ cm}}{f} + 1 = \frac{25\text{ cm}}{10\text{ cm}} + 1 = 3.50$$

PROBLEM 25-2 When a simple magnifier is positioned so that the image is at infinity, the resulting minimum magnification is 5. What is the focal length of this magnifier?

Solution Use Eq. (25-2):

$$f = \frac{25\text{ cm}}{M} = \frac{25\text{ cm}}{5} = 5\text{ cm}$$

PROBLEM 25-3 A simple magnifier has a focal length of 18.9 cm. What is the maximum magnification of this lens?

Solution Use Eq. (25-3):

$$M = \frac{25\text{ cm}}{18.9\text{ cm}} + 1 = 2.32$$

PROBLEM 25-4 A compound microscope has an objective lens with a focal length of 1.5 cm. The object is 1.8 cm from the objective. The focal length of the eyepiece is 2.5 cm. What is the total magnification of this microscope?

Solution You'll have to calculate the image distance from Eq. (24-1):

$$\frac{1}{q} = \frac{1}{f} - \frac{1}{p} = \frac{p - f}{fp}$$

$$q = \frac{fp}{p - f} = \frac{(1.5\text{ cm})(1.8\text{ cm})}{1.8\text{ cm} - 1.5\text{ cm}} = 9.0\text{ cm}$$

Now, use Eq. (25-4) to calculate the magnification:

$$M = \left(\frac{-q}{p}\right)\left(\frac{25\text{ cm}}{f_2}\right) = \left(\frac{-9.0\text{ cm}}{1.8\text{ cm}}\right)\left(\frac{25\text{ cm}}{2.5\text{ cm}}\right) = -50.0$$

PROBLEM 25-5 A compound microscope has an objective lens with a focal length of 2.0 cm. The eyepiece has a focal length of 1.5 cm. These two lenses are separated by a distance of 11.5 cm. Calculate the total magnification of this microscope.

Solution Calculate the image distance of the objective lens:

$$q = \ell - f_e = 11.5\text{ cm} - 1.5\text{ cm} = 10.0\text{ cm}$$

Then, calculate the object distance:

$$\frac{1}{p} = \frac{1}{f} - \frac{1}{q} = \frac{q - f}{fq}$$

$$p = \frac{fq}{q - f} = \frac{(2 \text{ cm})(10 \text{ cm})}{10 \text{ cm} - 2 \text{ cm}} = 2.5 \text{ cm}$$

Now, use Eq. (25-4):

$$M = \left(\frac{-q}{p}\right)\left(\frac{25 \text{ cm}}{f_2}\right) = \left(\frac{-10 \text{ cm}}{2.5 \text{ cm}}\right)\left(\frac{25 \text{ cm}}{1.5 \text{ cm}}\right) = -66.7$$

PROBLEM 25-6 An astronomical telescope has an objective lens with a focal length of 40 cm and an eyepiece with a focal length of 2 cm. What is the magnification of this telescope?

Solution Use Eq. (25-5):

$$M = -\frac{f_o}{f_e} = -\frac{40 \text{ cm}}{2 \text{ cm}} = -20.0$$

PROBLEM 25-7 A Galilean telescope has an objective lens with a focal length of 20 cm and an eyepiece with a focal length of −4 cm. What is the magnification of this telescope?

Solution Use Eq. (25-6):

$$M = -\frac{f_o}{f_e} = -\frac{20 \text{ cm}}{-4 \text{ cm}} = 5.00$$

PROBLEM 25-8 A positive lens and a negative lens are placed as close together as possible. The two focal lengths are 16 cm and −8 cm. What is the effective focal length of this combination of two lenses?

Solution Use Eq. (25-7):

$$\frac{1}{f_{\text{eff}}} = \frac{1}{f_1} + \frac{1}{f_2} = \frac{f_1 + f_2}{f_1 f_2}$$

$$f_{\text{eff}} = \frac{f_1 f_2}{f_1 + f_2} = \frac{(16 \text{ cm})(-8 \text{ cm})}{16 \text{ cm} - 8 \text{ cm}} = -16.0 \text{ cm}$$

PROBLEM 25-9 Two thin lenses are placed in contact. The first lens has a focal length of 20 cm and the effective focal length of the combination of the two lenses is 12 cm. Calculate the focal length of the second lens.

Solution From Eq. (25-7),

$$\frac{1}{f_2} = \frac{1}{f_{\text{eff}}} - \frac{1}{f_1} = \frac{f_1 - f_{\text{eff}}}{f_{\text{eff}} f_1}$$

$$f_2 = \frac{f_{\text{eff}} f_1}{f_1 - f_{\text{eff}}} = \frac{(12 \text{ cm})(20 \text{ cm})}{20 \text{ cm} - 12 \text{ cm}} = 30.0 \text{ cm}$$

Supplementary Exercises

PROBLEM 25-10 The angular size of the moon viewed from the surface of the earth is 0.52°. When the moon is observed with an astronomical telescope, its angular size is increased to 10.4°. What is the angular magnification of this telescope?

PROBLEM 25-11 A simple magnifier is used to examine a stamp placed at the focal point of the magnifier. The focal length of the magnifier is 5.0 cm. What is the magnification?

PROBLEM 25-12 A simple magnifier has a focal length of 4.0 cm. When a postage stamp is placed a little nearer the lens than 4.0 cm, the virtual image is at the nearpoint, 25 cm from the observer's eye. This arrangement produces the maximum magnification. What is the maximum magnification of this magnifier?

PROBLEM 25-13 A double-convex lens is made of flint glass with an index of refraction of 1.66 and has a focal length of 8.0 cm. What is the maximum magnification of this magnifier?

PROBLEM 25-14 A compound microscope has an objective lens with a focal length of 2.0 cm. The focal length of the eyepiece is 2.5 cm. A microscope slide is 2.5 cm in front of the objective. The final image of the eyepiece is at infinity. Calculate the total magnification of this microscope.

PROBLEM 25-15 The distance from the objective lens to the eyepiece of the microscope described in Problem 25-18 is 12.5 cm. What will the magnification be if this distance is reduced to 8.5 cm?

PROBLEM 25-16 An astronomical telescope has an objective lens with a focal length of 44 cm and an eyepiece with a focal length of 2 cm. What is the magnification of this telescope?

PROBLEM 25-17 A Galilean telescope has an objective lens with a focal length of 12 cm. The negative eyepiece has a focal length of −3 cm. What is the magnification of this Galilean telescope?

PROBLEM 25-18 A positive lens with a focal length of 8 cm is placed in contact with a negative lens whose focal length is −24 cm. What is the focal length of this lens combination?

PROBLEM 25-19 Two identical positive lenses are placed as close together as possible. The focal length of this lens combination is 7.5 cm. What is the focal length of each positive lens?

Answers to Supplementary Exercises

25-10: $M = 20.0$

21-11: $M = 5.00$

25-12: $M = 7.25$

25-13: 4.13

25-14: −40.0

25-15: −20.0

25-16: −22.0

25-17: 4.00

25-18: 12.0 cm

25-19: 15.0 cm

26 WAVE OPTICS

THIS CHAPTER IS ABOUT

☑ **Phase Difference**
☑ **Double-Slit Interference**
☑ **Single-Slit Diffraction**
☑ **Diffraction Gratings and Spectroscopes**
☑ **Phase Change upon Reflection**
☑ **The Air Wedge**
☑ **The Michelson Interferometer**
☑ **Polarization of Light**

26-1. Phase Difference

Electromagnetic waves behave in many ways like other kinds of waves, for example, they show *interference* and *diffraction*.

- **Interference** occurs when light waves from different sources overlap.
- **Constructive interference** occurs when the waves are *in phase* and, consequently, reinforce one another. [**recall:** Waves are in phase when they reach their peak at the same time and place. (Sec. 11-2)]
- **Destructive interference** occurs when the waves are *out of phase* and, consequently, partially or wholly cancel one another. Two waves are *exactly* out of phase when the maximum of one wave coincides with the minimum of the other.

When two light waves pass through one point in space, the resultant wave motion at that point depends on the **phase difference** between the two waves.

- If the phase difference is zero or an integral multiple of 2π radians, the resultant wave motion has a maximum value.
- If the phase difference is an odd integral multiple of π radians ($\pi, 3\pi, 5\pi, \ldots$), the resultant wave motion has a minimum value.

The difference between the distances that two light waves travel is their **path difference** Δs. If two waves with the same wavelength λ start from different points, they may not be in phase when they meet, but we can find the phase difference $\Delta\phi$ between them at any point with

RELATIONSHIP BETWEEN PHASE DIFFERENCE AND PATH DIFFERENCE

$$\Delta\phi = 2\pi\frac{\Delta s}{\lambda} \qquad (\Delta\phi \text{ in radians}) \qquad \textbf{(26-1)}$$

EXAMPLE 26-1: Radio waves are electromagnetic waves like light, but have longer wavelengths. Two sources of radio waves are broadcasting in phase at a wavelength of 1 cm and are separated by 2 cm. What is the phase difference at a point on the line joining the two sources and 0.75 cm from one of them?

Solution: If the point is 0.75 cm from one source it must be $(2 - 0.75)$ cm = 1.25 cm from the other. So the difference between the two path lengths is

$$\Delta s = (1.25 - 0.75)\text{ cm} = 0.5\text{ cm}$$

Now we can use Eq. (26-1) to find the phase difference.

$$\Delta\phi = 2\pi\frac{\Delta s}{\lambda} = 2\pi\frac{0.5 \text{ cm}}{1 \text{ cm}} = \pi \text{ rad} = 180°$$

Because the phase difference is 180° (an odd multiple of π radians) the two waves interfere destructively at this point. Notice that the waves interfere constructively at the point halfway between the sources, where $\Delta s = (1 - 1)$ cm $= 0$.

When light shines through a very small hole, the edges of the hole do not cast a sharp shadow. This is the result of another property of waves that light exhibits.

- **Diffraction** is the ability of a light wave to bend around the edge of an obstacle.

When light shines through a small hole, the combined effects of diffraction and interference result in an image that is a small bright disk with bright and dark fringes.

- At the center of a *bright* fringe, the waves incident on the screen are *in phase*.
- At the center of a *dark* fringe, the incident waves are *exactly out of phase*.

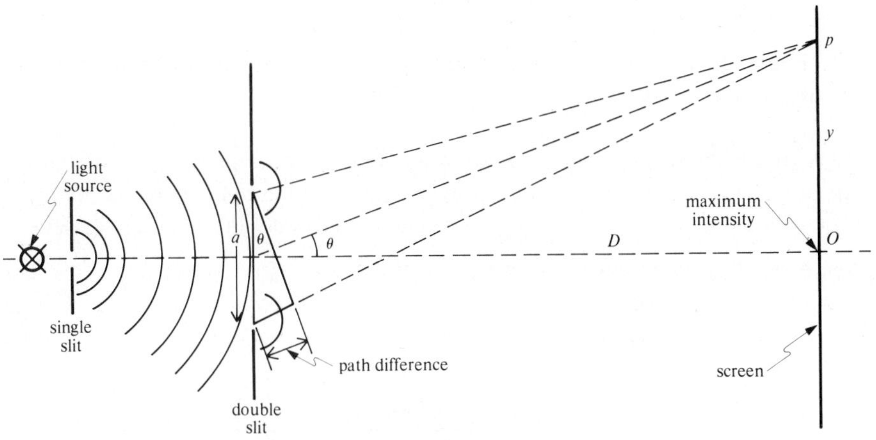

Figure 26-1

26-2. Double-Slit Interference

Figure 26-1 illustrates the interference pattern produced by light of a single wavelength (color) shining on two narrow parallel slits separated by a distance a. Because the single slit opposite the light source is equidistant from each of the double slits, the light waves emerging from the double slits are in phase. The point p on the screen is closer to the upper slit than to the lower one, so there is a difference in the path length of waves traveling from each slit to p. If there is a bright fringe at p, the path difference of the two rays reaching p is n, where n is an integer ($n = 0, 1, 2, 3, \ldots$) called the **fringe order**. Each fringe order corresponds to the center of a bright fringe at an increasing distance from point O. Point O is equidistant from each slit, so the two waves arriving there are in phase. Point O is the center of the fringe pattern and the intensity there is a maximum.

- The **intensity** of a light ray is the flow of energy per unit area.

note: The bright central band at point O (the line of maximum intensity) is called the *zero-order* fringe.

We can find the locations of the bright fringes in terms of the angle θ between a line from the midpoint between the slits to O and a line from the midpoint between the slits to p.

ANGULAR POSITIONS OF BRIGHT FRINGES

$$\sin\theta = \frac{n\lambda}{a} \qquad \textbf{(26-2)}$$

EXAMPLE 26-2: If the distance a between the slits in Figure 26-1 is 0.02 mm and the wavelength of the light is 500 nm, what is the angle between the central maximum and the second order bright fringe?

Solution: Directly from Eq. (26-2),

$$\sin\theta = \frac{n\lambda}{a} = \frac{2(5.00 \times 10^{-7}\ \text{m})}{2.00 \times 10^{-5}\ \text{m}} = 0.050$$

$$\theta = \arcsin(0.050) = 2.866°$$

EXAMPLE 26-3: Point p in Figure 26-1 is the center of the second-order fringe. Determine the phase difference between the light waves that arrive at p from the double slit.

Solution: Use Eq. (26-1). Note that the path difference Δs equals $n\lambda$, and that, for a second-order fringe, $n = 2$.

$$\Delta\phi = 2\pi\frac{\Delta s}{\lambda} = 2\pi\frac{2\lambda}{\lambda} = 4\pi \text{ radians}$$

The phase difference is an integer multiple of 2π, so the intensity at p is a maximum.

EXAMPLE 26-4: The distance D from the double slit to the screen is 60 cm, the slit distance a is 0.02 mm, and the wavelength from the light source is 500 nm. Calculate the distance y from point O to the second-order bright fringe at point p.

Solution: Use your result from Example 26-2.

$$\tan\theta = \frac{y}{D}$$

$$y = D\tan\theta = (60\ \text{cm})\tan 2.866° = 3.00\ \text{cm}$$

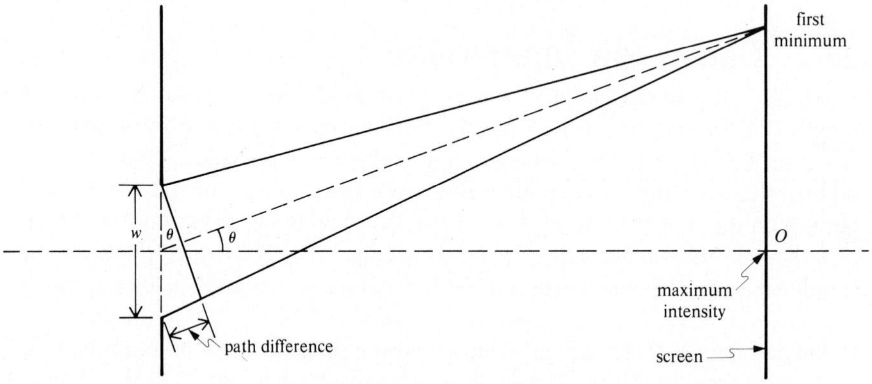

Figure 26-2

26-3. Single-Slit Diffraction

When a light wave is incident on a single narrow slit of width w, as shown in Figure 26-2, it also produces a diffraction pattern. The location of the *minimum* intensities, the *dark* fringes, in this diffraction pattern are given by

ANGULAR POSITIONS OF MINIMUM INTENSITIES $$n\lambda = w\sin\theta \qquad \textbf{(26-3)}$$

Here $n = 1$ represents the first minimum next to the central bright maximum. There is no $n = 0$ minimum.

EXAMPLE 26-5: Light with a wavelength of 600 nm is incident on a slit whose width is 0.01 nm. The screen is 80 cm from the slit.
Calculate the total width of the central maximum as shown in Figure 26-3.

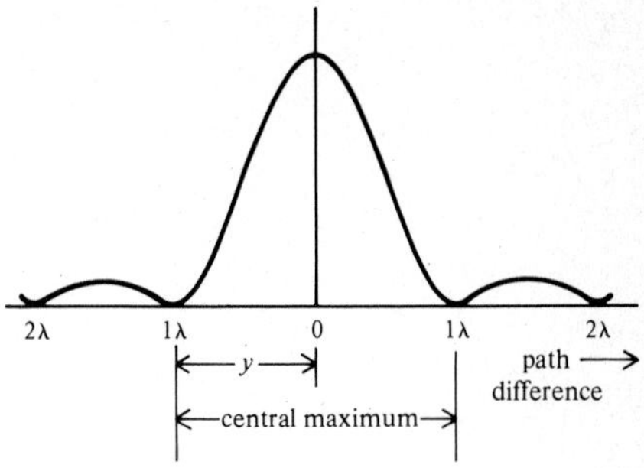

Figure 26-3

Solution: You can calculate the angular separation between the central maximum and the first minimum ($n = 1$) from Eq. (26-3):

$$n\lambda = w \sin\theta$$

$$\sin\theta = \frac{n\lambda}{w} = \frac{(1)600\ \text{nm}}{0.01\ \text{mm}} = \frac{6.0 \times 10^{-7}\ \text{m}}{1.0 \times 10^{-5}\ \text{m}} = 0.06$$

$$\theta = \arcsin(0.06) = 3.44°$$

Now, you can find y, the distance from the central peak to the first minimum:

$$y = D \tan\theta = (80\ \text{cm}) \tan 3.44° = 4.81\ \text{cm}$$

The total width of the central maximum is $2y = 9.62$ cm.

26-4. Diffraction Gratings and Spectroscopes

- A **diffraction grating** is a series of many equally spaced slits on a glass or metal plate.

These are often made by ruling very fine lines on the plate with a diamond tip. The unscratched spaces on the plate function as slits. Diffraction gratings often have about 10,000 lines per centimeter. When ordinary white light (all wavelengths) strikes a diffraction grating it produces an interference pattern similar to the pattern that a single wavelength produces when it strikes a double slit. The same equation (Eq. 26-2) that gives the angular positions of bright fringes produced by monochromatic light shining on a double slit gives the angular positions of fringes produced by a diffraction grating. But each bright fringe produced by white light shining on a diffraction grating is a rainbow pattern or **spectrum**. Since diffraction gratings separate light into its component wavelengths we can use them to determine what wavelengths any light source is producing.

- The **spectroscope** is a device used to measure wavelengths accurately.

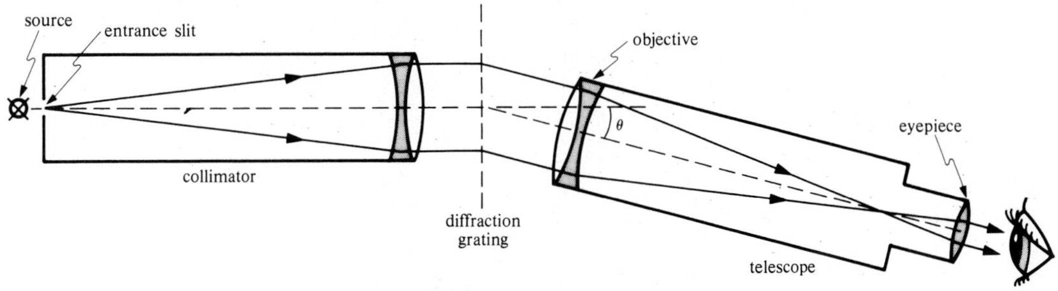

Figure 26-4. Grating spectroscope.

Either a prism or a diffraction grating may be part of a spectroscope (a prism is harder to use). The major parts of a grating spectroscope, as shown in Figure 26-4, are the **collimator**, which makes an image of light shining through a narrow slit, the *diffraction grating*, which disperses the light into spectra, and the movable *telescope*, through which the spectra can be viewed or photographed. You can use the angle at which the telescope must be placed to view a particular color to calculate the wavelength of the color.

EXAMPLE 26-6: A source emitting a red light and a violet light is viewed through a grating spectroscope. The diffraction grating in the spectroscope has 400 grating-lines per millimeter. The first-order angles at which the telescope must be placed to view these colors are 14.6° and 9.21°, respectively. Calculate the wavelength in nanometers of each color.

Solution: First, calculate the spacing between the lines on the grating:

$$a = \frac{1 \text{ mm}}{400} = 2.5 \times 10^{-3} \text{ mm}$$

Now you can use Eq. (26-2) to calculate the wavelengths. For the red light,

$$\sin \theta_1 = \frac{n\lambda}{a}$$

$$\lambda = \frac{a \sin \theta_1}{n} = \frac{(2.5 \times 10^{-3} \text{ mm})(\sin 14.6°)}{1} = 6.3 \times 10^{-4} \text{ mm} = 630 \text{ nm}$$

For the violet light,

$$\lambda = \frac{a \sin \theta_2}{n} = \frac{(2.5 \times 10^{-3} \text{ mm})(\sin 9.21°)}{1} = 4.00 \times 10^{-4} \text{ mm} = 400 \text{ nm}$$

26-5. Phase Change upon Reflection

Figure 26-5 illustrates two situations in which light is reflected.

- When light is traveling in air and strikes the surface of glass, the reflected wave undergoes a phase change of 180°.
- When light is traveling in glass and is reflected from the interface between the glass and air, there is no phase change.

A 180° change of phase occurs whenever light waves are reflected from a medium that has an index of refraction higher than the index of the medium in which the light waves are traveling.

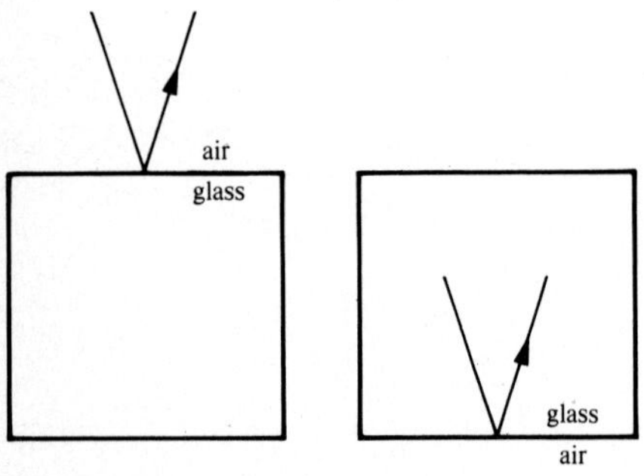

Figure 26-5. Reflection of light from a glass/air interface.

26-6. The Air Wedge

When two flat pieces of glass are placed in contact at one edge, forming a thin wedge of air, as shown in Figure 26-6, and are illuminated by a monochromatic light source (light of a single wavelength), an observer will see interference fringes. Because the path length gradually increases with increasing distance from the line of contact, the observer sees a series of alternately bright and dark bands.

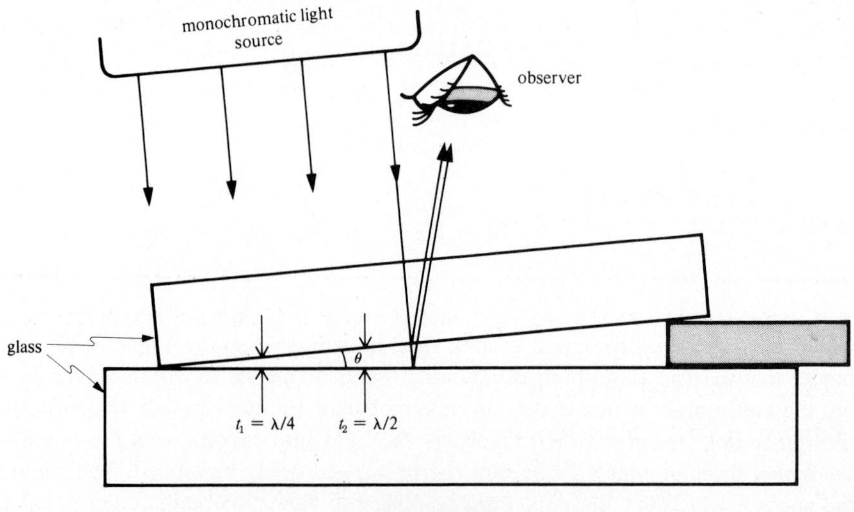

Figure 26-6

The area where the two pieces of glass are in contact appears dark because the light reflected from the lower surface is 180° out of phase with the light reflected from the upper surface. Where the separation of the two surfaces is one quarter of the wavelength of the light, the two reflected light waves will be in phase, resulting in a bright band. There will be another dark region where the thickness of the air wedge is one half the wavelength of the incident light. The thickness t of the air wedge at each dark band is

THICKNESS OF AN AIR WEDGE

$$t = \frac{m\lambda}{2} \tag{26-4}$$

where m is an integer counting the number of dark bands to the line of contact. We can use air wedges to measure the thickness of very small objects.

EXAMPLE 26-7: A very thin foil is placed between two glass plates, as shown in Figure 26-6, and light of wavelength 5.5×10^{-7} m is shining on the plates. When you look down on the plates, you see 25 dark bands. What is the width of the foil?

Solution: Use Eq. (26-4), and set $m = 24$ because the dark band at the contact between the two plates represents $m = 0$.

$$t = \frac{m\lambda}{2} = \frac{24(5.5 \times 10^{-7}\ \text{m})}{2} = 6.60 \times 10^{-6}\ \text{m}$$

EXAMPLE 26-8: What is the angle between the plates in Example 26-7 if the distance L along the top plate from the line of contact to the 25th dark band is 4 cm?

Solution: Simple trigonometry gives the answer to this problem.

$$\sin\theta = \frac{t}{L} \qquad \theta = \arcsin(t/L) = 0.00945°, \qquad \text{a small angle indeed!}$$

26-7. The Michelson Interferometer

- The **Michelson interferometer,** illustrated in Figure 26-7, is a device that uses the interference of light to measure distances, or changes of distances, with great accuracy.

The second surface of the glass plate M is "half-silvered" so that 50% of the light is reflected and 50% is transmitted. The two beams of light are returned to the observer's eye by the mirrors M_1 and M_2. When M_1 is moved a distance Δx, the path difference between the two light beams is $2\Delta x$. If n represents the number of fringes the observer sees when the movable mirror is moved a distance Δx, then the wavelength of the light emitted by the source can be calculated from the following equation:

WAVELENGTH OF LIGHT VIEWED WITH A MICHELSON INTERFEROMETER

$$\lambda = \frac{2\Delta x}{n} \tag{26-5}$$

EXAMPLE 26-9: When the movable mirror in a Michelson interferometer is moved a distance of 0.25 mm, a shift of 800 fringes is seen.
Calculate the wavelength of the light.

Solution: Directly from Eq. (26-5):

$$\lambda = \frac{2\Delta x}{n} = \frac{2(2.5 \times 10^{-4}\ \text{m})}{800} = 6.25 \times 10^{-7}\ \text{m}$$

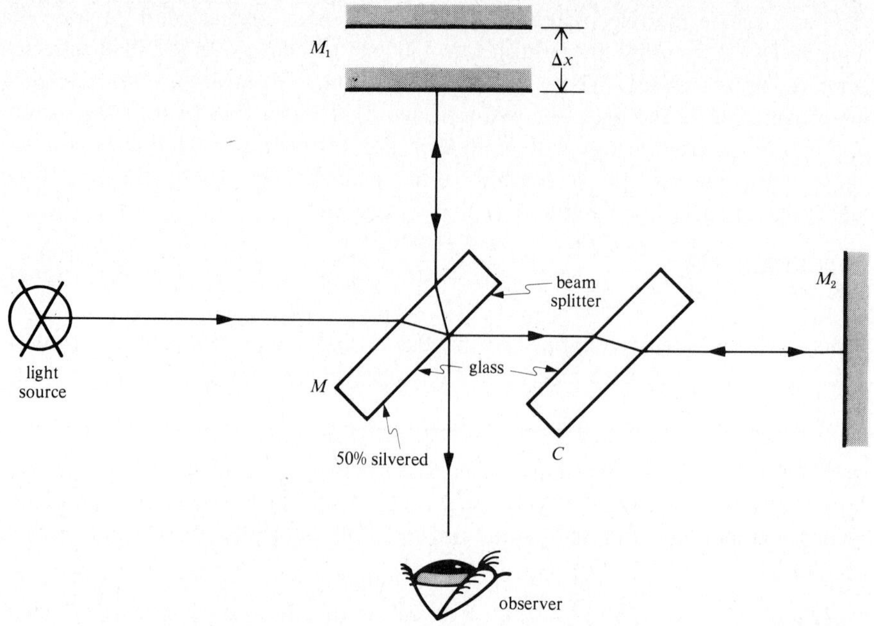

Figure 26-7. The Michelson interferometer.

EXAMPLE 26-10: The Michelson interferometer is illuminated with light whose wavelength is 546 nm. When the mirror is moved slowly through a small distance, a shift of 120 fringes is observed.
Calculate the distance the mirror is moved.

Solution: Use Eq. (26-5) and solve for the displacement:

$$\Delta x = \frac{n\lambda}{2} = \frac{(120)(5.46 \times 10^{-7}\ \text{m})}{2} = 3.28 \times 10^{-5}\ \text{m}$$

$$= 0.0328\ \text{mm}$$

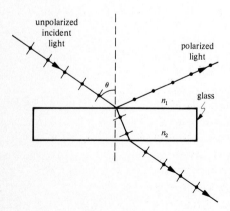

26-8. Polarization of Light

Light is a transverse wave (see Sec. 11-1), so the oscillations of the light wave are perpendicular to the direction of propagation. Most light sources produce unpolarized light. This means that the electric field vibrates in random directions—although always perpendicular to the direction of propagation—as illustrated in Figure 26-8a.

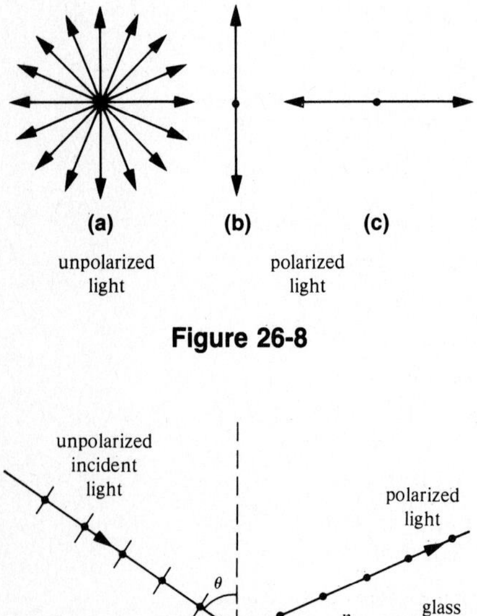

Figure 26-8

- In a **polarized** beam of light, the electric field vibrates in only one direction, as shown in Figures 26-8b and c.

If unpolarized light is incident upon a smooth surface at an angle called **Brewster's angle,** the reflected light is completely polarized with the **E**-field vibrating parallel to the reflecting surface, as illustrated in Figure 26-9. The angle of incidence that causes the reflected light to be 100% polarized is given by

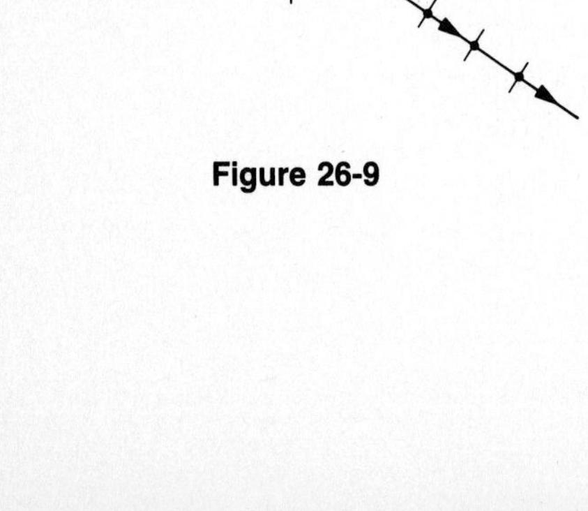

Figure 26-9

ANGLE OF INCIDENCE PRODUCING POLARIZED LIGHT

$$\tan \theta_1 = \frac{n_2}{n_1} \qquad \textbf{(26-6)}$$

where n_2 is the index of refraction of the surface, and n_1 is the index of refraction of the medium.

EXAMPLE 26-7: Light traveling in air ($n = 1.0003$) is incident upon a smooth glass surface whose index of refraction is 1.54. Find the angle of incidence that causes the reflected light to be completely polarized.

Solution: Directly from Eq. (26-6),

$$\tan\theta_1 = \frac{n_2}{n_1} = \frac{1.54}{1.0003} = 1.540$$

$$\theta_1 = \text{arc}\tan(1.540) = 57.0°$$

EXAMPLE 26-8: When unpolarized light is incident on a smooth water surface, the reflected light is 100% polarized when the angle of incidence is 53.05°. Find the index of refraction of the water.

Solution: From Eq. (26-6),

$$n_2 = n_1 \tan\theta_1 = (1.0003)\tan 53.05° = 1.33$$

SUMMARY

Relationship between phase difference and path difference	$\Delta\phi = 2\pi\dfrac{\Delta s}{\lambda}$ ($\Delta\phi$ in radians)	the phase difference is proportional to the ratio of path difference to wavelength
Angular positions of bright fringes (multiple slits)	$\sin\theta = \dfrac{n\lambda}{a}$	for light incident on two or more slits, gives the angles between the central maximum and bright fringes as measured from the midpoint between the slits as a function of slit spacing and wavelength
Angular positions of minimum intensities (single slit)	$n\lambda = w\sin\theta$	for light incident on one slit, gives the angles between the central maximum and dark fringes as measured from the center of the slit
Thickness of an air wedge	$t = \dfrac{m\lambda}{2}$ ($m = 0, 1, 2, \ldots$)	gives the thickness of an air wedge at each dark band
Wavelength of light viewed with a Michelson interferometer	$\lambda = \dfrac{2\Delta x}{n}$	gives the wavelength of the light emitted by a light source when n is the number of fringes an observer sees when the movable mirror is moved a distance x
Angle of incidence producing polarized light	$\tan\theta_1 = \dfrac{n_2}{n_1}$	gives the angle of incidence that causes reflected light to be 100% polarized

RAISE YOUR GRADES

Can you define . . . ?

- ☑ interference
- ☑ intensity
- ☑ diffraction
- ☑ polarized light

Can you . . . ?

- ☑ calculate the phase difference between two light waves when their path difference is $\lambda/2$
- ☑ write the equation for the angular position of the second-order bright fringe produced by two slits separated by distance a
- ☑ calculate the angular position of the first minimum in the diffraction pattern of a single slit of width w
- ☑ calculate the second-order angle of diffraction for a known wavelength
- ☑ calculate the angle of an air wedge between two flat pieces of glass if you know the distance between the dark fringes
- ☑ calculate the wavelength of the light illuminating a Michelson interferometer if you can count the fringes moving across the field of view when the movable mirror is displaced a known distance

SOLVED PROBLEMS

PROBLEM 26-1 The path difference between two light waves with the same wavelength is one quarter of the wavelength. What is the phase difference between these two waves?

Solution Directly from Eq. (26-1),

$$\Delta\phi = 2\pi\frac{\Delta s}{\lambda} = 2\pi\frac{\lambda/4}{\lambda} = \frac{\pi}{2}\text{ radians}$$

$$= \left(\frac{\pi}{2}\text{ rad}\right)\left(\frac{180°}{\pi\text{ rad}}\right) = 90°$$

PROBLEM 26-2 Two narrow parallel slits are 0.04 mm apart. The wavelength of the light falling on these two slits is 400 nm. The screen on which the interference fringes are seen is 60 cm from the double slit. Calculate the distance from the central maximum to the third-order bright fringe.

Solution Use Eq. (26-2) to calculate the angular position of the third-order maximum:

$$\sin\theta = \frac{n\lambda}{a} = \frac{3(4.0 \times 10^{-7}\text{ m})}{0.04 \times 10^{-3}\text{ m}} = 0.03$$

$$\theta = \arcsin(0.03) = 1.719°$$

Now, calculate the distance y from the central maximum to the third-order fringe:

$$\tan\theta = \frac{y}{D}$$

$$y = D\tan\theta = (60\text{ cm})\tan 1.719° = 1.80\text{ cm}$$

PROBLEM 26-3 A piece of photographic film is placed 80 cm from a pair of slits 0.03 mm apart. The distance from the central maximum to the first-order ($n = 1$) bright fringe is 1.60 cm. Calculate the wavelength of the light illuminating the double slit.

Solution First, calculate the angular separation of the central maximum and the first bright fringe:

$$\tan\theta = \frac{y}{D} = \frac{1.60\text{ cm}}{80\text{ cm}} = 0.02$$

$$\theta = \text{arc tan}(0.02) = 0.020\text{ rad}$$

(Notice that for small angles $\tan\theta \approx \theta$, so you can skip the step of finding the arc tangent.) Now, use Eq. (26-2):

$$\lambda = \frac{a\sin\theta}{n} = \frac{(0.03\times 10^{-3}\text{ m})\sin 0.02\text{ rad}}{1} = 6.00\times 10^{-7}\text{ m}$$

(The approximation for small angles applies to sines, too: $\sin\theta \approx \theta$).

PROBLEM 26-4 A single narrow slit with a width of 0.1 mm is illuminated by an unknown wavelength. The angular width of the central maximum, as measured from the positions of zero intensity on either side, is 9.0×10^{-3} radians. Calculate the wavelength falling on the single slit.

Solution Solve for the angular distance from the central maximum to the first minimum from Eq. (26-3), then calculate the wavelength:

$$\sin\theta = \frac{n\lambda}{w} \qquad n = 1$$

$$\lambda = w\sin\theta = (0.1\times 10^{-3}\text{ m})\sin 4.5\times 10^{-3}\text{ rad} = 4.50\times 10^{-7}\text{ m}$$

PROBLEM 26-5 A screen is 50 cm from the single slit described in Problem 26-4. Calculate the total width of the central maximum of this single-slit diffraction pattern.

Solution

$$2y = 2D\tan\theta = 2(50\text{ cm})\tan(9.0\times 10^{-3}\text{ rad}) = 0.900\text{ cm}$$

PROBLEM 26-6 A light source containing mercury produces a green radiation with a wavelength of 546.1 nm. The rulings of the grating have a spacing of 1.667×10^{-3} mm. Calculate the second-order diffraction angle for this wavelength.

Solution Directly from Eq. (26-2),

$$\sin\theta_1 = \frac{n\lambda}{a} = \frac{2(5.461\times 10^{-7}\text{ m})}{1.667\times 10^{-6}\text{ m}} = 0.655$$

$$\theta_1 = \text{arc sin}(0.655) = 40.9°$$

PROBLEM 26-7 The mercury light source of Problem 26-6 also produces a blue radiation with a wavelength of 435.8 nm. Calculate the second-order angle of diffraction for this wavelength.

Solution

$$\sin\theta_2 = \frac{2(4.358\times 10^{-7}\text{ m})}{1.667\times 10^{-6}\text{ m}} = 0.523 \qquad \theta_2 = \text{arc sin}(0.523) = 31.5°$$

PROBLEM 26-8 Two flat pieces of glass are in contact at one edge and slightly separated at the other edge. The parallel dark fringes are 2 mm apart. The wavelength of the light that produces the fringes is 540 nm. Calculate the angle between the two pieces of glass.

Solution Since the fringes are all 2 mm apart, the distance from the line of contact to the first fringe is 2 mm. So you can find the thickness of the wedge at the first fringe by using Eq. (26-4) for $m = 1$.

$$t = \frac{m\lambda}{2} = \frac{1(5.40\times 10^{-7}\text{ m})}{2} = 2.70\times 10^{-7}\text{ m}$$

$$\sin\theta = \frac{t}{L} = \frac{2.70\times 10^{-7}\text{ m}}{2.00\times 10^{-3}\text{ m}} = 1.35\times 10^{-4}$$

$$\theta = \text{arc sin}(1.35\times 10^{-4}) = 0.00773°$$

PROBLEM 26-9 If the two glass plates in Problem 26-8 are separated by a piece of paper whose edge is 3.4 cm from the contact between the two plates, calculate the thickness of the paper.

Solution Use your result from Problem 26-8.

$$\sin\theta = \frac{t}{L} \qquad t = L\sin\theta = (3.4\text{ cm})(1.35\times10^{-4}) = 4.59\times10^{-4}\text{ cm}$$

PROBLEM 26-10 When the movable mirror of a Michelson interferometer is displaced a distance of 0.2 mm, a shift of 625 fringes is seen. Calculate the wavelength of the light being used.

Solution Directly from Eq. (26-5),

$$\lambda = \frac{2\Delta x}{n} = \frac{2(2\times10^{-4}\text{ m})}{625} = 6.4\times10^{-7}\text{ m}$$

PROBLEM 26-11 When the mirror of a Michelson interferometer is moved a small distance, 280 fringes are observed. If the wavelength is 500 nm, what is the distance that the mirror is displaced?

Solution From Eq. (26-5),

$$\Delta x = \frac{n\lambda}{2} = \frac{280(5\times10^{-7}\text{ m})}{2} = 0.070\text{ mm}$$

PROBLEM 26-12 A ray of light traveling in air is incident on a piece of transparent plastic whose index of refraction is 1.40. Calculate the angle of incidence that will cause the reflected light to be completely polarized.

Solution From Eq. (26-6),

$$\tan\theta_1 = \frac{n_2}{n_1} = \frac{1.40}{1.0003} = 1.3996$$

$$\theta_1 = \arctan(1.3996) = 54.5°$$

Supplementary Exercises

PROBLEM 26-13 The phase difference between two waves at a point in space is 90° or $\pi/2$ radians. What is the corresponding path difference?

PROBLEM 26-14 Two narrow, parallel slits, separated by a distance of 0.25 mm, are illuminated by a light source whose wavelength is 480 nm. Calculate the angular separation of the central bright maximum and the first-order bright fringe.

PROBLEM 26-15 The fringes produced by the double slit in Problem 26-12 are observed on a screen that is 75 cm from the double slit. Calculate the linear separation of the central maximum and the first-order bright fringe.

PROBLEM 26-16 When the wavelength of the light source falling on the double slits in Problem 26-12 is increased, the linear separation of the fringes becomes 0.20 mm. Calculate the angular separation of these adjacent fringes when the screen is 75 cm from the double slits.

PROBLEM 26-17 Calculate the wavelength of the light falling on the double slit in Problem 26-14.

PROBLEM 26-18 A light source producing a wavelength of 620 nm shines on a diffraction grating. What is the first-order angle of diffraction for this wavelength if the spacing between the slits of the diffraction grating is 2×10^{-6} m?

PROBLEM 26-19 The grating spectroscope described in Problem 26-18 produces a second image of the entrance slit at a diffraction angle of 20.2°. This is a first-order image. What is the wavelength of this light?

PROBLEM 26-20 A single slit with a width of 0.10 mm is illuminated with light whose wavelength is 500 nm. Calculate the angular separation between the central maximum and the first minimum.

PROBLEM 26-21 The diffraction pattern produced by the narrow slit in Problem 26-20 is recorded on a photographic film placed 50 cm from the slit. Calculate the total width of the central maximum measured from the two points where the intensity is zero.

PROBLEM 26-22 Two flat pieces of glass form a narrow air wedge, which is illuminated by light whose wavelength is 620 nm. The distance from the first dark fringe to the second is 2.0 mm. Calculate the angle of the air wedge in radians.

PROBLEM 26-23 The upper piece of glass described in Problem 26-18 has a length of 5 cm. The left edge is in contact with the lower plate of glass and the right edge is held above the lower plate by a human hair. Calculate the thickness of the hair.

PROBLEM 26-24 The movable mirror of a Michelson interferometer is very slowly moved a distance of 0.12 mm. Five hundred fringes pass the field of view. Calculate the wavelength of the light.

PROBLEM 26-25 When the mirror of the Michelson interferometer described in Problem 26-24 is moved a very small distance, only 150 fringes are observed. Calculate the displacement of the mirror.

PROBLEM 26-26 Light traveling in air is incident on a piece of glass whose index of refraction is 1.60. Calculate the angle of incidence that will cause the reflected light to be completely polarized.

Answers to Supplementary Exercises

26-13: $\lambda/4$

26-14: 0.110°

26-15: 0.144 cm

26-16: 0.01528°

26-17: 6.67×10^{-5} mm

26-18 18.1°

26-19 691 nm

26-20: 0.2865°

26-21: 5.00 mm

26-22: 1.55×10^{-4} rad

26-23: 7.75×10^{-4} cm

26-24: 480 nm

26-25: 3.60×10^{-2} mm

26-26: 58.0°

27 SPECIAL RELATIVITY

THIS CHAPTER IS ABOUT

☑ **Time Dilation**
☑ **Relativistic Length Contraction**
☑ **Relativistic Mass Increase**
☑ **Relativistic Kinetic Energy**
☑ **Mass–Energy Relation**

Relativity is concerned with how we measure the distance and time interval between events occurring in different reference frames. Unfortunately, there is no "best" reference frame; that is, there is no reference frame—or object—that we can claim is at absolute rest. This means that we can't tell which one of two reference frames is moving and which is at rest. All we can say is that the two reference frames are moving with respect to each other. The theory of **special relativity** applies only to frames of reference in *uniform* relative motion—**inertial** frames—neither frame is accelerating with respect to the other. Because we cannot tell if any reference frame is at absolute rest, we can assign any inertial frame we like—usually the frame in which we are making our measurements—to being at relative rest, and measure physical quantities in other frames with respect to it.

Special relativity is based on two principles:

(**1**) The forms of the laws of physics are the same in all inertial frames of reference.
(**2**) The speed of light (in empty space) is the same for all observers, regardless of their velocity or the velocity of the source of light.

Many careful experiments have tested and verified these principles. These two principles lead mathematically to equations that predict phenomena that seem strange by the standards of our daily lives, but nevertheless also have been verified. The effects of special relativity are negligibly small, however, on objects whose velocities are small compared to the speed of light. If an object is moving at a velocity close enough to the speed of light that the effects of special relativity are substantial, we say the object is moving at a **relativistic velocity.**

27-1. Time Dilation

According to special relativity, there is no absolute time. The duration of a time interval between two events depends on the frame of reference in which the events are observed.

- Moving clocks run slower than clocks at rest, an effect of relativity known as **time dilation.**

If a spaceship with a clock on board moves past the earth at a high speed v, observers on the earth—"at rest" in their frame of reference—would conclude that the clock on the spaceship runs more slowly than an identical clock on the surface of the earth. Similarly, the crew of the spaceship—"at rest" in their frame of reference—would observe that the clock on the earth is slower than theirs. Both clocks, however, show the correct time for their respective frame of reference. If the time interval between two events on the spaceship is Δt_0, as measured by an observer on the spaceship, then the time interval Δt measured by observers on the earth is

TIME DILATION

$$\Delta t = \frac{\Delta t_0}{\sqrt{1 - v^2/c^2}} \qquad \textbf{(27-1)}$$

where c is the speed of light, approximately 3.0×10^8 m/s.

note: For most computations in this chapter, you'll find it easier if you solve for v/c first, and then square the result.

EXAMPLE 27-1: A spaceship passes the earth at a speed of 1.0×10^8 m/s. The time interval between two events that take place at the same point on the spaceship is 100 s.
What is the time interval between the events as recorded by observers on earth?

Solution: Begin by solving for v/c:

$$\frac{v}{c} = \frac{1.0 \times 10^8 \text{ m/s}}{3.0 \times 10^8 \text{ m/s}} = \frac{1}{3}$$

Now, directly from Eq. (27-1),

$$\Delta t = \frac{\Delta t_0}{\sqrt{1 - v^2/c^2}} = \frac{100 \text{ s}}{\sqrt{1 - 1/9}} = 106 \text{ s}$$

EXAMPLE 27-2: The time interval between two events occurring at the same spot on the earth is 200 s.
What time interval will be observed on a spaceship passing the earth at a speed of 1.0×10^8 m/s?

Solution: From Example 27-1, $v/c = 1/3$. Now, from Eq. (27-1),

$$\Delta t = \frac{200 \text{ s}}{\sqrt{1 - 1/9}} = 212 \text{ s}$$

EXAMPLE 27-3: A spaceship passes the earth at a high speed v. On board the spaceship, two events occur at the same place with a time interval of 300.0 s between them.Observers on the earth find that the time interval between these two events is 311.3 s. Calculate the speed of the spaceship relative to the earth.

Solution: Use Eq. (27-1) to solve for the spaceship's speed:

$$\sqrt{1 - v^2/c^2} = \frac{\Delta t_0}{\Delta t}$$

$$\frac{v^2}{c^2} = 1 - \left(\frac{\Delta t_0}{\Delta t}\right)^2$$

$$v = c\sqrt{1 - (\Delta t_0/\Delta t)^2}$$

$$= (3.00 \times 10^8 \text{ m/s})\sqrt{1 - (300/311.3)^2} = 8.01 \times 10^7 \text{ m/s}$$

27-2. Relativistic Length Contraction

- A moving object measures shorter than when it is at rest, an effect of relativity known as **length contraction.**

Suppose our spaceship, traveling at a high speed v, carries a 4.0-m section of pipe that is *parallel* to the spaceship's direction of motion. Physicists on the earth, using a meter stick that is at rest with respect to the earth, would observe a length of somewhat less than 4.0 m, according to the following equation:

LENGTH CONTRACTION

$$L = L_0\sqrt{1 - v^2/c^2} \qquad \textbf{(27-2)}$$

note: Length contraction occurs only in that dimension of an object *parallel* to its direction of motion.

EXAMPLE 27-4: A spaceship moves past the earth at a relative speed of 1.0×10^8 m/s. A 4.0-m length of pipe on board the spaceship is parallel to the spaceship's direction of velocity.
Calculate the length of this pipe as recorded on the earth.

Solution: Directly from Eq. (27-2),

$$L = L_0\sqrt{1 - v^2/c^2} = (4.0)\sqrt{1 - (1/3)^2} = 3.77 \text{ m}$$

EXAMPLE 27-5: A rigid object on a moving spaceship is parallel to the direction of motion. Its length, as measured on the spaceship, is 3.00 m. When measured by observers on the earth, its length is 2.598 m.
Calculate the relative speed of the spaceship.

Solution: From Eq. (27-2),

$$1 - v^2/c^2 = (L/L_0)^2$$

$$\frac{v^2}{c^2} = 1 - \left(\frac{L}{L_0}\right)^2$$

$$v = c\sqrt{1 - (L/L_0)^2}$$

$$= (3.00 \times 10^8 \text{ m/s})\sqrt{1 - (2.598/3.00)^2} = 1.50 \times 10^8 \text{ m/s}$$

27-3. Relativistic Mass Increase

- A moving object measures more massive than when it is at rest, an effect known as **relativistic mass increase.**

RELATIVISTIC MASS

$$m = \frac{m_0}{\sqrt{1 - v^2/c^2}} \qquad \textbf{(27-3)}$$

EXAMPLE 27-6: A proton with a rest mass of 1.67×10^{-27} kg moves at a speed of 1.2×10^8 m/s in an accelerator. Determine the relativistic mass of the proton.

Solution: First, calculate v/c:

$$\frac{v}{c} = \frac{1.2 \times 10^8 \text{ m/s}}{3.0 \times 10^8 \text{ m/s}} = 0.4$$

Now, use Eq. (27-3):

$$m = \frac{m_0}{\sqrt{1 - v^2/c^2}} = \frac{1.67 \times 10^{-27} \text{ kg}}{\sqrt{1 - (0.4)^2}} = 1.82 \times 10^{-27} \text{ kg}$$

EXAMPLE 27-7: An electron with a rest mass of 9.11×10^{-31} kg moves at a very high speed in a linear accelerator. The relativistic mass of the moving electron is 12.22×10^{-31} kg. Calculate the speed of the electron.

Solution: From Eq. (27-3):

$$\sqrt{1 - v^2/c^2} = \frac{m_0}{m} \qquad 1 - v^2/c^2 = \left(\frac{m_0}{m}\right)^2 \qquad \frac{v^2}{c^2} = 1 - \left(\frac{m_0}{m}\right)^2$$

$$v = c\sqrt{1 - (m_0/m)^2} = (3.0 \times 10^8 \text{ m/s})\sqrt{1 - (9.11/12.22)^2}$$

$$= 2.00 \times 10^8 \text{ m/s}$$

27-4. Relativistic Kinetic Energy

You have just seen that the mass of a moving body is different from that of the same body at rest. You also know that a moving body has kinetic energy. So it should not be surprising that mass and energy are connected concepts. The kinetic energy of a moving object depends on its relativistic mass m minus its rest mass m_0:

$$\text{KE} = mc^2 - m_0c^2 \qquad \textbf{(27-4)}$$

At velocities that are small in comparison to c, Eq. (27-4) reduces mathematically to the familiar equation for nonrelativistic kinetic energy, $\text{KE} = \frac{1}{2}mv^2$ (Eq. 6-5). Since we can use Δm to represent the increase in mass, $m - m_0$, we can write the equation for kinetic energy as

RELATIVISTIC KINETIC ENERGY

$$\text{KE} = (\Delta m)c^2 \qquad \textbf{(27-5)}$$

EXAMPLE 27-8: The rest mass of an electron is 9.11×10^{-31} kg. Calculate the kinetic energy of an electron moving at two thirds the speed of light.

Solution: First, use Eq. (27-3) to calculate the relativistic mass of the electron:

$$m = \frac{9.11 \times 10^{-31}\text{ kg}}{\sqrt{1 - (2/3)^2}} = 1.222 \times 10^{-30}\text{ kg}$$

Now, figure the increase in mass:

$$\Delta m = m - m_0 = (12.22 - 9.11) \times 10^{-31}\text{ kg} = 3.11 \times 10^{-31}\text{ kg}$$

Finally, calculate the KE from Eq. (27-6):

$$\text{KE} = (\Delta m)c^2 = (3.11 \times 10^{-31}\text{ kg})(3.00 \times 10^8\text{ m/s})^2 = 2.80 \times 10^{-14}\text{ J}$$

EXAMPLE 27-9: The rest mass of a proton is 1.67×10^{-27} kg. Calculate the kinetic energy of a proton moving through a linear accelerator at a constant speed of $0.429c$.

Solution:

$$m = \frac{1.67 \times 10^{-27}\text{ kg}}{\sqrt{1 - (0.429)^2}} = 1.849 \times 10^{-27}\text{ kg}$$

$$\Delta m = (1.849 - 1.67) \times 10^{-27}\text{ kg} = 1.79 \times 10^{-28}\text{ kg}$$

$$\text{KE} = (1.79 \times 10^{-28}\text{ kg})(3.00 \times 10^8\text{ m/s})^2 = 1.61 \times 10^{11}\text{ J}$$

27-5. Mass–Energy Relation

Not only is a change in relativistic mass equivalent to a change in kinetic energy, but a change in total mass is equivalent to a change in total energy. That is, the rest mass of a particle can be converted into energy, according to Einstein's famous equation

MASS–ENERGY RELATION

$$E = (\Delta m)c^2 \qquad \textbf{(27-6)}$$

This is the equation for mass–energy equivalence that is applied in nuclear reactors and atomic bombs. In these applications, reactions of atomic nuclei result in a slight loss of mass, which is transformed into energy. Conversely, in a process called **pair formation,** observed in experiments with subatomic particle accelerators, energy is transformed into two particles with a total mass m, determined by Eq. (27-6).

EXAMPLE 27-10: An atom of uranium loses 2.51×10^{-28} kg of mass during a fission reaction. Calculate the amount of energy produced when one mole of uranium (6.02×10^{23} atoms) undergoes this fission reaction.

Solution: First, calculate the amount of mass that is lost:

$$\Delta m = (2.51 \times 10^{-28}\ \text{kg/atom})(6.02 \times 10^{23}\ \text{atoms}) = 1.51 \times 10^{-4}\ \text{kg}$$

Then, use Eq. (27-4) to find the resulting energy:

$$E = (\Delta m)c^2 = (1.51 \times 10^{-4}\ \text{kg})(3.0 \times 10^{8}\ \text{m/s})^2 = 1.36 \times 10^{13}\ \text{J}$$

SUMMARY

Time dilation	$\Delta t = \dfrac{\Delta t_0}{\sqrt{1 - v^2/c^2}}$	time between two events on a moving object, as measured by an observer "at rest," as a function of the time between the events in the reference frame where the object is at rest
Length contraction	$L = L_0\sqrt{1 - v^2/c^2}$	length of a moving object, as measured by an observer "at rest," as a function of the object's length at rest
Relativistic mass	$m = \dfrac{m_0}{\sqrt{1 - v^2/c^2}}$	mass of a moving object, as measured by an observer "at rest," as a function of the object's mass at rest
Relativistic kinetic energy	$\text{KE} = mc^2 - m_0c^2$ or $\text{KE} = (\Delta m)c^2$	the kinetic energy of a moving particle is determined by its relativistic mass minus its rest mass
Mass–energy relation	$E = (\Delta m)c^2$	mass can be converted into an equivalent amount of energy, or energy into an equivalent amount of mass

RAISE YOUR GRADES

Can you define . . . ?

- ☑ special relativity
- ☑ time dilation
- ☑ relativistic length contraction
- ☑ relativistic mass increase

Can you . . . ?

- ☑ write the equation for time dilation
- ☑ write the equation for relativistic length contraction
- ☑ write the equation for the mass of an object moving at high speed
- ☑ calculate the speed of a moving proton whose observed mass is a given amount greater than its rest mass
- ☑ write the equation for the kinetic energy of a particle moving at a very high speed
- ☑ calculate the kinetic energy of a moving electron whose observed mass is a given amount greater than its rest mass

SOLVED PROBLEMS

PROBLEM 27-1 A spaceship moves past the earth at a speed of 0.5×10^8 m/s with respect to earth. Astronauts on the spaceship measure a 3-minute time interval between two events that occur at the same location on the ship. What is the corresponding time interval measured by observers on the earth?

Solution First, figure v/c:

$$\frac{v}{c} = \frac{0.5 \times 10^8 \text{ m/s}}{3.0 \times 10^8 \text{ m/s}} = \frac{1}{6}$$

Then, use Eq. (27-1) to calculate the time dilation:

$$\Delta t = \frac{\Delta t_0}{\sqrt{1 - v^2/c^2}} = \frac{180 \text{ s}}{\sqrt{1 - (1/6)^2}} = 183 \text{ s}$$

PROBLEM 27-2 Observers aboard a spaceship moving at high speed relative to earth measure the time interval between two events that require 4 minutes to complete at a fixed location on earth. The time interval as measured from the spaceship is 277.13 s. Calculate the speed of the spaceship relative to the earth.

Solution Use Eq. (27-1) and solve for the speed of the spaceship:

$$v = c\sqrt{1 - (\Delta t_0/\Delta t)^2} = (3.0 \times 10^8 \text{ m/s})\sqrt{1 - (240 \text{ s}/277.13 \text{ s})^2} = 1.50 \times 10^8 \text{ m/s}$$

PROBLEM 27-3 A spaceship 12.0 m long passes the earth at a speed of 0.6×10^8 m/s. Observers on earth find that the length of this spaceship is slightly less than 12 m. Calculate the length of the spaceship as observed from the earth.

Solution Use Eq. (27-2):

$$L = L_0\sqrt{1 - v^2/c^2} = (12.0 \text{ m})\sqrt{1 - (0.6/3.0)^2} = 11.8 \text{ m}$$

PROBLEM 27-4 A spaceship 10.0 m long passes the moon at a very high speed. Astronauts on the moon observe that the length of the spaceship is 9.165 m. Calculate the speed of the spaceship relative to the moon.

Solution Use Eq. (27-2) to solve for the speed of the spaceship:

$$v = c\sqrt{1 - (L/L_0)^2} = (3.0 \times 10^8 \text{ m/s})\sqrt{1 - (9.165 \text{ m}/10.0 \text{ m})^2} = 1.20 \times 10^8 \text{ m/s}$$

PROBLEM -27-5 An electron whose rest mass is 9.11×10^{-31} kg moves in a linear accelerator at a speed equal to 95% of the speed of light. Calculate the relativistic mass of this moving electron.

Solution Use Eq. (27-3) to calculate the relativistic mass of the electron:

$$m = \frac{m_0}{\sqrt{1 - v^2/c^2}} = \frac{9.11 \times 10^{-31} \text{ kg}}{\sqrt{1 - (0.95)^2}} = 2.92 \times 10^{-30} \text{ kg}$$

PROBLEM 27-6 An electron moves so rapidly in a linear accelerator that its relativistic mass is 5.0252 times its rest mass. Calculate the speed of this moving electron.

Solution Use Eq. (27-3) to solve for the speed of the electron:

$$v = c\sqrt{1 - (m_0/m)^2} = c\sqrt{1 - (1/5.0252)^2} = 0.98\ c = 2.94 \times 10^8 \text{ m/s}$$

PROBLEM 27-7 Calculate the relativistic kinetic energy of the high-speed electron described in Problem 27-6.

Solution First, calculate the increase in mass of this high-speed electron:

$$\Delta m = m - m_0 = 5.0252\ m_0 - m_0 = 4.0252(9.11 \times 10^{-31} \text{ kg})$$
$$= 3.667 \times 10^{-30} \text{ kg}$$

Now, use Eq. (27-5) to calculate its kinetic energy:

$$\text{KE} = (\Delta m)c^2 = (3.667 \times 10^{-30}\ \text{kg})(3.0 \times 10^8\ \text{m/s})^2 = 3.30 \times 10^{-13}\ \text{J}$$

PROBLEM 27-8 A proton in a linear accelerator is accelerated to a speed of 2.75×10^8 m/s. The rest mass of the proton is 1.67×10^{-27} kg. Calculate its relativistic mass.

Solution

$$m = \frac{1.67 \times 10^{-27}\ \text{kg}}{\sqrt{1 - (2.75/3.0)^2}} = 4.18 \times 10^{-27}\ \text{kg}$$

PROBLEM 27-9 Calculate the kinetic energy of the high-speed proton described in Problem 27-8.

Solution

$$\Delta m = 4.18 \times 10^{-27}\ \text{kg} - 1.67 \times 10^{-27}\ \text{kg} = 2.51 \times 10^{-27}\ \text{kg}$$

$$\text{KE} = (2.51 \times 10^{-27}\ \text{kg})(3.0 \times 10^8\ \text{m/s})^2 = 2.26 \times 10^{-10}\ \text{J}$$

PROBLEM 27-10 In a fusion reaction, four neutral hydrogen atoms are brought together to produce a neutral helium atom. The helium atom has a mass that is 4.764×10^{-29} kg less than the four hydrogen atoms. The mass of one mole of hydrogen is 1.008 g. How much energy will be produced if 500 g of hydrogen undergoes this fusion reaction?

Solution Calculate the number of hydrogen atoms:

$$n = (500\ \text{g})\frac{(6.02 \times 10^{23}\ \text{atoms})}{(1.008\ \text{g})} = 2.986 \times 10^{26}\ \text{atoms}$$

Now, divide by four to find the number of helium atoms produced:

$$\frac{2.986 \times 10^{26}}{4} = 7.465 \times 10^{25}\ \text{He atoms}$$

Calculate the mass lost by these helium atoms:

$$(4.764 \times 10^{-29}\ \text{kg/He atom})(7.465 \times 10^{25}\ \text{He atoms}) = 3.556 \times 10^{-3}\ \text{kg}$$

Finally, use Eq. (27-6) to find the amount of energy produced:

$$E = (\Delta m)c^2 = (3.556 \times 10^{-3}\ \text{kg})(3.0 \times 10^8\ \text{m/s})^2 = 3.20 \times 10^{14}\ \text{J}$$

Supplementary Exercises

PROBLEM 27-11 The time interval between two events that take place on a high-speed spaceship is 250 s. The speed of the spaceship with respect to earth is 1.0×10^8 m/s. Calculate the corresponding time interval as observed on the earth.

PROBLEM 27-12 The astronauts aboard a high-speed spaceship measure the time interval between two events that occur at the same location on the spaceship. They report to earth that the time interval is 400 s. Observers on earth measure the time interval as 536.66 s. Calculate the speed of the spaceship relative to the earth.

PROBLEM 27-13 The length of a spaceship measures 15.0 m when it is at rest on earth. When this spaceship moves past the moon at a speed of 1.2×10^8 m/s, observers on the moon observe that its length is less than 15.0 m. Calculate the length recorded by observers on the moon.

PROBLEM 27-14 A steel rod on a spaceship has a length of 2.50 m, as measured by observers on the ship. When this spaceship moves past earth, observers on earth discover that the length of the rod is 2.4495 m. Calculate the speed of the spaceship.

PROBLEM 27-15 An electron in a linear accelerator moves at a speed equal to 90% of the speed of light. Calculate the relativistic mass of this electron.

PROBLEM 27-16 A proton moving at a high speed has a mass that is 150% of its rest mass. Calculate the speed of this proton.

PROBLEM 27-17 In a nuclear fission experiment, there is a loss of 12 g of mass. Calculate the energy produced in this reaction.

PROBLEM 27-18 A nuclear fusion experiment produces an energy of 2.70×10^{12} J. Calculate the loss of mass that produced this energy.

PROBLEM 27-19 The rest mass of a proton is 1.67×10^{-27} kg. Calculate the kinetic energy of a proton moving at 75% of the speed of light.

PROBLEM 27-20 The rest mass of an electron is 9.11×10^{-31} kg. An electron moving at a high speed has a mass that is 105% of its rest mass. Calculate the kinetic energy of this electron.

Answers to Supplementary Exercises

27-11: 265 s

27-12: 2.00×10^8 m/s

27-13: 13.7 m

27-14: 6.00×10^7 m/s

27-15: 2.09×10^{-30} kg

27-16: 2.24×10^8 m/s

27-17: 1.08×10^{15} J

27-18: 3.00×10^{-5} kg

27-19: 7.69×10^{-11} J

27-20: 4.10×10^{-15} J

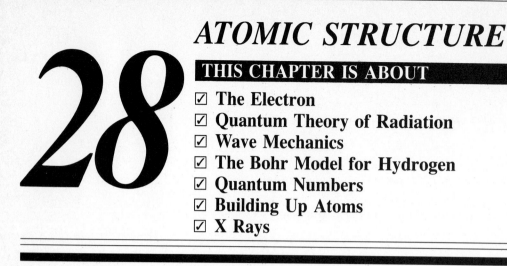

28 ATOMIC STRUCTURE

THIS CHAPTER IS ABOUT

- ☑ The Electron
- ☑ Quantum Theory of Radiation
- ☑ Wave Mechanics
- ☑ The Bohr Model for Hydrogen
- ☑ Quantum Numbers
- ☑ Building Up Atoms
- ☑ X Rays

You'll recall from Sec. 16-1A that an atom consists of a positively charged nucleus, containing protons and neutrons, surrounded by a cloud of negatively charged electrons. Most atoms are neutral, which means that the number of electrons is equal to the number of protons in the nucleus. (We'll look more closely at the nucleus in Chapter 29.)

28-1. The Electron

A. The charge on the electron

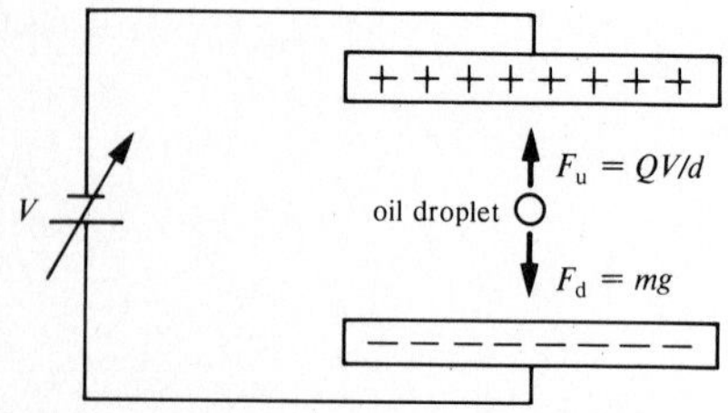

Figure 28-1. Millikan's oil-drop experiment.

We can measure the charge on an electron ($e = 1.60 \times 10^{-19}$ C) by introducing a tiny negatively charged droplet of mineral oil between the two plates of an air-filled capacitor, then adjusting the voltage V so that the droplet is in equilibrium between the two plates (Fig. 28-1). In equilibrium the upward force on the charged droplet QV/d (where d is the distance between the two capacitor plates and Q is the charge on the droplet) exactly balances the droplet's weight mg. So the charge on the droplet is

CHARGE ON AN OIL DROPLET

$$Q = \frac{mg}{V/d} \tag{28-1}$$

This experiment shows that the droplet can have only charges that are integer multiples of 1.60×10^{-19} C.

EXAMPLE 28-1: An oil droplet with a mass of 1.633×10^{-14} kg is in equilibrium between two capacitor plates. The voltage between the plates is 5000 V, and the distance between them is 2.0 cm.

Calculate the number of excess electrons on this droplet.

Solution: Use Eq. (28-1) to find the total charge on the droplet:

$$Q = \frac{mg}{V/d} = \frac{(1.633 \times 10^{-14}\ \text{kg})(9.8\ \text{m/s}^2)}{5 \times 10^3\ \text{V}}(2 \times 10^{-2}\ \text{m}) = 6.40 \times 10^{-19}\ \text{C}$$

Now, calculate the number of electrons:

$$6.40 \times 10^{-19}\ \text{C}\left(\frac{1\ \text{electron}}{1.60 \times 10^{-19}\ \text{C}}\right) = 4\ \text{electrons}$$

B. The mass of an electron

In a magnetic field, a charged particle will move in a circle because of the force due to the field, $\mathbf{F} = Q\mathbf{v} \times \mathbf{B}$ (Eq. 20-1). We can use this property to find the charge-to-mass ratio, and thus the mass, of an electron. The charge to mass ratio of the electron is given by

CHARGE-TO-MASS RATIO OF AN ELECTRON

$$\frac{e}{m} = \frac{E}{B^2 r} \tag{28-2}$$

where E is the magnitude of the electric field through which the electron is accelerated, B is the magnitude of the magnetic field accelerating the electron, and r is the radius of the circle.

EXAMPLE 28-2: An electron is accelerated by a magnetic field of 4×10^{-3} T through an electric field of 1.5×10^5 N/C. The radius of its orbit is 5.336 cm. Calculate the electron's (**a**) charge-to-mass ratio and (**b**) mass.

Solution:
(**a**) Directly from Eq. (28-2):

$$\frac{e}{m} = \frac{E}{B^2 r} = \frac{1.5 \times 10^5 \text{ N/C}}{(4 \times 10^{-3} \text{ T})^2 (5.336 \times 10^{-2} \text{ m})} = 1.757 \times 10^{11} \text{ C/kg}$$

(**b**) Divide the charge on the electron by its charge-to-mass ratio.

$$m = \frac{e}{e/m} = \frac{1.60 \times 10^{-19} \text{ km}}{1.757 \times 10^{11} \text{ C/kg}} = 9.11 \times 10^{-31} \text{ kg}$$

28-2. Quantum Theory of Radiation

- The **quantum theory** states that energy is not continuous, but rather is *quantized,* which means that energy exists in discrete packets or *quanta*.

Each atom in an object oscillates around its average position unless the temperature of the object is 0 K, and the energy of this oscillation is lost through heat radiation (Sec. 14-3C). The entire object radiates according to Stefan's law (Eq. 14-5), but each oscillating atom can emit energy only in discrete quanta. The energy E emitted by an oscillating atom is quantized according to **Planck's equation:**

PLANCK'S EQUATION

$$E = hf = \frac{hc}{\lambda} \tag{28-3}$$

where f is the frequency of the electromagnetic wave and h is a fundamental universal constant—**Planck's constant**—whose value is 6.626×10^{-34} J s. Each quantum of energy, called a **photon**, moves at the speed of light. [**recall:** λ is the wavelength of light and c is its speed. (Eq. 23-8)]

EXAMPLE 28-3: Determine the energy that an atom loses when it emits a photon with a wavelength of 6.0×10^2 nm.

Solution: You'll use Planck's equation, but first make sure that λ and c are expressed in compatible units:

$$\lambda = (6.0 \times 10^2 \text{ nm})(10^{-9} \text{ m/nm}) = 6.0 \times 10^{-7} \text{ m}$$

Then,

$$E = \frac{hc}{\lambda} = \frac{(6.626 \times 10^{-34} \text{ J s})(3.00 \times 10^8 \text{ m/s})}{6.00 \times 10^{-7} \text{ m}} = 3.31 \times 10^{-19} \text{ J}$$

A. The photoelectric effect

When light shines on a metal surface, some electrons in the metal absorb enough energy to escape from their atoms and be emitted from the surface of the metal, what we call the **photoelectric effect.** According to experiment:

(1) No electrons are emitted if the frequency of light is below a certain value, known as the *threshold frequency* f_0.
(2) Light of a higher frequency than f_0 causes the electrons to be emitted with greater kinetic energy. The maximum KE of the electrons ($\frac{1}{2}mv^2$) is directly proportional to frequency f.
(3) Increasing the intensity of the light increases the number of electrons emitted, but their maximum KE remains the same.

In explaining these results, Einstein extended the quantum concept to light itself. All electromagnetic radiation exists as tiny packets of energy—quanta or photons—whose energy depends on the frequency of the radiation, according to Planck's equation. Only photons with energy above hf_0 have enough energy to eject any electrons from the metal. The maximum kinetic energy that each electron can have is the energy of one photon, hf, minus the minimum energy needed to free the electron.

A photon has *zero* rest mass, and *always* moves at the speed of light. We can figure its momentum from the following equation:

MOMENTUM OF A PHOTON

$$p = \frac{h}{\lambda} \tag{28-4}$$

EXAMPLE 28-4: Calculate the momentum of a photon with a wavelength of 6.20×10^{-7} m.

Solution: Directly from Eq. (28-4):

$$p = \frac{h}{\lambda} = \frac{6.626 \times 10^{-34}\ \text{J s}}{6.20 \times 10^{-7}\ \text{m}} = 1.07 \times 10^{-27}\ \text{kg m/s}$$

B. The Compton effect

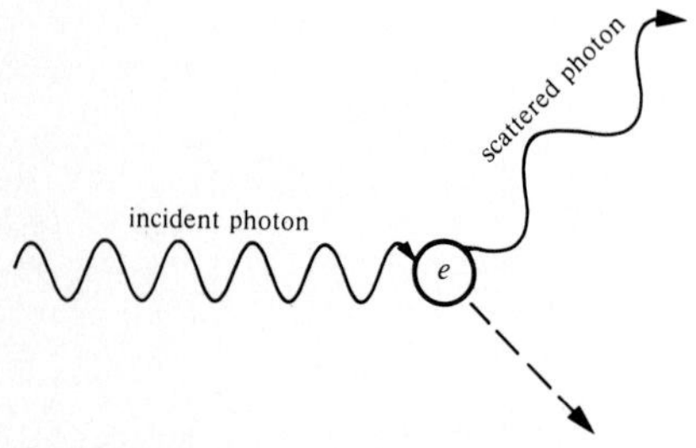

Figure 28-2. Compton effect.

We can devise experiments in which a photon can transfer some of its momentum to a particle of matter, just as if the photon were a particle too (Fig. 28-2).

- The **Compton effect** refers to the fact that scattered X-ray photons have a slightly lower frequency than the incident photons, indicating a loss of energy.

This effect is produced when a photon collides with an electron and some of the photon's energy is transferred to the electron. The energy hf' of the scattered photon is less than the initial energy hf of the incident photon. If we apply the laws of conservation of energy and momentum to the collision, we can write

KE TRANSFERRED TO AN ELECTRON

$$\text{KE} = \frac{1}{2}mv^2 = hf - hf' \tag{28-5}$$

EXAMPLE 28-5: An X-ray photon with a frequency of 4.225×10^{18} Hz strikes an electron. The frequency of the scattered photon is 4.149×10^{18} Hz. Calculate the kinetic energy transferred to the electron.

Solution: Directly from Eq. (28-5):

$$\text{KE} = h(f - f')$$
$$= (6.626 \times 10^{-34}\ \text{J s})(4.225 - 4.149) \times 10^{18}\ \text{Hz} = 5.036 \times 10^{-17}\ \text{J}$$

EXAMPLE 28-6: An X-ray photon with a wavelength of 8.0×10^{-11} m strikes an electron. The scattered photon has a wavelength of 8.8×10^{-11} m. Calculate the KE transferred to the electron.

Solution: Substitute c/λ for f in Eq. (28-5). [**recall:** The frequency of light is the ratio of its speed to its wavelength: $f = c/\lambda$. (Eq. 23-8)]

$$\text{KE} = hc\left(\frac{1}{\lambda} - \frac{1}{\lambda'}\right)$$
$$= (6.626 \times 10^{-34}\ \text{J s})(3.0 \times 10^{8}\ \text{m/s})\left(\frac{1}{8.0 \times 10^{-11}\ \text{m}} - \frac{1}{8.8 \times 10^{-11}\ \text{m}}\right)$$
$$= 2.26 \times 10^{-16}\ \text{J}$$

28-3. Wave Mechanics

Matter, as well as light, has both wave and particle properties. A particle of matter has a wavelength that is related to its momentum in the same way that a photon's wavelength is related to its momentum. The relationship of wavelength λ to momentum is given by the **de Broglie equation** for a particle of mass m traveling at speed v,

DE BROGLIE EQUATION

$$\lambda = \frac{h}{mv} \qquad \textbf{(28-6)}$$

where h is, once again, Planck's constant.

note: Any moving body may behave like a wave under certain conditions. However, wavelike properties are significant only when the size of an object is not much larger than the wavelength; the wavelength of any ordinary object is much too small to be measured or detected.

We can interpret this wave–particle duality of matter in terms of probabilities, the amplitude of the **wave function** associated with a particle representing the *probability* of finding a particle at a point. The wave–particle duality leads to a fundamental uncertainty in nature known as the **Heisenberg uncertainty principle:** The more precisely we determine a particle's position, the less precisely we can know its momentum, and vice versa. In quantitative terms, if we perform the best possible experiment, we can do no better than

HEISENBERG UNCERTAINTY PRINCIPLE

$$\Delta x\, \Delta p_x \geqq \frac{h}{4\pi} \qquad \textbf{(28-7)}$$

where, in relation to the x axis, Δx is the uncertainty in our measurement of the particle's position, Δp_x is the uncertainty of our measurement of its momentum in the x direction, and h is (you guessed it) Planck's constant 6.626×10^{-34} J s.

EXAMPLE 28-7: An electron moving at a speed of 1.5×10^{8} m/s has a linear momentum of 1.578×10^{-22} kg m/s. Calculate the electron's wavelength.

Solution: Directly from de Broglie's equation (28-6),

$$\lambda = \frac{h}{p} = \frac{6.626 \times 10^{-34}\ \text{J s}}{1.578 \times 10^{-22}\ \text{m/s}} = 4.20 \times 10^{-12}\ \text{m}$$

EXAMPLE 28-8: The x-axis position of a photon has been measured with an uncertainty of 2.0×10^{-8} m.
What is the uncertainty of the photon's linear momentum?

Solution: Use Eq. (28-7) and solve for the uncertainty of the momentum:

$$\Delta x\, \Delta p_x \geqq \frac{h}{4\pi}$$

$$\Delta p_x \cong \frac{h}{4\pi \Delta x} \cong \frac{6.626 \times 10^{-34}\ \text{J s}}{4\pi(2.0 \times 10^{-8}\ \text{m})} \cong 2.64 \times 10^{-27}\ \text{kg m/s}$$

28-4. The Bohr Model for Hydrogen

In Section 26-4 we saw how a grating spectroscope disperses a beam of light into a spectrum. If we equip a spectroscope with a device such as a camera or photovoltaic cell to record the spectrum, we call the resulting instrument a **spectrometer**. The study of spectra, **spectroscopy**, allows us to determine the structure of atoms and molecules. If we observe the light emitted by a hot gas, such as the flame of a Bunsen burner, through a spectroscope, we find that the spectrum is not continuous but consists of discrete bright lines of different wavelengths (colors) against a dark background. Each element emits its own characteristic combination of lines, so by observing the pattern of lines we can determine what elements emitted the light. The spectrum of the simplest element, hydrogen, is the key to understanding the spectra of the other elements, as well as the structure of the atom. The visible part of the hydrogen spectrum contains a series of lines called the **Balmer series.** The wavelengths corresponding to each line in the Balmer series are given by

BALMER SERIES

$$\frac{1}{\lambda} = R\left(\frac{1}{2^2} - \frac{1}{m^2}\right) \qquad \textbf{(28-8)}$$

where m is any integer greater than 2 and $R = 1.097 \times 10^7\ \text{m}^{-1}$, called the **Rydberg constant.**

EXAMPLE 28-9: What is the wavelength of the line in the Balmer series for $m = 3$?

Solution: Use Equation (28-8):

$$\frac{1}{\lambda} = R\left(\frac{1}{2^2} - \frac{1}{3^2}\right) = (1.097 \times 10^7\ \text{m}^{-1})(0.139) = 1.525 \times 10^6\ \text{m}^{-1}$$

$$\lambda = 6.56 \times 10^{-7}\ \text{m} = 656\ \text{nm}$$

The complete spectrum of hydrogen includes other series of lines at ultraviolet and infrared wavelengths. We can find the wavelengths of all lines in the hydrogen spectrum if we replace 2^2 in Eq. (28-8) with the square of any positive integer:

WAVELENGTHS IN THE HYDROGEN SPECTRUM

$$\frac{1}{\lambda} = R\left(\frac{1}{n^2} - \frac{1}{m^2}\right) \qquad \textbf{(28-9)}$$

where m is any integer greater than n.

Bohr was able to explain Eq. (28-9) by making three assumptions:

(1) Each electron may circle the nucleus only in orbits of certain discrete radii.
(2) The angular momentum of each integer is quantized and has a value of $nh/2\pi$ where n is a positive integer and h is Planck's constant.
(3) An electron may jump or fall from one allowed orbit (or *energy level*) to another by absorbing or emitting a photon. Each line in an element's spectrum corresponds to the emission of a photon as an electron falls from a particular energy level to another particular energy level.

- The **ground state** is the lowest energy orbit—the one in which the electron normally resides. Higher energy levels are **excited states**.

We can calculate these energy states for any hydrogen-like (single-electron) atom:

ENERGY LEVEL (HYDROGEN-LIKE ATOM)

$$E_n = -13.60 \text{ eV}\left(\frac{Z^2}{n^2}\right) \quad \textbf{(28-10)}$$

where Z is the **atomic number** of the atom (the number of protons in the nucleus) and n is the **principal quantum number,** an integer characterizing the energy state of the electron (see Sec. 28-5). The $n = 1$ state is the ground state of the H atom. The electron volt eV is the energy acquired by an electron in moving through a potential difference (Sec. 17-1) of one volt. Since 1 V = 1 J/C, 1 eV = $(1.60 \times 10^{-19}\text{ C})(1\text{ V}) = 1.60 \times 10^{-19}$ J.

EXAMPLE 28-10: Calculate the energy of a hydrogen atom when the quantum number is 3.

Solution: A hydrogen nucleus has one proton, so $Z = 1$.

$$E_3 = -13.60 \text{ eV}\left(\frac{Z^2}{n^2}\right) = -13.60 \text{ eV}\left(\frac{1}{3^2}\right) = -1.51 \text{ eV}$$

The radius of an allowed electron orbit in an atom is

RADIUS OF AN ELECTRON ORBIT

$$r = \frac{n^2h^2}{4\pi^2mkZe^2} \quad \textbf{(28-11)}$$

where n is the principal quantum number, h is Planck's constant, $m = 9.11 \times 10^{-31}$ kg (the mass of an electron), $k = 9.00 \times 10^9 \text{ N m/C}^2$ (the constant from Coulomb's law, Eq. 16-1), Z is the atomic number, and $e = 1.60 \times 10^{-19}$ C (the charge on an electron).

EXAMPLE 28-11: Determine the radius of the electron orbit for hydrogen when $n = 3$.

Solution: Directly from Eq. (28-11),

$$r = \frac{3^2h^2}{4\pi^2mkZe^2}$$

$$= \frac{(9)(6.626 \times 10^{-34}\text{ J s})^2}{4\pi^2(9.11 \times 10^{-31}\text{ kg})(9.0 \times 10^9\text{ N m}^2/\text{C}^2)(1)(1.6 \times 10^{-19}\text{ C})^2}$$

$$= 4.77 \times 10^{-10}\text{ m}$$

When an electron in a hydrogen atom drops from a higher-energy state E_m to a lower-energy state E_n, it emits a photon. The energy of the photon is

ENERGY OF A PHOTON EMITTED BY A HYDROGEN ATOM

$$hf = 13.6 \text{ eV}\left(\frac{1}{n^2} - \frac{1}{m^2}\right) \quad \textbf{(28-12)}$$

EXAMPLE 28-12: Calculate the (**a**) energy and (**b**) wavelength of the photon emitted by a hydrogen atom whose energy changes from E_3 to E_2.

Solution: (**a**) Directly from Eq. (28-10),

$$hf = 13.6 \text{ eV}\left(\frac{1}{2^2} - \frac{1}{3^2}\right) = 1.889 \text{ eV}$$

(**b**) Use Eq. (28-3) to find the photon's wavelength. (Remember to express E in units compatible with those of h and c.)

$$\lambda = \frac{hc}{E} = \frac{(6.626 \times 10^{-34} \text{ J s})(3.0 \times 10^8 \text{ m/s})}{(1.889 \text{ eV})(1.60 \times 10^{-19} \text{ J/eV})} = 6.58 \times 10^{-7} \text{ m}$$

EXAMPLE 28-13: A neutral helium atom ($Z = 2$) has a nucleus, which contains two protons and two neutrons, orbited by two electrons. If one of the electrons is removed, the helium atom has a positive charge. Calculate the energy level of a helium ion He^+ corresponding to the quantum number $n = 4$.

Solution: Use Eq. (28-8) to calculate the energy:

$$E_4 = -13.60 \text{ eV}\left(\frac{2^2}{4^2}\right) = -3.40 \text{ eV}$$

EXAMPLE 28-14: What is the wavelength of the photon emitted when the energy of a helium ion changes from E_4 to E_3?

Solution: First, use Eq. (28-8) to calculate the energy of the helium ion when $n = 3$:

$$E_3 = -13.60 \text{ eV}\left(\frac{2^2}{3^2}\right) = -6.044 \text{ eV}$$

The photon's energy equals the difference between E_4 and E_3.

$$\Delta E = E_4 - E_3 = (-3.40 \text{ eV}) - (-6.044) = 2.644 \text{ eV}$$

Finally, calculate the photon's wavelength from Eq. (28-3):

$$\lambda = \frac{hc}{E} = \frac{(6.626 \times 10^{-34} \text{ J s})(3.0 \times 10^8 \text{ m/s})}{(2.644 \text{ eV})(1.60 \times 10^{-19} \text{ J/eV})} = 4.70 \times 10^{-7} \text{ m}$$

EXAMPLE 28-15: Atoms can absorb photons, provided the energy of the photon is equal to the difference in two energy levels of the atom. Calculate the wavelength of the photon absorbed by a hydrogen atom when n changes from 2 to 3.

Solution: First, calculate the energy difference from Eq. (28-12). Remember that $\Delta E = E_2 - E_3 = hf = (hc/\lambda)$.

$$\frac{hc}{\lambda} = 13.6 \text{ eV}\left(\frac{1}{2^2} - \frac{1}{3^2}\right) = 1.889 \text{ eV}$$

Then, from Eq. (28-3):

$$\lambda = \frac{(6.626 \times 10^{-34} \text{ J s})(3.0 \times 10^8 \text{ m/s})}{(1.889 \text{ eV})(1.60 \times 10^{-19} \text{ J/eV})} = 6.58 \times 10^{-7} \text{ m}$$

28-5. Quantum Numbers

The picture of an electron orbiting a nucleus like a planet orbiting the sun is too simple. We cannot know the exact position of an electron (the Heisenberg uncertainty principle), only the *probability* of finding the electron in a certain volume of space, as determined by a wave function (Sec. 28-3). Each possible "orbit" of an electron is actually one wave function, which for the hydrogen atom is just a formula with sines and cosines. To tell one wave function from another, that is, to determine which energy level an electron is in, we need a set of four **quantum numbers.**

- The **principal quantum number** n corresponds to the allowed orbits in the Bohr model of an atom. n is a positive integer that can range from 1 to infinity, although values of 1 to 7 are sufficient to describe all atoms in the ground state. The value of n is the main factor determining the energy of an electron.
- The second quantum number l is the **angular momentum quantum number.** The value of l controls the general spatial distribution of electron probability and can take all positive integer values from 0 to $n - 1$, so there are n possible values of l associated with each value of n. For example, if $n = 4$, there are four l values: 0, 1, 2, and 3. We usually use the letters s, p, d, and f to represent l values of 0, 1, 2, and 3, respectively. Table 28-1 shows the use of this nomenclature for $n = 1$ to 5.
- The **magnetic quantum number** m_l describes the particular orientation of the spatial distribution described by l. The possible values of m_l are integers ranging from $-l$ to $+l$. If $l = 3$, m_l can be $-3, -2, -1, 0, 1, 2,$ or 3. There are $2l + 1$ possible values of m_l associated with each value of l.
- The **spin magnetic quantum number** m_s designates the possible spin orientations of the electron. It can have only two values: $+\frac{1}{2}$ and $-\frac{1}{2}$.

TABLE 28-1: Quantum Numbers

n	l	Letter description	m_l	m_s	Number of combinations
1	0	1*s*	0	$\pm\frac{1}{2}$	2
2	0	2*s*	0	$\pm\frac{1}{2}$	2 } 8
2	1	2*p*	$-1, 0, +1$	$\pm\frac{1}{2}$	6
3	0	3*s*	0	$\pm\frac{1}{2}$	2 } 18
3	1	3*p*	$-1, 0, +1$	$\pm\frac{1}{2}$	6
3	2	3*d*	$-2, -1, 0, +1, +2$	$\pm\frac{1}{2}$	10
4	0	4*s*	0	$\pm\frac{1}{2}$	2 } 32
4	1	4*p*	$-1, 0, +1$	$\pm\frac{1}{2}$	6
4	2	4*d*	$-2, -1, 0, +1, +2$	$\pm\frac{1}{2}$	10
4	3	4*f*	$+3, \ldots, 0, \ldots, +3$	$\pm\frac{1}{2}$	14
5	0	5*s*	0	$\pm\frac{1}{2}$	2 } 32
5	1	5*p*	$-1, 0, +1$	$\pm\frac{1}{2}$	6
5	2	5*d*	$-2, -1, 0, +1, +2$	$\pm\frac{1}{2}$	10
5	3	5*f*	$-3, \ldots, 0, \ldots, +3$	$\pm\frac{1}{2}$	14
5	4*				

*Atoms in which $l = 4$ do not occur in nature and have not yet been synthesized.

EXAMPLE 28-16: For an atom in which $n = 5$, determine **(a)** the permissible values of l and **(b)** the permissible values of m_l for $l = 2$.

Solution:

(a) For $n = 5$ the quantum number l may be an integer from 0 to $n - 1$, that is, 0, 1, 2, 3, or 4.

(b) For $l = 2$, m_l may be an integer from $-l$ to $+l$, that is, $-2, -1, 0, 1,$ or 2.

28-6. Building Up Atoms

A. The Pauli exclusion principle

The **Pauli exclusion principle** helps us to understand the way in which electrons are arranged to build up complex atoms—those atoms with atomic numbers Z greater than 1. This principle states that

- No two electrons in an atom can have exactly the same quantum numbers (n, l, m_l, and m_s).

This means that if you are given the first three out of the four quantum numbers for a particular atom, you know that there can be only two electrons possible in the atom with these particular quantum numbers, because there are only two possible values of m_s ($+\frac{1}{2}$ or $-\frac{1}{2}$).

- Each possible combination of n, l, and m_l specifies an **orbital**, a region in which an electron with a particular combination of quantum numbers may occur, and the energy that an electron in the orbital must have.

In a hydrogen atom, the energy sequence of orbitals depends *only* on the value of n: Larger values of n mean higher-energy (i.e., less stable) orbitals. In multielectron atoms, the values of *both* n and l affect the orbital energy level.

B. The electronic configuration

- The **electronic configuration** of an atom shows the way in which the electrons in an atom occupy, in order of increasing energy, the available orbitals and spin states.

Electrons with the same value of n are in the same **shell**. Electrons with $n = 1$ are in the K shell; those with $n = 2$ are in the L shell, and so on. Electrons with the same value of n and l are in the same subshell. We specify the electron configuration simply by giving the n value and the appropriate letter for l (see Sec. 28-5), with the number of electrons in each subshell given as a superscript. For example, the electronic configuration for sodium is written $1s^2 2s^2 2p^6 3s^1$, indicating that an electrically neutral sodium atom has two $1s$ electrons, two $2s$ electrons, six $2p$ electrons, and one $3s$ electron.

In the periodic table (at the back of this book), the elements are arranged in order of increasing atomic number. Each square contains the atomic number Z, the symbol for the element, and the atomic mass.

note: The periodic table also arranges the elements according to chemical properties—you can find these discussed in chemistry textbooks.

EXAMPLE 28-17: What is the maximum number of electrons in an atom that can possess the following sets of quantum numbers: **(a)** $n = 4$, $l = 2$; **(b)** $n = 5$, $l = 3$, $m_l = -1$; and **(c)** $n = 3$, $l = 3$?

Solution:

(a) m_l can be $-2, -1, 0, 1,$ or 2. For each m_l there can be two electrons, $m_s = +\frac{1}{2}$ or $-\frac{1}{2}$, so the maximum number of electrons is 10.

(b) Here $m_l = -1$ has been specified. The only other variable is m_s, which may be $+\frac{1}{2}$ or $-\frac{1}{2}$, so the maximum is 2.

(c) Trick question! If $n = 3$, l cannot equal 3. The l quantum number must be less than 3. The answer is zero.

EXAMPLE 28-18: What are the values of n and l that correspond to the following orbitals: $5d$, $3s$, $2p$? What is the maximum number of electrons that have these combinations of n and l?

Solution: The number is the value of n and the letter corresponds to the value of l (see Table 28-1). Once you find the value of l, use $2l + 1$ to find the number of electrons that can have different m_l values but the same n and l values. Then for each possible combination of n, l, and m_l, there may be two electrons, one with $m_s = +\frac{1}{2}$ and one with $m_s = -\frac{1}{2}$. So the total number of electrons that can have a particular combination of n and l is $2(2l + 1)$.

Orbital	n	l	Number of m values $(2l + 1)$	Total electrons
$5d$	5	2	5	10
$3s$	3	0	1	2
$2p$	2	1	3	6

28-7. X Rays

When an outer electron falls from a higher to a lower energy level, the photon it emits is often in the visible range, as in the hot gas studied through a spectroscope. When we remove an electron from the inner shell of an atom with many electrons, an electron from an outer shell drops to fill the vacancy. For example, if we remove a $1s$ electron from a zinc atom, a $2p$ or $3s$ electron might drop to the lower energy level. The change in energy is much greater than the change in energy of an excited outer electron, so the photon emitted by the zinc atom has a greater energy, in the X-ray range. If the electron falls from the L shell to the K shell ($n = 2$ to $n = 1$), we call the emitted photon a K_α X ray and its energy is

ENERGY OF A K_α X RAY

$$E_{K_\alpha} = 13.6 \text{ eV}(Z - 1)^2\left(\frac{1}{1^2} - \frac{1}{2^2}\right) \qquad \textbf{(28-13)}$$

The energy of a K_β X ray, which results from a $M \rightarrow K$ transition, is

ENERGY OF A K_β X RAY

$$E_{K_\beta} = 13.6\ \text{eV}(Z - 1)^2\left(\frac{1}{1^2} - \frac{1}{3^2}\right) \qquad \textbf{(28-14)}$$

EXAMPLE 28-19: Calculate the energy and the wavelength of a K_α X-ray photon produced by a molybdenum atom ($Z = 42$).

Solution: Use Eq. (28-13) to calculate the photon's energy:

$$E_{K_\alpha} = 13.6\ \text{eV}(42 - 1)^2\left(\frac{1}{1} - \frac{1}{4}\right)$$

$$= (1.71 \times 10^4\ \text{eV})(1.60 \times 10^{-19}\ \text{J/eV}) = 2.74 \times 10^{-15}\ \text{J}$$

Now, calculate the photon's wavelength from Eq. (28-3):

$$\lambda = \frac{hc}{E} = \frac{(6.626 \times 10^{-34}\ \text{J s})(3.0 \times 10^8\ \text{m/s})}{2.74 \times 10^{-15}\ \text{J}} = 7.25 \times 10^{-11}\ \text{m}$$

EXAMPLE 28-20: Calculate (**a**) the energy and (**b**) the wavelength of a K_β X-ray photon produced by a molybdenum atom.

Solution: (**a**) Directly from Eq. (28-14):

$$E_{K_\beta} = 13.6\ \text{eV}(42 - 1)^2\left(\frac{1}{1} - \frac{1}{9}\right) = 2.032 \times 10^4\ \text{eV}$$

(**b**)

$$\lambda = \frac{(6.63 \times 10^{-34}\ \text{J s})(3.0 \times 10^8\ \text{m/s})}{(2.032 \times 10^4\ \text{eV})(1.60 \times 10^{-19}\ \text{J/eV})} = 6.11 \times 10^{-11}\ \text{m}$$

SUMMARY

Charge on an oil droplet	$Q = \dfrac{mg}{V/d}$	the charge on an electron is measured by introducing a droplet of oil of mass m between two capacitors, then adjusting the voltage so that the oil remains in equilibrium
Charge-to-mass ratio of an electron	$\dfrac{e}{m} = \dfrac{E}{B^2 r}$	the mass of an electron is determined by forcing it to move through a magnetic field in a circular orbit
Planck's equation	$E = hf = \dfrac{hc}{\lambda}$	a photon's energy is equal to its frequency times Planck's constant
Momentum of a photon	$p = \dfrac{h}{\lambda}$	a photon's momentum is the ratio of Planck's constant and the photon's wavelength
KE transferred to an electron	$\text{KE} = \dfrac{1}{2} mv^2 = hf - hf'$	when a photon collides with an electron, it loses some momentum to the electron, decreasing its frequency

de Broglie equation	$\lambda = \dfrac{h}{mv}$	gives the relationship of the wavelength of a material particle to its momentum
Heisenberg uncertainty principle	$\Delta x\, \Delta p_x \geqq \dfrac{h}{4\pi}$	in trying to determine the position and momentum of a particle simultaneously, we can be no more precise than $h/4\pi$
Wavelengths in the hydrogen spectrum	$\dfrac{1}{\lambda} = R\left(\dfrac{1}{n^2} - \dfrac{1}{m^2}\right)$	the wavelength of a hydrogen emission line depends on the difference between the inverse squares of the quantum numbers of the final and initial energy levels
Energy level (hydrogen-like atom)	$E_n = -13.6\ \text{eV}\left(\dfrac{Z^2}{n^2}\right)$	gives the energy states for any single-electron atom as a ratio of its atomic number and its principal quantum number
Radius of an electron orbit	$r = \dfrac{n^2h^2}{4\pi^2 mkZe^2}$	gives the radius of an allowed electron orbit in an atom
Energy of a photon emitted by a hydrogen atom	$hf = 13.6\ \text{eV}\left(\dfrac{1}{n^2} - \dfrac{1}{m^2}\right)$	gives the energy emitted by a photon when an electron in a hydrogen atom drops from a higher- to a lower-energy state
Energy of a K_α X ray	$E_{K_\alpha} = 13.6\ \text{eV}(Z-1)^2\left(\dfrac{1}{1^2} - \dfrac{1}{2^2}\right)$	gives the energy of a K_α X ray resulting from a $L \rightarrow K$ transition
Energy of a K_β X ray	$E_{K_\beta} = 13.6\ \text{eV}(Z-1)^2\left(\dfrac{1}{1^2} - \dfrac{1}{3^2}\right)$	gives the energy of a K_β X ray resulting from a $M \rightarrow K$ transition

RAISE YOUR GRADES

Can you . . . ?

- ☑ write the equation for the charge on an oil droplet in equilibrium
- ☑ write the equation for the charge-to-mass ratio of an electron
- ☑ calculate the momentum of a photon whose wavelength is known
- ☑ calculate the energy of a photon whose wavelength is known
- ☑ write the equation for the KE of an electron struck by a photon
- ☑ calculate the de Broglie wavelength of a moving particle
- ☑ explain the Heisenberg uncertainty principle
- ☑ write the equation for the energy levels of a hydrogen-like atom
- ☑ write the equation for the orbital radius of an electron in a hydrogen-like atom
- ☑ write the equation for the energy of a photon emitted by a hydrogen atom
- ☑ write the equation for the energy of a K_α X ray and a K_β X ray

SOLVED PROBLEMS

PROBLEM 28-1 An oil droplet with excess of five electrons is in equilibrium in an electric field between two parallel plates of a capacitor. The voltage between the two plates is 6000 V and the distance between them is 1.5 cm. Calculate the mass of the oil droplet.

Solution Use Eq. (28-1) to solve for the mass:

$$Q = \frac{mg}{V/d} \qquad m = \frac{QV}{gd} = \frac{5(1.6 \times 10^{-19}\ \text{C})(6 \times 10^{3}\ \text{V})}{(9.8\ \text{m/s}^2)(1.5 \times 10^{-2}\ \text{m})} = 3.27 \times 10^{-14}\ \text{kg}$$

PROBLEM 28-2 An electron is accelerated by an electric field of 6.00×10^5 N/C through a magnetic field. The radius of its circular orbit is 4 cm, and its charge-to-mass ratio is 1.76×10^{11} C/kg. Calculate the magnitude of the magnetic field.

Solution Use Eq. (28-2) to solve for the magnetic field:

$$\frac{e}{m} = \frac{E}{B^2 r}$$

$$B = \sqrt{\frac{E}{r(e/m)}} = \sqrt{\frac{6.00 \times 10^{15}\ \text{N/C}}{(4 \times 10^{-2}\ \text{m})(1.76 \times 10^{11}\ \text{C/kg})}} = 9.23 \times 10^{-3}\ \text{T}$$

PROBLEM 28-3 Calculate the momentum of an X-ray photon with a wavelength of 0.0612 nm.

Solution Directly from Eq. (28-4),

$$p = \frac{h}{\lambda} = \frac{6.626 \times 10^{-34}\ \text{J s}}{6.12 \times 10^{-11}\ \text{m}} = 1.08 \times 10^{-23}\ \text{kg m/s}$$

PROBLEM 28-4 Calculate the energy of an X-ray photon whose wavelength is 0.072 nm.

Solution Use Planck's equation (28-3):

$$E = \frac{hc}{\lambda} = \frac{(6.626 \times 10^{-34}\ \text{J s})(3.0 \times 10^{8}\ \text{m/s})}{7.2 \times 10^{-11}\ \text{m}} = 2.76 \times 10^{-15}\ \text{J}$$

PROBLEM 28-5 A photon has an energy of 3.4×10^{-15} J. Calculate its frequency.

Solution Use Eq. (28-3) to find the frequency:

$$E = hf$$

$$f = \frac{E}{h} = \frac{3.4 \times 10^{-15}\ \text{J}}{6.626 \times 10^{-34}\ \text{J s}} = 5.13 \times 10^{18}\ \text{Hz}$$

PROBLEM 28-6 An X-ray photon with a wavelength of 0.60×10^{-10} m strikes an electron. The electron recoils with a KE of 2.56×10^{-16} J. Calculate the wavelength of the photon after its collision with the electron.

Solution Use Eq. (28-5) to calculate the energy of the photon after its collision with the electron:

$$\frac{1}{2}mv^2 = hf - hf'$$

$$hf' = \frac{hc}{\lambda} - \frac{1}{2}mv^2 = \frac{(6.626 \times 10^{-34}\ \text{J s})(3.0 \times 10^{8}\ \text{m/s})}{0.60 \times 10^{-10}\ \text{m}} - 2.56 \times 10^{-16}\ \text{J}$$

$$= 3.06 \times 10^{-15}\ \text{J}$$

Then, from Eq. (28-3),

$$\lambda = \frac{hc}{E} = \frac{(6.626 \times 10^{-34}\ \text{J s})(3.0 \times 10^{8}\ \text{m/s})}{3.06 \times 10^{-15}\ \text{J}} = 0.650 \times 10^{-10}\ \text{m}$$

PROBLEM 28-7 A proton with a mass of 1.67×10^{-27} kg travels at a speed of 4×10^5 m/s. Calculate the de Broglie wavelength of this proton.

Solution Use Eq. (28-6):

$$\lambda = \frac{h}{mv} = \frac{6.626 \times 10^{-34}\text{ J s}}{(1.67 \times 10^{-27}\text{ kg})(4 \times 10^5\text{ m/s})} = 9.92 \times 10^{-13}\text{ m}$$

PROBLEM 28-8 The linear momentum of a rapidly moving electron is measured with an uncertainty of $\Delta p_x = 5.5 \times 10^{-28}$ kg m/s. Calculate the uncertainty of the electron's position.

Solution Use Eq. (28-7) to calculate Δx:

$$\Delta x\,\Delta p \approx \frac{h}{4\pi} \qquad \Delta x \approx \frac{h}{4\pi\,\Delta p} \approx \frac{6.626 \times 10^{-34}\text{ J s}}{4\pi(5.5 \times 10^{-28}\text{ kg m/s})} \approx 9.59 \times 10^{-8}\text{ m}$$

PROBLEM 28-9 The single electron of an ionized helium atom is excited to $n = 3$. Calculate the radius of this electron's orbit.

Solution Use Eq. (28-11):

$$r = \frac{n^2h^2}{4\pi^2 mkZe^2} = \frac{3^2(6.626 \times 10^{-34}\text{ J s})^2}{4\pi^2(9.11 \times 10^{-31}\text{ kg})(9 \times 10^9\text{ N m}^2/\text{C}^2)(2)(1.6 \times 10^{-19}\text{ C})^2} = 2.38 \times 10^{-10}\text{ m}$$

PROBLEM 28-10 A single electron orbiting about a lithium ($Z = 3$) nucleus moves from $n = 4$ to $n = 3$. Calculate the frequency of the photon emitted by this hydrogen-like atom.

Solution From the energy of each level from Eq. (28-10):

$$E_4 = -13.6\text{ eV}\left(\frac{3^2}{4^2}\right) = -7.65\text{ eV} \qquad E_3 = -13.6\text{ eV}\left(\frac{3^2}{3^2}\right) = -13.6\text{ eV}$$

The energy of the photon is the difference between these two energy levels:

$$E = E_4 - E_3 = (-7.65\text{ eV}) - (-13.6\text{ eV}) = 5.95\text{ eV}$$

Then use Eq. (28-3) to calculate the photon's frequency:

$$f = \frac{E}{h} = \frac{(5.95\text{ eV})(1.6 \times 10^{-19}\text{ J/eV})}{6.626 \times 10^{-34}\text{ J s}} = 1.44 \times 10^{15}\text{ Hz}$$

PROBLEM 28-11 Calculate the wavelength of light emitted by a hydrogen atom when n changes from 5 to 3.

Solution Use Eq. (28-12) to calculate the photon's energy.

$$E = hf = 13.6\text{ eV}\left(\frac{1}{3^2} - \frac{1}{5^2}\right) = 0.967\text{ eV}$$

Now, use Eq. (28-3) to calculate the wavelength of this photon:

$$\lambda = \frac{hc}{E} = \frac{(6.626 \times 10^{-34}\text{ J s})(3.0 \times 10^8\text{ m/s})}{(0.967\text{ eV})(1.6 \times 10^{-19}\text{ J/eV})} = 1.28 \times 10^{-6}\text{ m}$$

PROBLEM 28-12 Very high energy electrons bombard a pure tungsten surface ($Z = 74$). Calculate the energy and the wavelength of the K_α X rays that are produced.

Solution Directly from Eq. (28-13),

$$E_{K_\alpha} = 13.6\text{ eV}(Z - 1)^2\left(\frac{1}{1^2} - \frac{1}{2^2}\right) = 13.6\text{ eV}(73)^2\left(1 - \frac{1}{4}\right) = 5.436 \times 10^4\text{ eV}$$

Then,

$$\lambda = \frac{hc}{E} = \frac{(6.626 \times 10^{-34}\text{ J s})(3.0 \times 10^8\text{ m/s})}{(5.436 \times 10^4\text{ eV})(1.6 \times 10^{-19}\text{ J/eV})} = 2.29 \times 10^{-11}\text{ m}$$

PROBLEM 28-13 When high-energy electrons bombard a tungsten target, both K_α and K_β X rays are produced. Calculate (**a**) the energy and (**b**) the wavelength of the K_β X-ray photons.

Solution (**a**) Use Eq. (28-14):

$$E_{K_\beta} = 13.6 \text{ eV}(Z - 1)^2\left(\frac{1}{1^2} - \frac{1}{3^2}\right) = 13.6 \text{ eV}(73)^2\left(1 - \frac{1}{9}\right) = 6.44 \times 10^4 \text{ eV}$$

(**b**)

$$\lambda = \frac{(6.626 \times 10^{-34} \text{ J s})(3.0 \times 10^8 \text{ m/s})}{(6.44 \times 10^4 \text{ eV})(1.6 \times 10^{-19} \text{ J/eV})} = 1.93 \times 10^{-11} \text{ m}$$

Supplementary Exercises

PROBLEM 28-14 An oil droplet with an excess of electrons is in equilibrium in an electric field. The weight of the droplet is 1.28×10^{-13} N and the voltage is 6000 V. The parallel plates of the capacitor are separated by 3 cm. Calculate the number of excess electrons on this oil droplet.

PROBLEM 28-15 Calculate the momentum of a photon whose wavelength is 0.8×10^{-10} m.

PROBLEM 28-16 A photon has an energy of 6.63×10^{-19} J. Calculate the wavelength of this photon.

PROBLEM 28-17 A photon with a wavelength of 3.0×10^{-7} m strikes an electron at rest. The scattered photon has a wavelength of 4.0×10^{-7} m. Calculate the KE of the electron.

PROBLEM 28-18 A proton moves along the x axis at a speed of 6×10^5 m/s. The mass of the proton is 1.67×10^{-27} kg. What is the de Broglie wavelength of this proton?

PROBLEM 28-19 A single electron orbiting a helium nucleus moves from energy level E_4 to E_2. Calculate the energy in joules of the photon emitted as a result of this energy change.

PROBLEM 28-20 Calculate the wavelength of the photon in Problem 28-19.

PROBLEM 28-21 Calculate the radius of an electron orbiting the hydrogen nucleus with $n = 1$.

PROBLEM 28-22 A single electron is in a circular orbit around a helium nucleus. What is the energy of this hydrogen-like atom if $n = 5$?

PROBLEM 28-23 The uncertainty of the position of an electron moving along the x axis is 2.5×10^{-6} m. What is the uncertainty of the electron's momentum?

Answers to Supplementary Exercises

28-14: 4 electrons

28-15: 8.28×10^{-24} kg m/s

28-16: 3.00×10^{-7} m

28-17: 1.66×10^{-19} J

28-18: 6.61×10^{-13} m

28-19: 1.63×10^{-18} J

28-20: 1.22×10^{-7} m

28-21: 5.30×10^{-11} m

28-22: -2.176 eV

28-23: 2.11×10^{-29} kg m/s

29 THE NUCLEUS

THIS CHAPTER IS ABOUT

☑ Nuclear Structure
☑ The Unified Atomic Mass Unit
☑ The Binding Energy of a Nucleus
☑ Radioactive Decay

29-1. Nuclear Structure

- The *nucleus* of an atom consists of neutrons and protons, collectively called **nucleons**.
- The **mass number** A is the total number of nucleons in a nucleus.
- The **atomic number** Z equals the number of protons in a nucleus.
- The number of neutrons in a nucleus is $A - Z$.
- **Isotopes** are nuclei of an element that contain the same number of protons, but different numbers of neutrons.

For example, the nucleus of the element chromium (symbol Cr) contains 24 protons. There are four different isotopes of this element, each with a different number of neutrons. They are, in order of abundance in nature, ${}^{52}_{24}Cr$ (83.8%), ${}^{53}_{24}Cr$ (9.5%), ${}^{50}_{24}Cr$ (4.3%), ${}^{54}_{24}Cr$ (2.4%), where the superscript is the mass number A and the subscript is the atomic number Z. Notice that ${}^{52}Cr$ contains 28 neutrons ($52 - 24 = 28$), ${}^{53}Cr$ contains 29 neutrons ($53 - 24 = 29$), and so on.

- A **nuclide** is any given combination of protons and neutrons.

The approximate radius of a nucleus (in meters) is given by

RADIUS OF A NUCLEUS

$$r \cong (1.2 \times 10^{-15}\ \text{m})\, A^{1/3} \tag{29-1}$$

EXAMPLE 29-1: Calculate the approximate radius of the most abundant isotope of chromium.

Solution: Directly from Eq. (29-1) with $A = 52$,

$$r \cong (1.2 \times 10^{-15}\ \text{m})(52)^{1/3} = 4.5 \times 10^{-15}\ \text{m}$$

29-2. The Unified Atomic Mass Unit

The most abundant isotope of carbon has six protons and six neutrons in its nucleus. One mole of neutral atoms of this isotope has a mass of 12.000 000 g.

- The **unified atomic mass unit** u is defined for neutral ${}^{12}_{6}C$ (nucleus plus six electrons) as exactly 12.000 000 u. There are Avogadro's number of atoms (6.02×10^{23}) in a mole (see note, Sec. 13-2C), so

$$1\ \text{u} = \frac{1}{12}\left(\frac{12.000\,000\ \text{g}}{6.02 \times 10^{23}}\right)\left(\frac{\text{kg}}{10^3\ \text{g}}\right) = 1.66 \times 10^{-27}\ \text{kg}$$

EXAMPLE 29-2: One million electron volts (1 MeV) equals 1.6×10^{-13} J and $1\ \text{u} = 1.66 \times 10^{-27}$ kg.
Calculate the energy equivalent of 1 u in units of MeV.

Solution: You can find the energy from Einstein's equation (27-4):

$$E = \Delta mc^2 = (1.66 \times 10^{-27} \text{ kg})(3.00 \times 10^8 \text{ m/s})^2$$

$$= 1.494 \times 10^{-10} \text{ J}\left(\frac{1 \text{ MeV}}{1.6 \times 10^{-13} \text{ J}}\right) = 934 \text{ MeV}$$

note: The generally accepted equivalent energy is 931 MeV, which we'll use after this.

29-3. The Binding Energy of a Nucleus

The total mass of a nucleus is always *less* than the sum of the masses of its constituent nucleons.

- The **binding energy** BE of a nucleus is the mass (or energy) difference between the mass of the nucleus and the sum of the masses of its constituent nucleons—and represents the amount of energy necessary to break apart the nucleus into its constituent nucleons.

The protons and neutrons in a nucleus are tightly bound together, so it takes a significant amount of energy to pull them apart. For example, the removal of a neutron from a carbon nucleus, written

$$^{12}_{6}\text{C} \rightarrow {}^{11}_{6}\text{C} + {}^{1}_{0}\text{n}$$

requires 18.7 MeV. You can calculate this energy from the masses of $^{12}_{6}\text{C}$, $^{11}_{6}\text{C}$, and the mass of the neutron, $^{1}_{0}\text{n}$, which are listed in Table 29-1. After the neutron is removed, the sum of the mass of the constituent nucleons is 11.011 432 u + 1.008 665 u = 12.020 097 u. The mass of the neutral carbon atom $^{12}_{6}\text{C}$ is 12.000 000 u.

$$(11.011\,432 + 1.008\,665) \text{ u} = 12.020\,097 \text{ u} + \Delta m$$

This means that there is an increase in mass of $\Delta m = 2.0097 \times 10^{-2}$ u.

TABLE 29-1: Properties of Light Nuclides

Symbol	Mass (u)	Abundance or Mode of Decay	Half-Life
e	0.000 549		
p	1.007 277		
$^{1}_{0}\text{n}$	1.008 665	β^-	10.6 min
$^{1}_{1}\text{H}$	1.007 825	99.985%	
$^{2}_{1}\text{H}$	2.014 102	0.015%	
$^{3}_{1}\text{H}$	3.016 050	β^-	12.3 y
$^{3}_{2}\text{He}$	3.016 030	0.0001%	
$^{4}_{2}\text{He}$	4.002 603	~100%	
$^{6}_{2}\text{He}$	6.018 893	β^-	0.80 s
$^{8}_{2}\text{He}$	8.0341	β^-	0.12 s
$^{9}_{6}\text{C}$	9.0312	β^+	0.13 s
$^{10}_{6}\text{C}$	10.016 81	β^+	19.5 s
$^{11}_{6}\text{C}$	11.011 432	β^+	20.3 min
$^{12}_{6}\text{C}$	12.000 000	98.9%	
$^{13}_{6}\text{C}$	13.003 354	1.11%	
$^{14}_{6}\text{C}$	14.003 242	β^-	5730 y
$^{15}_{6}\text{C}$	15.010 600	β^-	2.5 s
$^{16}_{6}\text{C}$	16.014 70	β^-	0.74 s

EXAMPLE 29-3: Calculate the binding energy BE associated with the increase in mass for the reaction $^{12}_{6}\text{C} \rightarrow {}^{11}_{6}\text{C} + {}^{1}_{0}\text{n}$.

Solution: BE = $(2.0097 \times 10^{-2} \text{ u})(931 \text{ MeV/1 u}) = 18.7$ MeV

EXAMPLE 29-4: Deuterium is an isotope of hydrogen whose nucleus contains one proton and one neutron.
How much energy is required to remove the neutron from the deuterium nucleus?

Solution: First, write the reaction:

$$^{2}_{1}\text{H} \rightarrow {}^{1}_{1}\text{H} + {}^{1}_{0}\text{n}$$

Now, set up the mass balance equation and find Δm, the increase in mass:

$$2.014\,102 \text{ u} + \Delta m = (1.007\,825 + 1.008\,665) \text{ u}$$

$$\Delta m = (2.016\,490 - 2.014\,102) \text{ u} = 2.388 \times 10^{-3} \text{ u}$$

Finally, you can calculate the BE:

$$\text{BE} = (2.388 \times 10^{-3} \text{ u})\left(\frac{931 \text{ MeV}}{1 \text{ u}}\right) = 2.223 \text{ MeV}$$

29-4. Radioactive Decay

- **Radioactivity** is the breakup or transformation (called **decay**) of unstable nuclei. Radioactivity occurs without external stimuli; it is caused by forces within the nucleus.
- A **disintegration** is the decay of one radioactive nucleus.

Unstable nuclei undergo radioactive decay by

- emitting an **α particle,** which consists of two protons and two neutrons and is the same as a helium nucleus ($^{4}_{2}He^{2+}$). In this **α decay** the nucleus decreases in Z by 2 and in A by 4.
- emitting an electron, $^{0}_{-1}e$. In this **β^- decay** a neutron changes into a proton (so that charge is conserved) and the decaying nucleus *increases* in Z by 1 but remains unchanged in A.
- emitting a **positron**, $^{0}_{+1}e$ (a particle with the same mass as an electron but bearing a *positive* charge). In this **β^+ decay** a proton changes into a neutron and the decaying nucleus *decreases* in Z by 1 but remains unchanged in A.
- capturing an electron. In this process a proton changes into a neutron and the decaying nucleus *decreases* in Z by 1 but remains unchanged in A (the net effect of electron capture is the same as that of positron emission).

Decays involving electron emission, positron emission, or electron capture are collectively called **β decay.**

- emitting photons of very high energy, called **γ rays.** In this **γ decay** Z and A both remain unchanged, but the nucleus drops from a higher to a lower internal energy state.
- **spontaneous fission,** breaking into two approximately equal fragments.

The type of radioactive decay that an unstable nuclide will undergo is determined by the nuclide's size and the ratio of protons to neutrons within it.

A. Half-life

- The **half-life** $T_{1/2}$ of radioactive decay is the time required for one half of the nuclei in any sample of a given isotope to decay.

The **disintegration constant** λ is related to the half-life as follows:

DISINTEGRATION CONSTANT

$$\lambda = \frac{0.693}{T_{1/2}} \quad \textbf{(29-2)}$$

The final number of nuclei N after time t is related to the original number of nuclei N_0 as follows:

RADIOACTIVE DECAY

$$N = N_0 e^{-\lambda t} \quad \textbf{(29-3)}$$

where e is the base of natural logarithms ($\sim$2.718) and λ is the disintegration constant. This equation indicates that the number of radioactive nuclei in a given sample decreases exponentially in time.

EXAMPLE 29-5: An isotope of carbon, $^{15}_{6}C$, undergoes β^- decay with a half-life of 2.5 s. If the original number of $^{15}_{6}C$ atoms is 2.0×10^{22},
(a) How many $^{15}_{6}C$ atoms will be present 5.0 s later?
(b) Into what nuclide will the other atoms have decayed?

Solution: **(a)** After 2.5 s have passed, one half of the original nuclei will have decayed, leaving only 1.0×10^{22} carbon atoms. After another 2.5 s have passed, there will be 0.5×10^{22} of the $^{15}_{6}C$ atoms. **(b)** In β^- decay, one neutron changes into a proton, increasing Z by one but leaving A unchanged. The element with seven protons in its nucleus is nitrogen, so $^{15}_{6}C$ decays into $^{15}_{7}N$.

EXAMPLE 29-6: The half-life of an isotope of carbon, ${}^{10}_{6}C$, is 19.5 s. If there are 3×10^{21} ${}^{10}_{6}C$ nuclei at time t, how many will there be 50 s after t?

Solution: First, calculate the disintegration constant (Eq. 29-2):

$$\lambda = \frac{0.693}{T_{1/2}} = \frac{0.693}{19.5 \text{ s}} = 0.035\,54 \text{ s}^{-1}$$

Now, find the final number of nuclei from Eq. (29-3):

$$N = N_0 e^{-\lambda t} = (3 \times 10^{21}) e^{-(0.035\,54 \text{ s}^{-1})(50 \text{ s})} = 5.07 \times 10^{20} \text{ atoms}$$

B. Activity

- The **activity** A of a radioactive nuclide is the *rate of decay* in disintegrations per second:

ACTIVITY OF A RADIOACTIVE NUCLIDE

$$A = \lambda N \tag{29-4}$$

Because the number of nuclei decreases with time, the activity also decreases. The equation for the activity at time t is

ACTIVITY AS A FUNCTION OF TIME

$$A = A_0 e^{-\lambda t} \tag{29-5}$$

EXAMPLE 29-7: Calculate the original activity of the carbon atoms described in Example 29-5.

Solution: First, you'll need the disintegration constant:

$$\lambda = \frac{0.693}{2.5 \text{ s}} = 0.2772 \text{ s}^{-1}$$

Then, you can calculate the activity at $t = 0$ from Eq. (29-4):

$$A_0 = \lambda N_0 = (0.2772 \text{ s}^{-1})(2.00 \times 10^{22}) = 5.54 \times 10^{21} \text{ s}^{-1}$$

EXAMPLE 29-8: A sample of the radioactive carbon isotope ${}^{10}_{6}C$ has a mass of 0.5 g. The disintegration constant of this carbon nuclide is 0.03554 s^{-1}. Calculate the activity of this sample after 100 s.

Solution: Calculate the mass in kg of a ${}^{10}_{6}C$ atom:

$$10.016\,81 \text{ u}\left(\frac{1.66 \times 10^{-27} \text{ kg}}{1 \text{ u}}\right) = 1.663 \times 10^{-26} \text{ kg/atom}$$

Now, calculate the number of carbon atoms present at the beginning:

$$\frac{0.5 \times 10^{-3} \text{ kg}}{1.663 \times 10^{-26} \text{ kg/atom}} = 3.007 \times 10^{22} \text{ atoms}$$

Then, calculate the number of carbon atoms left after 100 s (Eq. 29-3):

$$N = (3.007 \times 10^{22}) e^{-(0.035\,54 \text{ s}^{-1})(100 \text{ s})} = 8.60 \times 10^{20} \text{ atoms}$$

Finally, use Eq. (29-4) to calculate the activity after 100 s:

$$A = (0.035\,54 \text{ s}^{-1})(8.60 \times 10^{20}) = 3.06 \times 10^{19} \text{ s}^{-1}$$

EXAMPLE 29-9: The nucleus of tritium ${}^{3}_{1}H$ contains 2 neutrons and 1 proton. Its half-life is 12.3 y.
How many disintegrations per second occur in a 5-μg sample of tritium ${}^{3}_{1}H$?

Solution: Begin by calculating the mass of one tritium atom:

$$(3.016\,050 \text{ u})\left(\frac{1.66 \times 10^{-27} \text{ kg}}{1 \text{ u}}\right) = 5.007 \times 10^{-27} \text{ kg}$$

Then, calculate the number of atoms in the 5-μg sample:

$$N_0 = \frac{5 \times 10^{-9} \text{ kg}}{5.007 \times 10^{-27} \text{ kg}} = 9.986 \times 10^{17} \text{ atoms}$$

Now, calculate the disintegration constant:

$$\lambda = \left(\frac{0.693}{12.3 \text{ y}}\right)\left(\frac{1 \text{ y}}{365 \text{ d}}\right)\left(\frac{1 \text{ d}}{8.64 \times 10^4 \text{ s}}\right) = 1.787 \times 10^{-9} \text{ s}^{-1}$$

Finally, calculate the number of disintegrations per second:

$$A_0 = (1.787 \times 10^{-9} \text{ s}^{-1})(9.986 \times 10^{17}) = 1.78 \times 10^9 \text{ s}^{-1}$$

EXAMPLE 29-10: The half-life of tritium is 12.3 y. A sample of tritium ^3_1H contains 5×10^{16} atoms.
How many atoms of tritium will be in the sample after 20 years?

Solution: The number of seconds in 20 y is

$$(20 \text{ y})\left(\frac{365 \text{ d}}{1 \text{ y}}\right)\left(\frac{8.64 \times 10^4 \text{ s}}{1 \text{ d}}\right) = 6.307 \times 10^8 \text{ s}$$

You calculated the disintegration constant in Example 29-9: $\lambda = 1.787 \times 10^{-9} \text{ s}^{-1}$. Now, from Eq. (29-3),

$$N = (5 \times 10^{16})e^{-(1.787 \times 10^{-9} \text{ s}^{-1})(6.307 \times 10^8 \text{ s})} = 1.62 \times 10^{16} \text{ atoms}$$

SUMMARY

Radius of a nucleus	$r \cong (1.2 \times 10^{-15} \text{ m})A^{1/3}$	gives the approximate radius of a nucleus as a function of nuclear mass number
Disintegration constant	$\lambda = \dfrac{0.693}{T_{1/2}}$	gives the relation of the disintegration constant to the half-life of radioactive decay
Radioactive decay	$N = N_0 e^{-\lambda t}$	gives the number of radioactive nuclei remaining in a sample of a radioactive nuclide as a function of the number of nuclei at time $t = 0$
Activity of a radioactive nuclide	$A = \lambda N$	the number of disintegrations per second in a sample of a radioactive nuclide is the disintegration constant times the number of nuclei
Activity as a function of time	$A = A_0 e^{-\lambda t}$	gives the activity of a sample of a radioactive nuclide as a function of the sample's activity at time $t = 0$

RAISE YOUR GRADES

Can you define . . . ?

- ☑ an isotope
- ☑ a nuclide
- ☑ radioactivity
- ☑ α decay
- ☑ a β particle
- ☑ a positron
- ☑ a γ ray

Can you . . . ?

- ☑ calculate the approximate radius of a nucleus if you know the number of nucleons in the nucleus
- ☑ write the relationship between the disintegration constant and the half-life of a radioactive isotope
- ☑ given the half-life of a radioactive nuclide and the number of nuclei in a sample of the nuclide at time $t = 0$, calculate the number of radioactive nuclei in the sample at any other time
- ☑ given the half-life of a nuclide and the number of moles of the nuclide in a sample, calculate the sample's activity
- ☑ given the activity of a sample of a radioactive nuclide at time $t = 0$ and the nuclide's half-life, calculate the activity of the sample at any other time

SOLVED PROBLEMS

PROBLEM 29-1 The isotope of radium with the longest half-life has a mass number of 226. Calculate the approximate radius of the nucleus of $^{226}_{88}\text{Ra}$.

Solution Directly from Eq. (29-1),

$$r \approx (1.2 \times 10^{-15}\ \text{m})A^{1/3} \approx (1.2 \times 10^{-15}\ \text{m})(226)^{1/3} \approx 7.31 \times 10^{-15}\ \text{m}$$

PROBLEM 29-2 Calculate the binding energy of a neutron in a tritium nucleus ^3_1H.

Solution First, write the equation for this nuclear reaction:

$$^3_1\text{H} \rightarrow {}^2_1\text{H} + {}^1_0\text{n}$$

Now, set up the mass balance equation and find Δm. Use Table 29-1 to find the values.

$$2.014\,102\ \text{u} + 1.008\,665\ \text{u} = 3.016\,050\ \text{u} + \Delta m$$

$$\Delta m = .006\,717\ \text{u}$$

Then, calculate the binding energy:

$$\text{BE} = 6.717 \times 10^{-3}\ \text{u}\left(\frac{931\ \text{MeV}}{1\ \text{u}}\right) = 6.25\ \text{MeV}$$

PROBLEM 29-3 The isotope of carbon with $A = 11$ decays by emitting a positron β^+. The half-life of this nuclide is 20.3 min. Calculate the activity of a sample of $^{11}_6\text{C}$ with a mass of 2 mg.

Solution Calculate the mass in grams of a single $^{11}_6\text{C}$ atom:

$$(11.011\,432\ \text{u})\left(\frac{1.66 \times 10^{-24}\ \text{g}}{1\ \text{u}}\right) = 1.828 \times 10^{-23}\ \text{g}$$

Then, calculate the number of atoms of carbon present in the 2-mg sample:

$$N_0 = \frac{2 \times 10^{-3}\ \text{g}}{1.828 \times 10^{-23}\ \text{g}} = 1.094 \times 10^{20}\ \text{atoms}$$

Now, use Eq. (29-2) to calculate the disintegration constant:

$$\lambda = \frac{0.693}{T_{1/2}} = \left(\frac{0.693}{20.3\ \text{min}}\right)\left(\frac{1\ \text{min}}{60\ \text{s}}\right) = 5.69 \times 10^{-4}\ \text{s}^{-1}$$

Finally, you can use Eq. (29-4) to calculate the initial activity:

$$A_0 = \lambda N_0 = (5.69 \times 10^{-4}\ \text{s}^{-1})(1.094 \times 10^{20}) = 6.22 \times 10^{16}\ \text{s}^{-1}$$

PROBLEM 29-4 Calculate the activity of the sample of ${}^{11}_{6}\text{C}$ from Problem 29-3 after two hours.

Solution Use Eq. (29-3) to find out how many carbon atoms will be present after two hours:

$$N = N_0 e^{-\lambda t} = (1.094 \times 10^{20})e^{-(5.69 \times 10^{-4}\ \text{s}^{-1})(7200\ \text{s})} = 1.819 \times 10^{18}\ \text{atoms}$$

Then, from Eq. (29-4),

$$A = (5.69 \times 10^{-4}\ \text{s}^{-1})(1.819 \times 10^{18}) = 1.03 \times 10^{15}\ \text{s}^{-1}$$

PROBLEM 29-5 The initial activity of a sample of ${}^{11}_{6}\text{C}$ is $6.22 \times 10^{16}\ \text{s}^{-1}$. What will the activity be 10 days later?

Solution First, you'll have to calculate the number of seconds in 10 days.

$$t = (10\ \text{d})\left(\frac{24\ \text{h}}{1\ \text{d}}\right)\left(\frac{3600\ \text{s}}{\text{h}}\right) = 8.64 \times 10^{5}\ \text{s}$$

Then, use Eq. (29-5) to calculate the activity. From Problem 29-3, you know the disintegration constant is $5.69 \times 10^{-4}\ \text{s}^{-1}$:

$$A = A_0 e^{-\lambda t} = (6.22 \times 10^{16}\ \text{s}^{-1})e^{-(5.69 \times 10^{-4}\ \text{s}^{-1})(8.64 \times 10^{5}\ \text{s})} = 1.94 \times 10^{-197}\ \text{s}^{-1}$$

Such a low number, essentially 0, indicates that the last ${}^{11}_{6}\text{C}$ nucleus in the sample decayed long before the 10 days were up.

PROBLEM 29-6 The half-life of the nuclide ${}^{10}_{6}\text{C}$ is 19.5 s. A sample of this carbon nuclide has an activity of 5×10^{5} disintegrations per second. Calculate the number of atoms present in this sample.

Solution

$$\lambda = \frac{0.693}{19.5\ \text{s}} = 3.554 \times 10^{-2}\ \text{s}^{-1}$$

Then, from Eq. (29-4):

$$N = \frac{A}{\lambda} = \frac{5 \times 10^{5}\ \text{s}^{-1}}{3.554 \times 10^{-2}\ \text{s}^{-1}} = 1.41 \times 10^{7}\ \text{atoms}$$

PROBLEM 29-7 The sodium isotope ${}^{25}_{11}\text{Na}$ has a half-life of 60 s; the mass of one mole of this isotope is 25 g. Calculate **(a)** the number of atoms in a sample of 3 μg of this isotope and **(b)** the number of ${}^{25}_{11}\text{Na}$ atoms in the sample 30 minutes later.

Solution Use Avogadro's number (Sec. 13-2C) to calculate the initial number of sodium atoms:

$$N_0 = (3 \times 10^{-6}\ \text{g})\frac{6.02 \times 10^{23}\ \text{atoms/mole}}{25\ \text{g/mole}} = 7.224 \times 10^{16}\ \text{atoms}$$

Calculate the disintegration constant:

$$\lambda = \frac{0.693}{60\ \text{s}} = 0.01155\ \text{s}^{-1}$$

Then, use Eq. (29-3):

$$N = (7.224 \times 10^{16} \text{ atoms})e^{-(0.01155\,\text{s}^{-1})(1800\,\text{s})} = 6.76 \times 10^{7} \text{ atoms}$$

PROBLEM 29-8 What is the original activity of the 3-μg of sodium atoms described in Problem 29-7.

Solution Use Eq. (29-4):

$$A_0 = (0.01155 \text{ s}^{-1})(7.224 \times 10^{16}) = 8.344 \times 10^{14} \text{ s}^{-1}$$

PROBLEM 29-9 Calculate the number of disintegrations per second of the 3 μg of sodium described in Problem 29-7 after 5 minutes have passed.

Solution Use Eq. (29-5) with $t = 300$ s:

$$A = A_0 e^{-\lambda t} = (8.344 \times 10^{14} \text{ s}^{-1})e^{-(.01155\,\text{s}^{-1})(300\,\text{s})} = 2.61 \times 10^{13} \text{ disintegrations/s}$$

PROBLEM 29-10 A sample of sodium $^{25}_{11}$Na contains 5×10^{15} atoms. The half-life of this sodium nuclide is 60 s. How many atoms will remain after 6 minutes?

Solution Use the value of λ you calculated in Problem 29-7, 360 s for t, and Eq. (29-3):

$$N = N_0 e^{-\lambda t} = (5 \times 10^{15} \text{ atoms})e^{-(.01155\,\text{s}^{-1})(360\,\text{s})} = 7.82 \times 10^{13} \text{ atoms}$$

Another way to find the result is to remember that after one half-life, 1/2 of the nuclei remain, after two half-lives, $1/2 \times 1/2 = 1/2^2$ remain, etc. Here $t/T_{1/2} = 6$, so $1/2^6$ remain. Therefore

$$N = \frac{N_0}{2^6} = \frac{5 \times 10^{15} \text{ atoms}}{2^6} = 7.81 \times 10^{13} \text{ atoms}$$

Supplementary Exercises

PROBLEM 29-11 One of the isotopes of cobalt ($Z = 27$) has a mass number of 60. How many neutrons are there in a nucleus of this isotope?

PROBLEM 29-12 The common isotope of thorium has a mass number of 232. What is the approximate radius of a $^{232}_{90}$Th nucleus?

PROBLEM 29-13 What is the binding energy in MeV of a neutron in the nucleus of $^{4}_{2}$He?

PROBLEM 29-14 Calculate the amount of energy in MeV required to remove a neutron from the nucleus of $^{14}_{6}$C.

PROBLEM 29-15 The disintegration constant of the nuclide $^{24}_{10}$Ne is 3.417×10^{-3} s^{-1}. What is the half-life of this nuclide?

PROBLEM 29-16 A sample contains 4.0×10^{22} atoms of the sodium isotope $^{21}_{11}$Na. The half-life of this nuclide is 23 s. How many atoms of $^{21}_{11}$Na will be present in this sample after 10 minutes?

PROBLEM 29-17 The mass of the isotope $^{25}_{11}$Na is 24.99 u. How many atoms are in a sample of $^{25}_{11}$Na whose mass is 0.5 g?

PROBLEM 29-18 How many $^{25}_{11}$Na atoms will be present in a 0.5-g sample after 8 minutes? The half-life of this nuclide is 1.0 minute.

PROBLEM 29-19 A sample contains 1.205×10^{22} atoms of the sodium isotope $^{25}_{11}$Na. How many $^{25}_{11}$Na atoms will be in the sample after 500 s?

PROBLEM 29-20 What is the activity of the sample of $^{25}_{11}Na$ described in Problem 29-17?

PROBLEM 29-21 What is the activity of a 0.5-g sample of sodium $^{25}_{11}Na$ at time $t = 1000$ s if its activity at $t = 0$ is 1.392×10^{20} s^{-1}?

PROBLEM 29-22 The radioactive isotope of magnesium $^{23}_{12}Mg$ has a half-life of 12 s and a mass of 22.994 u. How many atoms are present in a sample whose mass is 8 mg?

PROBLEM 29-23 What is the original activity of the sample of magnesium described in Problem 29-22?

PROBLEM 29-24 What is the activity of 8 mg of the magnesium isotope $^{23}_{12}Mg$ after 200 s have passed?

Answers to Supplementary Exercises

29-11: 33

29-12: 7.37×10^{-15} m

29-13: 20.6 MeV

29-14: 8.17 MeV

29-15: 3.38 minutes

29-16: 5.63×10^{14} atoms

29-17: 1.20×10^{22} atoms

29-18: 4.71×10^{19} atoms

29-19: 3.74×10^{19} atoms

29-20: 1.39×10^{20} s^{-1}

29-21: 1.34×10^{15} s^{-1}

29-22: 2.09×10^{20} atoms

29-23: 1.21×10^{19} s^{-1}

29-24: 1.17×10^{14} s^{-1}

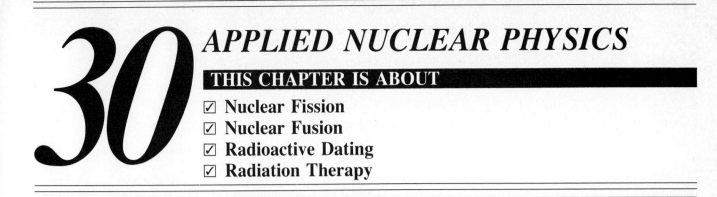

30 APPLIED NUCLEAR PHYSICS

THIS CHAPTER IS ABOUT

- ☑ Nuclear Fission
- ☑ Nuclear Fusion
- ☑ Radioactive Dating
- ☑ Radiation Therapy

30-1. Nuclear Fission

- **Fission** is a process in which the nucleus of an atom is *split* into large fragments.

For example, when a slow neutron strikes the nucleus of a ^{235}U atom, the nucleus splits apart, producing two or more fragments, additional neutrons, and energy. The neutrons cause neighboring uranium atoms to split apart. Two possible reactions for the fission of ^{235}U are

$$^{235}_{92}U + ^{1}_{0}n \rightarrow ^{97}_{38}Sr + ^{136}_{54}Xe + 3\,^{1}_{0}n \qquad \textbf{(a)}$$

$$^{235}_{92}U + ^{1}_{0}n \rightarrow ^{91}_{36}Kr + ^{142}_{56}Ba + 3\,^{1}_{0}n \qquad \textbf{(b)}$$

In order for the reaction to be possible, the total rest mass of the system must *decrease*. For both reactions, the mass of the uranium atom and the neutron before fission takes place is 235.0439 u + 1.0087 u = 236.0526 u. For reaction (**a**), the masses of the strontium atom, the xenon atom, and the three neutrons are

$$96.9261 \text{ u} + 135.9072 \text{ u} + 3.0260 \text{ u} = 235.8593 \text{ u}$$

a decrease of 0.1933 u, so this reaction is possible. Is reaction (**b**) possible? (See Example 30-2.) The mass lost in a fission reaction is converted to energy according to Eq. (27-6). Nuclear reactors use this energy to generate electric power. The rate at which a nuclear reactor produces energy is

POWER OF A NUCLEAR REACTOR

$$P = n\,\Delta mc^2 \qquad \textbf{(30-1)}$$

where Δm is the loss of mass in each reaction, c is the speed of light, and n is the number of reactions per second.

EXAMPLE 30-1: A large nuclear power plant generates 3000 MW of thermal power by the fission of ^{235}U. Assume that all the energy comes from reaction (**a**). How much ^{235}U undergoes fission each second?

Solution: You already found the mass converted to energy per reaction: 0.1933 u. Its energy equivalent is

$$E = mc^2 = (0.1933 \text{ u})(1.66 \times 10^{-27} \text{ kg/u})(3 \times 10^8 \text{ m/s})^2 = 2.888 \times 10^{-11} \text{ J}$$

The number of reactions per second is

$$n = \frac{P}{E} = \frac{3000 \times 10^6 \text{ J/s}}{2.888 \times 10^{-11} \text{ J/atom}} = 1.039 \times 10^{20} \text{ atoms/s}$$

The mass of this number of ^{235}U atoms is

$$(1.039 \times 10^{20} \text{ atoms})(1.66 \times 10^{-27} \text{ kg/u})(235.04 \text{ u/atom}) = 4.05 \times 10^{-5} \text{ kg}$$

EXAMPLE 30-2: Calculate the rate of energy production for a fission reactor using reaction (**b**) if 0.05 g of ^{235}U splits each second. The mass of one ^{91}Kr atom is 90.9234 u; the mass of one ^{142}Ba atom is 141.9164 u.

Solution: The total mass of the ^{235}U and neutron before fission is 236.0526 u. The masses of the ^{91}Kr, ^{142}Ba, and three neutrons are

$$90.9234\text{ u} + 141.9164\text{ u} + 3.0250\text{ u} = 235.8658\text{ u}$$

The number of uranium atoms that split apart each second is

$$n = \frac{5 \times 10^{-5}\text{ kg}}{(236.0526\text{ u})(1.66 \times 10^{-27}\text{ kg/u})} = 1.276 \times 10^{20}\text{ atoms}$$

The mass that is converted to energy during the fission of each ^{235}U atom is

$$\Delta m = 236.0526\text{ u} - 235.8658\text{ u} = 0.1868\text{ u}$$

and the rate of energy production is

$$P = n\,\Delta mc^2 = (1.276 \times 10^{20})(0.1868\text{ u})(1.66 \times 10^{-27}\text{ kg/u})(3 \times 10^8\text{ m/s})^2$$
$$= 3.56 \times 10^9\text{ W}$$

30-2. Nuclear Fusion

- **Fusion** is a process in which light nuclei are joined together.

When light nuclides traveling at very high speeds collide with each other, they may stick together. There is a loss of mass and a concomitant release of energy. Two possible fusion reactions are

$$^2_1\text{H} + {}^2_1\text{H} \rightarrow {}^3_1\text{H} + {}^1_1\text{H} \qquad \textbf{(c)}$$

$$^2_1\text{H} + {}^3_1\text{H} \rightarrow {}^4_2\text{He} + {}^1_0\text{n} \qquad \textbf{(d)}$$

These and other fusion reactions are the energy source of the sun and other stars. Nuclear power plants using fusion reactions have not yet been built but are under development.

EXAMPLE 30-3: In a hypothetical fusion reactor 0.02 g of deuterium undergoes fusion reaction (**c**) each second. Calculate the energy produced during each second of this fusion reaction.

Solution: For reaction (**c**),

mass before fusion: $2(2.014\,102\text{ u}) = 4.028\,204\text{ u}$

mass after fusion: $3.016\,05\text{ u} + 1.007\,825\text{ u} = 4.023\,875\text{ u}$

$$\Delta m = 4.028\,204\text{ u} - 4.023\,875\text{ u} = 0.004\,33\text{ u}$$

Next, find the number of reactions per second. Remember that two deuterium atoms take part in each reaction.

$$n = (2 \times 10^{-5}\text{ kg/s})\left(\frac{\text{u}}{1.66 \times 10^{-27}\text{ kg}}\right)\left(\frac{\text{reaction}}{2(2.014\text{ u})}\right) = 2.991 \times 10^{21}\text{ s}^{-1}$$

Now, calculate the energy produced each second

$$P = n\,\Delta mc^2 = (2.991 \times 10^{21}\text{ s}^{-1})(4.33 \times 10^{-3}\text{ u})(1.66 \times 10^{-27}\text{ kg/u})$$
$$\times (3 \times 10^8\text{ m/s})^2 = 1.93 \times 10^9\text{ J/s}$$

EXAMPLE 30-4: Suppose a fusion reactor contains deuterium $^2_1$H and tritium $^3_1$H at an extremely high temperature. During each second, 0.02 g of deuterium and tritium are fused. Calculate the rate at which energy is produced (in kW).

Solution: The masses of deuterium and tritium in reaction (**d**) before fusion are 2.014 10 u + 3.016 05 u = 5.030 15 u; the total mass after the fusion reaction is 4.002 60 u + 1.008 66 u = 5.011 26 u. The mass that is lost is

$$\Delta m = 5.030\,15\text{ u} - 5.011\,26\text{ u} = 0.018\,89\text{ u}$$

The mass (in kg) of the deuterium and tritium atoms is

$$m = 5.030\,15\text{ u}\left(\frac{1.66 \times 10^{-27}\text{ kg}}{1\text{ u}}\right) = 8.350 \times 10^{-27}\text{ kg}$$

and the number of fusion reactions that take place each second is

$$n = \frac{2 \times 10^{-5}\text{ kg/s}}{8.350 \times 10^{-27}\text{ kg}} = 2.395 \times 10^{21}\text{ s}^{-1}$$

The rate at which energy is produced is

$$P = n\,\Delta mc^2 = (2.395 \times 10^{21}\text{ s}^{-1})(1.889 \times 10^{-2}\text{ u})(1.66 \times 10^{-27}\text{ kg/u}) \times (3 \times 10^8\text{ m/s})^2 = 6.76 \times 10^9\text{ J/s} = 6.76 \times 10^6\text{ kW}$$

30-3. Radioactive Dating

Because the earth's atmosphere is bombarded by high-energy protons from space, called **cosmic rays**, the CO_2 in the atmosphere contains a very small amount of radioactive $^{14}_{2}C$. For every 10^{12} atoms of $^{12}_{6}C$, a stable isotope of carbon, there is one atom of carbon-14, which emits beta particles. This means that all living organisms contain one atom of $^{14}_{6}C$ for every 10^{12} atoms of $^{12}_{6}C$. When the organism dies, it stops exchanging CO_2 with its environment, so after death the ^{14}C lost by radioactive decay to the stable nitrogen ^{14}N is no longer replenished. The radioactivity due to ^{14}C in the remains decreases, and the age of the remains can be calculated from

DATING BY RADIOISOTOPES

$$t = \frac{\ln(A_0/A)}{\lambda} \qquad \textbf{(30-1)}$$

where t is the age, A_0 is the initial activity, A is the current activity, and λ is the disintegration constant.

EXAMPLE 30-5: An ancient piece of charcoal contains 12 g of carbon. The measured activity of this sample of carbon is 1.6335 s^{-1}. The half-life of carbon-14 is 5730 years.
(**a**) Calculate the original activity of this 12-g sample and then (**b**) determine the sample's age.

Solution: (**a**) The mass of one atom of carbon-12 is

$$m = (12.0000\text{ u})\left(\frac{1.66 \times 10^{-24}\text{ g}}{1\text{ u}}\right) = 1.992 \times 10^{-23}\text{ g}$$

The number of atoms of carbon-12 in the 12-g sample is

$$n = \frac{12\text{ g}}{1.992 \times 10^{-23}\text{ g}} = 6.024 \times 10^{23}\text{ atoms}$$

Now, divide by 10^{12} to determine the original number of atoms of carbon-14:

$$N_0 = \frac{6.024 \times 10^{23}\text{ atoms}}{1 \times 10^{12}} = 6.024 \times 10^{11}\text{ atoms}$$

You can calculate the disintegration constant from Eq. (29-2):

$$\lambda = \frac{0.693}{5730\text{ y}} = 1.2094 \times 10^{-4}\text{ y}^{-1}$$

Since one year contains 3.1526×10^7 s,

$$\lambda = \frac{1.2094 \times 10^{-4}\ \text{y}^{-1}}{3.1536 \times 10^7\ \text{s/y}} = 3.835 \times 10^{-12}\ \text{s}^{-1}$$

Use Eq. (29-4) to calculate the original activity:

$$A_0 = \lambda N_0 = (3.835 \times 10^{-12}\ \text{s}^{-1})(6.024 \times 10^{11}) = 2.310\ \text{s}^{-1}$$

(b) Use Eq. (30-1) to calculate the age of this 12-g sample of charcoal:

$$t = \frac{\ln(A_0/A)}{\lambda} = \frac{\ln(2.310/1.6335)}{1.2094 \times 10^{-4}\ \text{y}^{-1}} = 2.87 \times 10^3\ \text{y}$$

EXAMPLE 30-6: Several small bones from an extinct animal contain 16 g of carbon, with a measured activity of $0.770\ \text{s}^{-1}$.
Calculate the age of these bones.

Solution: You calculated the mass of one atom of carbon-12 in Example 30-5. Now, calculate the number of atoms of carbon-14 in 16 g of carbon:

$$N_0 = \left(\frac{16\ \text{g}}{1.992 \times 10^{-23}\ \text{g}}\right)\left(\frac{1}{10^{12}}\right) = 8.032 \times 10^{11}\ \text{atoms}$$

Calculate the original activity. From Example 30-5, the disintegration constant is $\lambda = 3.835 \times 10^{-12}\ \text{s}^{-1}$.

$$A_0 = \lambda N_0 = (3.835 \times 10^{-12}\ \text{s}^{-1})(8.032 \times 10^{11}) = 3.080\ \text{s}^{-1}$$

From Eq. (30-1),

$$t = \frac{\ln(3.080/0.770)}{1.2094 \times 10^{-4}\ \text{y}^{-1}} = 11.5 \times 10^3\ \text{years}$$

Carbon dating is useful for organisms that lived within the last 40,000 years. For minerals or older organic matter we can use the radioactive decay of other isotopes, such as $^{40}_{19}\text{K}$ and $^{238}_{92}\text{U}$. When ^{238}U decays, it starts a long chain of decays that ends with the stable isotope of lead ^{206}Pb. Every uranium atom that starts down the chain eventually ends up as lead. Thus the sum of the current number of ^{238}U atoms plus daughter ^{206}Pb atoms equals the original number of ^{238}U atoms. Because the half-life of ^{238}U is 4.468×10^9 years, only very old rocks can be dated in this way. The age of a sample is calculated by

AGE OF A URANIUM-BEARING SAMPLE

$$t = \frac{1}{\lambda_U} \ln\left(\frac{N_{Pb}}{N_U} + 1\right) \qquad \textbf{(30-3)}$$

where λ_U is the disintegration constant of ^{238}U and N_{Pb} and N_U are the numbers of ^{206}Pb and ^{238}U atoms, respectively.

EXAMPLE 30-7: A rock contains twice as many ^{238}U atoms as ^{206}Pb atoms. What is its age?

Solution: The disintegration constant of ^{238}U is

$$\lambda_U = \frac{0.693}{T_{1/2}} = \frac{0.693}{4.468 \times 10^9\ \text{y}} = 1.551 \times 10^{-10}\ \text{y}^{-1}$$

Now you can find the age with Eq. (30-3). The ratio $N_{Pb}/N_U = 0.5$

$$t = \frac{1}{\lambda_U} \ln\left(\frac{N_{Pb}}{N_U} + 1\right) = \frac{1}{1.551 \times 10^{-10}\ \text{y}^{-1}} \ln(0.5 + 1) = 2.61 \times 10^9\ \text{y}$$

30-4. Radiation Therapy

Energetic γ rays can be used to kill cancerous cells. A common source of γ rays for this purpose is cobalt-60 ($^{60}_{27}Co$). The mass of $^{60}_{27}Co$ is 59.9338 u and its half-life is 5.3 y.

EXAMPLE 30-8: Calculate the activity of one gram of $^{60}_{27}Co$.

Solution: Calculate the mass of one atom of cobalt-60 and the number of atoms contained in 1 g of cobalt:

$$m = (59.9338 \text{ u})\left(\frac{1.66 \times 10^{-24} \text{ g}}{1 \text{ u}}\right) = 9.949 \times 10^{-23} \text{ g}$$

$$N_0 = \frac{1.0 \text{ g}}{9.949 \times 10^{-23} \text{ g}} = 1.005 \times 10^{22} \text{ atoms}$$

Now, calculate the disintegration constant:

$$\lambda = \left(\frac{0.693}{5.3 \text{ y}}\right)\left(\frac{1 \text{ y}}{3.1536 \times 10^{7} \text{ s}}\right) = 4.146 \times 10^{-9} \text{ s}-1$$

And,

$$A_0 = (4.146 \times 10^{-9} \text{ s}^{-1})(1.005 \times 10^{22}) = 4.17 \times 10^{13} \text{ s}^{-1}$$

Radium is also used for radiation therapy. $^{226}_{88}Ra$ emits energetic α particles:

$$^{226}_{88}\text{Ra} \rightarrow {}^{222}_{86}\text{Rn} + {}^{4}_{2}\text{He}$$

The mass of radium-226 is 226.025 36 u. The mass of radon and the α particle are 222.017 53 u + 4.002 60 u = 226.020 13 u. The activity of a radioactive source is generally measured in curies (Ci), where one Ci equals $3.7 \times 10^{10} \text{ s}^{-1}$.

EXAMPLE 30-9: The loss of mass in the conversion of radium to radon causes the α-particle to acquire kinetic energy. Calculate the KE of the α particle.

Solution:

$$\Delta m = 226.025\,36 \text{ u} - 226.020\,13 \text{ u} = 0.005\,23 \text{ u}$$

$$\text{KE} = 0.005\,23 \text{ u}\left(\frac{931 \text{ MeV}}{1 \text{ u}}\right) = 4.87 \text{ MeV}$$

$$= (4.87 \times 10^{6} \text{ eV})(1.60 \times 10^{-19} \text{ J/eV}) = 7.79 \times 10^{-13} \text{ J}$$

EXAMPLE 30-10: Calculate the activity of a 5-g sample of radium-226 ($T_{1/2}$ = 1620 y) is curies.

Solution:

$$\lambda = \left(\frac{0.693}{1620 \text{ y}}\right)\left(\frac{1 \text{ y}}{3.1536 \times 10^{7} \text{ s}}\right) = 1.356 \times 10^{-11} \text{ s}^{-1}$$

$$N = \left(\frac{5 \text{ g}}{226.025 \text{ u}}\right)\left(\frac{1 \text{ u}}{1.66 \times 10^{-24} \text{ g}}\right) = 1.333 \times 10^{22} \text{ atoms}$$

$$A = \lambda N = (1.356 \times 10^{-11} \text{ s}^{-1})(1.333 \times 10^{22})$$

$$= (1.808 \times 10^{11} \text{ s}^{-1})\frac{1 \text{ Ci}}{3.7 \times 10^{10} \text{ s}^{-1}} = 4.89 \text{ Ci}$$

SUMMARY

Power of a nuclear reactor	$P = n\,\Delta mc^2$	the rate of energy production is the energy equivalent of the mass loss per reaction times the number of reactions per second
Dating by radioisotopes	$t = \dfrac{\ln(A_0/A)}{\lambda}$	the age of a sample containing a radioisotope can be determined from the sample's current activity, its original activity, and the isotope's disintegration constant
Age of a uranium-bearing sample	$t = \dfrac{1}{\lambda_U}\ln\left(\dfrac{N_{Pb}}{N_U} + 1\right)$	the age of old rocks can be determined from the amounts of ^{206}Pb and ^{238}U in them

RAISE YOUR GRADES

Can you define . . . ?

- ☑ fission
- ☑ fusion
- ☑ cosmic rays
- ☑ a curie

Can you . . . ?

- ☑ calculate the energy produced when a uranium atom splits apart
- ☑ calculate the number of uranium atoms contained in 1.0 kg of uranium
- ☑ calculate the rate of energy production if you know the number of uranium atoms undergoing fission each second
- ☑ determine how many neutrons are produced when an atom of uranium $^{235}_{92}$U splits into tellurium $^{133}_{52}$Te and zirconium $^{99}_{40}$Z
- ☑ calculate the energy produced when two deuterium atoms produce a tritium atom $^{3}_{1}$H and a hydrogen atom
- ☑ calculate the original activity due to ^{14}C of a carbon sample of known mass
- ☑ determine the age of a piece of charcoal if you know its mass and current activity due to ^{14}C
- ☑ calculate the age of a uranium-bearing rock if you know the numbers of ^{206}Pb and ^{238}U atoms in it
- ☑ calculate the activity of a given amount of ^{60}Co
- ☑ calculate the activity of a given amount of ^{226}Ra

SOLVED PROBLEMS

PROBLEM 30-1 A fission reaction of uranium-235 is

$$^{235}_{92}\text{U} + ^{1}_{0}\text{n} \rightarrow ^{141}_{56}\text{Ba} + ^{92}_{36}\text{Kr} + 3\,^{1}_{0}\text{n}$$

The mass of the uranium atom and the neutron is 235.0439 u + 1.0087 u = 236.0526 u. The masses of the barium atom, the krypton atom, and the three neutrons are

$$140.9139 \text{ u} + 91.9263 \text{ u} + 3.0261 \text{ u} = 235.8663 \text{ u}$$

Calculate the rate of energy production if 0.02 g of ^{235}U undergoes fission each second.

Solution Calculate the mass of one uranium atom and the number of uranium atoms that split apart each second:

$$m = (235.0439 \text{ u})\left(\frac{1.66 \times 10^{-27} \text{ kg}}{1 \text{ u}}\right) = 3.902 \times 10^{-25} \text{ kg}$$

$$n = \frac{2 \times 10^{-5} \text{ kg/s}}{3.902 \times 10^{-25} \text{ kg}} = 5.126 \times 10^{19} \text{ atoms/s}$$

Now, calculate the mass lost when the uranium atom splits apart:

$$\Delta m = 236.0526 \text{ u} - 235.8663 \text{ u} = 0.1863 \text{ u}$$

So, the rate of energy production in kW is

$$P = n\,\Delta mc^2 = (5.126 \times 10^{19} \text{ s}^{-1})(0.1863 \text{ u})(1.66 \times 10^{-27} \text{ kg/u})(3 \times 10^8 \text{ m/s})^2$$
$$= 1.43 \times 10^9 \text{ J/s} = 1.43 \times 10^6 \text{ kW}$$

PROBLEM 30-2 A fission reaction for uranium-235 is

$${}^{235}_{92}\text{U} + {}^{1}_{0}\text{n} \rightarrow {}^{138}_{54}\text{Xe} + {}^{90}_{38}\text{Sr} + 8\,{}^{1}_{0}\text{n}$$

The total mass of the uranium atom and the neutron is 236.0526 u. The masses of xenon, strontium, and the eight neutrons are

$$137.956 \text{ u} + 89.936 \text{ u} + 8.0696 \text{ u} = 235.962 \text{ u}$$

This reaction occurs in a reactor where 0.01 g of uranium is fissioned each second. Calculate the rate of energy production in kilowatts.

Solution The mass of one uranium-235 atom is 3.902×10^{-25} kg, so the number of uranium atoms fissioned each second is

$$n = \frac{1 \times 10^{-5} \text{ kg/s}}{3.902 \times 10^{-25} \text{ kg}} = 2.56 \times 10^{19} \text{ atoms/s}$$

and the mass lost is

$$\Delta m = 236.0526 \text{ u} - 235.9620 \text{ u} = 0.0906 \text{ u}$$

The power produced is

$$P = n\,\Delta mc^2 = (2.56 \times 10^{19} \text{ s}^{-1})(0.0906 \text{ u})(1.66 \times 10^{-27} \text{ kg/u})(3 \times 10^8 \text{ m/s})^2$$
$$= 3.47 \times 10^8 \text{ J/s} = 3.47 \times 10^5 \text{ kW}$$

PROBLEM 30-3 A fusion reaction is

$${}^{3}_{1}\text{H} + {}^{3}_{1}\text{H} \rightarrow {}^{4}_{2}\text{He} + 2\,{}^{1}_{0}\text{n}$$

The masses of these nuclides are given in Table 29-1. In a hypothetical fusion reactor, 0.02 g of tritium is fused each second. Calculate the power produced.

Solution Calculate the mass of the two tritium atoms and the number of fusion reactions per second:

$$m = 2(3.01605 \text{ u}) = (6.0321 \text{ u})\left(\frac{1.66 \times 10^{-24} \text{ g}}{1 \text{ u}}\right) = 1.001 \times 10^{-23} \text{ g}$$

The number of fusion reactions per second is

$$n = \frac{0.02 \text{ g}}{1.001 \times 10^{-23} \text{ g}} = 1.998 \times 10^{21} \text{ s}^{-1}$$

and

$$\Delta m = 6.0321 \text{ u} - (4.0026 \text{ u} + 2.0173 \text{ u}) = 0.0122 \text{ u}$$
$$P = n\,\Delta mc^2 = (1.998 \times 10^{21} \text{ s}^{-1})(0.0122 \text{ u})(1.66 \times 10^{-27} \text{ kg/u})(3 \times 10^8 \text{ m/s})^2$$
$$= 3.64 \times 10^9 \text{ J/s} = 3.64 \times 10^6 \text{ kW}$$

PROBLEM 30-4 Here is another fusion reaction:

$$^2_1H + ^3_1H \rightarrow ^1_0n + ^4_2He$$

If 0.02 g of deuterium and tritium fuses each second, calculate the power produced.

Solution

$$m = 2.014\,102\text{ u} + 3.016\,050\text{ u} = 5.030\,152\text{ u}$$

$$= (5.030\,152\text{ u})\left(\frac{1.66 \times 10^{-24}\text{ g}}{1\text{ u}}\right) = 8.350 \times 10^{-24}\text{ g}$$

$$n = \frac{0.02\text{ g}}{8.350 \times 10^{-24}\text{ g}} = 2.395 \times 10^{21}\text{ s}^{-1}$$

$$\Delta m = 5.030\,152\text{ u} - (1.008\,665\text{ u} + 4.002\,603\text{ u}) = 0.018\,884\text{ u}$$

$$P = n\,\Delta mc^2 = (2.395 \times 10^{21}\text{ s}^{-1})(1.8884 \times 10^{-2}\text{ u})(1.66 \times 10^{-27}\text{ kg/u})(3 \times 10^8\text{ m/s})^2$$

$$= 6.76 \times 10^9\text{ W}$$

PROBLEM 30-5 An ancient piece of charcoal contains 8 g of carbon. The measured activity of this charcoal is 0.7017 dis/s. One atom of $^{12}_{6}C$ has a mass of 1.992×10^{-23} g. Calculate the age of this piece of charcoal.

Solution Calculate the original number of atoms of carbon-14 in this 8 g of carbon:

$$N_0 = \left(\frac{8\text{ g}}{1.992 \times 10^{-23}\text{ g}}\right)\left(\frac{1}{10^{12}}\right) = 4.0161 \times 10^{11}\text{ atoms}$$

You calculated the disintegration constant of $^{14}_{6}C$ in Example 30-5:

$$\lambda = 1.2094 \times 10^{-4}\text{ y}^{-1} = 3.835 \times 10^{-12}\text{ s}^{-1}$$

Now, calculate the original activity of this sample of carbon from Eq. (29-4):

$$A_0 = \lambda N_0 = (3.835 \times 10^{-12}\text{ s}^{-1})(4.0161 \times 10^{11}\text{ atoms}) = 1.540\text{ s}^{-1}$$

Finally, use Eq. (30-1) to find the age of this piece of charcoal:

$$t = \frac{\ln(A_0/A)}{\lambda} = \frac{\ln(1.540/0.7017)}{1.2094 \times 10^{-4}\text{ y}^{-1}} = 6.50 \times 10^3\text{ years}$$

PROBLEM 30-6 A small rock contains some ^{238}U whose activity is 35 disintegrations per second. It also contains 1 mg of ^{206}Pb. Calculate the age of the rock.

Solution First find the number of ^{238}U atoms by using Eq. (29-4). You found the disintegration constant for ^{238}U in Example 30-7.

$$N_U = \frac{A}{\lambda} = \frac{(35\text{ s}^{-1})(3.1536 \times 10^7\text{ s/y})}{(1.551 \times 10^{-10}\text{ y}^{-1})} = 7.116 \times 10^{18}$$

Find the number of ^{206}Pb atoms by using the definition of the mole and Avogadro's number (Sec. 29-2).

$$N_{Pb} = (1 \times 10^{-3}\text{ g})\left(\frac{\text{mol}}{206\text{ g}}\right)\left(\frac{6.02 \times 10^{23}\text{ atoms}}{\text{mol}}\right) = 2.922 \times 10^{18}$$

Now you can find the age with Eq. (30-3):

$$t = \frac{1}{\lambda_U}\ln\left(\frac{N_{Pb}}{N_U} + 1\right) = \left(\frac{1}{1.551 \times 10^{-10}\text{ y}^{-1}}\right)\ln\left(\frac{2.922 \times 10^{18}}{7.116 \times 10^{18}} + 1\right) = 2.22 \times 10^9\text{ y}$$

PROBLEM 30-7 Some ancient bones containing 400 g of carbon were found in a cave in Colorado in 1980. From Example 30-5, the disintegration constant of carbon-14 is

$$\lambda = 1.2094 \times 10^{-4}\text{ y}^{-1} = 3.835 \times 10^{-12}\text{ s}^{-1}$$

The measured activity of these bones in 1980 was 55.55 s^{-1}. Calculate the age of these bones.

Solution In Example 30-5 you found the mass of one ^{12}C atom: 1.992×10^{-23} g. So the original number of ^{14}C atoms in the sample was

$$N_0 = \left(\frac{400 \text{ g}}{1.992 \times 10^{-23} \text{ g}}\right)\left(\frac{1}{10^{12}}\right) = 2.008 \times 10^{13}$$

$$A_0 = (3.835 \times 10^{-12} \text{ s}^{-1})(2.008 \times 10^{13}) = 77.007 \text{ s}^{-1}$$

$$t = \frac{\ln(A_0/A)}{\lambda} = \frac{\ln(77.007/55.55)}{1.2094 \times 10^{-4} \text{ y}^{-1}} = 2.70 \times 10^3 \text{ years}$$

PROBLEM 30-8 Three grams of ^{60}Co are used in an attempt to destroy a cancerous growth. The mass of a ^{60}Co atom is 59.9338 u and its half-life is 5.3 y. Calculate the activity of 3 g of $^{60}_{27}Co$.

Solution

$$\lambda = \frac{0.693}{T_{1/2}} = \frac{0.693}{5.3 \text{ y}} = 0.13075 \text{ y}^{-1} = \frac{0.13075 \text{ y}^{-1}}{3.1536 \times 10^7 \text{ s/y}} = 4.146 \times 10^{-9} \text{ s}^{-1}$$

Now, calculate the number of atoms of cobalt-60 contained in the 3 g of cobalt:

$$m = (59.9338 \text{ u})\left(\frac{1.66 \times 10^{-24} \text{ g}}{1 \text{ u}}\right) = 9.949 \times 10^{-23} \text{ g}$$

$$N_0 = \frac{3 \text{ g}}{9.949 \times 10^{-23} \text{ g}} = 3.015 \times 10^{22} \text{ atoms}$$

$$A_0 = \lambda N_0 = (4.146 \times 10^{-9} \text{ s}^{-1})(3.015 \times 10^{22}) = 1.25 \times 10^{14} \text{ s}^{-1}$$

PROBLEM 30-9 Calculate the activity of an 8-g piece of radium-226.

Solution You can calculate the disintegration constant of radium from Eq. (29-2). The half-life of radium-226 is 1620 years. The mass of a radium atom is 226.025 u.

$$\lambda = \left(\frac{0.693}{1620 \text{ y}}\right)\left(\frac{1 \text{ y}}{3.1536 \times 10^7 \text{ s}}\right) = 1.356 \times 10^{-11} \text{ s}^{-1}$$

Now, calculate the number of atoms of $^{226}_{88}Ra$ in 8 g of radium:

$$N = \left(\frac{8 \text{ g}}{226.025 \text{ u}}\right)\left(\frac{1 \text{ u}}{1.66 \times 10^{-24} \text{ g}}\right) = 2.132 \times 10^{22} \text{ atoms}$$

From Eq. (29-4),

$$A = (1.356 \times 10^{-11} \text{ s}^{-1})(2.132 \times 10^{22}) = 2.89 \times 10^{11} \text{ s}^{-1}$$

Supplementary Exercises

PROBLEM 30-10 A radioactive isotope of neon ($^{24}_{10}Ne$) has a half-life of 3.38 min. Calculate the disintegration constant of this isotope of neon.

PROBLEM 30-11 The mass of the neon isotope $^{24}_{10}Ne$ is 23.994 u. Calculate the mass (in kg) of one atom of neon-24.

PROBLEM 30-12 A nuclear fission reaction produces 172 MeV of energy. Calculate the loss of mass that produces this energy.

PROBLEM 30-13 In a nuclear fission reactor, 0.01 g of uranium-235 splits apart each second. The mass of a ^{235}U atom is 235.044 u. Calculate the number of ^{235}U atoms that split apart each second.

PROBLEM 30-14 The disintegration constant of a radioactive isotope is $\lambda = 8.387 \times 10^{-9}\ s^{-1}$. Calculate the half-life in years of this substance.

PROBLEM 30-15 In a fusion reaction, two deuterium nuclei produce a tritium nuclide and a proton. Calculate the energy produced by the reaction, ${}^2_1H + {}^2_1H \rightarrow {}^3_1H + {}^1_1H$.

PROBLEM 30-16 Suppose 0.06 g of deuterium undergoes a fusion reaction each second. Calculate the number of fusion reactions per second.

PROBLEM 30-17 Calculate the power in kilowatts produced by the fusion reaction described in Problems 15 and 16.

PROBLEM 30-18 Calculate the initial number of radioactive carbon-14 atoms in an ancient piece of charcoal containing 8 g of carbon. The mass of a carbon atom is 12.000 u.

PROBLEM 30-19 The disintegration constant of carbon-14 is $\lambda = 3.835 \times 10^{-12}\ s^{-1}$. Calculate the original activity of 8 g of carbon.

PROBLEM 30-20 The ancient piece of charcoal described in Problem 30-18 has an activity of 0.500 disintegrations per second. Calculate the age of this piece of charcoal.

Answers to Supplementary Exercises

30-10: $3.42 \times 10^{-3}\ s^{-1}$

30-11: 3.98×10^{-26} kg

30-12: 0.185 u

30-13: 2.56×10^{19} atoms/s

30-14: 2.62 years

30-15: 4.03 MeV

30-16: $8.97 \times 10^{21}\ s^{-1}$

30-17: 5.79×10^{6} kW

30-18: 4.02×10^{11} atoms of ${}^{14}_{6}C$

30-19: $154\ s^{-1}$

30-20: 9.30×10^{3} years

SECOND SEMESTER EXAM (Chapters 16–30)

1. A charge of 25 μC experiences an electrostatic force of 1.5×10^{-5} N. What is the magnitude of the electric field? **[Ch. 16]**

2. A point charge of 1.5×10^{-8} C is located at a point P. What is the magnitude of the electric field produced by this charge at a distance of 3 cm from P? **[Ch. 16]**

3. Point P is located 8 cm from a positive charge of 12×10^{-8} C. Point M is 5 cm from the same charge. What is the difference in potential between points M and P? **[Ch. 17]**

4. During an interval of 10 s, 2×10^{19} electrons pass through a small lamp. Calculate the current through this lamp. **[Ch. 16]**

5. A battery whose EMF is 1.5 V has an internal resistance of 1.2 Ω. This battery is connected to a lamp that has a resistance of 10.8 Ω. What is the potential difference across the terminals of this battery? **[Ch. 18]**

6. Calculate the rate at which electric energy is transformed into heat in the 10.8-Ω resistor of Problem 5. **[Ch. 18]**

7. A tungsten filament 2.5 cm long and 0.1 mm in diameter has a resistance of 0.1592 Ω. Calculate the resistivity of tungsten. **[Ch. 19]**

8. A galvanometer has an internal resistance of 1500 Ω. The full-scale current through this galvanometer is 200 μA. Calculate the resistance of a resistor in parallel with the galvanometer that will cause the full-scale current to be 5.0 A. **[Ch. 19]**

9. A charge of 8 μC is moving perpendicular to a magnetic field. The speed of the charge is 50 m/s and the magnetic force acting on it is 2.4×10^{-3} N. Calculate the magnitude of the magnetic field. **[Ch. 20]**

10. A straight conductor carrying a current of 6 A is 50 cm long. The conductor is located in a magnetic field whose magnitude is 4 T. The angle between the conductor and the magnetic field is 40°. Calculate the magnetic force acting on this conductor. **[Ch. 20]**

11. A conductor 10 cm long is moving perpendicular to a magnetic field of 5 T at a speed of 80 m/s. Calculate the EMF induced in this moving conductor. **[Ch. 20]**

12. An AC generator is placed in a circuit with a 20-Ω resistor. The maximum voltage of the generator is 120 V. What is the effective current through the resistor? **[Ch. 21]**

13. Calculate the rate at which electric energy is transformed into heat within the resistor in Problem 12. **[Ch. 21]**

14. An AC generator with a frequency of 50 Hz and an effective voltage of 50 V is connected in series with an 8-μF capacitor. Calculate the effective current through this capacitor. **[Ch. 21]**

15. A piece of wire has a resistance of 80 Ω at 0 °C. When it is heated to 60 °C its resistance becomes 105 Ω. Calculate the temperature coefficient of resistance of this wire. **[Ch. 22]**

16. The capacitance of a parallel-plate capacitor is 6 μF. The potential difference between the two plates is 400 V. Calculate the energy stored in this capacitor. **[Ch. 22]**

17. The average distance from the earth to the sun is 1.496×10^{11} m. How long does it take for electromagnetic waves to travel this distance? **[Ch. 23]**

18. A circuit contains an inductor (inductance, 4 mH) and a capacitor (capacitance, 8 μF) connected in parallel. Calculate the resonance frequency of this circuit. **[Ch. 21]**

19. The speed of propagation of visible light through a glass is 2×10^8 m/s. Calculate the index of refraction of this glass. **[Ch. 23]**

20. The index of refraction of air is 1.0003. The index of refraction of flint glass is 1.66. A ray of light strikes the surface of flint glass at an angle of incidence of 40°. Calculate the angle of refraction within the glass. **[Ch. 23]**

21. An object is placed 12 cm in front of a thin lens. The image is 24 cm behind the lens. What is the focal length of this lens? **[Ch. 24]**

22. An object is placed 10 cm in front of a lens whose focal length is 6 cm. At what distance behind the lens will the image be located?

23. A negative lens has a focal length of -8 cm. An object is placed 12 cm in front of this lens. Where is the image? **[Ch. 24]**

24. A thin lens consists of two convex surfaces. The radius of curvature of each surface is 10 cm. The index of refraction of the glass is 1.50. Calculate the focal length. **[Ch. 24]**

25. What is the magnification of the image described in Problem 21? **[Ch. 24]**

26. What is the maximum magnification of a double convex lens with a focal length of 5 cm? **[Ch. 25]**

27. The objective lens of a compound microscope has a focal length of 2.0 cm. The object is placed 3.0 cm in front of the objective lens. The eyepiece has a focal length of 2.5 cm. Calculate the total magnification of this microscope. **[Ch. 25]**

28. An astronomical telescope has an objective lens with a focal length of 40 cm. The focal length of the eyepiece is 2.0 cm. What is the magnification of this telescope? **[Ch. 25]**

29. The focal lengths of two thin lenses are 5 cm and 8 cm. What is the effective focal length of the two lenses when placed in contact? **[Ch. 25]**

30. Two light waves arrive at a point *P*. The path difference between these two waves is 1.5 λ. Calculate the phase difference between these two waves. **[Ch. 26]**

31. Two parallel slits 0.2 mm apart are illuminated with light whose wavelength is 600 nm. Calculate the angular position of the first-order bright fringe. **[Ch. 26]**

32. A single slit 0.12 mm wide is illuminated with light whose wavelength is 500 nm. Calculate the angular distance from the central peak to the first minimum. **[Ch. 26]**

33. Light traveling in air strikes a smooth glass surface whose index of refraction is 1.62. Calculate the angle of incidence that will cause the reflected light to be completely polarized. **[Ch. 26]**

34. A spaceship moves past the earth at a speed of 1.0×10^8 m/s. The interval between two events on the spaceship is 200 s. What is this interval as measured by observers on the earth? **[Ch. 27]**

35. The rest mass of the proton is 1.67×10^{-27} kg. What is the mass of a proton moving in an accelerator at a speed of 1.5×10^8 m/s? **[Ch. 27]**

36. In a certain nuclear reaction a mass of 1.2×10^{-5} kg is lost. Calculate the relativistic energy that results from this mass loss. **[Ch. 27]**

37. An oil droplet that is in equilibrium between the plates of a capacitor has a weight of 1.536×10^{-13} N. The voltage between the two plates is 4000 V and the distance between them is 2.5 cm. Calculate the number of excess electrons on this oil droplet. **[Ch. 28]**

38. What is the momentum of a photon whose wavelength is 300 nm? **[Ch. 28]**

39. What is the energy of a photon whose wavelength is 420 nm? **[Ch. 28]**

40. A photon with a frequency of 5.20×10^{18} Hz collides with an electron. The frequency of the scattered photon is 5.0×10^{18} Hz. Calculate the kinetic energy that is transferred to the electron. **[Ch. 28]**

41. Calculate the approximate radius of a uranium nucleus whose mass number *A* is 238. **[Ch. 29]**

42. Calculate the disintegration constant of ^{10}C, which has a half-life of 19.5 s. **[Ch. 29]**

43. If there are 3.4×10^{22} of the carbon atoms described in Problem 42 at time $t = 0$, how many atoms of this isotope will remain at $t = 2$ minutes? **[Ch. 29]**

44. What was the activity at time $t = 0$ of 3.4×10^{22} atoms of ^{10}C? **[Ch. 29]**

45. In a certain fission reaction the loss of mass is 0.12 u. Calculate the rate of energy production in kilowatts when 5.0×10^{24} of these reactions take place each second. **[Ch. 30]**

46. In a certain fusion reaction the loss of mass is 0.00433 u. If there are 5.0×10^{24} fusion reactions per second, calculate the rate of energy production in kilowatts. **[Ch. 30]**

47. An ancient piece of charcoal contains 15 g of carbon. The measured activity of this sample of carbon is $1.52\ s^{-1}$. What was the original activity of this piece of charcoal? The disintegration constant of carbon-14 is $3.835 \times 10^{-12}\ s^{-1}$. **[Ch. 30]**

48. Now calculate the age of this piece of charcoal. **[Ch. 30]**

49. **(a)** Find the location and magnification of the image produced by the lens combination shown in Figure E-5. Is the image real or virtual? Erect or inverted? **[Ch. 25]**

(b) The object in Figure E-5 is emitting light at a wavelength of 500 nm. What is the energy of each photon of this light? **[Ch. 28]**

(c) One of the photons strikes a free electron (mass $= 9.11 \times 10^{-31}$ kg) in the lens and loses some energy, so that its wavelength is now 600 nm. What is the electron's resulting velocity? **[Ch. 28]**

(d) The object has a mass of 1 g and is composed 100% of $^{60}_{27}Co$ at time $t = 0$. ^{60}Co decays by β^- decay with a half-life of 5.27 years. Predict the composition (what mass of what nuclides) of the object at time $t = 1$ year. **[Ch. 29]**

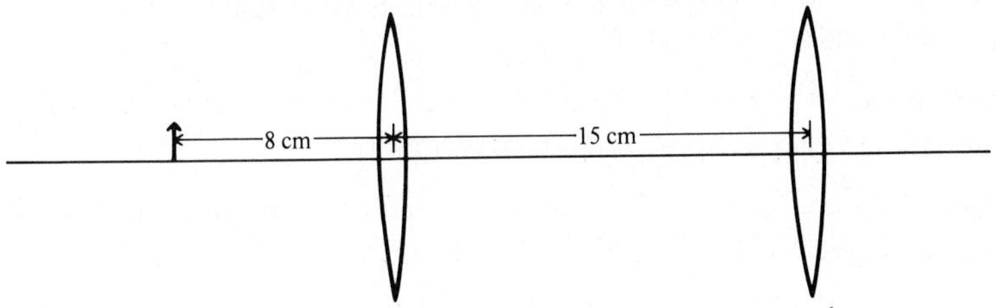

Figure E-5

Solutions to Second Semester Exam

1. The strength of an electric field at a point in space is the force exerted on a positive charge located at that point, divided by the magnitude of the charge.

$$E = \frac{F_{elec}}{Q} = \frac{1.5 \times 10^{-5}\ N}{25 \times 10^{-6}\ C} = 0.600\ N/C$$

2. The electric field due to a point charge is

$$E = \frac{kQ}{r^2} = \frac{(9 \times 10^9\ N\ m^2/C^2)(1.5 \times 10^{-8}\ C)}{(0.03\ m)^2} = 1.5 \times 10^5\ N/C$$

3. Calculate the potential at each point, then subtract to find the difference between them.

$$V_P = \frac{kQ}{r_P} = \frac{(9 \times 10^9 \text{ N m}^2/\text{C}^2)(12 \times 10^{-8} \text{ C})}{0.08 \text{ m}} = 1.35 \times 10^4 \text{ V}$$

$$V_M = \frac{kQ}{r_M} = \frac{(9 \times 10^9 \text{ N m}^2/\text{C}^2)(12 \times 10^{-8} \text{ C})}{0.05 \text{ m}} = 2.16 \times 10^{-4} \text{ V}$$

$$\Delta V = V_M - V_P = (2.16 - 1.35) \times 10^{-4} \text{ V} = 8.10 \times 10^{-3} \text{ V}$$

4.

$$I_{\text{ave}} = \frac{\Delta Q}{\Delta t} = \left(\frac{2 \times 10^{19} \text{ electrons}}{10 \text{ s}}\right)\left(\frac{1.60 \times 10^{-19} \text{ C}}{\text{electron}}\right) = 0.320 \text{ A}$$

5. First find the current by dividing the EMF by the total resistance. Then find the potential difference across the battery's terminals by subtracting the current times the internal resistance from the EMF.

$$I = \frac{\mathscr{E}}{R_{\text{tot}}} = \frac{1.5 \text{ V}}{(1.2 + 10.8)\ \Omega} = 0.125 \text{ A}$$

$$V = \mathscr{E} - IR_i = 1.5 \text{ V} - (0.125 \text{ A})(1.2\ \Omega) = 1.35 \text{ V}$$

6.

$$P = I^2R = (0.125 \text{ A})^2(10.8\ \Omega) = 0.169 \text{ W}$$

7. Use the equation relating the resistivity of a substance to the resistance of a resistor made of that substance, and solve for resistivity.

$$\rho = \frac{RA}{L} = \frac{(0.1592\ \Omega)\pi(1 \times 10^{-4} \text{ m}/2)^2}{2.5 \times 10^{-2} \text{ m}} = 5.00 \times 10^{-8}\ \Omega \text{ m}$$

8. First, calculate the voltage drop across the galvanometer. Second, find the current through the shunt. Finally, divide the voltage drop by this current to find the resistance.

$$V = I_G R_G = (200 \times 10^{-6} \text{ A})(1500\ \Omega) = 0.3 \text{ V}$$

$$I_S = I_{\text{tot}} - I_G = 5.0 \text{ A} - 200 \times 10^{-6} \text{ A} \approx 5.00 \text{ A}$$

$$R_S = \frac{V}{I_G} = \frac{0.3 \text{ V}}{5.00 \text{ A}} = 6.00 \times 10^{-2}\ \Omega$$

9. Solve the equation for the magnetic force on a moving charge for the magnitude of the magnetic field.

$$B = \frac{F_{\text{mag}}}{Qv \sin\theta} = \frac{2.4 \times 10^{-3} \text{ N}}{(8 \times 10^{-6} \text{ C})(50 \text{ m/s})(\sin 90°)} = 6.00 \text{ T}$$

10.

$$F_{\text{mag}} = I\Delta\ell B \sin\theta = (6 \text{ A})(0.5 \text{ m})(4 \text{ T})(\sin 40°) = 7.71 \text{ N}$$

11.

$$\mathscr{E} = B\ell v \sin\theta = (5 \text{ T})(0.1 \text{ m})(80 \text{ m/s})(\sin 90°) = 40.0 \text{ V}$$

12. First find the effective voltage in this purely resistive AC circuit. Then use Ohm's law to find the effective current.

$$V_{\text{eff}} = \frac{V_0}{\sqrt{2}} = \frac{120 \text{ V}}{\sqrt{2}} = 84.85 \text{ V}$$

$$I_{\text{eff}} = \frac{V_{\text{eff}}}{R} = \frac{84.85 \text{ V}}{20\ \Omega} = 4.24 \text{ A}$$

13.

$$P = I_{\text{eff}}^2 R = (4.24 \text{ A})^2(20\ \Omega) = 360 \text{ W}$$

14. First find the capacitor's reactance, then divide the effective voltage by the reactance to find the effective current.

$$X_C = \frac{1}{2\pi fC} = \frac{1}{2\pi(50 \text{ Hz})(8 \times 10^{-6} \text{ F})} = 398\ \Omega$$

$$I_{\text{eff}} = \frac{V_{\text{eff}}}{X_C} = \frac{50 \text{ V}}{398\ \Omega} = 0.126 \text{ A}$$

15. Solve the equation relating resistance at temperature T to resistance at 0 °C for the coefficient α.

$$\alpha = \left(\frac{1}{T}\right)\left(\frac{R_T}{R_0} - 1\right) = \left(\frac{1}{60\ °\text{C}}\right)\left(\frac{105\ \Omega}{80\ \Omega} - 1\right) = 5.21 \times 10^{-3}\ (\text{C}°)^{-1}$$

16. $$E_C = \tfrac{1}{2}CV^2 = \tfrac{1}{2}(6 \times 10^{-6}\ \text{F})(400\ \text{V})^2 = 0.480\ \text{J}$$

17. $$t = \frac{d}{c} = \frac{1.496 \times 10^{11}\ \text{m}}{3.00 \times 10^8\ \text{m/s}} = 499\ \text{s} = 8\ \text{min}\ 19\ \text{s}$$

18. $$f_{\text{res}} = \frac{1}{2\pi\sqrt{LC}} = \frac{1}{2\pi\sqrt{(4 \times 10^{-3}\ \text{H})(8 \times 10^{-6}\ \text{F})}} = 890\ \text{Hz}$$

19. $$n = \frac{c}{v} = \frac{3 \times 10^8\ \text{m/s}}{2 \times 10^8\ \text{m/s}} = 1.50$$

20. Use Snell's law.

$$n_1 \sin\theta_1 = n_2 \sin\theta_2$$

$$\theta_2 = \arcsin\left(\frac{n_1 \sin\theta_1}{n_2}\right) = \arcsin\left(\frac{1.0003 \sin 40°}{1.66}\right) = 22.8°$$

21. Solve the thin-lens equation for the focal length f.

$$f = \frac{pq}{q + p} = \frac{(12\ \text{cm})(24\ \text{cm})}{(12 + 24)\ \text{cm}} = 8.00\ \text{cm}$$

22. Solve the thin-lens equation for the image distance q.

$$q = \frac{fp}{p - f} = \frac{(6\ \text{cm})(10\ \text{cm})}{(10 - 6)\ \text{cm}} = 15.0\ \text{cm}$$

23. $$q = \frac{fp}{p - f} = \frac{(-8\ \text{cm})(12\ \text{cm})}{[12 - (-8)]\ \text{cm}} = -4.80\ \text{cm}$$

The minus sign indicates that the image is in front of the lens.

24. Use the lens-maker's equation and solve for the focal length f.

$$\frac{1}{f} = (n - 1)\left(\frac{1}{R_1} + \frac{1}{R_2}\right) = (1.50 - 1)\left(\frac{1}{10\ \text{cm}} + \frac{1}{10\ \text{cm}}\right) = 0.1\ \text{cm}^{-1} \qquad f = 10.0\ \text{cm}$$

25. $$m = -\frac{q}{p} = -\frac{24\ \text{cm}}{12\ \text{cm}} = -2.00$$

The minus sign indicates that the image is inverted.

26. $$M = \frac{25\ \text{cm}}{f} + 1 = \frac{25\ \text{cm}}{5\ \text{cm}} + 1 = 6.00$$

27. First calculate the image distance of the objective lens, to find the magnification of the objective. Then multiply the magnification of the objective by the maximum magnification of the eyepiece to find the total magnification of the microscope.

$$q = \frac{fp}{p - f} = \frac{(2\ \text{cm})(3\ \text{cm})}{(3 - 2)\ \text{cm}} = 6\ \text{cm}$$

$$M = \left(\frac{-q}{p}\right)\left(\frac{25\ \text{cm}}{f_e}\right) = \left(\frac{-6\ \text{cm}}{3\ \text{cm}}\right)\left(\frac{25\ \text{cm}}{2.5\ \text{cm}}\right) = -20.0$$

28. $$M = -\frac{f_o}{f_e} = -\frac{40\ \text{cm}}{2.0\ \text{cm}} = -20.0$$

29. $$\frac{1}{f_{\text{eff}}} = \frac{1}{f_1} + \frac{1}{f_2} \qquad f_{\text{eff}} = \frac{f_1 f_2}{f_2 + f_1} = \frac{(5\ \text{cm})(8\ \text{cm})}{(5 + 8)\ \text{cm}} = 3.08\ \text{cm}$$

30. $$\Delta\phi = 2\pi\frac{\Delta s}{\lambda} = 2\pi\frac{1.5\lambda}{\lambda} = 9.42 \text{ radians} = 540^\circ$$

31. $$\sin\theta = \frac{n\lambda}{a} \qquad \theta = \arcsin\left(\frac{n\lambda}{a}\right) = \arcsin\left(\frac{1(6\times10^{-7}\text{ m})}{2\times10^{-4}\text{ m}}\right) = 0.172^\circ$$

32. $$n\lambda = w\sin\theta \qquad \theta = \arcsin\left(\frac{n\lambda}{w}\right) = \arcsin\left(\frac{1(5\times10^{-7}\text{ m})}{1.2\times10^{-4}\text{ m}}\right) = 0.239^\circ$$

33. Recall that the index of refraction of air is 1.0003.

$$\tan\theta_1 = \frac{n_2}{n_1} \qquad \theta_1 = \arctan\left(\frac{n_2}{n_1}\right) = \arctan\left(\frac{1.62}{1.0003}\right) = 58.3^\circ$$

34. Use the equation for relativistic time dilation.

$$\Delta t = \frac{\Delta t_0}{\sqrt{1 - v^2/c^2}} = \frac{200\text{ s}}{\sqrt{1 - \left(\frac{1\times10^8\text{ m/s}}{3\times10^8\text{ m/s}}\right)^2}} = 212\text{ s}$$

35. $$m = \frac{m_0}{\sqrt{1 - v^2/c^2}} = \frac{1.67\times10^{-27}\text{ kg}}{\sqrt{1 - \left(\frac{1.5\times10^8\text{ m/s}}{3\times10^8\text{ m/s}}\right)^2}} = 1.93\times10^{-27}\text{ kg}$$

36. $$E = (\Delta m)c^2 = (1.2\times10^{-5}\text{ kg})(3\times10^8\text{ m/s})^2 = 1.08\times10^{12}\text{ J}$$

37. First find the total charge on the droplet, then divide by the charge on an electron to find the number of electrons.

$$Q = \frac{mg}{V/d} = \frac{1.536\times10^{-13}\text{ N}}{4000\text{ V}/0.025\text{ m}} = 9.6\times10^{-19}\text{ C}$$

$$9.6\times10^{-19}\text{ C}\left(\frac{1\text{ electron}}{1.60\times10^{-19}\text{ C}}\right) = 6\text{ electrons}$$

38. $$p = \frac{h}{\lambda} = \frac{6.626\times10^{-34}\text{ J s}}{300\times10^{-9}\text{ m}} = 2.21\times10^{-27}\text{ m}$$

39. $$E = \frac{hc}{\lambda} = \frac{(6.626\times10^{-34}\text{ J s})(3\times10^8\text{ m/s})}{420\times10^{-9}\text{ m}} = 4.73\times10^{-19}\text{ J}$$

40. $$\text{KE} = hf - hf' = (6.626\times10^{-34}\text{ J s})(5.2 - 5.0)\times10^{18}\text{ Hz} = 1.33\times10^{-16}\text{ J}$$

41. $$r \approx (1.2\times10^{-15}\text{ m})A^{1/3} = (1.2\times10^{-15}\text{ m})238^{1/3} = 7.44\times10^{-15}\text{ m}$$

42. $$\lambda = \frac{0.693}{T_{1/2}} = \frac{0.693}{19.5\text{ s}} = 0.0355\text{ s}^{-1}$$

43. $$N = N_0e^{-\lambda t} = (3.4\times10^{22})e^{-(0.0355\text{ s}^{-1})(120\text{ s})} = 4.80\times10^{20}$$

44. $$A_0 = \lambda N_0 = (0.0355\text{ s}^{-1})(3.4\times10^{22}) = 1.20\times10^{21}\text{ s}^{-1}$$

45. $$P = n\Delta mc^2 = (5\times10^{24})(0.12\text{ u})(1.66\times10^{-27}\text{ kg/u})(3\times10^8\text{ m/s})^2 = 8.96\times10^{10}\text{ kW}$$

46. $$P = n\Delta mc^2 = (5\times10^{24})(0.00433\text{ u})(1.66\times10^{-27}\text{ kg/u})(3\times10^8\text{ m/s})^2 = 3.23\times10^{9}\text{ kW}$$

47. If you recall that the mass of one ^{12}C atom is exactly 12.0 u, you can find the mass in kg of one ^{12}C atom. Then divide the mass of the charcoal by the atomic mass in kg to find the number of atoms in the sample. Next, multiply by the original ratio of ^{12}C to ^{14}C to find the original number of ^{14}C atoms. Finally, multiply by the ^{14}C disintegration constant to find the original activity.

$$m = (12.0\text{ u})(1.66\times10^{-27}\text{ kg/u}) = 1.992\times10^{-26}\text{ kg}$$

$$n = \frac{0.015\text{ kg}}{1.992\times10^{-26}\text{ kg}} = 7.53\times10^{23}$$

$$N_0 = 7.53\times10^{23}\left(\frac{1}{10^{12}}\right) = 7.53\times10^{11}$$

$$A_0 = \lambda N_0 = (3.835\times10^{-12}\text{ s}^{-1})(7.53\times10^{11}) = 2.888\text{ s}^{-1}$$

48.
$$t = \frac{\ln(A_0/A)}{\lambda} = \frac{\ln(2.888\ \text{s}^{-1}/1.52\ \text{s}^{-1})}{3.835 \times 10^{-12}\ \text{s}^{-1}} = 1.674 \times 10^{11}\ \text{s}\left(\frac{\text{y}}{3.156 \times 10^{7}\ \text{s}}\right) = 5.30 \times 10^{3}\ \text{y}$$

49. (**a**) First find the image distance and magnification produced by the first lens alone.

$$\frac{1}{p_1} + \frac{1}{q_1} = \frac{1}{f_1} \qquad q_1 = \frac{f_1 p_1}{p_1 - f_1} = \frac{(4\ \text{cm})(8\ \text{cm})}{8\ \text{cm} - 4\ \text{cm}} = 8\ \text{cm}$$

$$m_1 = -\frac{q_1}{p_1} = -\frac{8\ \text{cm}}{8\ \text{cm}} = -1$$

Now calculate the added effect of the second lens. The image produced by the first lens is the object for the second lens. The total magnification is the product of the two individual magnifications.

$$q_2 = \frac{f_2 p_2}{p_2 - f_2} = \frac{(6\ \text{cm})(15 - 8)\ \text{cm}}{(15 - 8)\ \text{cm} - 6\ \text{cm}} = 42.0\ \text{cm}$$

$$m_2 = -\frac{q_2}{p_2} = -\frac{42\ \text{cm}}{(15 - 8)\ \text{cm}} = -6 \qquad m_{\text{tot}} = (-1)(-6) = 6.00$$

The image distance q_2 is positive, so the image is real. The magnification m_{tot} is positive, so the image is erect.

(**b**) Use Planck's equation:

$$E = \frac{hc}{\lambda} = \frac{(6.626 \times 10^{-34}\ \text{J s})(3 \times 10^{8}\ \text{m/s})}{500 \times 10^{-9}\ \text{m}} = 3.98 \times 10^{-19}\ \text{J}$$

(**c**)
$$mv^2 = hf - hf' = hc\left(\frac{1}{\lambda} - \frac{1}{\lambda'}\right) \qquad v = \sqrt{\frac{hc}{m}\left(\frac{1}{\lambda} - \frac{1}{\lambda'}\right)}$$

$$v = \sqrt{\frac{(6.626 \times 10^{-34}\ \text{J s})(3 \times 10^{8}\ \text{m/s})}{9.11 \times 10^{-31}\ \text{kg}}\left(\frac{1}{500 \times 10^{-9}\ \text{m}} - \frac{1}{600 \times 10^{-9}\ \text{m}}\right)} = 2.70 \times 10^{5}\ \text{m/s}$$

(**d**) β^- decay is the conversion of a neutron to a proton and emission of an electron, resulting in the decaying nucleus increasing in atomic number Z by one and remaining unchanged in mass number A. A glance at the periodic table reveals that the element whose atomic number is 28 is nickel. So ${}^{60}_{27}\text{Co}$ decays into ${}^{60}_{28}\text{Ni}$. To find the composition of the object after 1 year, calculate the number of atoms in the object, the disintegration constant of ${}^{60}\text{Co}$, and the number of ${}^{60}\text{Co}$ atoms remaining after 1 year.

$$N_0 = (1\ \text{g})\left(\frac{\text{u}}{1.66 \times 10^{-24}\ \text{g}}\right)\left(\frac{\text{atom}}{60\ \text{u}}\right) = 1.004 \times 10^{22}\ \text{atoms}$$

$$\lambda = \frac{0.693}{T_{1/2}} = \frac{0.693}{5.27\ \text{y}} = 0.1315\ \text{y}^{-1}$$

$$N = N_0 e^{-\lambda t} = (1.004 \times 10^{22}\ \text{atoms})e^{-(0.1315\,\text{y}^{-1})(1\,\text{y})} = 8.803 \times 10^{21}\ \text{atoms}$$

The mass in grams of this number of ${}^{60}\text{Co}$ atoms is

$$m_{\text{Co}} = (8.803 \times 10^{21}\ \text{atoms})(60\ \text{u/atom})(1.66 \times 10^{-24}\ \text{g/u}) = 0.877\ \text{g}$$

So after 1 year the object consists of 0.877 g of ${}^{60}\text{Co}$ and 0.123 g of ${}^{60}\text{Ni}$.

APPENDIX A: SI Units

Quantity	Unit	Abbreviation	In terms of other SI units
Length	meter	m	
Mass	kilogram	kg	
Time	second	s	
Force	newton	N	$kg\ m/s^2$
Work or energy	joule	J	$N\ m = kg\ m^2/s^2$
Power	watt	W	$J/s = kg\ m^2/s^2$
Pressure	pascal	Pa	$N/m^2 = kg\ m^2/s^3$
Frequency	hertz	Hz	s^{-1}
Temperature	kelvin	K	
Amount of substance	mole	mol	
Electric charge	coulomb	C	A s
Electric current	ampere	A	C/s
Potential difference	volt	V	$J/C = kg\ m^2/(A\ s^3)$
Capacitance	farad	F	$C/V = A^2\ s^4/(kg\ m^2)$
Resistance	ohm	Ω	$V/A = kg\ m^2/(A^2\ s^3)$
Magnetic field strength	tesla	T	$N\ s/(C\ m) = N/(A\ m) = kg/(A\ s^2)$
Magnetic flux	weber	Wb	$T\ m^2 = V\ s = kg\ m^2/(A\ s^2)$
Inductance	henry	H	$V\ s/A = kg\ m^2/(A^2\ s^2)$

APPENDIX B: Commonly Used SI Prefixes

Prefix	Abbreviation	Factor
Mega	M	10^6
Kilo	k	10^3
Deci	d	10^{-1}
Centi	c	10^{-2}
Milli	m	10^{-3}
Micro	μ	10^{-6}
Nano	n	10^{-9}

APPENDIX C: Constants

Quantity	Symbol	Value
Acceleration of gravity at sea level	g	$9.8\ m/s^2$
Radian	rad	$180°/\pi = 57.3°$
Gravitational constant	G	$6.67 \times 10^{-11}\ N\ m/kg^2$
Atmospheric pressure at sea level		1.103×10^5 Pa
Speed of sound in air at sea level		343 m/s
Absolute zero		$0\ K = -273.15\ °C$
Avogadro's number		6.02×10^{23} atoms/mol
Universal gas constant	R	8.314 J/(mol K)
Calorie	cal	4.184 J
Specific heat of water		1.00 cal/(g °C)
Stefan–Boltzmann constant	σ	$5.67 \times 10^{-8}\ W/(m^2\ K^4)$
Charge on an electron		-1.602×10^{-19} C
Coulomb's constant	k	$9.0 \times 10^9\ N\ m^2/C^2$
Permittivity of free space	ε_0	$1/4\pi k = 8.854 \times 10^{-12}\ C^2/(N\ m^2)$
Permeability of free space	μ_0	$1.257 \times 10^{-6}\ N/A^2$
Speed of light in a vacuum	c	3.00×10^8 m/s
Mass of an electron		9.11×10^{-31} kg
Mass of a proton		1.67×10^{-27} kg
Planck's constant	h	6.626×10^{-34} J s
Rydberg constant	R	$1.097 \times 10^7\ m^{-1}$
Electron volt	eV	1.60×10^{-19} J
Unified atomic mass unit	u	1.66×10^{-13} kg = 931 MeV

APPENDIX D: Greek Alphabet

Α	α	Alpha	Ι	ι	Iota	Ρ	ρ	Rho
Β	β	Beta	Κ	κ	Kappa	Σ	σ	Sigma
Γ	γ	Gamma	Λ	λ	Lambda	Τ	τ	Tau
Δ	δ	Delta	Μ	μ	Mu	Υ	υ	Upsilon
Ε	ε	Epsilon	Ν	ν	Nu	Φ	ϕ	Phi
Ζ	ζ	Zeta	Ξ	ξ	Xi	Χ	χ	Chi
Η	η	Eta	Ο	o	Omicron	Ψ	ψ	Psi
Θ	θ	Theta	Π	π	Pi	Ω	ω	Omega

INDEX

Page numbers in italics refer to solved problems

C 2
D 3
E 4
F 5
G 6
H 7
I 8
J 9

Period number = n, the highest occupied electron level

Period	*s* Orbitals being filled: IA (Group numbers)	IIA	*d* Orbitals being filled (Transition elements): IIIB	IVB	VB	VIB	VIIB	VIIIB	VIIIB	VIIIB	IB	IIB	*p* Orbitals being filled: IIIA	IVA	VA	VIA	VIIA	Noble gases: VIIIA
	ns^1	ns^2	$(n-1)d^1ns^2$	$(n-1)d^2ns^2$	$(n-1)d^3ns^2$	$(n-1)d^4ns^{21}$	$(n-1)d^5ns^2$	$(n-1)d^6ns^2$	$(n-1)d^7ns^2$	$(n-1)d^8ns^2$	$(n-1)d^{10}ns^1$	$(n-1)d^{10}ns^2$	ns^2np^1	ns^2np^2	ns^2np^3	ns^2np^4	ns^2np^5	ns^2np^6
1	**H** 1 $1s^1$ 1.0079																	**He** 2 $1s^2$ 4.0026
2	**Li** 3 $2s^1$ 6.941	**Be** 4 $2s^2$ 9.01218											**B** 5 $2s^22p^1$ 10.81	**C** 6 $2s^22p^2$ 12.011	**N** 7 $2s^22p^3$ 14.0067	**O** 8 $2s^22p^4$ 15.9994	**F** 9 $2s^22p^5$ 18.9984	**Ne** 10 $2s^22p^6$ 20.179
3	**Na** 11 $3s^1$ 22.9898	**Mg** 12 $3s^2$ 24.305											**Al** 13 $3s^23p^1$ 26.9815	**Si** 14 $3s^23p^2$ 28.086	**P** 15 $3s^23p^3$ 30.9738	**S** 16 $3s^23p^4$ 32.06	**Cl** 17 $3s^23p^5$ 35.453	**Ar** 18 $3s^23p^6$ 39.948
4	**K** 19 $4s^1$ 39.098	**Ca** 20 $4s^2$ 40.08	**Sc** 21 $3d^14s^2$ 44.959	**Ti** 22 $3d^24s^2$ 47.90	**V** 23 $3d^34s^2$ 50.9414	**Cr** 24 $3d^54s^1$ 51.996	**Mn** 25 $3d^54s^2$ 54.938	**Fe** 26 $3d^64s^2$ 55.847	**Co** 27 $3d^74s^2$ 58.9332	**Ni** 28 $3d^84s^2$ 58.70	**Cu** 29 $3d^{10}4s^1$ 63.546	**Zn** 30 $3d^{10}4s^2$ 65.38	**Ga** 31 $4s^24p^1$ 69.72	**Ge** 32 $4s^24p^2$ 72.59	**As** 33 $4s^24p^3$ 74.9216	**Se** 34 $4s^24p^4$ 78.96	**Br** 35 $4s^24p^5$ 79.904	**Kr** 36 $4s^24p^6$ 83.80
5	**Rb** 37 $5s^1$ 85.4678	**Sr** 38 $5s^2$ 87.62	**Y** 39 $4d^15s^2$ 88.9059	**Zr** 40 $4d^25s^2$ 91.22	**Nb** 41 $4d^45s^1$ 92.9064	**Mo** 42 $4d^55s^1$ 95.94	**Tc** 43 $4d^55s^2$ (97)	**Ru** 44 $4d^75s^1$ 101.07	**Rh** 45 $4d^85s^1$ 102.905	**Pd** 46 $4d^{10}$ 106.4	**Ag** 47 $4d^{10}5s^1$ 107.868	**Cd** 48 $4d^{10}5s^2$ 112.40	**In** 49 $5s^25p^1$ 114.82	**Sn** 50 $5s^25p^2$ 118.69	**Sb** 51 $5s^25p^3$ 121.75	**Te** 52 $5s^25p^4$ 127.60	**I** 53 $5s^25p^5$ 126.904	**Xe** 54 $5s^25p^6$ 131.30
6	**Cs** 55 $6s^1$ 132.905	**Ba** 56 $6s^2$ 137.33	**La*** 57 $5d^16s^2$ 138.905	**Hf** 72 $4f^{14}5d^26s^2$ 178.49	**Ta** 73 $5d^36s^2$ 180.948	**W** 74 $5d^46s^2$ 183.85	**Re** 75 $5d^56s^2$ 186.207	**Os** 76 $5d^66s^2$ 190.2	**Ir** 77 $5d^76s^2$ 192.22	**Pt** 78 $5d^96s^1$ 195.09	**Au** 79 $5d^{10}6s^1$ 196.967	**Hg** 80 $5d^{10}6s^2$ 200.59	**Tl** 81 $6s^26p^1$ 204.37	**Pb** 82 $6s^26p^2$ 207.19	**Bi** 83 $6s^26p^3$ 208.980	**Po** 84 $6s^26p^4$ (209)	**At** 85 $6s^26p^5$ (210)	**Rn** 86 $6s^26p^6$ (222)
7	**Fr** 87 $7s^1$ (223)	**Ra** 88 $7s^2$ (226)	**Ac†** 89 $6d^17s^2$ (227)	**Rf** 104 (260)	**Ha** 105 (260)													

f Orbitals being filled

* Lanthanides ~ $4f^n5d^{0-1}6s^2$	**Ce** 58 $4f^15d^16s^2$ 140.12	**Pr** 59 $4f^35d^06s^2$ 140.907	**Nd** 60 $4f^45d^06s^2$ 144.24	**Pm** 61 $4f^55d^06s^2$ (145)	**Sm** 62 $4f^65d^06s^2$ 150.35	**Eu** 63 $4f^75d^06s^2$ 151.96	**Gd** 64 $4f^75d^16s^2$ 157.25	**Tb** 65 $4f^95d^06s^2$ 158.925	**Dy** 66 $4f^{10}5d^06s^2$ 162.50	**Ho** 67 $4f^{11}5d^06s^2$ 164.930	**Er** 68 $4f^{12}5d^06s^2$ 167.26	**Tm** 69 $4f^{13}5d^06s^2$ 168.934	**Yb** 70 $4f^{14}5d^06s^2$ 173.04	**Lu** 71 $4f^{14}5d^16s^2$ 174.97
† Actinides ~ $5f^n6d^{0-1}7s^2$	**Th** 90 $5f^06d^27s^2$ 232.038	**Pa** 91 $5f^26d^17s^2$ (231)	**U** 92 $5f^36d^17s^2$ 238.03	**Np** 93 $5f^46d^17s^2$ (237)	**Pu** 94 $5f^66d^07s^2$ (244)	**Am** 95 $5f^76d^07s^2$ (243)	**Cm** 96 $5f^76d^17s^2$ (247)	**Bk** 97 $5f^96d^07s^2$ (247)	**Cf** 98 $5f^{10}6d^07s^2$ (251)	**Es** 99 $5f^{11}6d^07s^2$ (254)	**Fm** 100 $5f^{12}6d^07s^2$ (257)	**Md** 101 $5f^{13}6d^07s^2$ (258)	**No** 102 $5f^{14}6d^07s^2$ (255)	**Lr** 103 $5f^{14}6d^17s^2$ (260)